Energy and the Environment:

Resources, Technologies, and Impacts

SECOND EDITION

May 2009

Reza Toossi, Ph.D., P.E.

Mechanical and Aerospace Engineering Department
California State University, Long Beach

VerVe Publishers, Inc.

11670 Sunset Boulevard, Suite 215
Los Angeles, CA 90049
United States of America

Registration Number TXu-270-020
Oct 1, 2005

ISBN 978-1-4276-1867-2
Library of Congress Control Number: 2006926865

Printing number: 9 8 7 6 5 4 3 2 1

Printed in the United States of America
on 100% recycled paper.

Hope has two beautiful daughters: anger and courage; anger at the way things are, and courage to change them. ~ St. Augustine

This book is dedicated to the people closest to my heart who inspired me, gave me the *energy* and provided me with *the environment* to complete this work.

My father Mr. Hossein Toossi Ardakan,

mother Safa,

and beautiful daughter Tara Safa

About the Author...

Reza Toossi is a professor of mechanical and aerospace engineering at CSULB. He received his B.S. degree from the Aryamehr (currently Sharif) University of Technology in Tehran, Iran and his M.S. and Ph.D. degrees from the University of California, Berkeley. He continued his Post Doctoral research studies in the Lawrence Berkeley Laboratory and joined the CSULB faculty in 1981.

Dr. Toossi has worked both as a research scientist and consultant on various projects related to aqueous aerosols and droplets in the atmosphere, nuclear safety, sensor design, air pollution dispersion modeling, flame propagation, fluid mechanics, and fiber optics. His current interests include conducting research and teaching courses in heat transfer, combustion, hybrid-electric vehicles, hydrogen storage, environmental engineering, and renewable energy sources.

Dr. Toossi is a member of ASME, ASEE, SAE, SPIE, AAPT, and ΤΒΠ *and a recipient of the 1995 CSULB Distinguished Faculty Scholarly and Creative Achievement, 1994/1995 Distinguished Faculty Teaching, and TRW Excellence in Teaching Awards.*

Chapter Summary

Table of Contents

Chapter 3. Wind Energy 45

Chapter 4. Hydro Energy 65

Chapter 5. Thermal Energy 85

Chapter 7. Fossil Fuels 143

Chapter 8. Air Pollution from Combustion Sources 173

Chapter 14. Transportation 333

Chapter 15. Economics of Energy 375

Energy and the Environment:

Sources, technologies, and impacts

Preface

Hope has two beautiful daughters: anger and courage; anger at the way things are, and courage to change them ~ St. Augustine.

Destiny is not a matter of chance, but of choice. Not something to wish for, but to attain. ~ William Jennings Bryan (1860-1925)

Energy is arguably the most fundamental concept in all of science and beyond. Chemical reactions are driven to a large extent by energy considerations. So are bowling balls and roller-coasters. Politicians worry about energy shortages and cosmologists suspect that an excess of energy is driving the expansion of the universe. The concept of energy is found everywhere.

Energy comes in many forms and, depending on whether or not it can replenish itself in a relatively short time, can be classified as renewable or non-renewable. Today's world, however, is powered mainly by nonrenewable fossil fuels— coal, oil and natural gas. Oil is particularly important because it is liquid and thus conveniently transported. It is also a key component of numerous clothing, pharmaceutical, and construction products. Only 1% of all energy consumption is fulfilled by renewable sources: solar, wind, waves, and geothermal; 88% comes from fossil fuels, while nuclear and hydroelectric account roughly equally for the remaining 11%. Fossil fuels are, however, of limited supply, and unless huge new deposits are discovered, they will soon deplete. It is estimated that at the current rate of consumption, we will run out of oil in 40 years, natural gas in 60, and coal in 180 years. If the growth in demand is considered, especially that in many developing countries, these numbers will be even lower.

The United States, the most industrialized nation, is particularly dependent on oil. Oil consumption in this country accounts for roughly 26% of the world's. Today, about 60% of the US oil comes from abroad, as compared to about 40% in the early 1990s. As more countries become industrialized, the rate of consumption grows rapidly. Consumption in China and India in particular is expected to grow by two-thirds in the next quarter century.

Although this book is written at a level suitable as a textbook for introductory course in energy and environment for undergraduate students with very little mathematics or science background, it can be read by *anyone* interested in technical, political, environmental, and economical issues related to energy. To provide material appropriate for engineering and science students, additional topics are included in the appendices and in various boxes.

This manuscript is divided into seventeen chapters. In Chapter 1, we introduce general concepts related to energy and its sources, reserves, production and patterns of use. Chapter 2 deals with mechanical forms of energy and the concepts of work, power, and simple machines. Wind power and hydro power are two aspects of mechanical energy and are discussed in Chapters 3 and 4, respectively. In Chapter 5, we introduce the reader to concepts of temperature, heat, and principles of thermodynamics. Furthermore, we show how these concepts can be used to design practical machines ranging from internal combustion engines and electricity-generating power plants to refrigerators, air conditioners, and heat pumps. Next, we introduce the chemical forms of energy, and its storage as biomass through photosynthesis (Chapter 6), or as fossil fuels (Chapter 7). The environmental consequences of fossil fuel combustion are discussed in Chapter 8. Geothermal energy is one source that provides thermal energy directly without the need to burn biomass or fossil fuel (Chapter 9). The radiation from the sun can also be used as a source of heat and light to use in our residential and commercial buildings, or to generate electricity; these topics are covered in Chapter 10. The energy released by nuclear reactions and the benefits and dangers of radiation exposure are discussed in Chapters 11 and 12 respectively.

In Chapter 13, we cover the basics of electricity generation, storage, transportation, and distribution. A short review of transportation technologies is given in Chapter 14. A brief overview of modern economic theories is given in Chapter 15, and is used to evaluate merits of different energy technologies. The economics of the environment and costs associated with environmental damage and cleanup technologies are given in Chapter 16. In the final chapter (Chapter 17) we define sustainability from economic and environmental perspectives, address issues associated with globalization, and offer a possible roadmap for a sustainable future. Each chapter is concluded with a list of additional resources, exercises, and one or more mini-projects. For easy reference, all essential data, formulas, and tables of unit conversions are compiled in four appendices and a glossary of terms at the end of the book.

The book is structured in a highly modular form applicable to a wide range of audiences including engineering and science majors, liberal arts students, and the general public. Each chapter stands alone and the text can be taught in any order that the instructor deems suitable. Widely different curricula can be designed and tailored to any audience simply by selecting to focus on sections from different chapters without loss of generality. For example, an environmental engineering course may include a summary of various energy technologies, with emphasis on air pollution, radiation, and environmental economics; a science curriculum may emphasize various technologies and incorporate some engineering designs.

In the course of writing the text, great care was taken to avoid excessive detail while emphasizing the fundamentals of energy and important environmental and technological concepts. Except for a few very simple equations, all equations are given in boxes labeled "Mathematical Interlude…" can be excluded by non-science, non-engineering majors. Information thought to be advanced or of interest to a specific group is placed between two icons a and ▪, indicating the beginning and end of such material. Additional information is also provided in capsulated forms in separate boxes titled "Digging Deeper...", "FYI...", "Point/Counterpoints...", and "Did You Know That...?".

Writing a textbook is no easy task and is rarely, if ever, accomplished without help from others. I wish to take the opportunity to thank my many colleagues and friends who made valuable suggestions, edited the manuscript, or assisted in design and layout. Dr. Hamid Hefazi offered constructive comments for the chapter on wind energy. A big thanks to Dr. Abas Goodarzi, who contributed valuable insights on the chapters dealing with electricity and transportation systems, to Dr. Adel Alapour for review and helpful comments of chapter on nuclear energy, and to Professors Darwin Hall and Ali Toossi Ardakani for their critique on chapters regarding the economics of energy and the environment. I also had many helpful discussions with long-time colleague and a coauthor on an upcoming physics text, Dr. Igor Glozman, professor of physics and astronomy, about the organization and presentation style. Mses. Emilie Ryan and Whitney Graves proofread the manuscript, Ms. Shiori Amano prepared the bulk of the graphics, Ms. Nicky Mohd helped in design of the book layout, and Ms. Joyce Kwon designed the book cover. Thanks also to Dr. Babak Mikhak for his artistic talents for design of a companion website. Finally, this book would not have been possible if it were not for valuable suggestions of numerous students who took a GE course with the same title in recent years at California State University, Long Beach.

As a final note, any manuscript is prone to have a certain number of typographical errors, at least in the first edition. In the case of this publication, the typographical errors have actually been planted on purpose, and that purpose is suggested by the famous American cartoonist Randy Milholland: *"Typos are very important to all written form. It gives the reader something to look for so they aren't distracted by the total lack of content in your writing."*

Reza Toossi
May 5, 2006
Long Beach, California

Preface to the Second Edition

Since the first edition of the text was published in May 2006, there have been major changes in the world energy situation. The market price of petroleum has spiked from $36 a barrel to $130 at the time of this writing in May 2008; retail prices of gasoline in the US have increased from $2.78 a gallon to $4.00 a gallon in the past two years. Indonesia, one of the major suppliers of oil in the last few decades, has become a net importer of oil and is cancelling its membership in OPEC. Over the past decade China has nearly doubled its oil consumption and is expected to double it again in the next decade. Similarly, other non-OECD countries are demanding higher and higher percentages of the total world petroleum output. During these happenings, the political instability in the Middle East and other oil rich countries are spreading. The US occupation of Iraq continues with no end in sight, and now rumors of an impending war with Iran are being circulated. In Nigeria, despite a cease fire signed by the government and various rebel groups, pipeline sabotage and kidnappings of foreign workers are routinely interrupting oil operation. As a result, the Nigerian oil production has decreased by 15 percent in the past two years. Similarly, in Algeria, terrorist attacks targeting oil facilities have sparked concern among foreign oil companies operating in North Africa. Thanks to skyrocketing oil prices, oil companies are enjoying record profits. Only three oil companies, Chevron, Exxon-Mobil, and Royal Dutch Shell, reported combined net profits of 90 billion dollars in 2007, the single largest annual profit in US corporate history.

As the world is waking up to the realities of energy scarcity, there are works in progress to develop new and alternative sources of energy. Progresses in generating electricity from waves, solar sources, and wind are being reported. Hybrid vehicles are slowly gaining traction, although fuel cell cars still remain a distant goal. New findings show that, contrary to original beliefs, biofuels do not reduce greenhouse emissions and may in fact divert vast resources away from food production. The recent shortages in food supply around the world is partly due to the fact that subsidies many farmers receive to plant corn and other crops to make biofuels is more than what they would earn from cultivating food crops. The situation is expected to get worse in near terms.

A new UN report on the effects of global warming on climate patterns, land resources, and vegetation is sobering. The trends point to a complicated blend of changes including violent shifts in weather patterns, disruption in patterns of food production, and restrictions on the availability of fresh water resources. As of April 2008, 178 countries have ratified the Kyoto Protocol, leaving the United States as the only major country refusing to abide by the treaty. There have been failures and successes, but the final verdict on the success of the protocol is still out.

This second edition is being prepared out of necessity to reflect the dynamic nature of energy resources, emerging technologies, and associated environmental implications. In addition to substantial editorial changes, the new edition will differ from the previous edition in the following ways:

1. The course content has been expanded and reorganized to reflect a more natural flow of the subjects.
2. All graphs, charts, and tables are updated to reflect the latest data available from reputable resources.
3. New scientific findings, technological advances, and economical theories are incorporated and their merits are discussed.
4. Additional examples are provided throughout the text, and supplementary exercises and mini-projects are given at the end of each chapter.

Reza Toossi, May 2009

Energy: Past, Present, and Future

CHAPTER 1

Forests to precede civilizations, deserts to follow ~
Francois Rene Chateaubriand (1768-1848)

The future ain't what it used to be ~ Pogo

Since the introduction of internal combustion engines, petroleum has played an ever-increasing role in our lives. The estimates of petroleum resources and its remaining life have been very crude at best. Following WWI, US geologists estimated that US petroleum reserves would only last for 30 to 40 years, but in the mid 1920s the predictions changed after the discovery of huge oil deposits in Texas and Oklahoma. In fact, the huge petroleum surplus forced oil companies to sharply lower their prices, which made Middle Eastern oil largely irrelevant for many decades.[1]

With the mass production of internal combustion engines, the rapid rise in consumption of petroleum, and the nationalization of oil industries by several major oil-producing countries in the 1950s and 1960s, new concerns arose over the long-term availability of oil reserves. In 1970, the Club of Rome, an international think-tank of scientists, economists, businessmen and political leaders commissioned a study which focused on the earth's physical limits to investigate the long-term economical and environmental consequences of the prevalent patterns of population and consumption on the depletion of natural resources. The result of the study was published in a book entitled The *Limits to Growth*[2], and concluded that unless major steps were taken to limit population and slow industrialization, natural resources would quickly deplete, resulting in global economic crises, famine, and irreversible environmental damage.[3] Based on their model, at the prevailing rate of growth, within a time span of less than 100 years, the world's population is expected to increase to 10 billion, per capita food production to drop by three quarters, pollution to increase tenfold, and all of the oil and gas supplies would deplete.[4]

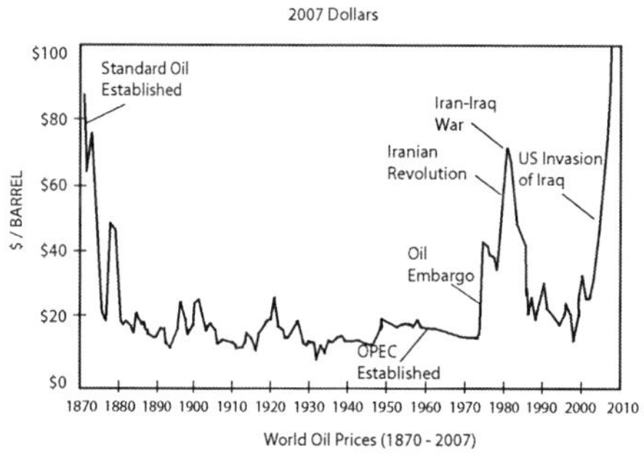

Figure 1-1
Crude oil prices since 1870 adjusted to 2007 values.
Source: Data prior to 2006 was adapted from Transportation Energy Data Book, 2007.

In 1973, following the Arab-Israeli War and the oil embargo by the Organization of Petroleum Exporting Countries (OPEC), and again in 1979, as a result of the Iranian Revolution, the doomsday scenarios predicted by earlier forecasts seemed to become even more authentic, and within a decade the price of oil increased sharply (Figure 1.1). The

[1] Mann, Charles C., Getting Over Oil, MIT Technology Review, January/February 2002.
[2] Meadows D., The Limits to Growth: A Report for the Club of Rome's Project on the Predicament of Mankind, Universe Books, 2nd Ed, 1972.
[3] The model assumed exponential growth for the rate of population increase and consumption of depletable resources, but a fixed source of supply, which necessitates an eventual collapse. Models such as this are widely referred to as "pessimist", or "neo-Malthusian" models.
[4] It should be noted that this point of view had its own critics. For example, Kahn, H., et al [The Next 200 years: A Scenario for America and the World, William Marrow, 1976.] argue that human ingenuity will, at the critical moments, intervene to devise technologies that can assure continuous development and better use of resources. Therefore, according to these technological optimists, resources are not finite but grow as needs arise. As we will argue throughout the book, we believe even with technological innovation and, at this time, additional unknown resources, the resources can eventually run out and we will, sooner or later, have to face the unwanted consequences of our resource mismanagement.

higher price of oil brought a profit windfall for oil companies, but also a major recession throughout the world. As the oil scare spread, so did calls for development of alternative sources of energy and higher efficiency. It also accelerated the search for new petroleum resources, which eventually led to the discovery of new oil fields in the Caspian Sea, Africa, the North Sea and Mexico, which caused oil prices to drop once again.

To prevent prices from falling, OPEC countries cut production, putting an even greater burden on its member countries that heavily depended on oil revenues for their economic livelihood. To boost their revenues, some countries actually sold their oil below market prices and at volumes above the limit set by the organization. The problem became so widespread that OPEC lifted the quotas by the mid-1980s. Since then, OPEC has tried several times to manipulate the prices by limiting production. Their attempts have failed every time, however, because they could not agree on a unified pricing policy and because developed countries made the transition to other resources such as coal, gas, and nuclear power.

As we begin the twenty-first century, energy has once again become a primary concern. A six-fold population increase during the past century and higher energy demands by developing countries, along with the rapid depletion of fossil resources have resulted in new tensions, that began with the 1991 invasion of Kuwait by Iraq resulting in the First Gulf War and continued with the US invasion and occupation of Iraq in 2003. It appears that unless concrete steps are taken to decrease the dependence of the industrialized world on foreign oil imports, many more wars will be fought and regional conflicts will continue to destabilize the world for many decades to come.

It is impossible to consider issues related to energy without looking at its effect on the environment. As we put more and more pollutants into our atmosphere, primarily as a result of human activity and the increasing use of fossil fuels, we are also affecting the earth's climate. When burned, fossil fuels produce carbon dioxide, nitric oxides, carbon monoxide and other harmful gases. Most of these gases adversely affect human and animal health and degrade material property. They are also the primary cause of depletion of ozone in the stratosphere and formation of acid rain and acid fog in the troposphere. Carbon dioxide is not considered a pollutant by most; however, it has a blanketing effect that prevents heat from escaping the atmosphere, causing the environment to warm up.

As petroleum resources are used up, we must switch to alternative sources of energy. Unfortunately, the public so far has shown little interest because the cost of utilization of these resources is higher than those of conventional fuels. How quickly we will be able to meet challenges posed by depleting non-renewable energy resources and the deteriorating environment depends on the way we measure the cost to society and on how far we are willing to go to level the playing field by transferring these costs to energy producers. For example, costs associated with petroleum are not limited to the transactions borne by the buyers and sellers. The true cost must include military costs associated with oil security, various subsidies to oil producers and customers in term of tax breaks and other exemptions, and environmental and health damage resulting from fossil fuels. If the polluters are forced to pay for these damages and if the many fossil fuel subsidies are eliminated, prices will have to be raised, which may help to encourage better efficiency through the use of more environmentally friendly "green" products.

Different approaches have been suggested to internalize these costs. Some economists propose the passage of additional "nuisance laws" that allow the public to sue for damages caused by reckless release of noxious emissions and toxic pollutants into the atmosphere. Two examples that they often cite and have had some success within the last few years are the costs incurred by tobacco companies to pay for health damages caused by cigarette smoking and the clean-up of the Exxon Valdez oil spill in Alaska.

Environmental activists are pushing for a "carbon tax" to account for future costs of climate change and to cover other health costs and damage to the environment. How much the cost will be and how much tax should be levied is very difficult to predict. No matter which approach is adopted, polluting companies are far from being supportive of these measures; they spend a lot of money to fight against lawsuits and new regulations that limit their activities, ultimately passing the costs back to the customers and making the battle a difficult one to win.

What is Energy?

Energy is generally regarded as the single most important concept in all of science. Human cultures are defined largely by their uses of energy resources. Every interaction in our universe involves a transfer of energy between two points or a transformation of energy from one form to another. The word "energy" is bandied about so often that we normally take its meaning for granted. We hear people described as being "full of energy," the world as having "an energy crisis," and the need to "conserve energy." Despite the immense impact of energy on our lives, the term "energy" does not lend itself to a simple definition. The difficulty in producing a precise definition for energy is derived from the fact that energy is a concept, an abstract idea, and not a material object such as an apple or a tree. As Heinberg points out in his book The Party's Over,[5] "few understand exactly what energy is and yet we know that it exists. Indeed, without it, nothing would exist". We can extend this description to define energy as what is required to make anything happen. Given that, energy

[5] Heinberg, Richard, The Party's Over: Oil, War and the Fate of Industrial Societies, New Society Publishers, Gabriola Island, BC, Canada, 2003

must be defined indirectly, in terms of what it does or is capable of doing. In fact, the presence of energy is revealed only when changes take place.

Defining Energy

In physics, energy is defined as *the ability or potential to perform work or cause changes*. As this definition implies, a system with energy need not do any work or cause a change, it only has to have the capacity to do so. A piece of wood sitting at rest some place does not do any work even though it has the capacity to do so (for example by burning it in a furnace to boil water to steam which then turns a turbine). The piece of wood therefore contains energy. In fact, all objects have energy by virtue of their own mass.

Origin of Energy

The scientific theory of the creation of the universe is based on a huge explosion (big-bang), which took place some 12-15 billion years ago. The explosion resulted in the birth of a cosmic nuclei "egg" consisting of a very condensed mass of elementary particles. This nucleus led to the formation of our stars, galaxies, and planets. These processes were of a type which, today, we call thermonuclear, i.e. matter converted into energy. Hydrogen was the first (and simplest) element that was formed and is still the most common element in the universe. It also makes up the core of our sun, which is estimated to be at a temperature in excess of 15 million degrees and a pressure of many billions of atmospheres. Every hour approximately 16 billion tons of solar mass is converted into energy through fusion of hydrogen to heavier atoms. The cores of heavier and bigger stars are at even higher temperatures and pressures. This allows the fusion of other atoms to form heavier elements such as carbon, oxygen, silicon, and eventually iron.

Most of the energy known to us has its source in the sun, i.e. solar energy. The sun radiates energy in all directions in the form of electromagnetic waves, including heat and light. A tiny fraction of this energy is beamed toward the earth, of which one-third is reflected back to the sky before ever reaching the ground. Of the remaining two-thirds, some is absorbed by the atmosphere, causing it to heat. The rest is intercepted by the earth in the form of light and heat, of which a mere 0.03% is transformed into biomass through photosynthesis. Biomass can be used as food and consumed by animals and humans, directly burned, or used to produce secondary fuels such as methanol and ethanol. If biomass is shielded from air, it decays and under suitable conditions, can eventually turn into fossil fuel-- coal, oil, and natural gas.

Because of different vegetation and rock formations, solar radiation does not heat the land and the ocean uniformly. Differential heating of land and water surfaces results in wind patterns and the ocean waves, two major sources of renewable energy. The earth is also being affected by the gravitational forces of its neighboring celestial bodies, most notably the moon, which is the cause of tides and much of the wave patterns in open channels. Tidal power plants are constructed to exploit the variation in the tides' potential energy in the same way that hydroelectric power stations use the potential energy from falling water.

Both nuclear fuel and geothermal energy have roots in the earth's early stages of development. Uranium fuel is the remains of the radioactive decay of the heavy materials and isotopes formed during the early stages of cosmic nuclei cooling. Uranium can be extracted from earth and processed for use as fuel in a nuclear reactor. What remains inside the earth continues to produce heat and form magma that can find its way to the earth's surface. Along the way it heats rocks, minerals, and water reservoirs which together constitute our geothermal resources.

Forms of Energy

There are many different ways that we can classify energy. For example we can classify all forms of energy as potential or kinetic, external or internal, primary or secondary, renewable or non-renewable, or in terms of its application as thermal, mechanical, chemical, electrical, nuclear, etc.

Potential energy is stored energy. This energy can be stored in atomic nuclei, molecular bonds, or gravitation. **Kinetic** energy is energy in action such as microscopic motion of atoms and molecules (heat), or gross motion of matter (falling rock, wind or river).

External energy is defined with respect to some outside frame of reference. Energies contained in winds, waves, and falling waterfalls -- commonly referred to as *mechanical energy* -- are of this type. **Internal** energy is energy locked within the internal structure of atoms and molecules. Examples are mass, chemical, thermal, light, and nuclear energy.

Mass is energy of matter by virtue of its own existence. Until 1905, when Einstein formulated the general theory of relativity, mass was not considered a form of energy. Einstein's famous formula $E = mc^2$ (where m is the mass and c is the speed of light in vacuum) gives a relationship between mass and the energy associated with it. It expresses the amount of energy that is given off if matter is completely annihilated. *Chemical energy* is the energy locked in the molecules of various substances. The energy stored in a molecule of carbon dioxide is that energy which holds two atoms of oxygen with one atom of carbon together. Biomass and fossil fuel store their energy as chemical energy. *Thermal energy,* or heat, is the energy associated with the random motion of individual molecules of matter. It can be considered as kinetic energy at the microscopic level. Geothermal and a big portion of solar energy are in the form of thermal energy. *Light energy* (also called radiant or electromagnetic energy) is the energy associated with a quantum of energy called a "photon" as it travels through outer space. The sun is the source of light energy. *Nuclear energy* is the energy trapped in the nucleus of an atom. This energy can be liberated by splitting apart a large nucleus to form two or more lighter atoms (fission) or by combining two light atoms to form a heavier atom (fusion).

Depending on its source, energy can also be divided into primary or secondary. **Primary** energy sources are those sources that can be directly converted to heat or mechanical work. The human intervention is limited only to extraction, cleaning, and separation, without changing the physical or chemical characteristics of the sources.[6] There are five main sources of primary energy that we use today; these include fossil fuel, nuclear, geothermal, solar, and tidal.[7] **Secondary** energy comes from transformation of primary energy. Electricity is a secondary source of energy that can be produced from any of the primary sources mentioned above. (Table 1-1).

Table 1-1. Energy Forms

Primary	Secondary
Fossil fuel Solar (Wind, Hydro, Biomass) Tides Geothermal Nuclear	Town gas (from coal); Gasoline, heating oil, diesel, and jet fuel (from crude); Alcohol, synfuel, and charcoal (from coal, oil shale, biomass) Hydrogen (from fossil, nuclear, solar) Electricity (from anything)

Depending on their long-term availability, energy resources can also be classified as either renewable or non-renewable. **Renewable** resources are those that will replenish themselves naturally in a relatively short period. Examples of renewable energy are solar, wind, hydroelectric, ocean thermal, waves, and tides. **Nonrenewable** resources are fossil fuels and uranium. Geothermal hot water, steam reservoirs and some material such as peat (decayed vegetable matter) and wood can be considered renewable if the rate of usage is small enough so as to allow their natural replenishment over time.

Units of Energy

There are two major systems of measurement in the world today: The United States Customary System (USCS) and the International System of Units (SI or metric). The metric system was adopted by the General Conference on Weights and Measures as the international system of units to be used in all international commerce. The scientific community, for the most part, has adopted this system to report findings and carry out scientific calculations. The United States is the last major industrial nation which has not converted fully to the metric system, a delay that has put us at a disadvantage in world trade.

[6] OECD/IEA/Eurostat, Energy Statistics Manual, 2005. http://www.iea.org/Textbase/publications/free_new_Desc.asp?PUBS_ID=1461

[7] Some consider waste as a primary source, as it is a surplus from any other process that has no further use for that particular process.

Each system has its own fundamental units and other quantities (including energy) can be expressed in terms of them. The fundamental units of the USCS are foot (for length), pound (for weight) and second (for time). All other units can be reduced to these units. For example, velocity is given in feet per second (ft/s), work and energy are given in pound-feet (lb-ft), and power is given in horsepower – equal to 550 lb-ft/s.

In the SI system of units, fundamental units of measurements are kilogram (for mass), meter (for length), and second (for time). Weight is a derived quantity defined as the force of gravity acting on an object of a given mass:

$$W = m \cdot g \qquad (1\text{-}1)$$

The unit of weight is newton, defined as the force of acceleration of a mass of 1 kg by 1 m/s^2 (1 N = 1 kg. m/s^2). In this equation, m is mass in kilogram and g is the gravitational acceleration. On earth g = 9.8 m/s^2 = 9.8 N/kg. In other words, objects on the influence of the earth gravity will accelerate at the rate of 9.8 meters per second (22 miles per hour) for each second they fall. Alternatively, it can be concluded that objects experience a force of 9.8 newtons per each kilogram of mass they posses.

In Chapter 2, we will define work as the energy needed to displace a force by a certain distance; it is therefore reasonable to conclude that a newton-meter commonly referred to as joule, is an appropriate unit for both work and energy (1 J = 1 N.m). Other units that have been derived and are used in this text are the watt (1 W = 1 J/s) for power, the ohm (Ω) for electrical resistance, the volt (V) for electromotive force, and the ampere (A) for electric current.[8]

A Note on Notations...

FYI ...

It is customary that units are represented by their letters in lower case (except at the beginning of a sentence) when they are spelled out -- kilograms, pascals, newtons, and meters. When abbreviated, they remain in lowercase unless they are proper names-- kg, Pa, N, and m. When we are speaking of individuals, we use the capital letters in accordance with English nomenclature-- Newton, Volt, or Pascal. Prefixes follow the notations shown in the table on the right.

Example: According to *Newton*'s second law of motion, the weight of an object (in *newtons*) is measured as the product of its mass (in *kilograms*) and the gravitational acceleration constant (in *meters per second squared*). Calculate the weight of an average-sized apple with a mass of 0.1 *kg* at a point with a gravitational acceleration constant of 10 m/s^2.

Solution: Weight is found from equation (1-1) as:

W = m.g = (0.1 *kg*)(10 m/s^2) = 1 *N* (or 1 *newton*)

Power of ten notations

	Prefix	Symbol
10^{-12}	pico	p
10^{-9}	nano	n
10^{-6}	micro	µ
10^{-3}	mili	m
10^{3}	kilo	k
10^{6}	Mega (million)	M
10^{9}	Giga (billion)	G
10^{12}	Tera (trillion)	T
10^{15}	Peta	P
10^{18}	Exa	E
10^{21}	Zeta	Z
10^{24}	Yotta	Y

[8] The NIST Reference on Constants, Units, and Uncertainty (*http://physics.nist.gov/cuu/Units/index.html*).

The fact that the SI system is based on 10 and is much easier to use than US customary units is not the reason the SI has found the popularity that it receives today. The main difference is the choice of weight in USCS and mass in the SI system as the fundamental units. A truck that has a mass of 2,000 kilograms in Detroit is still 2,000 kilograms in Los Angeles, or in Paris. In fact, if you load the truck and take it to the moon, it still has the same mass. On the contrary, a truck which weighs 4,000 pounds in Detroit will have a slightly different weight in Paris. Even the weight in Detroit varies by day, as the temperature and pressure changes. If you ship the same truck to space, it will have no weight at all.

In addition to the USCS and SI units, other units have been traditionally used (and unfortunately are still in use today). In the case of energy, these are BTU, therms[9], quad[10], barrels of oil[11], and even tons of TNT[12]. In order to conform to the rest of the world, we will concentrate primarily on SI units in this text. Since many people, especially in the United States, are most familiar with the USCS units, we will include these units when we feel the digression aids in comprehension. The unit conversion tables are given in Appendix B.

Example 1-1: The latest data gives the US energy consumption in 2007 as 101 Quads. Express this in BTU, joules, barrels and cubic kilometers of oil.
Solution: From the table of unit conversion given in Appendix B:

$$1\ \text{Quad} = 1\ \text{Quad} \times \frac{10^{15}\ \text{BTU}}{\text{Quad}} \times \frac{\text{barrels of oil}}{5.80\text{x}10^{6}\ \text{BTU}} \times \frac{0.159\ \text{m}^3}{\text{barrel}}$$

Or:
101 Quads = $1.01\text{x}10^{17}$ BTU = $1.07\text{x}10^{20}$ J = $1.74\text{x}10^{10}$ barrels of oil = 17.4 billion barrels of oil (bbo) = 27.6 cubic kilometers of oil

Example 1-2: Express the weight of a 180-pound person in newtons.
Solution: A person weighing 180 pounds (we mean pound -force or lbf) has a mass of 180 pounds (we mean pound-mass or lbm). The same person has a mass of 180/2.2 = 82 kilograms and a weight of 82x9.8 = 800 newtons.

Energy Supply and Energy Demand

Throughout history, humans have depended on energy to meet their needs for cooking, heating, transportation, and other daily activities. The primitive man depended on energy in the form of food for his survival.

[9] Therm is used mainly in the United States for metering household natural gas (1 therm=100,000 BTU).
[10] A quad is 10^{15} BTU or 1.055 EJ.
[11] A barrel of crude oil: 5.8 million BTU.
[12] Trinitrotoluene commonly known as TNT is a major constituent of many explosives, with the energy equivalent of one million food calorie per ton (strange but true!). TNT has a lower energy density than gasoline (32.2 MJ/kg), or even sugar (17 MJ/kg). The main attractiveness of TNT is that it does not have to be mixed with air to burn and in its ability to release this energy rapidly.

Table 1-2. Daily Per Capita Consumption of Energy (x1000 kcal)

	Primitive man 1,000,000 years ago	Hunting man 100,000 years ago	Early agricultural man 5,000 years ago	Advanced agricultural man 1,000 years ago	Industrial man 100 years ago	Modern technological man today
Transport-ation				1	14	63
Machinery*			4	7	24	91
Heating		2	4	12	32	66
Food**	2	3	4	6	7	10
Total	**2**	**5**	**12**	**26**	**77**	**230**

* Agricultural and industrial
** Including animal feeds
Source: Cook, E., "The Flow of Energy in an Industrial Society," Scientific American, *p. 135, 1971.*

The hunting man depended on energy 2.5 times more than the primitive man, as he learned to burn wood for cooking and heating. The demand increased even more when early human used animals to harvest crops. As agricultural society became more developed, wind and wave energies were harnessed to complement animal power to provide energy needed for farming and for transportation. The industrial revolution brought about by invention of the steam engine required increasingly more energy, mainly coal (See Table 1-2). By mid twentieth century, as society became more technological, and as demand for energy grew, coal was supplemented by petroleum, natural gas, and nuclear fuel. As Figures 1-2 and 1-3 indicate, the demand for energy has increased dramatically in the last few decades and is expected to continue to increase for the next few decades.

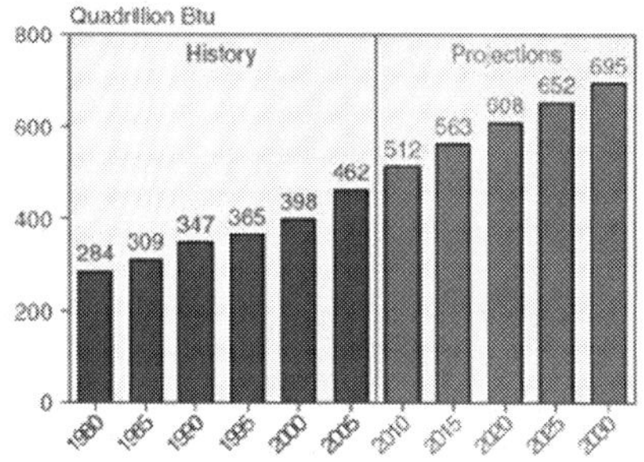

Figure 1-2
World Energy Consumption, 1990-2030.
Source: International Energy Outlook, 2007, Report No. DOE/EIA-0484.

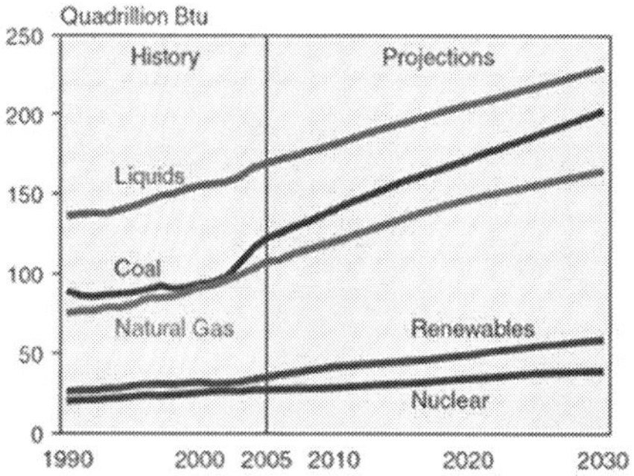

Figure 1-3
Historical and projected trends in energy demand between 1990-2030
Source: International Energy Outlook, 2007, Report No. DOE/EIA-0484.

Because of various economical and technological factors, most of the energy used today is from non-renewable sources – mainly fossil fuels. Worldwide, petroleum makes up most of the world energy production, followed by natural gas, coal, hydroelectric and nuclear resources. Excluding hydroelectric, renewable energy has been almost completely ignored and constitutes around 1-2% of the total energy production.[13]

Energy Use in the United States

Currently, energy consumption in the United States is one of the highest in the world and is expected to remain so in the foreseeable future.[14] The US, however, has long passed its peak oil production, and the reduction in petroleum had to be offset with other fuels, mainly coal and natural gas, or by foreign imports.[15] What is clearly evident is that energy resources are continuing to dwindle as both population and per capita consumption increase.

> **Question:** At the current rate of consumption, every American uses energy at about twice the rate of Japanese or Europeans, and

[13] US DoE, Energy Information Administration, Monthly Energy Review, July 2003.
[14] As we shall see later, energy cannot be consumed, only transformed from one form to another.
[15] "Annual Energy Outlook with projection to 2025," Energy Information Administration website (*http://www.eia.doe.gov/oiaf/aeo*).

US Per Capita Energy Consumption

Did You Know That ...?

Every year, an average American consumes energy equivalent to:

- 10 tons of coal, or
- 10,000 m³ of natural gas, or
- 10,000 liters of petroleum.

compared to Indians, the per capita consumption is 28 times greater. What contributes to the Americans' higher rate of energy use?

Answer: The large discrepancy is not only due to the large economies of the United States (and Canada), but is also a reflection of North Americans' appetite for large cars, luxury items, and personal comfort.

Table 1-3. Total U.S. Energy Consumption by Source, 2003

Energy Source	*Total Energy (Quad)*	*Percent of Total*
Petroleum	38.2	39.0
Natural Gas	23.5	24.0
Coal	22.5	23.0
Nuclear	7.8	8.0
Hydro	2.7	2.7
All others	3.3	3.3
Total	**98**	**100**

Source: EIA, Annual Energy Outlook, Report No. DOE/EIA-0383 (2005).

Primary energy consumption in the United States by sources is given in Table 1-3 and Figure 1-4. About 86% of all energy used in the United States is from fossil fuels. All forms of renewable energy (biomass, hydroelectric, solar, wind, etc) make up 6% of energy needs. Because of its portability, convenience of use, and relatively greater energy density, oil is the fuel of choice for a variety of applications. The United States, as the biggest consumer of energy, uses one quarter of the total 85 million barrels per day of petroleum produced around the world. Petroleum consumption is limited mainly to transportation (Figure 1-5) and certain industrial processes (primarily petrochemical). Figure 1-6 gives the US total energy consumption by sectors. Each of the transportation, industrial processes, and residential and commercial sectors consume roughly a third of all energy. The pattern of energy use described here is not limited to the US alone; similar trends are observed in other regions of the world.

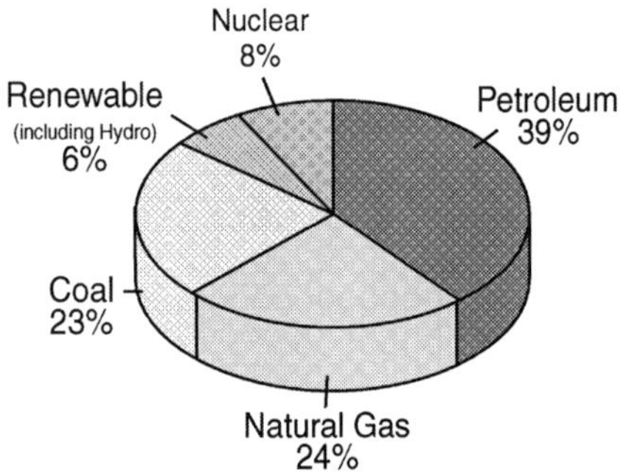

Figure 1-4
US total energy use by source.

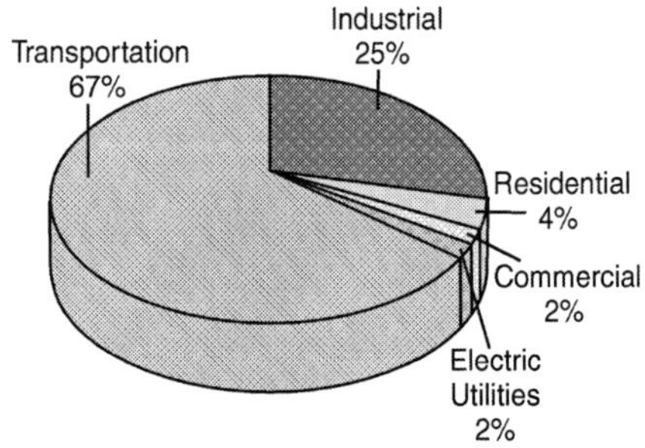

Figure 1-5
US petroleum consumption by sector.

Energy Use in Other Parts of the World

The recent data on energy consumption suggests moderate to rapid increases in the rates of fossil fuel consumption for various countries (Table 1-4). Russia's growth was small due to instability following the breakup of the Soviet Union. US, Canada, and other industrial countries have had moderate growths in the past 20 years, whereas energy use in many developing countries increased significantly. For example, for the same 20 year period, the energy consumptions in China and India were tripled. Even Iran, a country with vast petroleum reserves used much of its production for domestic consumption. The rapid increase is explained by a relatively faster rate of population growth and a higher per capita consumption reflected in higher total industrial outputs of these countries.

It is expected that in a near future, many under-developed countries will demand a greater portion of the world's petroleum resources. The increase will be higher in developing countries that aggressively push for industrialization and more rapid economical growth.

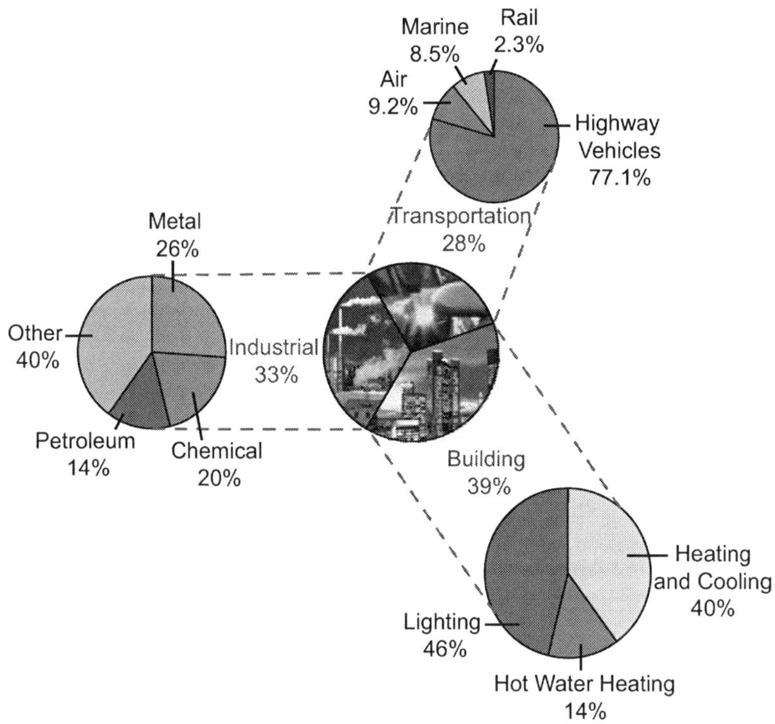

Figure 1-6
US total energy use by sector

Energy Reserves

It is very difficult to predict the total amount of energy reserves in the world, as by definition, renewable energy sources will be infinite. What is certain is that we will sooner or later run out of nonrenewable resources, primarily fossil fuels. The time that it will take to use up the remaining fossil resources is closely related to the accuracy of estimates of proven energy reserves and the rate at which new resources are found and extracted. The rate of consumption of fossil fuels is equally hard to predict, as it is affected by numerous factors such as population growth, economic activity, cost, environmental concerns, and the availability of other resources. Many studies suggest that within the next 30 to 50 years, we will have depleted most, if not all, of the natural gas and oil in the world. Even coal has a limited life and cannot be the answer to our long-term energy needs.

Table 1-4. Energy Consumption for selected Countries (Quads)

Country	*1984*	*1994*	*2004*	*% rise per year*
Canada	9.8	12.0	13.6	1.6
United States	76.8	89.3	100.4	1.3
England	8.4	9.5	10.0	1.0
Japan	15.7	20.2	22.6	1.8
Russia	NA	29.3	30.0	0.2
China	20.5	34.0	59.6	5.3
India	5.5	10.0	15.4	5.2
Iran	2.2	3.7	6.4	5.3
World Total	**299.9**	**357.3**	**446.4**	**2.0**

Source: EIA, Annual Energy Outlook, Report No. DOE/EIA-0383 (2006).

Doubling Time and Exponential Growth

Many phenomena (such as population growth, savings in a bank, and the spread of an epidemic) behave in a nonlinear fashion such that their changes vary in an exponential manner from one year to the next. An important characteristic of exponential growth is that a quantity grows by a certain percentage each year. This might not seem to be much at the beginning, but the cumulative amount can become quite staggering over time. For example, consider a city of 100,000 inhabitants. Because of immigration, increased birth and decreased death rates, its population expands at a rate of 5% annually. After the end of the first year, population will grow to 105,000 and then to 110,250 at the end of the second year.

The Incredible Power of the Power of Two[i]

FYI ...

As legend has it, chess was invented by a mathematician who served the king of Persia. When the king asked what the inventor wanted as a reward for his great game, the man simply asked for one grain of rice in the first square of the chess board, and double the number in the next square until all the squares are filled. The king laughed at this humble request; little did he know that the country would be bankrupt before long. As you might have guessed, the number becomes staggering as it climbs to 2 in the second square, 4 in the third, 8 in fourth, and $2^{63} \sim 10^{19}$ grains of rice in the last square, equal to the empire's total rice production for roughly two thousand years.

Although we cannot attest to the accuracy of this story, this example should be a reminder of how exponential growth can quickly lead to amazingly large numbers in a relatively short time.

[i] Some contend that chess was adapted from the India game "Chaturanga", but archaeologicall evidences point to the origin to Sasanid Dynasty in Persia (224-651 CE). See http://www.cais-soas.com/CAIS/Sport/chess.htm.

A simple calculation shows that this city will double to 200,000 people in about 14 years, double again to 400,000 in 28 years, reach 3 million after the passing of one generation (70 years) and grow to over 13 million after a century.

Old French Riddle:
At first there is only one lily pad in the pond, but the next day it doubles, and thereafter each of its descendants doubles. The pond completely fills up with lily pads in 30 days. When is the pond half full?

Answer: on the 29th day.

It is constructive to give a convenient method to express the growth of quantities that vary in this fashion (that is, those which grow by a fixed percentage every year) in terms of their doubling time. Doubling time is defined as the time it takes a quantity to double, and can be reasonably approximated by dividing 70 by the percentage growth rate:[16]

$$T_2 = \frac{70}{Percentage\ \ Rate\ \ of\ \ Growth} \tag{1-2}$$

As we will see in Chapter 11, the same equation can be used to calculate the time which it takes a sample which decays in population (i.e. has a negative growth rate) to reduce in half. This is known as half-life, an example of which is the activity of radioactive material or any other sample which decreases yearly by a fixed percentage.

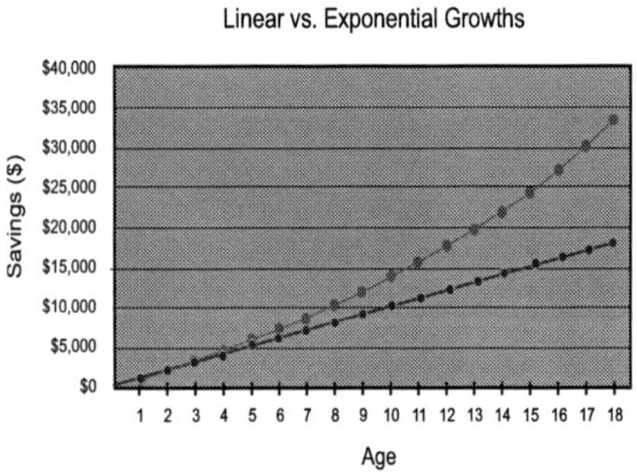

Ex. 1-3

Example 1-3: A couple is planning to save money to pay for their new-born son's college tuition by putting aside $1000 a year. How much would the child have when he enters college at 18-years of age, if his parents decided to deposit the money in a) a safe deposit box, or b) a saving account that paid 7% interest annually?
Solution: If the money is put in a deposit box, the growth is linear—the annual increase is always the same. The saving is $1,000 after the first year, $2,000 after the second year, etc. At age 18, the child will have $18,000 in savings.

If the money is invested in a saving account, the percentage rate of

[16] Equation 1-2 assumes that growth is taking place continuously. This is definitely true in the case of population growth and energy consumption. In many other instances, such as interest paid by a bank, growth comes in steps. For example a bank that pays 3% interest per annum compounded monthly pays 3/12=0.25% each month which results in a slightly different growth rate. As we will see in Chapter 15, banks and other financial institutions often apply the "Rule of 72" - divide 72 by the interest rate to find the time in which a saving deposit doubles in value.

increase is the same each year but the total increase is exponential. Let's say the parents deposit $1,000 on the child's first birthday. The bank pays 7% ($70) in interest for the first year. With the additional $1000 deposit at the end of the first year, the total money saved after the second birthday is 1,000(1.07)+1,000 = $2,070, and 2,070(1.07)+1,000 = $3,215 after the third. By his 18th birthday, the total deposit has grown to nearly $34,000 instead of $18,000 he would have had if no interests were paid.[17] The difference between the linear and exponential growths becomes substantial by the time the boy is ready for college.

Example 1-4: It is estimated that the world's petroleum production will peak sometime around 2005-2010, at which time we will have used up one half of all our petroleum reserves. If the total cumulative amount of oil which has been produced from the early days of discovery in the 1860s until today is estimated at 800 billion barrels, how long would it take before we deplete all our petroleum reserves? According to the latest projections,[18] the world energy consumption will increase by 57% between 2002 and 2025.
Solution: What may come immediately to mind is that since it took us 145 years to use half of our reserves (1860-2005), we will have oil for another 145 years. This is unfortunately far from the truth! In fact we will use as much oil as we have previously used in only one doubling time. Assuming demand continues to grow at the rate of 2.47% a year (57% in 23 years), the doubling time is calculated from Equation 1-2 as 70/2.47 = 28.3 years. Unless we cut consumption or find new reserves of oil and new sources of energy, we expect to run out of oil approximately 28 years after its peak -- by 2033-2038.

Example 1-5: According to a US Census, in 1973, the national birth and death rates were 15.6 and 9.4 per thousand. Furthermore, the Immigration and Naturalization Services (INS) data showed that 400,000 people or roughly 1.5 persons for every 1,000 people immigrated into the United States during that year. Calculate the population growth rate in 1973. If the population were to increase at the same rate, when would the US population reach twice that of 1973?
Solution: The growth per 1,000 persons was 15.6-9.4+1.5 = 7.7 or 0.77%. If growth were to continue at this rate, the US population would double the 1973 population in 70/0.77 = 91 years, in year 2064.

Example 1-6: What is the rate of population increase in a community in which there are only two children per family?
Answer: The community has zero rate of population increase. Essentially, each parent replaces their children for themselves.

[17] The cumulative value of all payments can be calculated by applying Equation (15-4) directly.
[18] "World Energy and Economic Outlook", Energy Information Administration, DOE/EIA-0484, July 2005.

Example 1-7: A bacterial colony fills up a jar at 12:00 pm. If the bacteria double in population every minute, when is the jar 1/16 full?
Answer: The jar is half full one minute before noon at 11:59, ¼ full at 11:58, 1/8 full at 11:57, and 1/16 full at 11:56.
The data on the use of petroleum shows that the world had used only 1/8 of its reserves in 1973, and the consumption has doubled every ten years for the last few decades. As of the writing of this text, roughly half of all petroleum reserves are used up; we are at 11:59 hours. The clock ticked 11:56 am during the oil crisis of early 1970s.

Example 1-8: A popular brand of antibacterial cleaner advertises that it kills 99.9% of all kitchen bacteria. Assuming that bacteria can double their population every 24 hours, how often must the disinfectant be applied to maintain kitchen bacteria in check?
Answer: Assuming the company's claim is accurate, after the antibacterial application 0.1% (or 1 in every 1000) bacteria remain. For the population to reach its original level, it must increase 1000 fold or about 10 doubling times. The disinfectant must be applied at least once every 10 days!

Energy and the Environment

As nonrenewable energy sources are depleted, the level of pollution in the environment is rising. Because fossil fuels make the bulk of our energy use today, most pollution is combustion generated. The harmful effects of these pollutants on the environment are becoming more pronounced, as evidenced by increased level of toxins in our atmosphere, rise in global warming, and depletion of ozone in the stratosphere. In addition to fossil fuels, the burn-up of nuclear fuel has resulted in pile up of tremendous amount of nuclear waste. Unlike other pollutants that can become ineffective in periods of many months or years, the nuclear waste material remains radioactive for a very long time and practically permanent relative to the span of human life.

Technology Options

Options for averting future energy crises exist on both the demand and supply sides. **Demand-side management**[19] implies a reduction of consumption by consumers of energy, either voluntarily or mandated by the government. It often implies promoting technologies that increase efficiency without sacrificing comfort.[20] Examples of demand-side management are better insulation, fluorescent lighting, energy-efficient appliances, hybrid vehicles, and using waste energy to produce heat (cogeneration).

[19] Many consider the demand-side management as tools and strategies for reducing energy use at a particular time as well as total consumption, irrespective of its impact on the environment or inconvenience that it causes on consumers. Here we concentrate only on technologies that are sustainable, i.e. do not harm the environment or the people.
[20] According to this definition, reducing the speed limit from 65 mph to 55 mph and turning down the thermostat from 70°F to 65°F in winter are not considered efficiency measures because they are accompanied by a certain degree of sacrifice.

Supply-side management requires finding new resources and processes that increases the chance of discovery, facilitates extraction, and allows more efficient processing (higher yields). New exploration techniques are available that can detect deeper and more out-of-reach locations in mines and reservoirs. Advances in engineering have made it possible to exploit vast hydro and solar resources with costs fast approaching those of conventional energy sources. Nuclear breeder technology is extending the lifetime of our nuclear supply by order of magnitudes. We are finding unconventional sources of energies (hydrates, tar sands, oil shale, etc.) that may open the door to vast sources of energy. Finally, in the not-too-distant future, fusion technology may provide us with the ultimate source of clean and cheap energy.

Status of Energy Technologies

Until the advent of the industrial revolution in the 18th century, humans used mainly animal power and mechanical power provided by wind and water. During the eighteenth century wood became the primary energy source used for cooking, space heating, and to fuel the industrialization. As forests diminished, coal became the primary fuel for powering steam locomotives. New applications were also found in the form of coke in casting and smelting, and as town gas for lighting. Coal consumption increased throughout the nineteenth century and peaked as petroleum was being discovered as a more convenient fuel that could be transported in pipelines and used directly in internal combustion engines (Figure 1-7). As new petroleum resources were discovered and better techniques were developed for refinement, petroleum found more and more applications in a wide variety of technologies (including pharmaceutical, petrochemical, and other industrial sectors) and still dominates the energy market as the fuel of choice. As a result of the rapid industrial growth, it is estimated that 30% of all stocks of nonrenewable energy (including fossil fuels) was used in the twentieth century. Although most of nonrenewable energy resources remain, the reign of fossil fuel is not expected to last much longer. It is believed that world oil has already peaked or is soon to peak in its production, and will be depleted by the end of this century. Use of nuclear fission is closely following the trend of fossil fuel as uranium reserves deplete, nuclear waste piles up, and problems associated with reactor safety make nuclear technology increasingly less attractive.

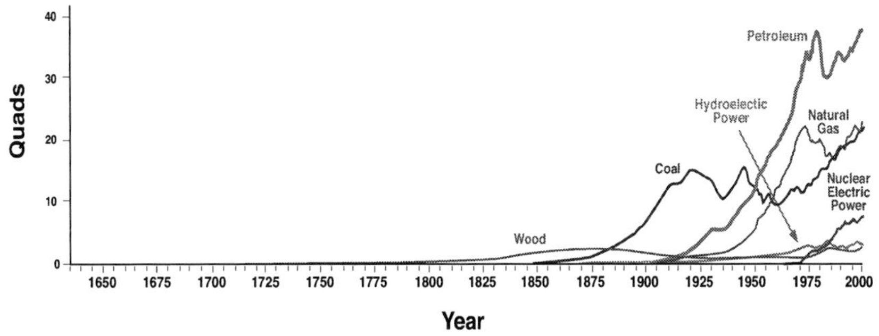

Figure 1-7
US Energy Consumption (Quads) by Source, 1635-2000.
Source: Energy Information Agency website, http://www.eia.doe.gov/emeu.aer/eh/intro.html.

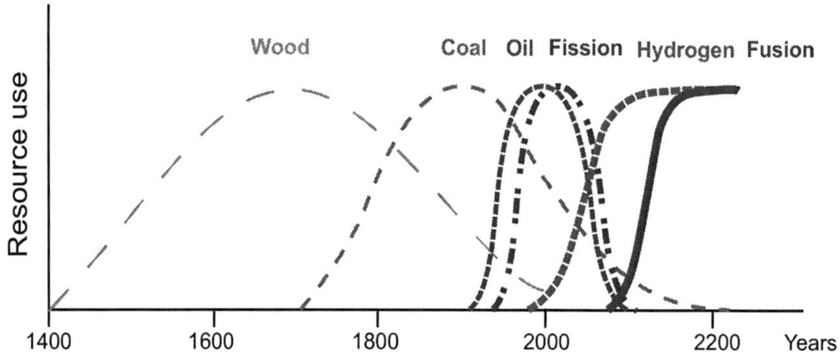

Figure 1-8
Energy sources: past, present, and future.

As energy becomes more scarce, the pressure to find alternative sources and switch to more efficient technologies increases. Although there have been major efforts to promote renewable sources of energy, most cannot compete economically with fossil and nuclear energy sources - yet. However, much research is being done and many have great hopes that renewable resources will play an ever-increasing role in meeting our energy needs during this century. Nuclear fusion remains a dream for the future and is not expected to be commercially plausible until the last decades of this century (Figure1-8).[21]

In the following chapters we will study some of these resources and how they will play a role in reaching our energy goals. New technologies are being developed to harness wind, biomass, tidal energy, and ocean currents, as well as thermal gradients in the oceans and under the ground. New materials are being found to produce cheaper photovoltaic, thermo-photovoltaic, and fuel cells. The next decade may prove to be one of the most interesting eras for energy enthusiasts. In fact, some economists proclaim, energy and the environment (as well as biotechnology and information technology) will become the most dominant sector affecting the economy in this century. Among the changes occurring in various energy sectors are:

Hydropower is going through major changes. The traditional methods of hydropower generation were waterwheels and hydroelectric plants. Although hydroelectric energy has already been exploited to its maximum potential in the United States and much of the industrialized world, small-scale hydro projects may serve as valuable resources in most of the developing countries. Research efforts are also underway, and some demonstration facilities have been built, to show the potential to harness the rise and fall of tides, underwater currents, and the motion of ocean waves.

Wind is another significant source of power. New advances in manufacturing processes have allowed the manufacture of lighter and stronger blades. Power electronics have also developed to a stage where variable-speed wind turbines with sophisticated control strategies have

[21] "Careers in renewable energy", DOE/GO-102001-1130, January 2001 (*http://www.nrel.gov/docs/fy01osti/28369.pdf*).

resulted in wind turbine designs with a wider range of operation and higher efficiencies.

Biomass is not only the source of all our food, but can also be used to generate energy. Although biomass is a renewable resource, it is not clean and, if burned, produces the same type of pollutants as fossil fuels. However, unlike fossil fuel, biomass does not produce greenhouse gases, and the overall emission is lower. Newer technologies allow higher productivity and cleaner burning. Genetically modified crops can be produced that are tailored to yield high heating values and burn with considerably less pollution.

Geothermal resources include hot water, steam, magma, and dry rock. With today's technology only hot water and steam reservoirs can be economically exploited, but works are underway to exploit the earth's natural thermal gradients. Geothermal heat pumps are available that use the relatively uniform temperature in shallow ground to provide winter heating and summer cooling with a payback time of only a few years.

Solar is another potentially huge, clean, and renewable source of energy, capable of producing both heat and electricity. Flat panel solar collectors are becoming increasingly more efficient and economical. For higher temperatures applications concentrators such as parabolic and dish collectors are manufactured with lighter yet durable materials. For electrical power generation, solar energy can be concentrated by means of a large array of mirrors and dishes (called a heliostat) that can heat up water in a receiver or be used to run a Stirling engine. Alternatively, electricity can be generated directly by photovoltaic cells which are becoming more and more efficient and at a cost of only a fraction of what they were just a few years ago. Thin films and organic plastics have been found that could substitute for silicon semiconductors -- the material used to manufacture most photovoltaic chips. Spherical solar cells are also being introduced in the market, which collect light from all directions, resulting in significantly better collection efficiency.

Hydrogen is considered by many to be the fuel of the 21st century, as much as wood was in the eighteenth, coal was in the nineteenth, and fossil fuel was in the twentieth century. Hydrogen is the cleanest of all fuels, whether it is burned or used to power a fuel cell. The only byproducts are heat and water vapor. The hydrogen technology is still in its infancy, and may still face unanticipated problems.

Nuclear Fission is the most controversial source of energy. Although power generated is clean and the cost is comparable to electricity generated by fossil fuels and hydroelectric plants, the unresolved issues of nuclear waste disposal and safety associated with these plants are daunting. In light of the terrorist attacks of September 11, 2001, these safety issues have taken an even higher priority. Work is being done to design safer reactors. One such design is called a Pebble Bed Reactor, designed originally by Germans and perfected by South Africans. Small grains of uranium, 0.5–mm in diameter, are placed in microspheres covered by layers of porous carbon and silicon graphite coating. Thousands of these microspheres are packed into pebbles, and then the pebbles are arranged in modular forms. The design is considered inherently safe, because fuel can never reach the melting temperature. Thus, the danger of meltdown scenarios similar to those in Three-Mile-Island and Chernobyl are considered highly unlikely. Furthermore, the graphite will act as a casket to contain the radioactive material even after the fuel is burned, eliminating the possibility of any radiation leak.

Nuclear Fusion is still considered the ultimate source of energy. About forty years ago, many had predicted that by the year 2000 fusion reactors would be operational producing cheap, clean, inexhaustible energy. Today, we are still coping with many of the same problems we were facing in the 1960s. The current prediction is that fusion may become practical only in the second half of this century. Time will tell whether we have to revise our predictions again.

Summary

After food, air and water, energy is the most basic need of human beings for survival. Without it we could not cook, provide heating or cooling, move from place to place, and could not have achieved any of the technological advances of the present time. The unprecedented economic growth of the past century was made possible only by the availability of vast resources of cheap petroleum. At the present time we are relying primarily on the following sources for satisfying our energy needs:

- Fossil fuels provide roughly 80% of total global consumption. Fossil fuels are, however, associated with many environmental problems such as global warming, ozone depletion, acid rain, and toxic air pollutants such as oxides of nitrogen, sulfur, and carbon.
- Nuclear energy is responsible for about 6.6% of the world's total consumption. However, it generates radioactive and toxic wastes that can remain for a very long time.
- Hydroelectric and biomass energy constitute 7% of total consumption. Hydroelectricity has relatively minor environmental impact and is renewable. Biomass is mainly in the form of wood and is renewable as long as it does not result in deforestation of existing forests.
- All the other forms of energy including solar, wind, and geothermal make up only between 1-2% of the total energy consumption.

With an expected increase in consumption of fossil fuels, we would also expect an increase in the rate of carbon emission into the atmosphere. As petroleum and natural gas reserves dwindle, coal will probably be the ultimate source for their replacement. When burned, coal, peat, and other solid fuels produce more emissions per unit amount of heat than petroleum or natural gas. They are also the fuel of choice for many developing countries that have the largest expected percentage increase in consumption, and who have limited financial resources to buy cleaner petroleum and natural gas. In the next couple of decades, the environmental consequences of fossil fuel consumption will undoubtedly be a major source of contention among various industrial and developing countries.

As we are depleting our valuable fossil fuel resources, we must find solutions to our increasing energy needs, which must take one or more of these three forms: to stop or even reverse the population growth, to conserve energy and reduce consumption, or to find new energy supplies, mainly in the form of renewable resources. How the future looks will largely depend on how we act to achieve these goals today.

Additional Information

Books

1. Meadows, D. H., et al., The Limit to Growth, Universe Books, 1972. Also see, The Limit to Growth: The 30-Year Update, Chelsea Green Publishing Company, 2004.
2. Diamond, J., Collapse: How Societies Choose to Fail or Succeed, Penguin Group, USA, 2004.
3. Cleveland, C. J., Encyclopedia of Energy, Elsevier Direct Science, 2004.

Periodicals

1. *The International Journal of Energy*, Science Direct Elsevier Publishing Company.
2. *Applied Energy,* Elsevier Publishing Company.
3. *Journal of Energy Resource Technology*, ASME International.
4. *The Energy Journal,* The quarterly journal of the IAEE's Energy Education Foundation, *(http://www.iaee.org/en/publications/journal.aspx).*

Government Agencies and Websites

1. Energy Citation Database, US Department of Energy (*http://www.osti.gov/energycitations*).
2. Environmental Protection Agency (*http://www.epa.gov*).
3. US Department of Energy *(http://www.doe.gov).*
4. The NIST Reference on Constants, Units, and Uncertainty *(http://physics.nist.gov/cuu/Units/index.html).*

Non-Government Organizations and Websites

1. The Club of Rome (*http://www.clubofrome.org*).
2. The Sierra Club *(http://www.sierraclub.org).*

Exercises

I. Essay Questions:

1. Why has per capita energy use in technological society increased so much? Do you expect this trend to continue in the 21st century?

2. What are the main factors responsible for the rapid rise in energy consumption in the developing world?

3. Imagine that China and India were to enjoy the same high standard of living as people in Canada and the United States. What percentage of the global energy supply would these countries demand? How would this affect the energy supply to the rest of the world? Discuss political and environmental implications.

4. Describe why natural gas is a cleaner fuel than heating oil. Why is coal considered to be the dirtiest fossil fuel?

5. Some argue that, if external costs associated with different fuels were considered, renewable energy sources would not necessarily be more expensive than fossil fuels. Explain and give examples.

6. What are the major problems with using biomass as a fuel?

7. What are the sources of nuclear, wind, wave, and geothermal energies?

8. What forms of energy sources constitute renewable resources, and what forms constitute non-renewable energy? How is the distinction made?

9. What is the difference between internal and external forms of energy? Give examples of each.

10. Indicate whether each unit represents energy (E), power (P), or neither (N)
    a. joule
    b. megawatt
    c. newton
    d. kilowatt-hour
    e. joule/day
    f. watt/day
    g. newton-meter
    h. horsepower
    i. ton of coal
    j. billion barrels of oil per year

II. Problems

1. Express the weight of a 220-lb man in newtons. What is his mass in kilograms?

2. A candy bar has 100 Calories. What is its food energy content in joules?

3. If the total energy consumption in the US was about 100 quads in 2006 and assuming the doubling time is 20 years, calculate the expected energy consumption in 2046.

4. According to the US Department of Energy, the total US energy consumption was 100 Q in 1996. During the same period, US crude oil production was 5 million barrels per day. Another 12 million barrels of crudes and 3 million barrels of petroleum products were imported daily. Express the US daily petroleum consumption in quads. What percentage of total US needs was provided by petroleum?

5. How much does an apple with a mass of 100 grams weigh in newtons?

6. According to the data, there are 22 births and 9 deaths annually for every 1000 persons living in the world. What is the annual rate of growth in the world's population?

7. The US used roughly100 quads of energy in 2006. Express the US annual energy consumption in:
   a. BTU
   b. EJ
   c. kWh
   d. Barrels of oil

III. Multiple Choice Questions:

1. According to the predictions by the Club of Rome scientists, the world supply of petroleum
   a. Will diminish in the next 10 years

b. Will continue to increase for the next 20 years
c. Will diminish by 2100
d. Should have already been depleted
e. Would never run out, because new reserves would always be found

2. The rise in oil prices of the mid-1970s can be attributed to
   a. The Arab-Israeli war
   b. The Arab oil embargo
   c. The energy crisis
   d. Additional investments for discovery of new oil reserves
   e. All of the above

3. Compared to gasoline, natural gas produces a lower amount of greenhouse gases because
   a. Gasoline is a byproduct of distillation, which involves processes that produce carbon dioxide
   b. Gasoline contains additives that produce carbon dioxide
   c. Gasoline has a smaller number of carbon atoms per molecule than natural gas
   d. Gasoline has a higher ratio of H/C than natural gas
   e. Gasoline has a lower ratio of H/C than natural gas

4. Energy is needed to
   a. Exert a force through a distance
   b. Lift a mass to a higher elevation
   c. Move an electric charge through a wire
   d. Accelerate a mass from rest to a higher velocity
   e. All of the above

5. Which of the following is considered to be a secondary form of energy?
   a. Geothermal
   b. Tides
   c. Food
   d. Electricity
   e. None of the above

6. What sector of the US economy consumes the greatest volume of petroleum?
   a. Residential
   b. Commercial
   c. Industrial
   d. Transportation
   e. Military

7. Which of the following statements are correct?
   a. Hydro energy refers not only to the energy from falling water from dams, but also to energy contained in tides, waves and ocean currents.
   b. Nuclear fission refers to the energy released from the breaking an atom apart. Fusion is the energy released from recombining two lighter atoms to a heavier atom.
   c. Heat is the thermal energy associated with the random motion of molecules.
   d. Solar thermal energy refers to harnessing the sun's rays to produce heat.
   e. All of the above.

8. Nuclear is the most controversial source of energy because
   a. Its cost is high compared to fossil fuels
   b. It will be a long time before the technology is matured
   c. The danger from the release of radioactive gases and the storage of radioactive wastes is too large
   d. Uranium, the fuel for nuclear reactions, is depleting quickly
   e. All of the above

9. The source of nuclear fuel can be traced back to
   a. The thermonuclear reaction in the sun
   b. The reaction between atomic particles in interstellar space
   c. The early stages of the earth's formation
   d. Photosynthesis
   e. All of the above

10. Some examples of renewable sources of energy are
    a. Nuclear, fossil, and geothermal
    b. Wind, ocean waves, solar, and nuclear
    c. Fossil, ocean thermal, biomass, and tidal
    d. Ocean currents, waves, solar, and biomass
    e. Mechanical, chemical, nuclear, and electrical

11. Chemical energy is
    a. Energy locked in molecules
    b. Energy by the virtue of an object's own mass
    c. Energy locked in a nuclei
    d. Energy associated with the motion of molecules
    e. Energy associated with the position of molecules

12. The SI system is preferable to the USCS because the
    a. SI system is easier to use
    b. SI units are independent of the location of the measurement
    c. SI system is finally adopted by the United States
    d. SI system has been universally used for a very long time
    e. Both a and b

13. Two convenient units for expressing a very large amount of energy are
    a. BTU and calorie
    b. Therms and Calorie
    c. Quad and TJ
    d. Tons of TNT and kWh
    e. All of the above

14. What is the approximate weight of a bag containing ten medium-sized apples?
    a. ~1 N
    b. ~10 N
    c. ~100 N
    d. ~ 1 kg
    e. ~ 1 lbm

15. A million people each have one million dollars. Their total assets are
    a. $ 2 M
    b. $ 1 B
    c. $ 1 G
    d. $ 1 T
    e. $ 1 P

16. One thousandth of one microsecond is
    a. 1 ms
    b. 1 μs
    c. 1000 μs
    d. 1 ns
    e. 1 ps

17. Per capita energy consumption in a modern technological society is roughly ____________ times that in the primitive society.
    a. 5
    b. 10
    c. 50
    d. 100
    e. 500

18. Which of the following countries had the highest rate of per capita energy consumption in the last 20 years?
    a. Canada
    b. The United States
    c. England
    d. China
    e. Russia

19. A 1200-W hair dryer has been operating for 20 minutes. What is the power and energy consumed by the dryer?
    a. 400 W, 400 J
    b. 400 W, 0.4 kWh
    c. 1200 W, 400 J
    d. 1200 W, 0.4 kWh
    e. 400 W/hr, 1200 J

20. Electricity can be generated using
    a. Mechanical energy
    b. Chemical energy
    c. Radiant energy
    d. Nuclear energy
    e. All of the above

21. In a modern nuclear power plant, energy is produced by
    a. Combining uranium atoms
    b. Burning uranium atoms
    c. Splitting uranium atoms
    d. The reaction between plutonium and uranium atoms
    e. All of the above

22. Worldwide, which source provides the most energy?
    a. Petroleum
    b. Coal
    c. Natural gas
    d. Hydroelectric
    e. Nuclear

23. Which source of energy is used most by Americans?
    a. Petroleum
    b. Coal
    c. Natural gas
    d. Hydroelectric
    e. Nuclear

24. Which of the following gases is considered most responsible for global warming in the atmosphere?
    a. Sulfur dioxide
    b. Carbon dioxide
    c. Water vapor
    d. Nitrous oxide
    e. Ozone

25. Solar, wind, hydropower, biomass, and geothermal are all renewable sources of energy because they
    a. Are free and abundant
    b. Can be converted directly into electricity
    c. Can be replenished by nature in a relatively short period of time
    d. Are relatively clean and pollution free
    e. All of the above

26. What percentage of the US electricity generation comes from renewable energy?
    a. 1%
    b. 6%
    c. 15%
    d. 30%
    e. 50%

27. Which of the following renewable energy sources currently provides the US with the most energy?
    a. Wind
    b. Waves
    c. Geothermal
    d. Solar
    e. Hydropower

28. Electricity is the movement of
    a. Photons
    b. Atoms
    c. Molecules
    d. Electrons
    e. Protons

29. Kinetic and potential energy are energies of ____________ and ____________, respectively.
    a. Springs and gravity
    b. Solids and liquids
    c. Position and motion
    d. Motion and position
    e. Velocity and height

30. __________ followed by __________ has the largest per capita consumption of energy.
    a. US, Canada
    b. Canada, US
    c. US, Russia
    d. US, Japan
    e. US, England

31. In comparison to an individual living in Japan, every American is using roughly ______________ energy.
    a. Half as much
    b. The same amount
    c. Twice as much
    d. 5 times as much
    e. 10 times as much

32. Every American on the average consumes energy equivalent to
    a. 10,000 liters of petroleum
    b. 10,000 liters of natural gas
    c. 10,000 barrels of oil
    d. 10,000 tons of coal
    e. 10,000 kilograms of TNT

33. The average American consumes most energy for
    a. Lighting
    b. Heating water
    c. Heating and cooling rooms
    d. Refrigerating foods

e. Transportation

34. Although the US has only ________ percent the world's population, it uses roughly _____ percent of the world energy.
    a. 1, 10
    b. 1, 25
    c. 5, 5
    d. 5, 25
    e. 25, 25

35. Which of the following renewable energy resources is currently the most heavily used in the world?
    a. Solar
    b. Biomass
    c. Geothermal
    d. Hydro
    e. Wind

36. Which of the following nonrenewable energy resources is currently the most heavily used in the United States?
    a. Natural gas
    b. Petroleum
    c. Coal
    d. Nuclear
    e. Hydro

37. Which of the following energy resources is currently the most produced in the United States?
    a. Natural gas
    b. Petroleum
    c. Coal
    d. Nuclear
    e. Hydro

38. Which of the following is not considered a source of energy?
    a. Fossil fuels
    b. Wind
    c. Geothermal
    d. Electricity
    e. Nuclear

39. As developing countries become increasingly industrialized
    a. They will need to import more oil
    b. There will be less oil available for industrialized countries
    c. There will be stiff competition for control of the oil resources
    d. The tension between third world and industrial countries will increase
    e. All of the above

40. The US's (and the world's) total output of energy in the order of decreasing resources are
    a. Coal, natural gas, and petroleum
    b. Petroleum, wind, and nuclear
    c. Petroleum, natural gas, and coal
    d. Petroleum, nuclear, and solar
    e. Petroleum, hydro, and nuclear

41. The sun gets its energy from
    a. Burning hydrogen
    b. Burning biomass
    c. Nuclear fission
    d. Nuclear fusion
    e. All of the above

42. Which of the following resources are considered primary?
    a. Biomass
    b. Alcohol
    c. Hydrogen
    d. Town gas
    e. None of the above

43. Most renewable energy sources can be traced back to
    a. Solar energy
    b. Hydro energy
    c. Nuclear fission
    d. Nuclear fusion
    e. All of the above

44. Which of the following statements is correct?
    a. Geothermal energy meets a large portion of today's energy demand.
    b. Geothermal energy includes steam and hot water reservoirs, magma, and dry rock.
    c. Geothermal energy has its roots in the interaction between earth and the moon.
    d. Dry rock plays a significant part of geothermal energy exploited today.
    e. All of the above.

45. Which fuel is used to generate the most electricity in the US?
   a. Coal
   b. Petroleum
   c. Natural gas
   d. Nuclear
   e. Wood

46. Which of the following sets of numbers represents a linear growth?
   a. 5, 8, 11, 14, 17
   b. 5, 8, 12, 17, 20
   c. 4, 8, 16, 32, 64
   d. All of the above
   e. None of the above

47. Which of the following sets of numbers represents an exponential growth?
   a. 5, 8, 11, 14, 17
   b. 5, 8, 12, 17, 23
   c. 4, 6, 9, 13.5, 20.25
   d. All of the above
   e. None of the above

48. For several decades, American petroleum consumption doubled every decade. This corresponds to an annual percent growth rate of
   a. 2%
   b. 5%
   c. 7%
   d. 10%
   e. Not sufficient data

49. It is expected that by the end of the century, the primary source of energy will be
   a. Hydrogen
   b. Fission
   c. Coal
   d. Petroleum
   e. Wood and other biomass

50. For large high-temperature applications such as those requiring superheated steam, a more convenient method of using solar energy is
   a. Photovoltaic cells
   b. Flat-plate collectors
   c. Dish collectors
   d. Heliostat
   e. All of the above

IV. True or False

1. The main cause of global warming is the burning of fossil fuels.

2. The higher the H/C ratio in the fuel, the dirtier the combustion will be.

3. Many environmentalists advocate the imposition of a "carbon tax" as a mechanism for reducing the budget deficit.

4. Fuel cells are devices in which hydrogen and fuel react to produce steam.

5. The rate of increase in energy consumption per capita in India is larger than in the US.

6. Pebble-bed nuclear reactors are widely used in Europe.

7. More than 80% of all the energy used in the world is from fossil fuels.

8. A person weighs less on the moon than on the earth because the gravity is weaker on the moon.

9. Electricity is a primary source of energy.

10. All forms of energy can be divided into kinetic or potential energy.

V. Fill-in the Blanks

1. The _____________ is still commonly used in the petroleum industry. It is equal to 42 gallons.

2. ___________ is used to express large quantities of energy. It is equal to 10^{15} BTU.

3. __________ will likely replace fossil fuels as the main source of energy within the next two centuries.

4. \$120 invested at 7% compound interest will grow to \$240 in about ______ years.

5. __________ and __________ are the two countries with the highest rate of energy consumption in the world.

6. There is a good chance that within the next hundred years __________ is going to be the most widely used fuel.

7. Natural gas is transported mainly by__________.

8. The next most logical number in the series 3, 9, 15, 21 is _______.

9. The next most logical number in the series 2, 6, 18, 54 is _______.

10. A person with a mass of 60 kilogram, has a weight of __________ newtons.

VI. PROJECT I - US Population Projection

The US population marked its 300,000,000th birth sometime during October 2006. The figure was derived from the 2000 population census, and assumed there are one live birth every seven seconds, one death every 13 seconds, and one net increase in immigration per 30 seconds. Assuming that this trend continues for a foreseeable future, calculate:

1. The percentage annual growth rate
2. The doubling time
3. The year when population reaches 600,000,000
4. Using equation C.1 (p.479), find the projected US population in 2050

PROJECT II - Energy and the Affluence

From the discussion given in this chapter it appears that industrialized countries consume considerably more energy than developing and under-developed countries. In fact many economists see a strong correlation between a country's wealth as measured by Gross Domestic Product (GDP), or Gross National Product (GNP) and the per capita energy consumption. In this project, you are asked to prepare a spreadsheet tabulating the latest published data for several developed and developing countries. Search the internet to find the latest population and economic data. Good sources for this information are the World Bank and the Energy Information Agency. CIA's Fact Book may also be consulted.

1. Plot the per capita energy consumptions of these countries versus their GDP and GNP on a log-log scale.
2. Calculate the ratio of the total energy consumption and GDP for the United States, China, India, and the country of your birth. Which country uses energy more efficiently?
3. Discuss the pattern, and whether such a correlation exists or not. If not, explain the discrepancy.

PROJECT III - Energy and the Environmental Literacy

In this exercise you are to evaluate the students' understandings of current issues related to energy and the environment. Prepare a questionnaire with 20 or more questions asking students about topics of importance to energy and the environment. Sample questions of great importance are (but definitely not limited to) population growth, the availability of fossil fuels, patterns of use, renewable energy sources, nuclear safety, environmental impacts, regional conflicts, policy issues, and economics.

You need to interview as many students as needed (among those who are taking the course with you and from randomly selected students in your college or university) to have a minimum of 10 respondents to each of the questions.

Tabulate responses and write a short essay summarizing your findings. From these results, discuss whether these students benefit if they would enroll in a similar course? Do you see any statistically significant differences between male and female students? Between native and foreign-born students?

Mechanical Energy

CHAPTER 2

Give me a point to lean on and I'll lift the world up
~ Archimedes (287-212 B.C.)

Mechanical energy, in the form of muscle power, was the first energy source that humans utilized. Hunters used muscle power to chase and wrestle animals or escape them when in danger. The use of tools for hunting and agriculture opened a new era in human development. The bow and arrow was used in hunting and in warfare, and wind was harnessed to navigate sailboats along rivers and on seas. As animals became domesticated, humans learned to exploit animal power - first oxen, then mules and horses - for pulling their plows and carts. The great civilizations of China, Egypt, Persia, Greece, and Rome relied on forced labor and animals to plow their farms; at the same time, their soldiers fought for the acquisition of more land and slaves.

As the empires crumbled and slaves became scarce, the work of humans and animals was increasingly supplemented by power from watermills and windmills. Although the invention of windmills and watermills can be traced back to ancient Persians, Greeks and Romans, their use was limited primarily to grinding grains. They became widespread only at the end of the first millennium when Europeans used mills to operate such tools as saws and looms and to pump water from wells.

The work from muscles, the work performed by windmills and watermills, and the work done by simple tools are all examples of mechanical energy. In this chapter we will define mechanical energy and how it manifests itself in the form of kinetic and potential energies. Hydro and wind energies will be discussed in Chapters 3 and 4, respectively.

Work and Power

In Chapter 1, we defined energy as the capacity to do work. This definition implies that work and energy are interrelated -- one cannot be defined without the other. Let us first introduce the more tangible concept - work - and then define energy in terms of it.

Work

Intuitively, **work** is the change in an object's state of motion that results from the application of a force; no work is performed unless the object is displaced in the same direction as the applied force. Sideway displacements do not contribute, so no work can be done by a force that is perpendicular to the displacement. When force is at an angle, only the component of force in the direction of motion must be considered. To move an object twice the distance, with the same amount of force, twice the work must be done. Likewise, if it requires twice the effort to move an object the same distance, the work performed will still be two times as much. It is therefore understandable to define work as the product of the force (F) times the distance (d).

$$W = F \cdot d \qquad (2\text{-}1)$$

When force is at an angle, only the component of force in the direction of motion must be considered.

In SI units, force is given in **newtons** (N) and distance in **meters** (m). The work is therefore given in **newton-meters** (N.m) or **joules** (J).

Question: Which of the following represents the performance of work?

a. An apple falls off an apple tree
b. A horse pulls a carriage
c. A balloon ruptures and air rushes out
d. A woman carries a basket of fruits over her head
e. A boy pushes against a wall until exhausted

Answer: In each of the first three instances, a force has caused a movement in the direction of the force. In case a, the force of gravity causes the apple to fall. In case b, the horse applies a force in the direction of motion. In case c, the higher pressure in the balloon forces air out of the balloon. Instances d and e, however, do not constitute work. The force that the woman applies to the basket is upward and perpendicular to the direction of her motion. Similarly, when the boy pushes on a wall, there is no displacement and he performs no work. This might be quite confusing, as in both instances the person can become very tired. In fact, the body does continuous work to pump the blood to various organs of the body, to expand and contract various muscles in hands and feet, and force air in and out of our lungs; all of these actions serve to maintain the normal body functions. Although no mechanical work was performed in cases d and e, both the woman and boy use up a lot of chemical energy via metabolism.

Question: A boy is moving a heavy box along a hallway. If the boy wants to move the box with the smallest amount of effort (putting in the least amount of work), which approach should he choose?

a. Pull the box with a rope parallel to the floor.
b. Push the box parallel to the floor.
c. Pull the box with a rope at an angle.
d. All of the above require the same amount of effort.

Answer: The answer is d. Although the boy might be a bit more comfortable in one situation over another, the work done is exactly the same in all cases. Work is needed because of friction between the two surfaces; the amount of work is equal to the magnitude of the friction force multiplied by the distance traveled. The component of force perpendicular to the ground is not doing any work. The smart thing to do is probably to put the box on a wheeled cart or a dolly to reduce the friction.

Question: You inflate a balloon by blowing into it. Do you perform work? By how much?
Answer: Yes, you do work because you are applying a force (moving your diaphragm to push the air out) through a distance (expanding the balloon outward). The amount of work is equal to the stored elastic potential energy; that is, the amount of energy needed to stretch (deform) the balloon.

Having defined work in terms of forces and displacements, we can now turn to defining mechanical energy as the energy acquired by an object upon which work is performed. Alternatively, it can be said that an object which possesses mechanical energy is capable of doing work. Mechanical energy can be described as either kinetic energy or potential energy.

Kinetic Energy

Kinetic energy is the energy of motion. Since any displacement can be viewed as a combination of a linear motion and a rotation, there are two forms of kinetic energy-- linear (the energy of motion from one location to another) and rotational (the energy due to rotational motion about the center of mass). The energy from wind and rotating flywheels are two examples of kinetic energy. In this book we concern ourselves mainly with the kinetic energy of a body in linear motion, which for an object of mass (m) and velocity (v) is:

$$E = \frac{1}{2} mV^2 \qquad (2\text{-}2)$$

When mass is expressed in kilograms and velocity in meters per second, kinetic energy will be in joules.

EX 2-1

Example 2-1: Comets are time capsules that offer clues about the formation and evolution of our solar system, some 4.5 billion years ago. To better understand how the outer solar system was formed a team of NASA scientists designed a probe called "Deep Impact" which was to smash into a nearby comet (actually 83 million miles away!) at very high speeds and for the first time ever look into the comet. The 820-lb probe collided with the comet at 23,000 mph on July 4, 2005, making a hole the size of a football field. What was the kinetic energy of the probe at the time of impact?
Solution: Converting the data into SI units, the probe's mass is 820/2.2 = 372 kg; this mass collides with the comet at (23,000x1.609)/3600 = 10.28 km/s = 10,280 m/s. The kinetic energy of impact is:

$$E = \frac{1}{2} mV^2 = \frac{(372).(10{,}280)^2}{2} = 1.97x10^{10} J = 19.7\ GJ$$

Which is equivalent to $1.9x10^{10}/4.2x10^6$ = 4,680 kg = 4.68 tons of TNT.

Potential Energy

As its name implies, potential energy is energy at rest that is waiting, capable of changing in the shape or position of an object in a force field. Energy from a hydroelectric dam, from tides, and from a slingshot are all examples of potential energy. There are two types of potential energy: 1) *Potential energy of form* is the result of twisting, compressing, stretching, and bending matter out of its natural shape; 2) *Potential energy of position* is the energy of a mass caused by its higher elevation relative to its surroundings. The changes in potential energy due to deformation of bodies are not considered in this book.

Potential energy of position (also called gravitational energy) can be calculated as the work needed to move an object from one location to a new location at a higher elevation. It is the product of weight (mg) and the height (h) to which the object was raised:

$$E_{pot} = mgh \qquad (2\text{-}3)$$

In this equation, g is a constant signifying the gravitational acceleration between two objects (equal to 9.81 m/s^2 on earth), m is mass in kilograms, h is height in meters, and E_{pot} is potential energy in joules.

Gravitational energy exists between any two objects separated by a distance. In the following example, the earth constitutes the other object.

Example 2-2: Calculate the gravitational energy released as a result of the collapse of the 110-story World Trade Center in New York City on September 11, 2001. Each tower had a mass of about 0.5 million tons and a height of 415 m.[1]
Solution: If it is assumed that the mass of the tower was uniformly distributed between floors, then the average height of the fall was ½(415) = 207.5 m and the potential energy of the combined towers was $E = 2x5.10^8x9.81x207.5 = 2x10^{12}$ J.

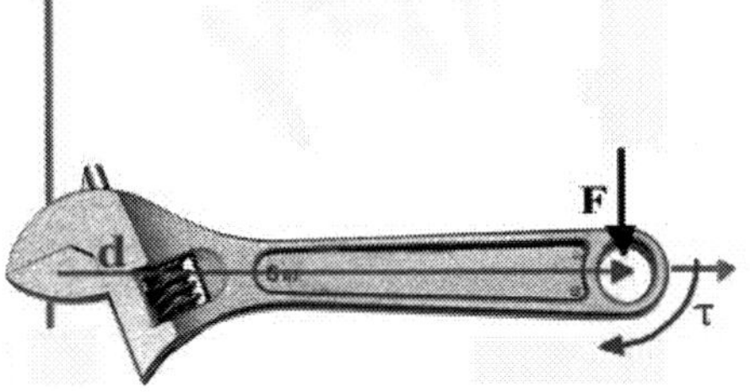

Figure 2-1
Torque

Torque

Torque is a measure of work (energy) required to rotate an object. Intuitively, torque can be interpreted as "force with a twist," since it results in rotation of the object upon which the force is applied. To tighten a screw or to pedal a bicycle we must apply torque. Torque can be calculated by multiplying the force and perpendicular distance between the axis of rotation (or pivot) and the "line of action" of the force. This distance is called "lever arm" or the "arm of the force." (See Figure 2-1).

$$\tau = F \cdot d \qquad (2\text{-}4)$$

[1] See FEMA Evaluation Report at *http://www.fema.gov/library/wtcstudy*.

Work and Power

FYI ...

Given enough fuel there is, in principle, no limit to how much work a machine can perform. However, the rate at which the work is performed is certainly limited.

In comparing this with equation (2-1), we can see that torque has the same unit as work (N·m).

Power

A rock climber and a trail hiker of similar mass moving to the top of a cliff use the same amount of work. They each have to lift their own weight the entire height of the cliff. The hiker, however, can reach the top of the cliff much faster than the rock climber. The difference is the rate at which the rock climber and hiker are doing the work, or their rate of energy consumption or power. Power is defined as:

$$P = \frac{W}{t} \qquad (2\text{-}5)$$

We know work is force times distance; therefore, power must be force times distance traveled per second. Since distance per second describes velocity, power equals force times velocity:

$$P = F \cdot V \qquad (2\text{-}6)$$

The rock climber performs the same task at a slower rate, using less power. For the same reason, you cannot carry a heavy load as quickly as a light load because your power output is too low.

Just as we showed that power for motion along a straight line is equal to force times velocity, we can show that rotational power is calculated as torque times angular velocity:

$$P = \tau \cdot \omega \qquad (2\text{-}7)$$

In this equation $\omega = 2\pi N$ is the angular velocity and N is the shaft speed in revolutions per second.

From the definition, we see that the SI unit of power is J/s, commonly called a watt (W) in honor of James Watt, the inventor of the steam engine. Because a watt is a relatively small unit of power, a kilowatt (1 kW = 1,000 W) is often used instead. In the US, power is customarily expressed in horsepower, which is equivalent to 746 watts (1 h.p. = 746 W). Table 2-1 compares the power output of an average man and horse with that of several machines.

Table 2-1. Power Outputs of Few Basic Machines

Laptop Computer	10 W
Man	100 W
Horse	746 W (1 horsepower)
Automobiles	100 kW
Water/Gas Turbines	1 MW
Power Plants	1 GW
Space Shuttle Main Engines	30 GW

Example 2-3: How much energy does it take an 80 kg man to run up a flight of stairs 6-m high? What is the power expenditure if it takes 10 seconds to climb the stairs? Assume $g = 10\ m/s^2$.

Solution: The energy consumed is E = mgh = 80x10x6 = 4800 J.

Watt and Horsepower

FYI ...

It is ironic that Watt coined the term "horsepower" to incite coal miners to buy his steam engine to replace their pit ponies.

The power expenditure is P = E/t = 4,800/10 = 480 W (0.63 h.p.). Obviously, it takes the same person a bit longer than 20 seconds (maybe 25 s) to climb two flights of stairs. In this case, although the amount of energy spent is exactly twice as much (9600 J), the power reduces to 9,600/25 = 384 W (0.5 h.p.). Although humans can demonstrate impressive muscle power for a short time, even the best of athletes cannot sustain a power output of more than one-tenth of a horsepower.

Example 2-4: To accelerate an automobile from 0 to 100 km/hr in 10 seconds, how much total energy must be provided by the fuel? What is the power rating? Would it make a difference if the car were accelerated to 100 km/hr in 20 seconds instead? Assume that the automobile has a mass of 1200 kg.
Solution: The final velocity of the car is v = 100,000/3600 = 27.78 m/s. The kinetic energy is then equal to $0.5 \times 1200 \times 27.78^2$ = 463,963 J. Whether this acceleration is carried out in 10 s or 20 s is irrelevant. The power rating, however, depends on the rate at which energy is consumed. If the car accelerates in 10s, power will be P = E/t = (463,963 J)/(10 s) = 46,396 J/s (46.4 kW or 62.2 h.p.). If the car's acceleration takes place in 20 s, the power requirement would be reduced by half (23.2 kW or 31.1 h.p.).

Example 2-5: It is expected that the world population will reach 10 billion by the year 2050. If, on the average, each person requires approximately 2 kW of electric power for his needs, how much total electric power must be produced in 2050? If 40% of all the electric power is produced by burning oil (as is done today), how much oil do we need annually to meet the demand? Each barrel of oil has an energy equivalent of 1,700 kWh.
Solution: The total electric output must exceed (10×10^9 persons) x (2×10^3 W) = 2×10^{13} W = 20,000 GW of electricity. As we will see later, power plants produce electricity with efficiencies of around 33%; we need to use three units of petroleum energy to produce one unit of electricity. If 40% of the electric power is produced by oil, we will need (0.4)(60,000 GW) x (365x24 h) = 2.1×10^8 GWh = 2.1×10^{14} kWh of thermal energy. The amount of oil needed would be $2.1 \times 10^{14}/1{,}700$ = 1.24×10^{11} barrels = 124 billion barrels of oil (bbo). The current world total petroleum reserves are estimated at 3021 bbo (See Table 6-5).

Example 2-6: An electric motor runs at 1500 rpm and delivers 5.0 hp. How much torque does it deliver?

Power of Words

FYI ...

Sound waves generally carry very little energy. While talking, a person normally emits a maximum of 10^{-4} watts of sound energy. This means that the average sound energy involved in giving a one-hour lecture is less than a joule. If such energy could be used to power a light 100 watt bulb, about a million people would be needed. Enlisting the services of the same crowd, about 10 minutes of casual conversation would be required in order to heat up an 8 ounce (200 gram) cup of coffee. So in terms of energy production, talk is really cheap!

Solution: The angular velocity of an electric motor and the power delivered are:

$\omega = 2\pi N = (2\pi \text{ rad/rev})(1500 \text{ rev/min})(1 \text{ min}/60\text{s}) = 157 \text{ rad/s}$

$P = (5.0 \text{ hp})(746 \text{ W/hp}) = 3730 \text{ W}$

Torque is calculated using equation 2-7,

$\tau = P/\omega = 3730 \text{ W} / 157 \text{ rad/s} = 23.76 \text{ N.m}$

Question: A cart filled with bricks is to be loaded onto a truck. Three ramps of different lengths are available (See figure).

a. Which situation requires the least amount of force?
b. Which situation requires the least amount of work?
c. Which situation requires the least amount of power?

Answer: Situation (a) requires the least amount of force, but the force must be applied over the longest distance. The product of force times distance, or work, is the same in all cases. The power expenditure depends on how long it takes to perform this work. If you pull faster, the time is shorter and more power is needed. Assuming speeds are the same in all cases, it takes longest for the cart to move up the longest ramp (a), and the least power is consumed. Can you verify these conclusions from equations 2-1 through 2-3?

(a)

(b)

(c)

Question: Two identical cars are leaving Los Angeles at the same time. The first car arrives in San Francisco in 5 hours, while it takes 10 hours for the second car to reach the same destination. Which car consumes more energy? Which car puts out more power?

Answer: Assuming that the cars' efficiencies are equal at all speeds [in reality, cars are designed to give a better gas mileage at about 80-100 km/h (50-60 mph)], both cars use the same amount of fuel (energy), i.e. work the same amount. The first car, however, uses up fuel twice as fast, therefore putting out twice the power of the second car.

Question: Suppose a small passenger sedan with a power rating of 80 hp can accelerate from 0-60 mph in 12 seconds. How long would it take a sports car with a power rating of 240 hp to accelerate by the same amount? Which car performs more work?

Answer: Both cars accelerate to the same speed (and gain the same amount of kinetic energy), therefore they perform the same amount of work. The second car uses three times the power—which means it does the same work in one third of the time, or 4 seconds.

The Golden Rule of Mechanics

The "golden rule" of mechanics essentially states that **whatever you lose in power you gain in displacement**. We commonly use pulleys and ramps to lift things which would otherwise be too heavy for muscle alone. In both cases, a smaller force achieves the necessary work, but over a greater distance.

> **Question:** Consider a bottle opener. In lifting the handle, you do work on the opener, which in turn does work on the cap it is lifting. Which one is greater, your work or the work of the bottle opener?
> **Answer:** Because the opener cannot be a source of energy, the output work (work of opener) can never exceed the input work (your work). The opener simply aids in the transfer of energy from you to the bottle cap. In fact, nearly all of your work will go into deforming rather than into lifting the bottle cap. Of course, energy was not really lost; it was just not used in the way you intended.

Although the output work is always less than the input work, output force does not have to be smaller than input force. This is exactly what a machine is supposed to do - *transform part of the energy (work) into a form most convenient -- that requiring less force.*

Simple machines

Machines are an important part of our daily existence; they vary in complexity from simple tools like screwdrivers and scissors, to more complex machines such as cranes and automobiles. Whether powered by engines or by people, machines make our tasks easier. A machine eases the load by changing either the magnitude or the direction of a force. Simple machines are used to reduce the amount of force required to do particular types of work, but the trade off is that you must apply this force over a greater distance. Others, such as bicycles, are considered complex (or compound) because they are made out of a number of simple machines. There are six types of simple machines (See Figures 2-2a-2-2f):

1. **Inclined plane** (any slanted surface used to raise a load from a lower level to a higher level). Work is carried out using less effort by moving a smaller force over greater distances. Examples include ramps for wheelchairs, ramps to load luggage onto a plane, and escalators. It is believed that Egyptian pyramids were built using similar ramps.

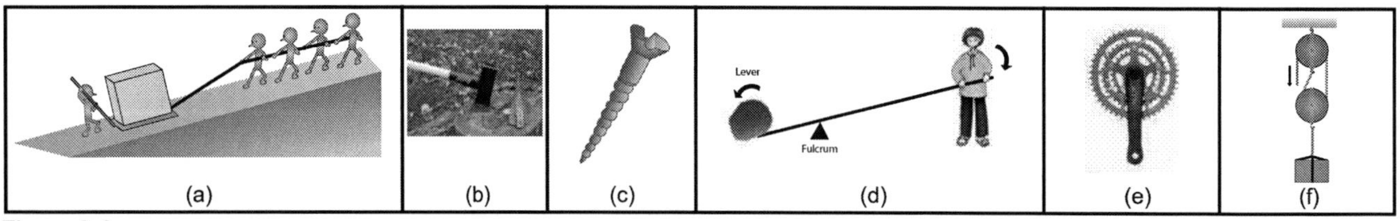

Figure 2-2
Simple Machines - (a) ramp, (b) wedge, (c) screw, (d) lever, (e) wheel and axle, (f) pulley

2. **Wedge** (essentially two back-to-back ramps or inclined planes turned on their sides). Wedges are used mainly to distribute force on both sides. It works by applying a smaller force over a larger area. Nearly all cutting machines (chisels, axes, scissors, and knives) use different variations of the wedges.

3. **Screw** (the circular version of the inclined plane). A screw is really an inclined surface wrapped around a cylinder or a cone. Some examples are ship propellers, corkscrews, meat grinders, ordinary screws, and jar lids.

4. **Lever** (a bar or a rigid object that rests and rotates about a point called a fulcrum). Levers make work easier by allowing a small force to be applied over a long distance, instead of a large force to act over a short load arm. A lever is characterized by a fulcrum or pivot, a force (effort) arm and a weight arm. Depending on the position where force and weight apply relative to the pivot, there are three kinds of levers. In a **first class lever**, the fulcrum is between the force and the weight (pliers, seesaw, and scissors). When the weight is between the fulcrum and the force, the lever is called a **second class lever** (nutcracker, and wheelbarrow). In **third class levers** the force is located between the fulcrum and the weight (baseball bat, shovel, tongs, and human arm).

5. **Wheel and axle** (a wheel or spoke attached to an axle or a lever). Here, a smaller force applied to the longer motion at the wheel results in a shorter more powerful motion at the axle. Examples are windmills, gears, doorknobs, and steering wheels.

6. **Pulley** (a chain, belt or rope wrapped around a wheel). A pulley is often used to lift a heavy weight with less force. A pulley makes work easier because it changes the direction of force. Instead of lifting up, you will pull down on the load using your body weight. Like other simple machines, the trade-off is the longer distance that the smaller force must travel at the end of the rope. Examples are the bicycle chain and the pulley used to move a curtain up and down.

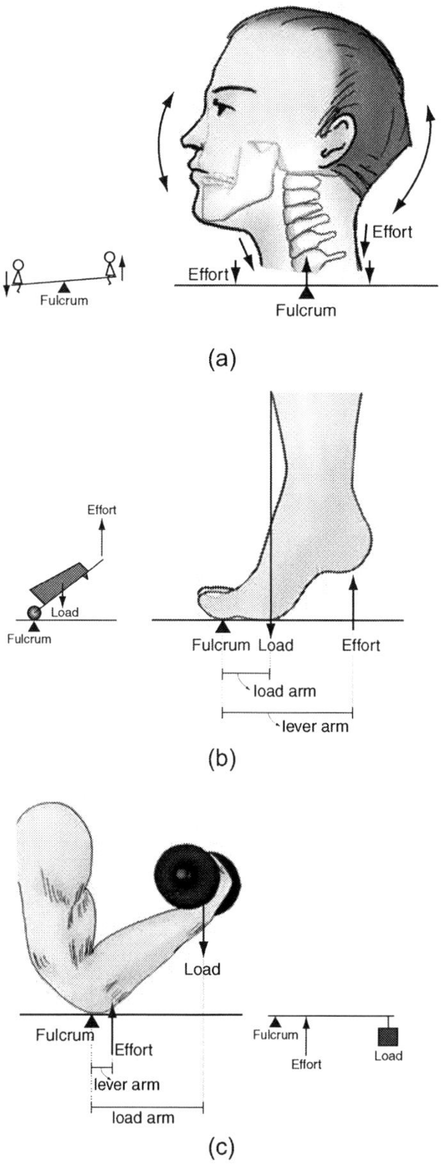

Figure 2-3
Examples of levers; first class (a), second class (b), and third class (c).

Question: Which of the six simple machines reduce the amount of work needed to raise a load?
Answer: None of them. Machines do not change the amount of work required. They merely make the work easier.

Question: How can a small child lift a heavy load?
Answer: In all the examples shown in Figure 2-3, the force required to carry out the tasks is reduced by making the lever arm bigger. In fact, we can show that the product of:

force in x lever arm = force out x load arm

Mechanical Efficiency and Mechanical Advantage

Two factors are important in selecting an appropriate machine for performing a given task: mechanical efficiency and mechanical advantage.

Mechanical efficiency gives an indication of the losses that may be incurred while performing a certain job. It is the fraction of the input work W_i that is available for carrying out the desired output work W_o:

$$\eta_m = \frac{W_o}{W_i} \qquad (2\text{-}8)$$

In a real machine, some of the input work is inevitably lost as heat, so all real machines have efficiencies that are less than 1.

> **Question:** Suppose you hook two machines together, one with efficiency of 40% and the other with an efficiency of 60%. What is the efficiency of the two machines operating together?
> **Answer:** If you are tempted to say 100%, try to resist. The first machine will feed only 40% of the original input energy into the second machine, which then puts out 60% of the 40%. In other words, each machine loses some energy and the net efficiency is 40%x60% = 24%.
>
> **Question:** A transmission gearbox transmits power from the engine to the axle through a number of interlocking gears. To get more power we need to shift the transmission to a lower gear, which will cause the speed to decrease. Can we design transmissions that increase power and at the same time increase velocity?
> **Answer:** No. A machine can increase the magnitude of the effort (input) force or increase the velocity of the object to be moved, but not both. If a machine did both at the same time, the output work would exceed the input work, which is impossible.

Mechanical advantage (MA) is the amount by which the machine multiplies the force. It is the ratio of the force exerted by a machine (or the resistance force F_r) to the force exerted up on a machine (or the effort force F_e). In an ideal machine no energy is wasted so the output work equals the input work ($W_o = W_i$), i.e; $F_r \cdot d_r = F_e \cdot d_e$. Therefore,

$$MA = \frac{F_r}{F_e} = \frac{d_e}{d_r} \qquad (2\text{-}9)$$

where d_r and d_e are corresponding displacements by F_r and F_e (lever arms). Forces are not, however, the only things that are magnified by levers. Since the longer arm of the lever has to travel a greater distance than the smaller arm of the lever in the same amount of time, it travels faster.

$$\frac{V_e}{V_r} = \frac{d_e}{d_r} \qquad (2\text{-}10)$$

Lifting the Earth*

FYI ...

"Give me a point to stand and I will lift up the earth!" proclaimed the great Archimedes, the genius of antiquity who discovered the laws of leverage. He realized that a lever would allow one to lift a very heavy object while applying a very small force. In principle, his assertion about lifting the earth, though somewhat ambitious, does seem perfectly reasonable. By using a very long lever, for instance, you could attempt to balance one object weighing as much as the earth with an applied force on the other end. Let's suppose, for argument's sake, that Archimedes can apply a force roughly equal to his own weight. Clearly, the much lighter Archimedes would have to position himself much farther from the pivot than the heavy object. In fact, the ratio of the two lever arms would be precisely the inverse ratio of their respective weights (or masses). Since the mass of the earth is $M_E \sim 6 \times 10^{24}$ kg and the mass of Archimedes was probably around 100 kg (to keep the numbers nice and round), the ratio of the arms would be roughly 6×10^{22}. A big number, but so what? Well, the difficulty arises when one considers the time involved in lifting the "earth" by any noticeable amount. For example, in order to lift the earth even by 1 millimeter, Archimedes would have to move his end through a giant arc 6×10^{19} m long. Just how long do you think that would take? Working against a force equal to his weight (1000 N), Archimedes would generate about 1.3 horsepowers(1 horsepower = 746 W) if he moved his end at the respectable rate of 1 m/s. At this rate, the task would take 6×10^{19} s or about 2 trillion years. Even if Archimedes were to move his end at the speed of light — nature's fastest — the task would still take him over 6 thousand years.

* Excerpts from a physics text currently under preparation jointly by this author and Professor Igor Glozman, Department of Physics, Highline Community College, Des Moines, Washington 98198.

You can infer from equations 2-9 and 2-10 that the gain in velocity or displacement is inversely proportional to reduction in effort, and is therefore inversely proportional to mechanical advantage. Hence you may use a lever to gain either mechanical advantage or speed, but not both.

Question: What is the mechanical advantage of a hammer driving a nail through a block of wood?
Answer: The energy used to carry the hammer through its flight (from the time of raising your hand until the time that hammer hits the nail) is used to push the nail only a few millimeters into the wood. For example, if we lower our hand by 50 cm, but the nail is pushed down only by 0.5 cm, then we have reduced the displacement by100 times, and at the same time multiplied the force by a factor of 100.

Question: Using the pulley arrangement shown in Figure 2-2, how much force is required to lift a 200-pound package? What is the mechanical advantage?
Answer: The load is divided between two cables holding the pulley at the right. The force is one half (100 pounds) and the mechanical advantage is two.

Unlike mechanical efficiency, mechanical advantage may or may not be greater than 1. Machines are designed to make a task simpler by increasing the force applied; therefore, for most machines mechanical advantage is greater than one. The muscles in our bodies also act as levers. Unlike machines, however, in many instances they do not make a particular task (such as lifting a weight) simpler. In fact, the human body is not designed for strength, but for agility, speed and wide ranges of motion.

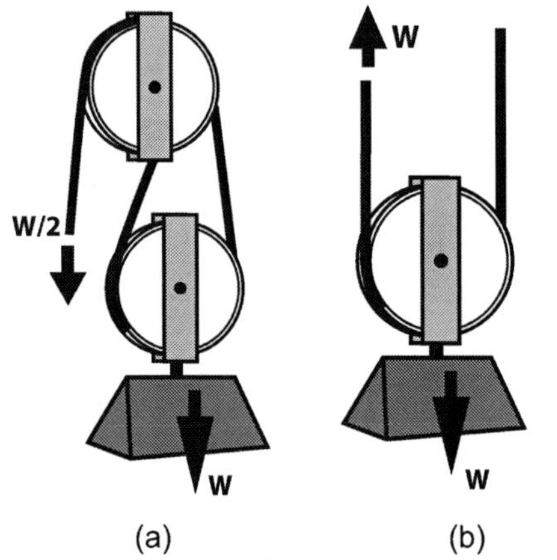

EX 2-7

Question: Our muscles are much stronger than they seem. For instance, the bicep is attached to the forearm about 8 times closer to the fulcrum (the elbow) than one's hand is when lifting a weight. What is the advantage of this seemingly inefficient arrangement?
Answer: The advantage is speed (also a factor of 8), which is frequently more important in the animal world.

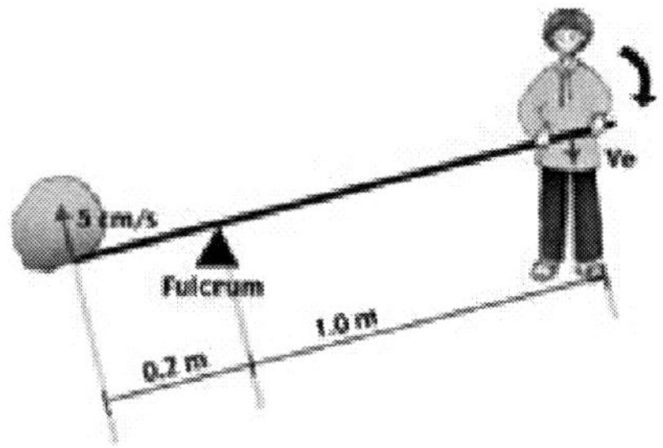

EX 2-8

Example 2-7: What is the mechanical advantage of the system of pulleys shown in figure Ex 2-7?
Solution: The mechanical advantage of a pulley system is approximately equal to the number of supporting ropes or strands. A single pulley can only change the direction of the pull to lift a load. Therefore, pulley shown in (a) has a mechanical advantage of 1 (MA = 1). The pulley system in (b) has two supporting pulleys and therefore has a MA = 2.

Example 2-8: In figure Ex 2-8, what is the force required for moving a 200 kilogram stone? What is the mechanical advantage? If we want to lift the stone at a rate of 5 cm a second, what is the speed at which we need to push down on the lever?
Solution: From equation (2-9) we have (200x9.8)x0.2 = F_ex 1.0, or F_e = 392 N. The mechanical advantage is:

$$MA = \frac{F_r}{F_e} = \frac{(200)\,(9.8)}{392} = \frac{1.0}{0.2} = 5.0$$

Using equation 2-10, we have

$$V_e = V_r \cdot \frac{d_e}{d_r} = 5 \cdot \frac{1.0}{0.2} = 25\ cm/s$$

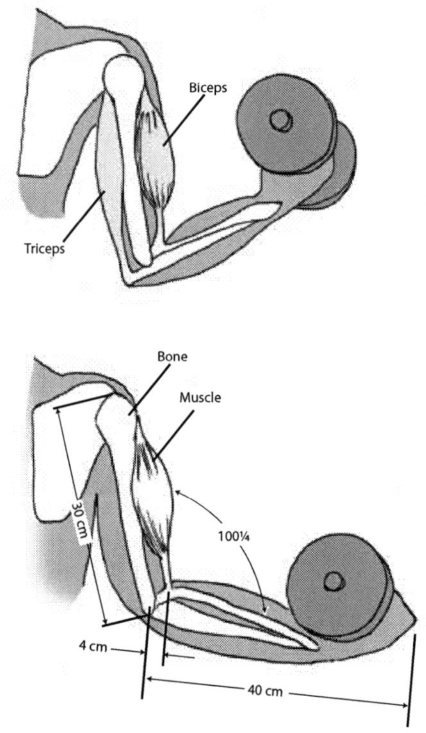

Figure 2-4
A human elbow acts as a third class lever during lifting a weight.

Human Body as a Machine

The human body can be seen as a machine that consumes food and converts it to mechanical work and heat (See Chapter 5). Power is transmitted through the action of forces on the muscles, bones, and joints that make body's lever system. As muscles contract, they pull a series of tendons and bones (levers) across joints (fulcrums, pivots, or hinges) to perform various tasks. Muscles always act as pairs, so when one contracts, its antagonist extends. Examples of the first, second and third class levers are the joint between skull and vertebrae (neck joints), the Achilles tendons, and the elbow joints, respectively. Most of the movements of the body are produced by third class levers, where the force is between fulcrum and weight. This design lends itself to speed of movement rather than force. For example, when we lift a weight, the biceps and triceps work as a pair; when the biceps flex the arm (effort) or lift a weight (load or resistance), the triceps relax to extend the arm (Figure 2-4).

> **Question:** A man is lifting a heavy weight by contracting the bicep muscle, and at the same time relaxing his triceps muscle. The elbow acts as the fulcrum. What is the force needed to lift a 10 pound weight?
> **Answer:** Force needed is higher by ratio of the arms (40 to 4); we need ten times the weight of the object to support the weight. In the discussions above we assumed that there are no losses involved. When losses are present, mechanical advantage will be reduced by a factor equal to the mechanical efficiency.

Summary

Mechanical energy is the total energy an object has excluding the thermal and internal energies associated with the structure of its atoms and molecules. In other words, it is the sum of its potential, kinetic, and rotational energies. Mechanical energy is what it takes to lift, move, turn, and twist an object. Examples of mechanical energy are energy possessed by a fast ball, energy stored in a spring or a rubber band, and the energy it takes to move a heavy object up a stairs. Simple machines are devices that are at our disposal to carry out the difficult tasks with less effort; they allow us to do the same amount of work by applying a smaller force (effort) but over longer distances. In performing any task there are always some frictional forces that we must overcome. For this reason, some mechanical energy is always converted to heat. In other words, in real systems, mechanical efficiencies are always smaller than "one".

Exercises

I. Essay Questions:

1. Which action requires more work: carrying a weight to the second floor from the ground using a ramp, using stairs or using an elevator?

2. A bicyclist in a road race like the Tour de France rides at about 20 mph (9 m/s) for over 6 hours a day. About 250 watts of power go into moving the bike against the resistance from air, gears, and tires. Another 750 watts are lost as heat. How many calories must the racer ingest per day in order to sustain such a rigorous regimen? (Use 1 calorie = 4.18 J)

3. How can levers and pulleys help in reducing the force needed to lift heavy objects?

4. Can a small child lift an adult with the help of a lever? How?

5. Suppose a tree root is under a section of cement sidewalk and over a ten-year period it lifts up the heavy slab. On the other hand, a strong worker with a crowbar could lift an identical section of sidewalk within a matter of seconds. Compare the work and power for both cases.

6. A car with a mass of 2,000 kg is traveling at 36 km/hr. What is the car's kinetic energy?

7. How much work do you do while climbing to the roof of a 30 meter tall building? Does it matter if you walk up the stairs slowly or quickly? If you skip every other step?

8. In the example above, what is the power you need to walk up the stairs if the vertical speed on the stairs is 0.5 m/s?

9. If the weight of an object being lifted is 200 kg and the number of supporting ropes the pulley system has is four, what would be the system's MA? How much effort weight would you actually be lifting?

10. Determine whether each of the tools mentioned below are first class, second class, or third class levers.
    a. Scissors
    b. Hammer used to insert a nail
    c. Hammer used to pull up a nail
    d. Fishing rod
    e. Nutcracker
    f. Screw
    g. Finger-nail clippers
    h. Wheelbarrow
    i. Achilles tendon, pulling or pushing across the heel of the foot
    j. Crowbar
    k. Elbow joint when lifting a load

II. Multiple Choice Questions:

1. Which of the following statements are correct?
    a. Whenever forces are involved, work is performed.
    b. Work is performed only when forces are normal to the direction of displacement.
    c. Work is involved when an object is displaced in the direction of the force, but not in the opposite direction.
    d. Work is involved when an object is displaced in the direction of a force.
    e. None of the above.

2. Work is by definition
    a. The product of force multiplied by the time during which it is exerted
    b. The product of force multiplied by the distance through which it acts
    c. Equivalent to power/time
    d. Equivalent to energy multiplied by time
    e. None of the above

3. In the language of mathematics
    a. d = W/F
    b. F = W.d
    c. F = d/W
    d. W = F/d
    e. W = d/F

4. Which of the following objects possess mechanical energy?
    a. A car traveling on a flat surface
    b. A car going downhill

c. A drawn bow ready to be released
d. A book sitting on the top of a bookshelf
e. All of the above

5. After a skydiver has reached his terminal speed (constant speed) in the atmosphere, he
   a. Gains kinetic energy at the expense of potential energy
   b. Gains potential energy at the expense of kinetic energy
   c. Releases heat at the expense of potential energy
   d. Releases heat at the expense of kinetic energy
   e. None of the above

6. What happens to the kinetic energy of a moving car as it comes to a complete stop?
   a. It is lost.
   b. It is converted to heat.
   c. It is stored in the flywheel.
   d. It is stored in the battery.
   e. None of the above.

7. Which of these statements are true?
   a. Power is the rate of energy expenditure per unit time.
   b. Power can be calculated as the product of force times velocity.
   c. Power can be expressed in units of watts or horsepower.
   d. All of the above.
   e. None of the above.

8. There are 100 steps to get from the ground-floor of a building to its rooftop. Person A walks up to the roof in 100 steps, while person B climbs up skipping every other step. It takes person A two minutes to reach the top, while person B can do it in half the time. From the information provided one can deduce that
   a. Both A and B perform the same amount of work, but A has more power output than B
   b. Both A and B perform the same amount of work, but B has more power output than A
   c. A required more work because he has to walk 100 steps
   d. B requires less work because it takes him a shorter time to make the climb
   e. B requires less work, because he only has to work half the time

9. Which of the following statements are true?
   a. Pulleys and ramps allow for the lifting of heavy objects because less work is needed.
   b. Pulleys and ramps allow for the lifting of heavy objects because the same work can be performed over larger distances.
   c. Pulleys and ramps allow for the lifting of heavy objects because the same work can be performed over a longer period of time.
   d. Pulleys and ramps allow for the lifting of heavy objects because they perform some of the work for you.
   e. All of the above.

10. The "golden rule" of mechanics implies that
    a. Whatever you lose in power, you gain in displacement
    b. Whatever you lose in energy, you gain in power
    c. Whatever you lose in power, you gain in time
    d. Whatever you lose in energy, you gain in displacement
    e. All of the above

11. Which of the following is an example of kinetic energy?
    a. An automobile lifted by a hydraulic jack
    b. Water sitting on a reservoir on top of a mountain
    c. A rubber ball as it is hitting a solid wall
    d. A turbine being turned by a falling waterfall
    e. None of the above

12. The main function of simple machines is to multiply
    a. The force
    b. The work
    c. The power
    d. The energy
    e. All of the above

13. A fork is an example of a
    a. Pulley

b. Ramp
c. Wedge
d. Wheel and axle
e. None of the above

14. A roller skate is a an example of a
    a. Lever
    b. Ramp
    c. Wedge
    d. Wheel and axle
    e. None of the above

15. A seesaw is an example of a
    a. Lever
    b. Ramp
    c. Wedge
    d. Wheel and axle
    e. None of the above

16. The lever arm is the
    a. Displacement from the pivot to the line of action of the force
    b. Displacement from the pivot to the point of application of the force
    c. Parallel displacement from the pivot to the line of action of the force
    d. Perpendicular displacement from pivot to line of action of the force
    e. Equal but opposite displacement from pivot to line of action of the force.

17. When a hammer is used to pull a nail out of a piece of wood, it acts as a
    a. Wedge
    b. Lever
    c. Ramp
    d. Pulley
    e. None of the above

18. Nearly all cutting machines use
    a. Ramps
    b. Levers
    c. Wedges
    d. Wheels and axles
    e. All of the above.

19. Which of the following can be represented by the same units?
    a. Work, torque, and acceleration
    b. Work, torque, and kinetic energy
    c. Force, torque, and energy
    d. Force, power, and torque
    e. Work, torque, and angular velocity

20. The relationship of energy to power is like
    a. miles to miles per hour
    b. force to velocity
    c. distance to time
    d. kilowatt to killowatt-hour
    e. oil to electricity

III. True or False?

1. Whenever a force undergoes a displacement work is performed.

2. Mechanical efficiency is the percentage of input work that can be converted to output work.

3. Unlike mechanical efficiency, mechanical advantage may or may not be greater than 1.

4. Torque is defined as the amount of force that is required to rotate an object.

5. When there is no friction, the sum of potential and kinetic energies remains constant.

6. In the metric system of units, torque measurements are usually given in newton-meters.

7. Torque is calculated by multiplying the force by the lever arm it is acting upon.

8. A wedge can be seen as a two-sided ramp with surfaces to distribute the load.

9. The sharper the wedge and the closer the threads on the screw, the greater the mechanical advantage.

10. A baseball bat is an example of first-class lever.

11. The human body can be seen as a machine that consumes food and converts it to mechanical work and heat.

12. Mechanical efficiency is defined as the ratio of input to output works.

13. In an ideal machine the output work and the input work are equal.

IV. Fill-in the Blanks

1. Work is the change in the state of motion that results from the application of___________.

2. No work can be done by a force that is ___________to the displacement.

3. Torque has the same unit as _________ and is expressed in __________ in SI units.

4. Tweezers are an example of a class _________ lever.

5. A pulley system with two supporting ropes would have the mechanical advantage of____________.

V. PROJECT I - The Mountain of Pharaoh

The pyramids of Egypt were built to enshrine the pharaoh after their death. The ancient Egyptians believed that the pharaohs held the key to the afterlife for everyone, and the pyramids symbolize this strong belief. The most impressive of the Egyptian pyramids and one of the Seven Wonders of the World is the Great Pyramid of Giza. Its construction was commissioned around 2,550 B.C. by King Khufu of the Fourth Dynasty. The Pyramid was constructed using around two million blocks of stone each weighing an average of 2.5 tons. The heaviest blocks weighed 40 to 60 tons and were used in constructing the ceiling of the burial chamber. It has been estimated that it took 20,000 to 25,000 slaves to build the pyramid.

Possibly the oldest structure in existence, the Great Pyramid took about 20 years to build. Its base covers 13.6 acres (55,000 meters squared); at 145 m tall, it was the tallest structure in the world until the Eiffel Tower was built in 1889.

You are asked to research the probable methods that Egyptians used to build the Pyramids. Carry out a simple calculation to determine the amount of energy that was expended to build the Great Pyramid of Giza. Assuming that an average man could put out 100 watts of power continuously, estimate the number of slaves it took to work 12 hours a day over the entire year to finish the job in 20 years.

Explain any discrepancy with data you may find in other literatures. Discuss different theories on achieving this task, and comment on their feasibilities.

Hint: Energy can be estimated as the amount of mass which had to be lifted from the ground to its center of mass -- the weight of the pyramid times the height of the centroid above the base. The volume and the height of the center of mass of pyramid are: $V = a^2 \cdot h/3$; and $z = h/4$.

Pyramid of Giza

PROJECT II - Energy Consumption During a Climb

In this project, you are asked to calculate the amount of work done and energy consumed during a climb to the top of a six-story building. To perform this assignment, you are asked to climb up from the first floor to the roof of a six-story building twice, once at a slow rate and the second time at a faster rate.

<u>Note the following data:</u>

1. Your mass in kilograms
2. The number of stairs and the average height of each stair (in meters)
3. The time it took you to climb the stairs
4. The time it took you to descend to the first floor (take the average of the two runs)

<u>Now, answer the following questions:</u>

1. What was the total height that you climbed?
2. What was your average speed?
3. What was the force used to perform the climb?
4. What is the work needed to overcome the gravitational potential energy?
5. How much energy did you burn during the climb? Note that the actual work is a lot more (by about a factor of 4) than the work of gravity. Additional

energy is necessary to power your muscles, nourish and repair cells, and to maintain your body temperature.

6. Calculate the average power consumption during each run by dividing work performed by time it took you to climb (or descend) the stairs.
7. Calculate the average power consumption during each run by multiplying your force by your average velocity.
8. In which instance did you perform more work, climbing up slowly or quickly?
9. In which instance did you put out more power, climbing up slowly or quickly?
10. Assuming the work performed in climbing up the stairs is used to generate electricity to turn on a 100-W fluorescent light, how long would the light stay on?

Work Sheet for Project II

Your name: ________________

Your mass: ________________ kg Your weight: ________________ N

Number of stairs: _____ Average height of each stair: __ m Total rise: ____________ m

	Slow Climb	Fast Climb	Descent
Force (N)			
Time to climb, t (s)			
Average climb velocity, v (m/s)			
Total work (N.m)			
Energy consumed (kJ)			
Average Power output (W) $P = W/t$			
Average Power output (W) $P = F{\cdot}v$			

Answer to question 7

Answer to question 8

Answer to question 9

Answer to question 10

CHAPTER 3

Wind Energy

Of all the forces of nature, I should think the wind contains the greatest amount of power. ~ Abraham Lincoln

In the last quarter of a century, wind energy technology has gone through revolutionary changes making wind the fastest growing source of electricity in the world. Worldwide, 120 gigawatts (120,000 megawatts), or about 1.5 percent of the total electricity, is generated by wind turbines, though numbers are growing rapidly. In the last decade, the average annual growth of the installed capacity of wind power plants has been increasing by 32%.[1] Many European countries, in particular Denmark, Germany, and Spain, have been heavily investing in wind energy. Germany, with more than 18,400 MW, is the leader in wind generation capacity with Spain, the United States, India, and Denmark ranking next. In terms of the percentage of total electricity production, Denmark ranks first; it currently produces 20% of its electricity from wind and is planning to raise this figure to 40% by 2030.

In the United States, only three states - North Dakota, Kansas, and Texas - have sufficient wind that, if harnessed, could satisfy national electricity needs. Though these states have the most wind, they are sparsely populated and there are not many transmission lines. Most wind turbines are located in three areas, all of them in California. Today, the US generates about 9,100 MW, or a little more than half a percent of its annual electrical generation capacity from wind, but is expected to increase its capacity to about 5% of its total electricity demand by 2025.[2]

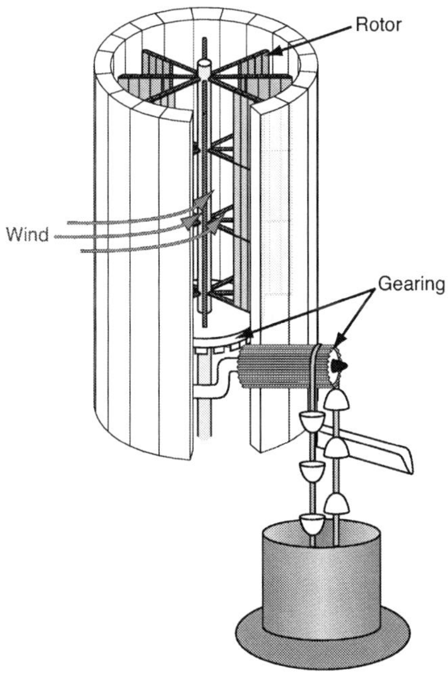

Figure 3-1
Persians invented the first practical windmills for grinding wheat and pumping water.

Overview

Wind is air in motion. We can't see it, but we can see its effect all around us. Just like moving water, the wind represents a tremendous source of natural energy. Like waterwheels, windmills were among the original prime movers that replaced human beings as a source of power. Wind power has been used as early as 5000 years ago by the Egyptians to sail ships across the Nile. The first windmills were invented by Persians around 200 BC to pump water from wells and to grind grain. They were constructed by fastening bundles of reeds onto wooden frames mounted on vertical shafts (hence called vertical windmills) housed in brick or clay walls. Wind entered through an opening at the side and was caught between the spokes radiating from the shaft (Figure 3-1). The technology was exported to China after Genghis Khan imprisoned Persian millers and forced them to build windmills to power irrigation systems in China.

[1] Kenisarin, M. M., "Worldwide State of Wind Power Engineering," *Applied Solar Energy*, Vol. 38, No. 4, 2002, Allerton Press, Inc., New York.
[2] American Wind Energy Association Website, *http://www.awea.org*.

It was not until the twelfth century that windmills found their way to Europe, where their use became increasingly widespread until the early 19th century. These mills used vanes that looked like huge paddles mounted on a horizontal pole (hence called horizontal windmills). The Dutch used windmills mainly for draining water from their low-lying land (hence Netherlands), which was quite prone to flooding (Figure 3-2).

Figure 3-2
Dutch horizontal wind turbine

Figure 3-3
Brush windmill in Cleveland

Figure 3-4
Two-bladed wind turbines

The first use of a large windmill to generate electricity was in 1888 by Charles Brush in Cleveland, Ohio.[3] The Brush machine shown in Figure 3-3 had a rotor 17 meters in diameter with 144 blades and provided enough electricity to light 350 incandescent lamps. The device worked for 20 years until the advent of steam engines and the popularization of low-cost, seemingly inexhaustible fossil fuels made windmills less and less attractive. Later works by aerodynamicists showed that the most efficient number of blades is 2-4, a far smaller number than those found in earlier windmills (Figure 3-4).

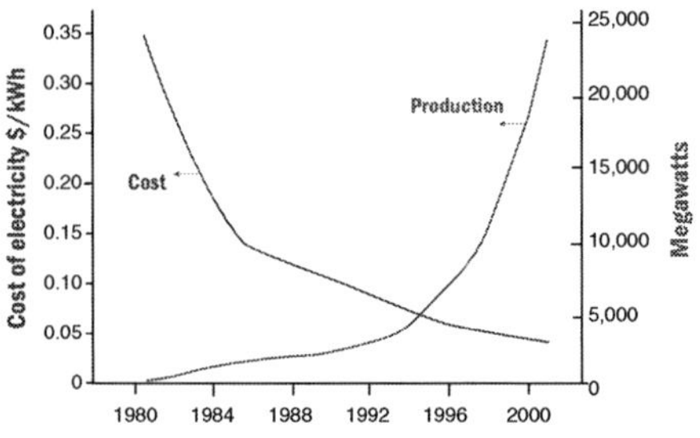

Figure 3-5
As the electricity output from wind energy has increased, cost of electrical generation has decreased dramatically.

The world's wind electric energy generation capacity has been climbing steadily since 1980, reaching 60,000 megawatts in 2005, and 120,000 megawatts in 2008; this is largely due to new technological innovations that have reduced production cost from $1.00 to around $0.04 for one kilowatt-hour (Figure 3-5). In 1980, the United States was the leading producer of electricity from wind and accounted for 80% of the world's capacity. Its share, however, has been declining ever since and accounts for only 15% of the world's capacity today.[4]

Global and Local Wind Patterns

Winds are generated as a result of two factors, the non-uniform heating of the earth by the sun and the rotation of the earth. Equatorial regions receive the most radiation, whereas polar regions receive the least. The difference in ground temperatures between the equator and the poles induces a global circulation pattern where hotter (and lighter) air rises

[3] "Mr. Brush's Windmill Dynamo," *Scientific American*, December 20, 1890.
[4] Global Wind Energy Council, "Global Wind Power Continues Expansion," press release, 17 February 2006.

near the equator, and colder (and heavier) air sinks at the poles. As a result, overall wind flow direction is from the poles toward the equator close to the surface and from the equator toward the poles in the upper atmosphere. The detailed flow pattern, however, is much more complicated than this. The upper atmosphere wind (called geostrophic wind) is largely driven by the earth's rotation and temperature (and thus pressure) differences. Close to the earth's surface, other factors such as mountains, valleys, and shorelines are important in establishing the local wind patterns.

Sea breezes occur during the daytime when landmasses are heated more quickly than the sea. Sand has a lower heat capacity than water and cannot hold solar heat as effectively, resulting in its temperature rising above that of the water nearby. As the air rises above the hotter land, air from the cooler sea moves to replace it, resulting in a sea breeze. At night, the land gives off heat more quickly and its temperature drops faster than the surrounding sea, resulting in **land breezes**. At dusk, there is often a period of tranquility when the temperatures of land and sea are equal (Figure 3-6a).

Valley and mountain breezes are due to a combination of both differential heating and local topography. As the sun rises, it hits the mountain tops first and, as the day progresses, the mountain slopes, causing differential heating between the two. As warmer air rises off the slopes, cool valley air moves up to fill the vacuum (**valley breeze**). In the afternoon, as the sun sets, the opposite occurs and we have **mountain breezes** (Figure 3-6b).

Wind Rose

Selection of a proper site for installing wind farms requires detailed meteorological data at different times. Wind data are routinely collected by wind anemometers installed on top of towers in the direction of the prevailing wind. Because speeds and directions are continually changing, 10-15 minute average values are recorded. The data is plotted on a wind rose, which is a single, graphical representation of speed, direction, and frequency of occurrence. Data are often normalized to the total period of observation in order to indicate frequencies. One simple way to

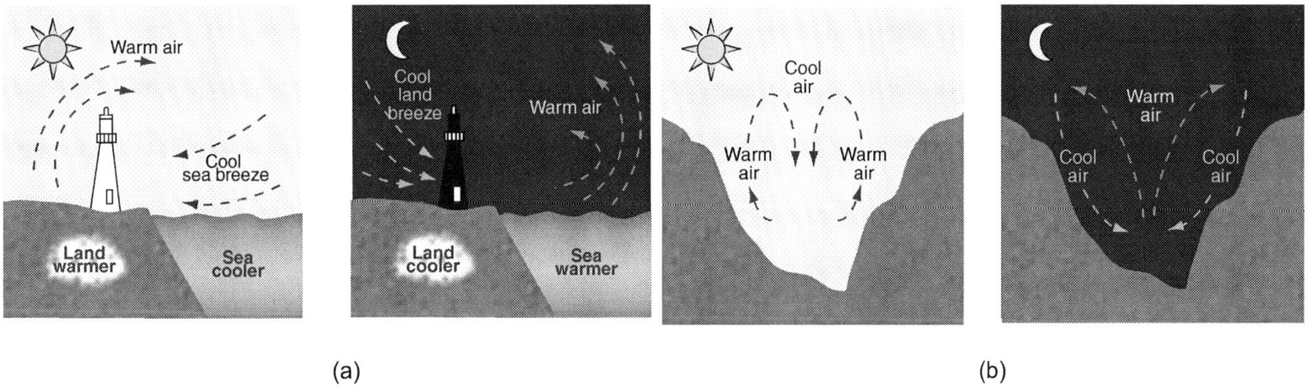

Figure 3-6
Wind formation along a) the coast lines, and b) mountain ranges.

Global Circulation Pattern

Digging Deeper ...

If the earth were not rotating, wind near the earth's surface would flow directly from the poles towards the equator and, in the upper atmosphere, from the equator to the poles, completing the loop. Since the globe is spinning, observers at the equator have a higher tangential velocity than observers farther north. To observers in the northern hemisphere, air moving south is turning to the right, acquiring a westward component and appearing to lag behind the earth. This deflection can be viewed as if a force (called **Coriolis force**) were pushing the wind toward the west. The opposite is true for the southern hemisphere. The result is that the wind flows horizontally, more or less along the isobars (paths of constant pressure). These winds are known as *geostrophic winds*.

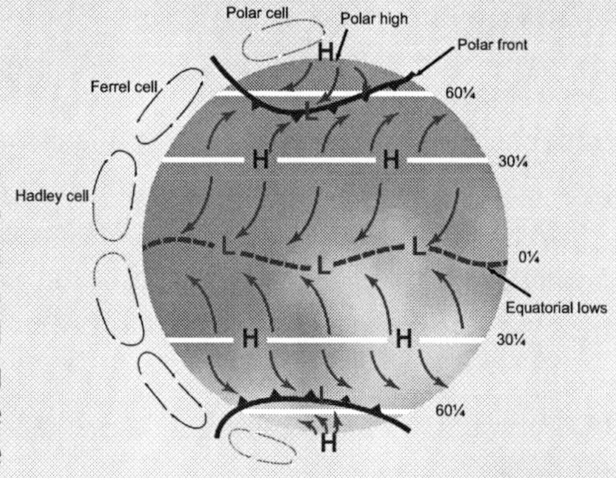

One factor that complicates the general circulation pattern is the radiative cooling of the upper atmosphere; this causes flow to become unstable and air to subside at 30° north and south latitudes, thus falling short of either pole. Air parcels sinking at 30° latitude will be pushed both north and south, towards the poles and the equator. These streams will collide at 60° latitude, producing new lows and polar fronts-- breaking up the large cells into two smaller Polar and Ferrell cells. The result is the prevailing winds and a global circulation pattern resembling that of the figure shown.

The geostrophic wind pattern described above occurs at altitudes above 1,000 meters. At regions below, we have synoptic winds that are associated with large-scale movements of warm and cold fronts. Hurricanes, tornadoes, and typhoons are of this kind. At ground level, surface effects dominate. Surface obstacles such as tall buildings, hills, and valleys, as well as proximity to the sea and the heating by local sources (such as factories, power plants and freeways), will determine the local wind pattern.

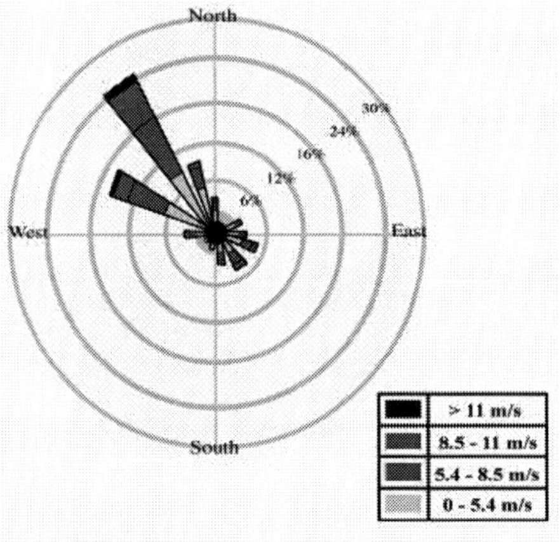

Figure 3-7
Wind rose for Fresno, California for the month of April. Concentric circles represent the frequency, and different colors indicate ranges of wind velocity.
Source: Natural Resources Conservation Service, US department of Agriculture, http://wcc.nrcs.usda.gov/climate/windrose.html)

graphically indicate both duration and direction is to draw bars extending radially from the centre of the rose in the direction of the wind; magnitude is represented proportionally to the time the wind spent in that direction at a given speed. The direction of the wind is traditionally taken as the direction it is coming from. The thickness or the color of the bars can be used to indicate the range of speeds in a given direction. A typical wind rose is shown in Figure 3-7. The data were accumulated over a period of ten years in Fresno, California, for the month of April. From the figure, it can be deduced that over 50% of the time the wind blows to the northwest at speeds of 1.8-3.3 m/s (16% of the time), 3.3-5.4 m/s (20% of the time), 5.4-8.5 m/s 11% of the time, 8.5-11 m/s 2% of the time and greater than 11 m/s less than 1% of the time.

Depending on the wind power density and speed, different geographical areas have been divided into seven classes. Class 1 regions are not suitable for wind energy development, and class 2 regions may become only marginally acceptable. Class 3 areas will be suitable in the future, as technology matures. Class 4 and higher are considered suitable for wind power with existing technology.

Figure 3-8 shows the yearly electricity production in W/m^2 of rotor area for different locations within the continental United States. As the data indicates, the strongest winds are found along coastlines, along ridges, and on the Great Plains.[5,6] Furthermore, it is estimated that up to 6% of US land is suitable for development of wind farms and has the potential to generate 1.5 times its electricity needs.[7]

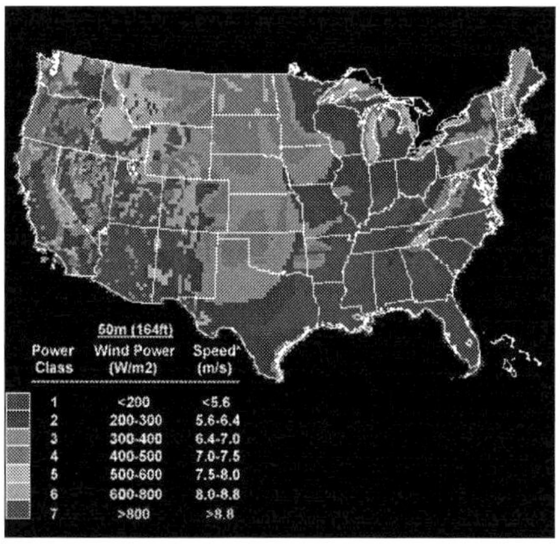

Figure 3-8
United States Wind Resource Map.
Source: National Wind Technology Center, http://www.nrel.gov/wind/wind_map.html.

Wind Turbines

Wind mills are devices that convert the kinetic energy of wind into the mechanical energy of rotating blades. The more kinetic energy, the greater the potential for producing work and power. The total energy transferred depends on the wind speed, the size of the rotor, and the density of the air. In practical terms, there are limits to the wind velocities that wind turbines can utilize. If the wind velocity is too low (cut-in speed), the kinetic energy is not sufficient to overcome friction at the bearings. At velocities above about 50 mph (cut-out speed), wind power is strong enough to knock the blades off, rendering the turbine inoperable and creating a real danger to nearby buildings, traffic, and people.

> **Question:** What is the difference in operation of an electric fan and a wind turbine?
> **Answer:** A fan uses electricity to produce wind, whereas a wind turbine uses wind to make electricity.

In ancient windmills and other early vertical axis windmills, the wind pushed and rotated the sail with a force called a *drag*. These windmills were inherently less efficient than later horizontal axis wind turbines that have been put into operation in the past four hundred years.

Modern wind turbine operation is similar to that of the wings of an aircraft. Because of the shape and angle of the airfoil (blade) cross-section, the air changes direction as it approaches the airfoil. The change in airstream direction results in a change in its momentum. As we have learned from Newton's second law, a change in momentum requires a force. The force of the airfoil on the air is counteracted with an equal and opposite force on the airfoil (remember action and reaction), which is called *lift*. In airplanes, lift causes the plane to become airborne, whereas in wind turbines, lift forces the blades' rotation about the hub, which in turn drives a generator. Perpendicular to the lift force, a drag force impedes rotor rotation.

Unlike helicopter blades and aircraft wings, which are designed for the greatest lift, turbine blades are designed to reduce drag. High lift is essential in aircrafts to prevent stall. Stall refers to a condition at which the angle of attack is so steep that no lift is produced. Even a small dent in the blade or airfoil can trigger a stall. In wind turbine applications, some stalls are welcome in high wind conditions to slow the turbine down and prevent

[5] Wind Energy Resources of United States, U.S. Department of Energy (DOE), can be found at the National Wind Technology Center web site at *http://www.nrel.gov/wind/*.
[6] DOE "Wind Powering America," web site at *http://www.eren.doe.gov/windpoweringamerica/*.
[7] US Department of Energy (*http://www.doe.gov)*

damage. In new designs, the blade angles can be adjusted to change the lift-to-drag ratio and to optimize the turbine's energy output for different wind speeds and directions.

> **Question:** Most rotor blades have twisted surfaces. Why?
> **Answer:** To optimize the angle of attack and prevent stall as wind flows from the tip (with a very high rotational speed) toward the hub (with a low rotational speed).

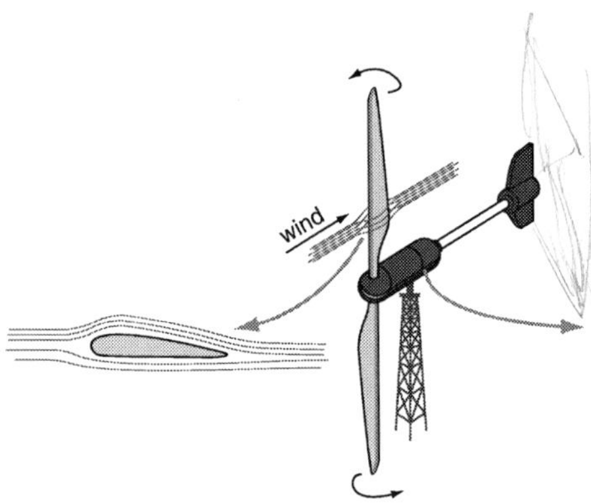

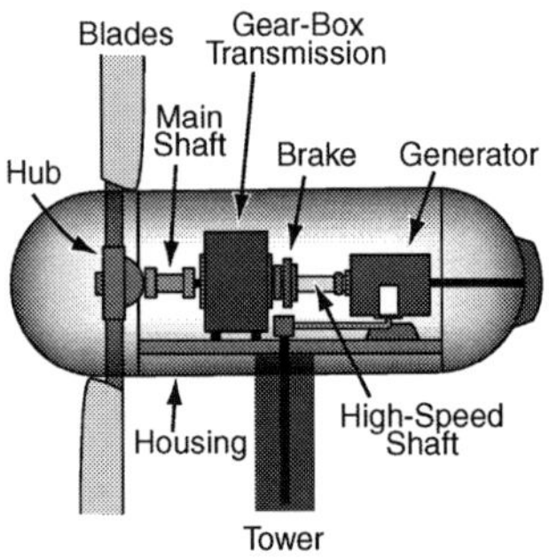

Figure 3-9
Principle of operation of a wind turbine.

The major components of a wind power generator are the hub (on which rotor blades are attached), a gearbox, an electrical generator, and a controller with associated cooling units. Except for the hub, the entire assembly is housed inside an enclosure called a nacelle that is mounted on top of a tower (Figure 3-9). In addition, modern wind generators have anemometers for measuring wind speed, vanes for determining direction, and a yaw mechanism that can tilt the rotor in the direction of the wind. The vane assures that the turbine continuously faces into the wind. The tower carries the weight and raises the turbine above trees, buildings and other nearby obstructions. Wind turbines are more efficient when installed on tall towers because they face faster winds. In addition, most turbines are equipped with automatic governors that protect rotors from spinning out of control in gusts and during high wind speeds.

Basic operation of wind power generation is as follows. Wind drives the blades mounted on a shaft rotating at about 10 to 50 revolutions per minute. The shaft is connected to a transmission or gearbox and changes the speed to the 3,000 rpm needed to generate utility grade electricity. An electrical generator converts the rotational speed of the shaft to electricity, at which point it is transmitted through an underground cable to a field transformer and, eventually, to a utility substation. Here, electricity is added to the grid and combines with electricity generated by a variety of other sources such as coal, nuclear, and hydroelectric power plants.

Not all kinetic energy can be converted to useful work. As incoming wind collides with the blades, some reflects off of the rotor surface, and some is lost as a result of friction at the generator, limiting the wind turbine's efficiency to about 40% at best.[8]

Wind turbines are available in a variety of sizes. Machines vary in size from 0.6 m in diameter (rated at about 500 W) to 60 m in diameter (rated at about 3 MW). **Rated power** is the power that a wind turbine produces when running at its optimum, i.e. at the highest wind speed.

Wind Power

Not all the kinetic energy present in winds can be extracted by a wind

[8] It can be shown that the theoretical maximum efficiency of wind turbines is 59%, known as the Betz limit.

turbine. In addition to various losses, such as slippage and drag, some of this energy must also be carried downstream of the turbine in order to maintain the air flow.

The power generated by a wind turbine is proportional to the kinetic energy of the mass of air swept through its rotor. The mass itself increases with the air density, the size of the rotor, and the speed of the wind. As a result, total power generated by a wind turbine increases with the air density, the cube of the wind velocity, and rotor area, that is, the square of the rotor diameter (See Box "Power and Torque").

Wind speed increases with height above the ground, so the height of the tower on which the rotor is mounted indirectly affects power production. Figure 3-10 shows the increase in power with the height of the tower. For example, a wind turbine mounted on a tower 120' (36 m) tall will be exposed to wind speeds twice that of a tower 30' (9 m) tall. This alone results in the production of eight times more power.

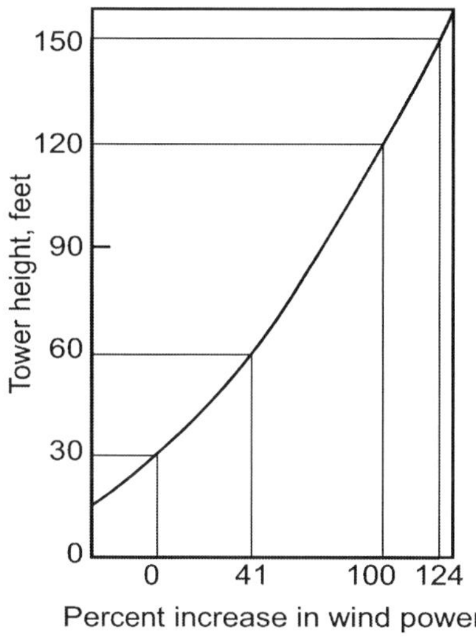

Figure 3-10
Effect of tower height on wind power

Air density (and power output) decreases with altitude at the rate of roughly 2.5% for every 1000' above sea level. Temperature affects the power derived from wind energy in two ways. First, they affect wind patterns and intensity. Second, they affect the air density and thus the mass of air drawn through the turbine. In desert and open areas, where most wind turbines are installed, temperature differences as high as 60°C are possible between summer and winter seasons. As a result, as much as 20% more power can be extracted in winter when the air is denser.

a Design Considerations

There are two types of wind turbines, horizontal and vertical. In horizontal turbines, the axis of the rotor is parallel to the ground, whereas in vertical turbines, the axis of the rotor is perpendicular to the ground. **Horizontal turbines** are usually of a propeller type, while **vertical turbines** have C-shaped blades and look like eggbeaters. These turbines (called Darrieus machines) are made with 2-3 blades and are designed to capture wind from any direction; they have the advantage that the heavy components (the gearbox, generator, and controllers) can be placed on the ground where wind is weaker. Because these turbines are mounted close to the ground, wind speeds are relatively low so they are generally less efficient than horizontal-axis machines. As a result, most modern wind turbines are of the horizontal type.

Turbines can also be designed such that either the front side (upwind) or rear side (downwind) of the rotor faces the wind. Most commercial turbines use an upwind design. The major advantage of **upwind** turbines is that there are no wake losses. The primary advantage of **downwind** machines is that the blades can be made from cheaper and more flexible materials because there is no danger of them hitting the tower. The main

Power and Torque

Mathematical Interlude ...

As the air rushes through the rotor, it follows an imaginary stream tube, slowing down to convert its kinetic energy into the rotational energy of the blades. Since the total volume of the air must remain nearly the same, it must expand as it crosses the rotor. The volume swept is proportional to the cross-sectional area of the blades and the velocity of the wind perpendicular to the rotor:

$$Q/t = A.V = \frac{\pi d^2}{4} V \qquad \text{(i)}$$

where:

Q is the air volume(m^3)
V is the wind velocity(m/s)
t is the time(s)
A is area of the disk rotor(m^2)
d is the rotor diameter(m)

The kinetic energy of this parcel of air is given as

$$E = \frac{1}{2} mV^2 = \frac{1}{2} \rho Q V^2 = \frac{\pi}{8} d^2 V^3 t \qquad \text{(ii)}$$

Power is energy per unit time, or

$$P = \frac{\pi}{8} \rho d^2 V^3 \qquad \text{(iii)}$$

This equation shows that wind velocity is the most important factor affecting power output. As the wind speed doubles, kinetic energy increases 4-fold, while twice as much air sweeps past the blades per unit time, resulting in a power increase of 2x2x2 = 8 times.

Equation (iii) assumes that all the energy contained in the approaching wind is converted to useful work. In practice, there are a number of factors that limit the energy utilization and rate of power production. Some of the wind will spill over and be deflected off of the rotor blades. Also, the velocity of the air after passing through the rotor is not negligible. Furthermore, there are losses at the generator, gearbox, converter, and other power conditioning devices. Overall efficiencies of 20-50% are reasonable. Including losses, equation (iii) is modified to give electrical power output as:

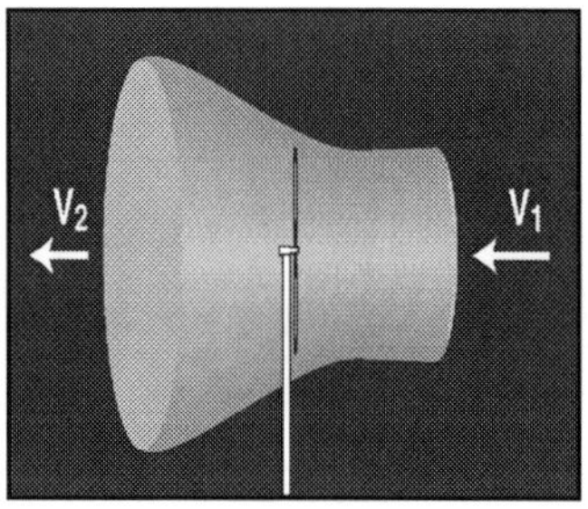

Stream tube.

$$P_e = \frac{\pi}{8} \rho d^2 V^3 \eta \qquad \text{(iv)}$$

Where η is the turbine efficiency.

Torque is calculated by dividing power by angular velocity of the shaft.

$$\tau = \frac{P}{\omega} \qquad \text{(v)}$$

where ω is the angular velocity of the rotor.

Example: Calculate the maximum power output and torque delivered by a 12-m diameter wind turbine when wind blows at 10 m/s. Assume rotor spins at 200 rpm and air density of 1.16 kg/m^3.

Solution: The maximum power is delivered when $\eta = 1$

$$P = \frac{\pi}{8}(1.16)(12)^2(10)^3 = 65{,}596\ W = 65.6\ kW \quad \text{and,} \quad t = \frac{65.6}{2\pi(200/60)}(1.16)(12) = 3.1\ kJ$$

disadvantage of downwind turbines is that wind speed suddenly drops behind the tower, so there is additional wind shade where the blade passes behind the tower which causes vibration of the rotor and results in more fatigue.

How many blades are optimal? Theoretically a wind turbine needs only one blade, but most have two and a few have three or more. The greater the number of blades, the more stable the turbines are, but also the heavier and more expensive. Multi-bladed rotors used in older designs are less efficient, but provide more torque and have therefore been used traditionally for pumping water even at low wind speeds. Two- and three-bladed rotors are cheaper, lighter, and run at faster speeds; however the increased tip speed can also make them noisier. They also have moderate starting torque which makes their operation difficult at very low wind speeds. One especially important factor in two-bladed wind turbines is wind shear. Because air has a greater velocity further from the ground, wind puts greater force on the top blade than the bottom blade. To correct this, wind turbines are equipped with a teetering mechanism, which equalizes the forces by allowing the blades to tilt slightly around their central pin. In some other designs, blades have a fixed pitch so instead of pivoting at the hub, these turbines flex as wind speed picks up. To guard against over-speeding, each blade has a small tip brake that at high speeds tips into the wind, stalling the turbine. Figure 3-11 compares performance of various rotor designs at different wind speeds.

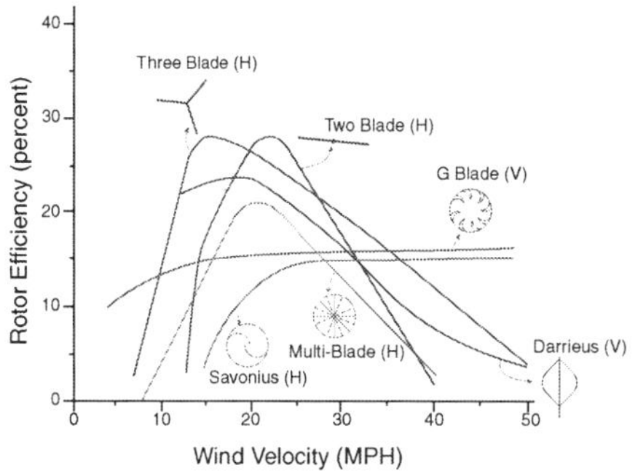

Figure 3-11
Effect of rotor design on performance
(Adapted from Wind Energy Systems, by G. L. Johnson, Prentice Hall, 1985.)

> **Question:** Windmills traditionally used for pumping water have small solid rotors with many blades. What is the main advantage of these designs?
> **Answer:** The many blades assure operation at very slow wind speeds, allowing continuous operation all year. These windmills are, however, very inefficient at high speeds and must be shut down to prevent damage.

Size

The size of wind turbines varies greatly and, depending on application, they produce power from a few watts to many megawatts. Micro turbines produce power in the range of 20-500 watts and are used in such applications as battery chargers and recreational vehicles. Turbines generating less than 50 kW are considered small, those producing between 50 kW and 1 MW are medium-sized, and turbines producing above 1 MW are large (Figure 3-12). Small turbines are used in applications such as pumping water, whereas medium and large wind turbines are used for producing electricity.

Figure 3-12
The world's largest wind turbine is now the Enercon E-126. This turbine has a rotor diameter of 126 meters (413 feet). The turbine being installed in Emden, Germany and will produce 7 megawatts (or 20 million kilowatt hours per year). That's enough to power about 5,000 households of four in Europe.
(Image courtesy of Enercon Corporation).

For larger metropolitan areas, either a large number of small turbines or a few large turbines must be installed. There are advantages and disadvantages to each option. The choice of using small or large rotors

Angle of Pitch and Angle of Attack

FYI ...

Some mistake the pitch and the angle of attack as being the same. The pitch is the angle the airfoil chord makes with the rotor plane of rotation. The angle of attack is the angle the chord line makes with the direction of flight or the relative wind direction. They are only the same in the absence of induced flow; that is, when the aircraft is in horizontal flight or, in the case of a wind turbine, when wind blows parallel to the rotational axis of the rotor. Generally, the two angles are different.

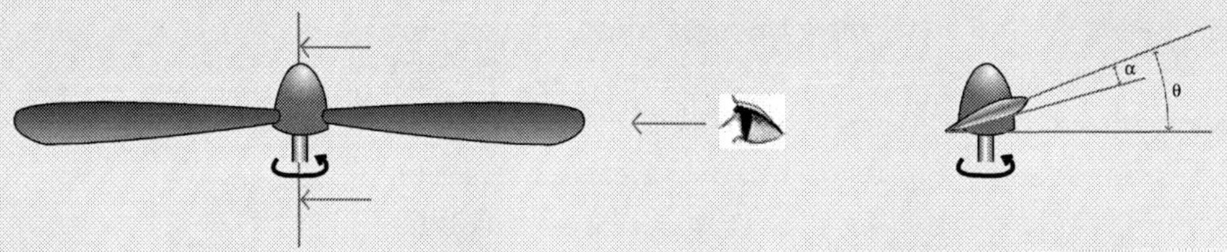

Top and side views of a wind turbine rotor and propeller blades. V_1 is the wind velocity, V_0 is the component of wind in the axial direction (or in the case of the aircraft, its forward speed), θ is the blade pitch angle, and α is the angle of attack.

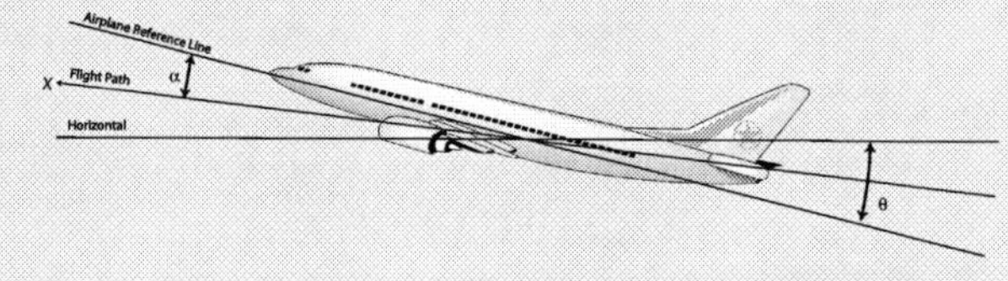

Pitch and angle of attack of an airplane in flight.

and generators depends on the application and on the distribution of wind energy throughout the year. For the same total capacity, larger turbines occupy less space and are cheaper to install. For example, a 400-kW turbine costs considerably less than four 100-kW turbines. The cost of delivering electricity is also lower for larger turbines than smaller ones. In addition, larger turbines can utilize the energy contained in high-speed winds and have greater efficiency. The drawback is that they cannot produce power at low speeds. The main advantage of small turbines is that they can produce continuous power for most of the year, as they require only slow to moderate wind speeds to operate. However, much of the energy in high-speed wind is wasted.

Taller towers are generally better than shorter towers, as wind velocity rapidly decreases near the ground. As shown earlier, the total energy in any wind stream is proportional to the cube of the wind speed, so turbines installed on taller towers could, in principle, be much more efficient. The cost of tower construction, however, could be much higher and may not justify the extra gain in efficiency. As a rule of thumb, tower heights are roughly equal to rotor diameter and cost one-fifth of the total price of the turbine. Stand-alone turbines are usually more economical, less complicated, and require less maintenance. They are used mainly in remote areas and small cities where local electrical grids are not able to support the power generation from larger wind turbines.

Power Control

To prevent failure, the system must have safety features that control sudden increases in power. Modern wind turbines are equipped with mechanisms that automatically adjust the rotor speed as wind speed and direction change. This can be done by controlling the pitch, controlling the angle of yaw, or initiating a stall.

The best way to control the power is by changing the blades' pitch. Pitch is the angle that the airfoil chord makes with the rotor plane of rotation (also called the angle of incidence). With variable pitch designs, bearings are inserted between the blades and rotor hub. When the wind speed becomes too great, rotor blades are turned along their longitudinal axis and slightly out of the wind to allow some wind to pass by without increasing lift. At low wind speeds, blades are turned into the wind. The advantage of a **pitch-controlled** scheme is that relatively constant rotor speeds can be maintained within a large range of wind speeds.

Another method of controlling rotor speed is to turn the rotor out of the wind by adjusting its yaw angle and tilting it either into or away from the direction of the wind. To optimize efficiency, turbine blades must face the wind at all times (they must be in a plane perpendicular to the direction of the wind). A large wheel called a yaw bearing turns the nacelle with the rotor into the wind. At very high speeds, the rotor is intentionally turned away from the direction of the trailing wind, reducing the total volume of air passing through the rotor and reducing the power. Since yawing bends the rotor, turbines running with **yaw-** or **tilt-control** experience large fatigue loads.

In **stall-controlled** wind turbines, blades are designed to ensure stall when wind velocity reaches a set value. The blades are shaped such that excess wind speed creates sufficient turbulence on the backside and the rotor stalls. The basic advantage of stall control is that the rotors remain fixed to the hub, which removes many of the complex control-mechanisms necessary in pitch and yaw control methods. Most large wind turbines use this approach.

Generator

In order to generate electricity, the wind turbine must be coupled with a generator. Most generators are directly connected to the grid and produce power at a nearly constant frequency; they are therefore constrained to operate within a narrow range of wind speeds. Sudden gusts of wind can speed up the rotation, so the drive train and tower must absorb the extra torque. Normally, generators produce 690 V, three phase alternating currents at frequencies of 60 (for the US) or 50 (for most other places in the world) hertz. The current is amplified through transformers in order to increase the voltage to the 10-30 kV used in most local transmission lines.

In indirect grid connections, wind turbines operate independently of the generator and the system operates in a variable speed mode determined instantaneously by wind speed. This setup has the advantage of being able to run a generator at all wind speeds. The cost is, however, considerably higher. In this case, voltage from the generator must be modified to match that of the neighboring electrical power grid. This must be done in several steps. First, the variable frequency alternating current must be converted into a constant frequency alternating current. Next, the alternating current must be converted to direct current using a rectifier. Finally, using a DC-AC inverter, the direct current must be converted to an alternating current at a frequency matching the grid line. Some filtering may also be required to eliminate unwanted frequencies, thus producing clean electric power with the desired frequency.

Power Train

The power generated by the wind turbine must be transferred to a generator. If the generator and turbine rotor are connected directly, then the rotor must turn with the same rotational speed as the generator. For a 50 Hz generator, this translates to 3,000 rpm. For a large diameter rotor, the tip speed will be extremely high, reaching several times the speed of sound. This causes an extreme torsional load and an excessive amount of noise which is not acceptable in populated areas. Rotational speed of the generator can be lowered in several ways.

The first option is to use a slow-moving AC generator with a large number of poles. For example, if the number of magnets in the stator of a synchronized generator is doubled, the magnetic field must rotate only half a revolution before it changes direction and the generator will run at 25 Hz (or 1,500 rpm). In theory, we can keep increasing the number of poles until the generator revolves at the same speed as the wind turbine rotor. The more practical approach is to use a gearbox. A **gearbox** is a device that converts slow-speed, high-torque power from the turbine into high-speed, low-torque power required to run the generator. An alternative approach is to use direct-drive generators that can operate at the rotational speed of the rotor, thus eliminating the need for a gearbox.

Hybrid Systems

According to World Bank estimates, as many as two billion people, 40% of the world's population, live in villages that are not tied to a utility grid. For these villages, a hybrid of energy sources including wind, solar, and diesel-powered generators would be most suitable. In the United States, a typical stand-alone hybrid system involving photovoltaic cells and wind is advantageous over either system by itself because it takes advantage of longer and brighter sunlight in summers when wind speeds are lower. In winters, the opposite is true; winds are strong when there is less sunlight. During periods of peak demand or when neither wind nor sunlight is sufficiently available, an auxiliary diesel can produce the additional energy required. In either case, to assure power is available at nights or during periods of low winds, a battery storage system is needed (Figure 3-13a).

When grid connection is available, it is possible to sell extra power to the utility companies and buy back electricity when demand exceeds the capacity of the system. In this case, there is no need to store the electricity in batteries or have an auxiliary diesel power generating plant (See Figure 3-13b).

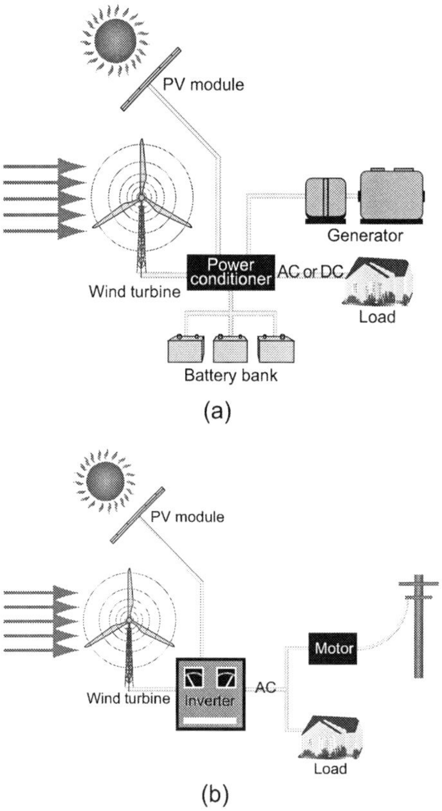

Figure 3-13
Hybrid systems involving a variety of energy sources can be used to power remote villages and generate income for their owners. (a) Stand-alone, (b) Grid-connected.
Adapted from "Small Wind Electric Systems: A U.S. Customer's Guide", Office of Energy Efficiency, U.S. Department of Energy, DOE/GO-102001-1293, [October 2002.]

Safety

To make wind energy economically viable, wind turbines must endure a wide range of hazardous conditions and have a long lifetime. Modern designs deploy various sensors that protect turbines, gearboxes, and generators against excess vibration, overheating, and high speed wind gusts.

Depending on wind speed and direction, the wind turbine produces fluctuating torque and varying forces on the blades, causing the rotor and tower to swing back and forth. The blades are made to be flexible and are able to vibrate. The frequency of oscillation depends on the height of the tower and on the material and weight of the rotor and nacelle. If the rotor spins at a synchronized speed with other vibrational frequencies, the oscillation can amplify and the tower could sway out of control. Wind turbines are often equipped with a controller system which starts the turbine when wind reaches a certain velocity (around 3-6 m/s) and shuts the machine off when wind velocity exceeds 20-30 m/s. Most wind turbines are designed to have a maximum output at wind speeds around 15 m/s. Higher wind speeds are rare, and lower wind speeds cannot produce sufficient power.

Noise can also be a problem. There are two sources of noise: mechanical and aerodynamic. Major sources of mechanical noise are the gearbox, drive shaft, and turbine blades. The primary source of aerodynamic noise is the air flow from the trailing edges of the blades.

One of the interesting features of electric motors and generators is that they require some electrical load to operate. If they overheat or a load is removed, the rotors will accelerate out of control and the devices will fail. To avoid overheating, most generators are equipped with either air or water cooling systems. To protect them from overspeeding and accidental disconnection from electrical grids, the common practice is aerodynamic braking, in which rotor blades are rotated 90 degrees along their longitudinal axis. Once the danger is over, a built-in hydraulic system returns the blades to their original orientation.

In addition to the control strategies described above, large turbines are

Figure 3-14
The Kappel wind farm in Denmark is a good example of how the technology can be integrated with the environment in an aesthetically pleasant manner.

Figure 3-15
Conceptual architectural design of a twin-tower building with three integrated 35-m diameter, 250 kW horizontal axis wind turbines. This arrangement is expected to provide 20% of the building's electricity.
Source: Compbell, N. et al., Wind Energy for the Built Environment Project (Project WEB), PF4.11, EWEC 2001, Copenhagen, 2-6 July2001.

Figure 3-16
The world's first offshore wind farm north of the island of Lolland in Denmark, consists of 11 turbines each producing 450 kW of energy.

equipped with complex, computer-operated control systems. These monitor environmental parameters (such as wind speed, direction, atmospheric temperature, and pressure), operating parameters (rotational speed of the rotor, torque, and the yaw angle), and a large number of devices such as pumps, actuators, and valves.

Siting

One of the important considerations in the design of any wind power generation station is its location. In general, wind generators must be installed in areas with an open view and follow the altitude contours of the prevailing winds. Accessibility to the site is also important. The site must be chosen so that the wind is mostly clear of obstacles, such as trees and tall buildings. Pattern and distances between wind turbines must be chosen so that the wake of one turbine does not interfere with the operation of adjacent turbines. The practical guideline is that turbines facing the prevailing wind direction should be spaced between 5 and 9 rotor diameters apart; turbines facing perpendicularly to prevailing winds should be from 3 to 5 rotor diameters apart.[9] Hilltops have the added advantage that they can pick up drafts, and the wind speeds are generally higher. Valleys are also suitable because tunnel effects result in higher wind speeds in valleys than those found in open spaces. When wind energy is chosen to service a small community or a local region, a site must be selected near a 10-30 kV power line; otherwise, the additional cost of extending the power grid can be prohibitive.

Because of their appearance, noise, and their potential adverse effects on the price of surrounding properties, wind farms can be objectionable to nearby communities. Modern planning procedures and sensitive site selection can help to minimize visual impact. Turbines can be obscured from view by planting trees or constructing other, similar screens. Some consider large wind turbines more aesthetically pleasing than smaller ones because they generally have lower rotational velocities. The selection of certain turbine colors, structures, and layouts can also help to minimize intrusiveness. Urban architects and city planners need to study these effects and design sites to match the local landscape or provide interesting additions to the nearby structures (Figures 3-14 and 3-15).[10]

Offshore Wind Farms

Wind turbines do not always need to be placed on land and, if they are installed offshore, may actually benefit from generally cooler and smoother lake and sea surfaces. Furthermore, because noise is not as much of a concern as with onshore facilities, turbines can be designed to operate at higher rotational speeds. As a result, a 20% greater efficiency is achieved at offshore wind farms (Figure 3-16). The main disadvantages of offshore wind farms are potential interferences with shipping routes and

[9] Danish Wind Industry Association, (http://www.windpower.org).
[10] Palmer, A., "Towers of Power," Popular Science, Vol 259, Iss 6, Dec 2001.

Is wind energy practical for you?

FYI ...

A small wind electric system may work for you if:

- Your house is thermally well-insulated. Making your house more energy efficient saves energy and money, thus requiring a smaller wind generator
- There is enough wind where you live
- You live in a house or ranch over one acre in size
- Tall towers are allowed in your neighborhood
- You have enough space
- You are comfortable with long term investments
- Your electricity cost is over $150 a month

Reference: "Small Wind Electric Systems: A US Customer's Guide", Office of Energy Efficiency, U.S. Department of Energy, DOE/GO-102001-1293, October 2002.

the additional cost of undersea cabling needed to connect the generator to the main electrical grid.

Summary

Wind energy is a clean, renewable source of energy. Only 1% of the total wind energy available is enough to fulfill all global energy needs, meaning wind is a potential source of an enormous amount of energy. The main drawback to wind energy is that it is noisy, intermittent, and to some unsightly. Other disadvantages cited are potential danger to birds, which may result in changes in migration patterns of some bird species, and shadow flicker – when moving blades cast shadow on nearby residences. However, wind turbine technology is undergoing rapid development: new composites have allowed substantial weight reduction, turbine noise is continually reduced, and new manufacturing technologies have allowed larger and larger blades-- making construction of multi megawatt turbines possible. At present, the average cost of production of electricity from wind energy is slightly higher than that from fossil fuel, but as demand for energy increases and fossil fuel sources deplete, the cost is rapidly becoming competitive with or even cheaper than those of nonrenewable sources of energy. Wind energy is expected to be a major energy source for most developing and island nations, as well as many developed European countries in the 21st century.

Additional Information

Books

1. Gipe P., Wind Energy Basics, —A comprehensive guide to modern small wind technology. AWEA (*http://www.awea.org*).
2. Elliott, D. et al., Wind Energy Resource Atlas of the United States, by American Wind Energy Association (*http://rredc.nrel.gov/wind/pubs/atlas*).
3. Khennas, S., Small wind systems for rural energy services, London: ITDG Pub., 2003.

Periodicals

1. *Solar Energy*, Direct Science Elsevier Publishing Company, the official journal of the International Solar Energy Society ®, is devoted to the science and technology of solar energy applications, and includes the indirect uses such as wind energy and biomass.
2. *Home Power Magazine*—bimonthly magazine for farm and home wind turbines (*http://www.homepower.com*).

Government Agencies and Websites

1. Energy Efficiency and Renewable Energy Clearinghouse (*http://www.eren.doe.gov*).
2. National Wind Technology Center, National Renewable Energy Laboratory (*http://www.nrel.gov/wind*).
3. Energy Efficiency and Renewable Energy: Wind Energy Technologies, US DOE (*http://www.eere.energy.gov/windandhydro/wind_technologies.html*).

Non-Government Organizations and Websites

1. American Wind Energy Association (*http://www.awea.org*).

Exercises

I. Essay Questions:

1. Airports are rarely located near big mountains. What concerns prevent building an airport near mountains?

2. Discuss the principles of operation of wind turbines. What are the major components of a wind generator? How do the wind turbine blades differ from those used in helicopters?

3. Define the pitch and angle of attack. Under what conditions are they the same?

4. Name the components of a modern wind power plant and describe their functions.

5. How does the height of a tower on which a wind turbine is installed affect its operation? What are the advantages and disadvantages of taller versus shorter towers?

6. What is the difference between a horizontal and a vertical wind turbine? How do they vary in shape? What is the advantage and disadvantage of an upwind design over a downwind design?

7. What are the conditions under which wind power generation can be economical for an average family? Discuss tax and other economical ramifications of switching from fossil fuel power plants to wind in your neighborhood.

8. Which one generates more power, a wind turbine 10-m in diameter at a wind speed of 10 m/s, or a wind turbine 20-m in diameter at a wind speed of 5 m/s?

II. Multiple Choice Questions:

1. People from which country invented the first practical wind turbine?
   a. US
   b. Netherlands
   c. Egypt
   d. Persia
   e. China

2. Which country utilizes the most wind energy today?
   a. US
   b. Germany
   c. Denmark
   d. India
   e. Iran

3. Since 1980, the US percentage share of wind capacity has been
   a. Declining
   b. Increasing
   c. Remaining about the same

4. What percentage of the US's electrical capacity comes from wind?
   a. Less than 1%
   b. About 5%
   c. About 15%
   d. About 50%
   e. About 99%

5. How much does the energy content of the wind change if wind speed suddenly doubles?
   a. About one-half
   b. About the same
   c. About twice as much
   d. About four times as much
   e. About eight times as much

6. How much does the energy content of wind change if the rotor size suddenly doubles?
   a. About one-half
   b. About the same
   c. About twice as much
   d. About four times as much
   e. About eight times as much.

7. How much does the energy content of wind change if the rotor size suddenly doubles, while at the same time the wind speed reduces by one half?
   a. About one-half
   b. About the same
   c. About twice as much
   d. About four times as much.
   e. About eight times as much

8. The rated power of a wind turbine is
   a. The maximum power a turbine can produce
   b. The average between maximum and minimum power
   c. The average between powers produced at the highest and lowest possible speeds under which wind turbine can operate
   d. The average power a turbine produces over a period of one day
   e. The average power a turbine produces over a period of one year

9. Two wind mills 10 feet and 30 feet in diameters are standing side-by-side against wind blowing at 5 knots. The smaller wind turbine produces 10 kW of electric power. The larger turbine is expected to produce
   a. 10 kW_e
   b. 30 kW_e
   c. 60 kW_e
   d. 90 kW_e
   e. 270 kW_e

10. A windmill system which produces 5 kilo watts of electrical power when the wind has a velocity of 3 m/s will produce ______ kilo watts when the wind velocity is 9 m/s.
    a. 1 2/3
    b. 15
    c. 45
    d. 135
    e. None of the above

11. The angle between the blade's chord and the direction of the incoming relative wind is called
    a. Yaw
    b. Pitch
    c. Roll
    d. Angle of attack
    e. None of the above

12. The angle between the blade's chord and the axis of rotation is called
    a. Yaw
    b. Pitch
    c. Roll
    d. Angle of attack
    e. None of the above

13. The main function of the gearbox is
    a. To increase the turning force (torque) of the wind turbine rotor and use a smaller generator
    b. To reduce the turning force (torque) of the wind turbine rotor and use a smaller generator
    c. To increase the turning force (torque) of the wind turbine rotor and use a larger generator
    d. To increase both the speed and the torque of the wind turbine
    e. To reduce both the speed and the torque of the wind turbine

14. Which of the following statements is <u>not</u> correct?
    a. Theoretically, wind turbines need only one blade.
    b. Generally, multi-bladed rotors are less efficient than rotors with only 2-3 blades.
    c. Two and three-bladed rotors have moderate starting torques which make them unsuitable at low wind speeds.
    d. Two-bladed rotors are lighter and run at faster speeds, which make their operation quieter and with considerably less vibration.
    e. The greater the number of blades, the more stable the turbines are, but also the heavier and more expensive.

15. Which of the following statements is correct?
    a. The power generated by a wind turbine varies linearly with the kinetic energy of the mass of air passing through its rotor.
    b. For the same wind velocity and rotor size, more power is generated in summers than in winters.
    c. Total power generated by a wind turbine increases with the square of the rotor area.
    d. Total power generated by a wind turbine increases with the square of the wind speed.
    e. All of the above statements are correct.

III. True or False?

1. Today, wind energy furnishes less than 1% of the

total energy needs in the world.

2. The greater the number of blades, the higher the efficiency of the wind turbines.

3. Land breezes occur during the daytime when land warms up much faster than the surrounding water.

4. Valley and mountain breezes are affected strongly by local topography.

5. The wind rose is a mechanical device for measuring the wind velocities.

6. Lift is the force perpendicular to the surfaces of the wind turbine blades.

7. Rated power is the power that a wind turbine produces at its optimum operating conditions.

8. Modern wind turbines vary in size and can produce power ranging from a few watts to several megawatts.

9. Wind velocity decreases with height above the ground.

10. The best location for installing a wind farm is in wide open fields, away from the populated areas but close enough to the national power grid.

IV. Fill-in the Blanks

1. Winds are generated as a result of two factors, the non-uniform heating of the earth by the sun and the _________________________.

2. Winds flowing on a path along the lines of constant pressure are called __________ wind.

3. The power a turbine produces when it is run at its optimum operating condition is called __________ power.

4. The total power generated by a wind turbine is not constant and varies with wind speed and direction, rotor diameter, and air ____________.

5. For best efficiency, most wind turbines are designed with _____ to _____ blades.

6. The power output of a wind turbine __________ with altitude at the rate of 2.5% for every _______ feet above sea level.

7. The direction of the lifting force is roughly __________ to the local flow field.

8. To modify torque from a wind turbine to run a generator, wind turbines are usually equipped with _____________ .

9. The best way to control the power is by changing the blades' ___________ angle.

10. Only ________ percent of the total available wind energy is enough to satisfy all of the world's energy needs.

V. Project – Designing a Wind Farm

A small village is planning to install a wind farm to meet its energy demand. In this project you are asked to evaluate various wind turbine technologies to meet the electricity needs of the community. The design options considered are the G-mill, two-blade, three-blade, multi-blade, Darrieus (egg-beater), and Savonius turbines. For comparison, assume that all wind designs must provide 60 kW of power when wind speed is around 33 mph.

1. Calculate the average wind speed and wind power factor for each speed range. Wind power factor is the ratio of power produced at a given wind speed to that at nominal speed (in this case 33 mph).
2. Use data given in Figure 3-11 to estimate the rotor efficiency. Calculate the power (kW) produced by each turbine when operated at a given wind velocity.
3. Find the total wind energy for each wind turbine by multiplying power times the number of hours of operation. Sum up capacities to find the annual cumulative value (kWh/year).
4. Assuming 50% conversion efficiency from mechanical to electrical power, determine electrical power for the most efficient turbine.

5. Assuming that the average cost of electricity generation is $0.12/kWh, determine the total annual cost.

The following wind data are available from local meteorological offices:

Wind Velocity	0-5	6-10	11-15	16-20	21-25	26-30	31-35	36-40	>40
Hours per Year	1200	1600	1950	1700	1480	960	280	50	10

Project Work Sheet

Wind velocity (mph)	**0-5**	**6-10**	**11-15**	**16-20**	**21-25**	**26-30**	**31-35**	**36-40**	**>40**	**kWh/year**
Hours per year	**1200**	**1600**	**1950**	**1700**	**1480**	**960**	**280**	**50**	**10**	-----
Average wind velocity				18			33			-----
Wind factor				0.16*			1			-----
G-Mill	η (%)									-----
	kW						60			-----
	kWh									
Two-Blade	η (%)									-----
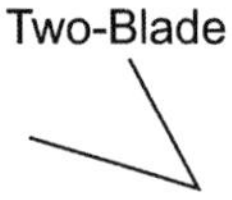	kW						60			-----
	kWh									
Three-Blade	η (%)			27%			18%			-----
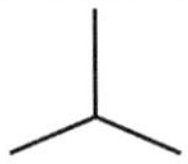	kW			14.58**			60			-----
	kWh			24,786						
Darrieus	η (%)									-----
	kW						60			-----
	kWh									
Multi-Blade	η (%)									-----
	kW						60			-----
	kWh									
Savonius	η (%)									-----
	kW						60			-----
	kWh									

The Most efficient turbine: _________
Total mechanical energy: _____________ kWh
Total electricity generation capacity by one turbine: _____________ kWh
Total annual operating cost of one turbine: US $ __________

* Wind factor = $(V/V_{nominal})^3 = (18/33)^3 = 0.162$

** Power calculated by scaling the nominal power (60 kW), corrected for wind velocity and efficiency,
P (18 mph) = 60 kW x 0.162 x (0.27/0.18) = 14.58 kW

Hydro Energy

CHAPTER 4

The ocean is a body of water occupying about two-thirds of a world made for man - who has no gills.
~ Ambrose Bierce (1842 - 1914)

Oceans cover more than 70% of the earth's surface, making them the world's largest source of hydro energy. There are many different ways to extract energy from water. Seawater is the source of deuterium, the ideal fuel for nuclear fusion. Surface water stores a massive amount of solar energy that can be exploited to design thermal power plants. In addition, water contains mechanical energy that can be converted to useful work in the form of the kinetic energy of river streams, or the potential energy of waterfalls, tides, and ocean waves. According to some estimates, these resources have the potential to produce 1-2 terawatts of electricity. This amount is enough to cover the energy demands of the entire globe, but tapping into most of that potential is not yet economically feasible. In this chapter, we will investigate how the mechanical energy stored in water can be used to design practical devices. Thermal and nuclear conversion options are discussed in Chapters 5 and 11 respectively.

Hydroelectric Energy

Egyptians harnessed energy from flowing water about 2,000 years ago by turning waterwheels to grind their grain. These primitive devices allowed the force of falling water to act on a waterwheel and provide rotational energy or shaft power. Through the centuries, mechanisms were designed to facilitate many other applications beyond the simple grain mills of the Egyptians. By the time of the Industrial Revolution, water power was driving tens of thousands of waterwheels. Today, hydropower is the most widely available renewable energy. It is exploited almost exclusively for generating electric power, providing about 20% of all electricity used around the world and 10% of the US electrical capacity.

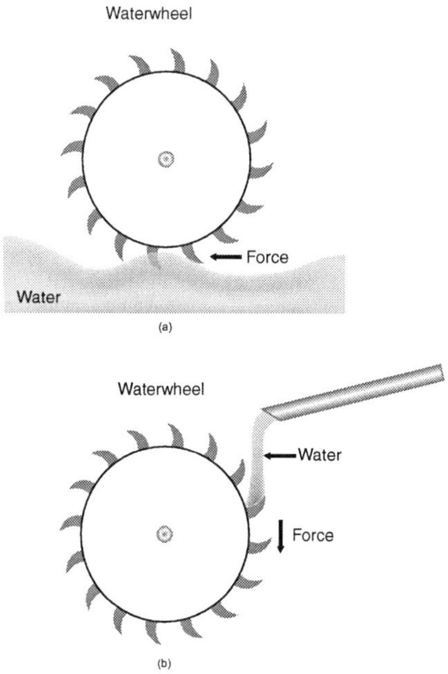

Figure 4-1
Vertical water mills; (a) undershot, and (b) overshot

Question: Two medieval varieties of waterwheels were undershot and overshot wheels[1,2] (Figure 4-1). Undershot refers to a paddle wheel fixed to the bank of a river or hung from an overhead bridge. It is turned by the impulse of the water current. Overshot water mills work by bringing a stream of water through a pipe or canal and pouring it onto the wheel from above. Which of the designs seems to have a higher efficiency?
Answer: Overshot wheels. The weight of the water falling on the blades (gravity) forces the wheel to turn at a faster speed. Furthermore,

[1] Hodges, H., "Technology in the Ancient World," Barnes & Noble, N.Y., 1992.
[2] Reynolds, J. Windmills and Watermills, Praeger, New York, 1970.

undershot wheels require a steady wave current, so their operation could be interrupted when water levels drop during the summer months.

Question: Watermills come in vertical and horizontal configurations (Figure 4-2). What are the advantages and disadvantages of each configuration?
Answer: Most early watermills were used solely for grinding wheat, barley and other seeds using a flat rock (stone mill). Horizontal mills were easiest to operate as they could directly turn the stone mill. The vertical designs required gears to transfer power to the stone mill and thus were not very convenient.

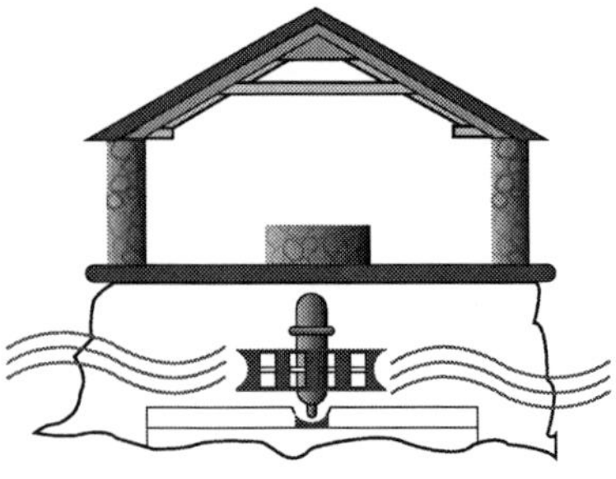

(a)

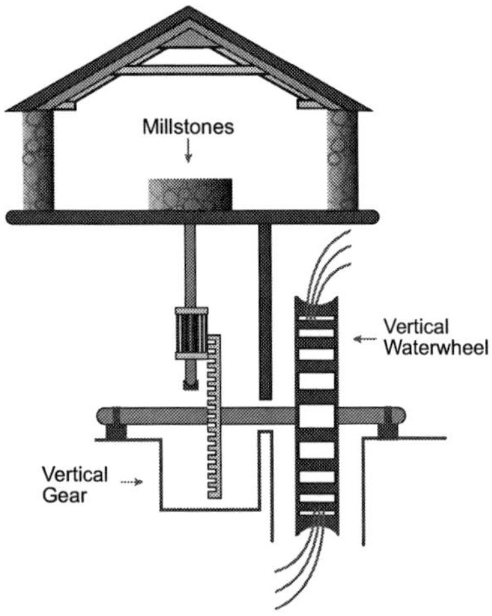

(b)

Figure 4-2
Water wheel configurations: (a) horizontal and (b) vertical

Power Generation

Currently there are about 680,000 MW of hydroelectric capacity installed throughout the world, which fulfills about a quarter of the world's electricity demand and supplies more than one billion people with power. Worldwide, the amount of hydroelectric power that can be developed is about five times greater than current capacity. With 332 billion kWh and 12.5% of the world's production, Canada is the leading hydropower producer; it satisfies 60% of its electricity needs by hydropower. After Canada, Brazil, and China, the United States is the fourth largest producer of hydroelectricity in the world.[3] Norway relies on its vast hydropower resources for 99% of its electricity, making it the cleanest country in the world in terms of energy use.[4]

Hydroelectric plants range in scale from large falling water plants used in developed nations to small river runoff plants with no dams or water storages used for rural electrification in less-developed countries. China has the most hydropower resources and has installed more than 40% of the entire world's small hydro (50 kW_e-10 MW_e) capacity.[5] For now, the largest hydroelectric plant in the world is located at Itaipú along the Parana River between Paraguay and Brazil. It has a power output of 12,600 MW and supplies 80% of the electricity consumed in Paraguay and a quarter of Brazil's total supply (Figure 4-3). The Itaipu plant will, however, concede its ranking to the Three Gorges plant being built over the Yangtze River in China. Slated for completion by 2010, it will produce 18,200 MW. The tallest waterfall in the world is Angel Falls in Venezuela, with a total drop of 980 m. The largest hydroelectric plant in the United States (and the third largest in the world after Itaipu in Brazil and Guri in Venezuela) is the Grand Coulee on the Columbia River in Washington State. Its three power plants can collectively produce 6,800 MWe. The power plant in Niagara, at 1,950 MWe capacity, is only of modest size.

[3] 2002 data can be found at *http://www.eia.doe.gov/pub/international/iealf/table15.xls.*
[4] Energy Information Agency (*http://www.eia.doe.gov*)
[5] Please note the difference between MW_e denoting the electrical power output and MW, which refers to thermal power output. As we will see later, only 30-35% of the thermal energy can be converted to electricity. For hydroelectric plants we can use kW and kW_e interchangeably.

Although the majority of large hydroelectric plants (>10 MWe) have already been constructed where it made economic sense, small-scale hydro plants have been largely ignored. Figure 4-4 shows the exploited and potential hydroelectric resources for various continents.[6] As this data indicates, most of the potential hydro sources for industrial nations have been used up. However, a huge hydro potential remains to be utilized by developing countries.

(a)

(b)

Figure 4-3
The hydroelectric generating station in Itaipu with a capacity of 12,600 MWe boasts the largest facility of this kind in the world (a). The tallest waterfall in the world is the Angel Falls in Venezuela (b).

Plant Types

There are three kinds of hydropower plants: impounded plants, pumped storage plants, and run-of-the-river plants.

Impounded plants, in which water is impounded in a reservoir behind a dam, are the most common. The water storage and release cycles can be relatively short (storing water at night for daytime power generation), or long (storing spring runoff for power generation in the summer). In these plants, water always flows downward from a storage reservoir behind a dam to run a turbine. Turbines are devices that convert the energy of a moving fluid (usually water, steam, or air) into the rotational energy of a shaft. During peak demands, where sufficient electricity cannot be generated by conventional means, enough water is released from the reservoir to meet additional power requirements. The primary issue with these plants is that the water flow rate downstream from the dam can change greatly, causing a sudden power surge. This often involves dramatic environmental consequences including soil erosion, degrading shorelines, crop damage, disruption of fisheries and other wildlife, and even flooding or droughts.

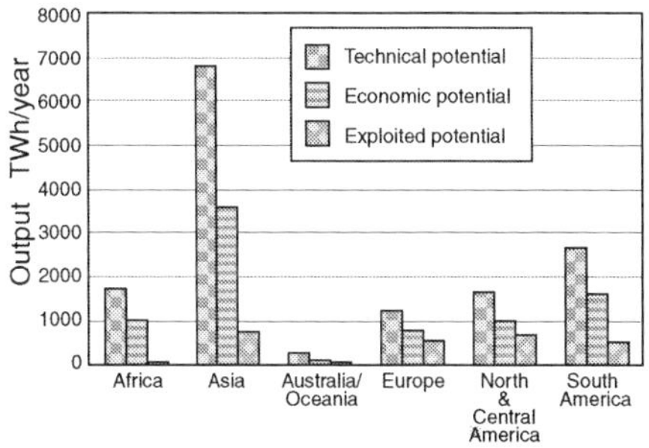

Figure 4-4.
Exploited hydro potential by continent.

Pumped storage plants (PSP) reuse water after its initial use to generate electricity. This is accomplished by pumping water back into a storage tank at a higher elevation during off-peak hours when the need for electric power is low. During peak demands or when there is an unexpected spike in the electrical load, water is allowed to flow back into the lower reservoir to produce more electricity. An important advantage of PSPs is the quick delivery of power during emergencies and power surges. In comparison, a typical coal- or natural gas-fired power plant takes many hours to start. In the United States, about one quarter of all hydropower generated is from pumped storage plants.

Instead of water being pumped to the storage reservoir at a higher elevation, during the peak demand, water can be allowed to fall in underground reservoirs dug out in hard rocks. When demand falls, the base load can pump the water back to the reservoir. Since there is smaller fluctuation in the level of water in the storage reservoir, the system is more stable.

In modern pumped storage plants, the same turbine-generator that

[6] Paish, O., "Small hydro power: technology and current status", *Renewable & Sustainable Energy Reviews*, 6 (2002) 537-556.

Power from Falling Water

Mathematical Interlude ...

A certain mass of water, m, has a certain amount of potential energy given by

$$E = mgH \qquad \text{(i)}$$

where H is the effective pressure head of water across the turbine (roughly equal to the height of the waterfall) and g is the gravitational acceleration equal to 9.81 m/s² at sea level. The power produced is calculated by replacing the mass, m, by mass flow rate $\dot{m}$[kg/s] in the above equation:

$$P_e = \dot{m}gH \qquad \text{(ii)}$$

Not all the power can be converted to electricity, as there are losses in the turbine and in the generator. If η_T and η_G are the efficiencies of the turbine and the generator, ρ is the water density and Q is the volumetric flow rate of water, then electric power generated is written as:

$$P_e = \eta_T \eta_G \rho g Q H \qquad \text{(iii)}$$

Modern hydroelectric plants employ turbines and generators with efficiencies as high as 90%. This gives an overall plant efficiency of $\eta = \eta_T . \eta_G = 0.90 \times 0.90 \sim 80\%$.

Example: A hydroelectric plant is supplied with water from a dam located 50 meters above the inlet to a turbine. As the water falls, its potential energy is converted into kinetic energy before it flows through the turbine at the rate of 10 cubic meters per second. Assume water has a density of 1,000 kg/m³ and that both the turbine and generator efficiencies are $\eta_T = \eta_G = 1.0$. Estimate the electrical power in megawatts that this plant can produce.
Solution: Power produced is calculated from equation (iii) as:

$$P_e = (1.0)(10.0\, m^3/s)(1000\, kg/m^3)(10.0\, m/s^2)(50m) = 5x10^6\, W_e = 50\, MW_e$$

generates electricity from falling water can also be used to pump the water back into the storage tank. In this case, the generator changes the direction of the electric field, forcing the turbine to rotate in the reverse direction and act as a motor which runs the pump.

Run-of-River Plants are typically low dams where the amount of water running through the turbine varies with the flow rate of water in the river. The flow rate of water and the amount of electricity that is generated in the run-of-river plants is generally smaller than in pumped storage plants, and changes continuously with seasons and weather conditions. Since these plants do not block water in a reservoir, their environmental impact is minimal.

Plant Design

Water used by a hydroelectric plant is stored behind a dam at a certain elevation (*head*) above a turbine. The water flows through a penstock and through the blades of the turbine, causing the turbine to rotate. Depending on amount of electric energy required, the flow through the turbine is adjusted by series of gates and valves. The turbine shaft is normally coupled to the generator shaft to produce electricity.

In a typical small hydro scheme, a portion of the water is diverted from a river or stream through an intake valve. It is then passed through a metal screen into a settling chamber where stones, timbers, and other debris are removed and suspended particles of dirt settle before water enters the turbine. Since no reservoir is blocking the flow of water, the impact on the river and its habitat is minimized.

Depending on its application, either an impulse or a reaction turbine is used. In an **impulse turbine**, the blades are fixed to a rotating wheel and available head is converted into kinetic energy by a contracting nozzle. The high velocity jet then impinges on the blades and turns the turbine. The most common impulse turbines are of the *Pelton* type, where a series of cupped buckets are set around its rim. A high-speed jet of water enters the wheel tangentially (Figure 4-5a) and is deflected 180 degrees by the cups. Nearly the entire momentum of the water is used to impart an impulse that forces the wheel to turn. Impulse turbines are used most often with falling mountain streams with large heads and relatively small flow rates.

A **reaction turbine** uses the pressure difference across the blades for producing a hydrodynamic lift that propels the blades. To produce the highest lift, blades must be highly aerodynamic and shaped like airfoils. Unlike the impulse turbine, the rotors (also called runners) of reaction turbines are fully immersed in water. Two common reaction turbines are the Kaplan and Francis turbines. In *Kaplan (propeller)* turbines, water impinges on a propeller -- similar to a ship's propeller-- fitted inside the penstock (Figure 4-5b); in *Francis* turbines, water flows radially inward to the runner housed in a spiral casing and is exited through a diffuser (Figure 4-5c). Kaplan turbines are common for the run-of-river projects with low head and large flow rates. Francis turbines are suitable in dams where high flow rates and relatively large heads are available.

The main advantage of reaction turbines over impulse turbines is that the blades can be adjusted to match the rotor speed closely to the generator speed. Recent advances in power electronics allow turbines and generators to run at varying speeds. This allows simpler and cheaper propeller turbines like Pelton to be used instead of the more advanced and expensive Kaplan and Francis.

(a) (b) (c)

Figure 4-5
a) Pelton; b) Kaplan; and c) Francis Turbines.
Photo Courtesy of Gugler Hydro Energy GMBH.

Tidal Energy

Tides are the vertical rise and fall of the sea level as a result of gravitational attraction between the earth, the moon and, to a lesser degree the sun. The gravitational attraction of the moon causes the water facing toward the moon to bulge. On the other side of the earth, the moon pulls the earth from the water, and we have a second bulge and another high tide. At the same time, low tides occur at the right angle to the moon's pull. Other factors that affect the force, shape, and height of the tides are the time of the year (position of the sun), size, depth, and shape of the ocean basin.

The highest highs and the lowest lows are called the spring tides. **Spring tides** (they come from the Saxon word *springen* meaning the swelling of water and have nothing to do with the spring season) occur a few days after the full and the new moons, when sun, moon, and earth are aligned. Spring tides take place about twice a month. When the sun-earth line is at the right angle to the earth-moon line (i.e during half moons), the gravitational effects of the sun and the moon tend to cancel each other, tides are lower than average range and we have the **neap tides** (neap means scanty or lacking). Since with respect to the moon, the earth rotates about its axis once every 24 hours 50 minutes, we expect an interval of 12 hours and 25 minutes between two high or two low tides; each *tidal day* is therefore 1.035 times a *solar day*.

Accompanying the vertical rise and fall of water, there are also horizontal or lateral movements. As the tides channel between islands or into bays and estuaries **tidal currents** occur. During the high tide, the tidal current known as the *flood tide* flows towards the shore. About six hours later, the current reverses and flows away from shore. This is the *ebb tide.*

Using tides for power generation is not a new concept. Tidal mills operated off the western coasts of Europe for many centuries before being replaced by cheaper methods of producing energy. These mills took advantage of the natural rise and fall of coastal tides by allowing water to fill up a pond during flood tides and then emptying that water over a watermill during the ebb tide.[7] Modern versions of tide mills are similarly designed. A barrier is built across an estuary and is allowed to be filled by water during the flood tide. The barrier is equipped with gates that are closed as the basin fills with water. As the tide recedes into the open sea during its ebb, the released water drives a number of turbines which in turn drive generators and produce electricity. Upon emptying, the basin gates are closed to allow the water in the sea to rise. This creates a head against the basin during the flood which can be emptied for a second time, turning the turbines in the opposite direction and generating power (Figure 4-6).

[7] Gies, F. J., Cathedral, Forge, and Waterwheel: Technology and Invention in the Middle Ages, Harper Perennial, New York, 1994.

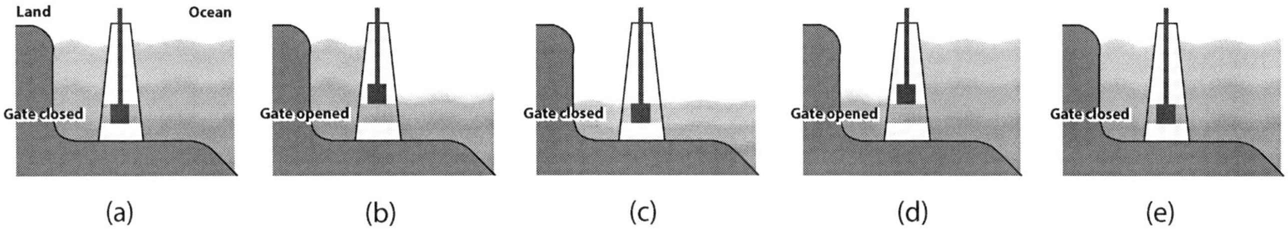

Figure 4-6
Operation of a tidal plant
(a) During period of high-tides water fills the basin. (b) At low tides the gate is opened and water is allowed to empty into the open sea while turning a turbine. (c) Once all the potential energy is exhausted gates are closed again, allowing water rise in the open sea. (d) The flow is from the open sea to the basin and turbine runs in the opposite direction. (e) The gate is closed until water rises again and repeats the operation.

The largest tidal power station that was ever constructed, a 240-MW plant built in 1966, is still operating today in La Rance estuary in Brittany, France (Figure 4-7). The annual generation is some 640 million kWh. Two other plants are operating commercially. The first, with an electrical generating capacity of 18-MW, is the Annapolis Royal Station in Nova Scotia off the Bay of Fundy in Canada; the second is a small 1.8 MW tidal plant in Kislaya Bay near Murmansk in the Russian Arctic.[8] Currently, there are no tidal power plants in the United States and none are planned.

Figure 4-7
Tidal power plant at the Bay of Rance estuary in France.

The potential energy of tidal power has been shown to increase with the basin area and the square of the average range between the low and high tides.[9,10] To be practical, tidal ranges of at least 5 meters (16.4 feet) are recommended. Oceans have tidal ranges between 5 to 10 feet; in narrow passages and estuaries ranges as high as 50 feet are common. Unfortunately, there are only about 40 sites on the planet that meet this requirement.[11]

Tides are clean, entirely predictable and renewable sources of energy. The major drawbacks are the large distances of most suitable sites from the population centers and the high capital cost of building the barrage across the estuaries. In addition, tidal power can be used only a few hours a day when the tide is moving in or out of the barrage.

Waves

A **wave** is a disturbance that travels through space and time. Unlike electromagnetic (optics) waves that can propagate through vacuum, mechanical waves (acoustics, ocean surface waves) need a medium to travel. Waves are characterized by their crests (highs) and troughs (lows), and may be classified as *traverse waves* with vibrations perpendicular to the direction of propagation (violin strings, light) or *longitudinal waves* with vibrations parallel to the direction of the propagation (horn, radar). Ripples on the surface of water are a combination of both traverse and

[8] Charlier, R. H., "Sustainable co-generation from the tides: A review," Renewable & Sustainable Energy Reviews, 7, 187-213, 2003.
[9] Bernstein, L. B., "Central tidal-power stations in contemporary energy production," State Publishing House, Moscow, 1961.
[10] Frau, J. P., "Tidal energy: promising projects: La Rance, a successful industrial-scale experiment," *IEEE Transactions on Energy Conversion*, Volume: 8, Issue: 3, September 1993. pp. 552-558.
[11] See the Department of Energy website at *http://www.eere.energy.gov/consumerinfo/factsheets/nb1.html*.

Tides

Digging Deeper ...

Tides are vertical displacement of water. The action of tides is best understood by examining the law of gravity as stated by the Newton's Law of Universal Gravitation. According to Newton, any two masses m_1 and m_2, separated by a distance r, are attracted to each other with a force of

$$F = G\frac{m_1 \cdot m_2}{r^2} \quad \text{(iv)}$$

where G is the Universal Gravitation constant equal to $6.67x10^{-11}$ $N.m^2/kg^2$.

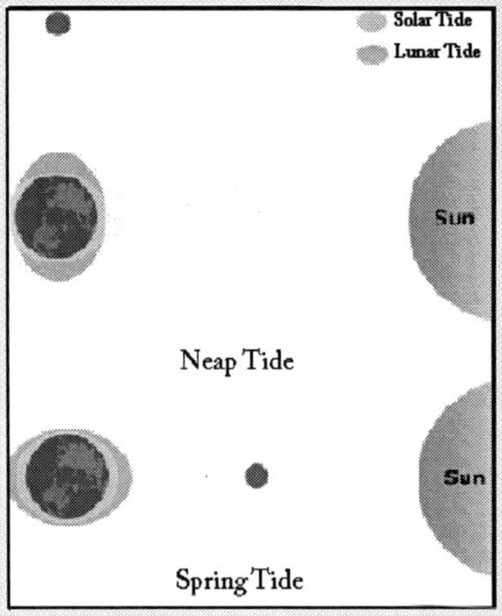

According to this law, as the moon orbits the earth it is pulled by the earth as though it were connected by a rope to the earth's center. The pull of gravity becomes weaker at distances farther away from the moon. As result of progressively declining gravity, the earth is stretched along the earth/moon line. Liquids have weaker internal bonds than solids, so ocean water stretches more than land masses. The points on the surface of the earth perpendicular to a line extending between the earth and moon feel less force than the earth's center. The result is that water is pulled along the moon-earth line and is pushed in the perpendicular direction. The net effect is that water accumulates at both ends along the moon-earth line (high tides). A little more than six hours later the earth has rotated by 90 degrees relative to the moon and the situation reverses (low tides).

But what about the pull of the sun? It just so happens that, although the gravitational pull of the sun is much stronger than that of the moon, because the sun is so much farther than the moon, its pull does not change very much from one side of the earth to the other, and its effect is only half that of the moon. The sun, however, gives a helping hand during the new and full moons when it is in line with the earth and the moon. Under these conditions, we have very strong tides which are called spring tides. When the sun, earth and the moon forming a right angle, as is the case during quarter moons, the effect is smallest and we have neap tides.Because the positions of the earth, sun, and moon change every day, tides vary from day to day, and from one location to another.

Question: Why do celestial bodies tend to be round?
Answer: Gravity pulls all the mass toward the center.

Example: Which system experiences a stronger attraction force, the earth-sun or the earth-moon? The sun, earth, and moon have masses of $2x10^{30}$, $6x10^{24}$, and $7.4x10^{22}$ kilograms respectively. The mean distance between the earth and the sun is $1.5x10^{11}$ m while the distance between the earth and the moon is $3.84x10^{8}$ m.
Solution: Substitute into equation (iv), we get:

$$F_{sun\text{-}Earth} = G\frac{M_S \cdot M_E}{r_{S\text{-}E}^2} = 6.67x10^{-11}\frac{(2x10^{30})\,(6x10^{24})}{(1.5x10^{11})^2} = 3.56x10^{22}\,N$$

$$F_{Moon\text{-}Earth} = G\frac{M_M \cdot M_E}{r_{M\text{-}E}^2} = 6.67x10^{-11}\frac{(7.4x10^{22})\,(6x10^{24})}{(3.84x10^{8})^2} = 2x10^{20}\,N$$

This calculation shows that the gravitational force of attraction of the sun on the earth is over 177 times that of the moon. As we mentioned above, it is not the absolute force of gravity, but the relative strength of gravitational forces across opposite points on the earth that determines the size of tidal range.

longitudinal waves.

Waves are mathematically represented by their *wavelength* (λ), the distance between two consecutive crests (or troughs), and *amplitude,* the maximum height between crests and troughs. *Period,* T, is the time for one complete cycle of a wave oscillation.

Wave Energy

As wind blows across the ocean, some of its energy is transferred to the water in the form of waves. Since wind itself is a result of non-uniform heating of the earth's surface by the sun, waves can be considered a stored form of solar energy (Figure 4-8). The rougher the water, the larger the ripples and the easier it is for the wind to transfer its energy. As wind loses its energy, waves become calmer and eventually reach their steady-state oscillations. As waves get closer to the coast and enter shallower water, they grow taller, their lengths decrease, and water particles move forward. The particles closer to the bottom fall behind until the wave eventually breaks on to the shore.

Figure 4-8
As wind blows over large bodies of water it continuously supplies energy to waves.

Waves have form (potential energy) and motion (kinetic energy). Potential energy is that which is expended to distort a flat sea surface into the shape of the wave. Kinetic energy is the energy put into wave by water motion in and beneath the wave, causing molecules of water to rotate in a circular fashion, higher in diameter at the surface and decreasing exponentially with depth.[12] Contrary to popular belief, waves do not move along the water surface – the water simply cycles around in small circles without any mass being transported forward. These are called stationary or transverse waves and are very different than longitudinal waves such as sound waves and waves generated by compressing a spring.

Surface Waves

Surface waves result from the superposition of many waves of varying wavelengths and heights. The wavelengths are, however, limited to about twice the average depth of the ocean. Ocean floors are, at maximum, 4 kilometers deep, so waves can be as long as 8 kilometers. Longer surface waves travel faster and further than shorter ones, but the energy content of a wave is related solely to its height. The surface currents flow clockwise in the northern and counter-clockwise in the southern hemispheres.

Wave power is measured in kW/m of wave crest (the highest point on a wave). The maximum power a wave can carry is about 100 kW per meter of wave front in the high seas, about 30-60 kW/m around the Atlantic coasts of Europe, and 10 kW/m off the western coast of the United States (See box "Power of a Wave"). Due to their greatest potential, the west coast of Europe is the most suitable for developing wave energy schemes (Table 4-1).[13] For example, it is estimated that waves in the British Isles could be tapped for more than 20% of British and 75% of Irish electricity needs.[14] At this time, most efforts remain in the research stage, although the few demonstration projects promise to play a significant role for wave technology in the near future. The European Union projects that by 2010 up to 1,000 megawatts of electricity could come from ocean waves.

Table 4-1. Potential for wave energy for major European countries

Country	Near Shore (TWh/year)	Offshore (TWh/year)
UK	14-21	43-64
Ireland	7-11	21-32
Portugal	4-6	12-18
Spain	3-5	10-16
France	3-5	12-18

[12] Mayo, N., "Ocean Waves -- Their Energy and Power," *Physics Teacher,* Vol. 35, September 1997.
[13] Thorpe, T. W., "A Brief Review of Wave Energy," Report no. ETSU-R120, The UK Department of Trade and Industry, May 1999.
[14] Edwards, R."The Big Break," New Scientist, pp. 30-34, October 1998.

Power of a Wave

Mathematical Interlude ...

The power delivered by a wave is obtained by dividing the total energy transported by its period:

$$P = \frac{\rho g^2 T H^2}{32\pi} \quad \text{(v)}$$

In this equation:

- P is power of wave front [W/m]
- g is the local gravitational acceleration [m/s²]
- ρ is the water density (kg/m³)
- T is period (time it takes two successive crests or troughs pass the same point [s], and;
- H is the wave height (amplitude) [m]

Example: A deep-water wave with a wavelength of 20 m and amplitude of 3 m travels at a speed of 5 m/s. Calculate the maximum power that can be exploited from such a wave.
Solution: Taking water density $\rho = 1030$ kg/m³, $g=9.81$ m/s², and a period of $T = 20/5 = 4$ s, we have:

$$P = \frac{(1030)(9.81)^2(4)(3^2)}{32\pi} = 35{,}500\ W/m = 35.5\ kW/m \text{ of wave front}$$

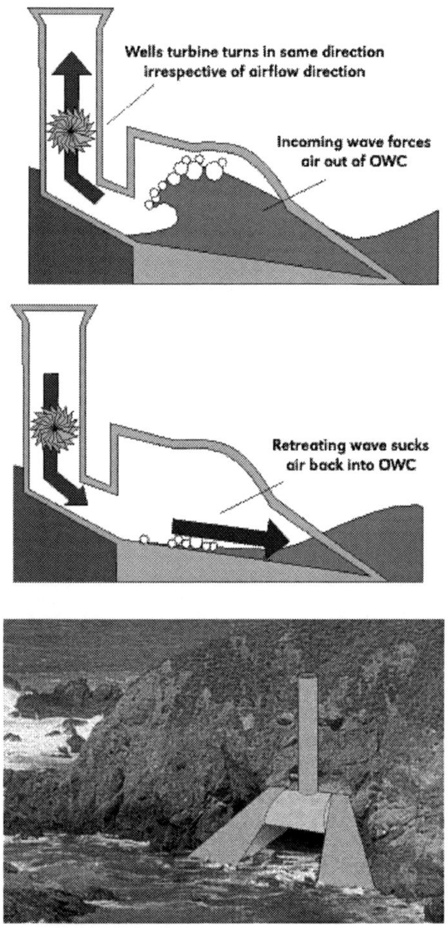

Figure 4-9
Schematic of an Oscillating Water Column.
Image courtesy of Fujita Research

Underwater (Marine) Currents

Just as wind is affected by local terrain, coastal waves are affected by underwater topography. Narrow channels between islands and around the coastal edges of oceans are best suited for power generation and can provide as much as one megawatt of power per square kilometer of seabed. To increase the velocity of the water and thus its kinetic energy, some propose building barriers across channels and narrow straights. The best locations are near the shore at depths of between 20 to 30 meters, where wave velocities are approximately 2-3 meters per second.

Power Generation

Wave power generation plants are of either the fixed or floating types. Fixed generating devices are located along the shore or are fastened to the seabed and are generally simpler to maintain and operate. Floating devices are installed on floating platforms. Examples of fixed energy conversion systems are oscillating water columns, tapered channel systems, and underwater turbines. Examples of floating systems are the Archimedes Wave Swing and the Salter's Duck. These devices work directly by activating a generator or pushing a working fluid (water or air) to drive a turbine and generator. No large-scale commercial wave power plants have been built yet, although major research is underway and several prototype systems have been built in Norway, Japan, India, and Scotland.

Onshore Systems

Most wave power machines use the **Oscillating Water Column.** They work by trapping air over the surface of water in a chamber that then moves a piston up and down. This can be achieved by the up and down motion of water in open seas, or by the back and forth motion of waves as they slam on the shoreline. In one design, waves coming toward the shore

a Power from Underwater Currents

Mathematical Interlude ...

The power of a current follows the same formulation that was given for wind power:

$$P_e = \frac{\pi}{8} \rho d^2 V^3 \eta \qquad \text{(vi)}$$

where:

- P is delivered power in watts,
- ρ is the density of seawater,
- d is the turbine diameter, and
- V is the current speed. For tidal currents close to the shoreline in estuaries is a sinusoidal function in time, and
- η is the turbine efficiency.

push water through a channel and act as a pump to compress a column of air that is then passed through a turbine-generator system to produce electricity. As the waves recede, air is sucked back, causing the turbine to continue its operation (Figure 4-9). Wells turbines are the key to the successful operation of the oscillating wave column system (Figure 4-10). These turbines always turn in one direction, independent of the direction of flow. This feature allows them to provide continuous operation whether waves are moving toward the coastlines or moving away from them. The first operational wave power station, called LIMPET (for Land-Installed Marine-Powered Energy Transformer), was installed on the Scottish island of Islay and generates 500 kW of electric power (Figure 4-11).[15]

Figure 4-10
The Wells Turbine continues to rotate in one direction even as the direction of flow reverses
Photo courtesy of Wavegen Corporation (http://www.wavegen.co.uk)

Another onshore wave technology device is the **Tapered Channel**, which consists of a tapered channel and a reservoir constructed on a cliff a few meters above sea level (Figure 4-12). Due to the narrowing of the channel, waves rise and water pours into and fills a reservoir. The reservoir provides the necessary head to run a turbine and generate power. Tapered channel systems are especially suitable during peak demands since they store energy in the reservoir until it is needed.

Figure 4-11
The Limpet 500.
Image Courtesy of Wavegen Corporation

Offshore Systems

In an effort to develop practical viable floating wave power devices for offshore applications, a prototype floating-type version of the oscillating water column called the **Mighty Whale** was designed by the Japan Marine Science and Technology Center. The 50 x 50 m platform was anchored to the bottom of the sea near Japan, operated from 1998-2002, and produced about 110 kW of electricity (Figure 4-13).[16] Mighty Whale also acts as a wave breaker to calm water for the fisheries.

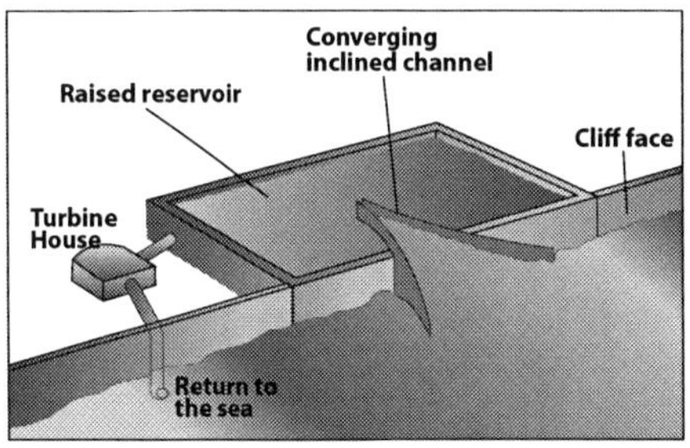

Figure 4-12
TapChan wave energy device.
A demonstration plant has been installed off a remote Norwegian island and operating since 1985.

The **Archimedes Wave Swing** (AWS) consists of a number of air-filled chambers submerged below the sea surface and is connected by movable floats that oscillate up and down as waves move over them. A series of linkages convert the vertical oscillation of the platform into rotational motion, which is in turn used to generate electricity. A pilot project off

[15] Islay LIMPET Wave Power Plant, Company website at *http://www.wavegen.co.uk.*
[16] Japan Marine Science and Technology Center, *http://www.jamstec.go.jp/jamstec/MTD/Whale/.*

Figure 4-13
The Mighty Whale
Image courtesy of The Marine Science & Technology Centre of Japan (http://www.jamstec.go.jp/)

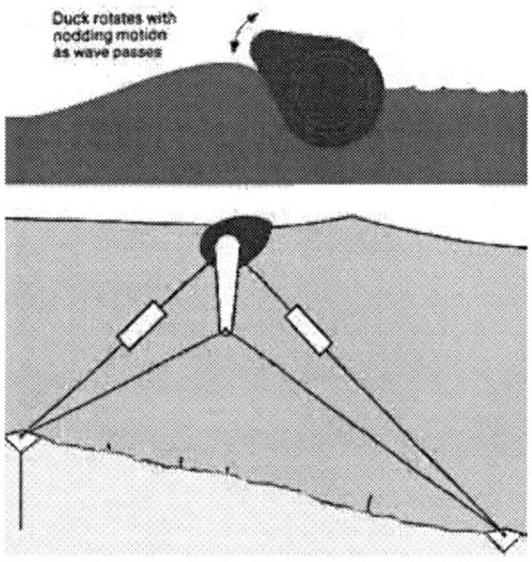

Figure 4-14
Salter's Duck

Figure 4-15
Underwater turbines
Image courtesy of TidalStream Ltd., UK.

the coast of Portugal is constructed and produces 8 MW of electricity.

Salter's Duck (Figure 4-14) operates on a similar principle with the exception that it is installed on a floating chamber connected to a fixed platform that swings up and down as a wave passes. The bobbing motion is converted to rotational motion (for example, using it to pump a hydraulic fluid through the blades of a turbine) which then runs a generator.[17]

Underwater turbines are similar to wind turbines, except that the kinetic energy in water is converted into rotational energy by underwater axial turbines (Figure 4-15). In contrast to wind velocity, currents in deep waters have relatively steady speeds, and therefore no energy storage system is necessary.

Siting

Like wind turbines, wave energy devices can be clustered in "wave farms" near the shorelines or moved off-shore. The main advantages of near-shore plants are that they are much cheaper to construct and there is less distance for electrical cables to be laid on the ocean floors. Their main disadvantage is that waves break up and lose much of their energy as they approach the coastlines. Shore-based stations are mostly considered for small-scale electricity generation. Offshore devices have the advantage of being more efficient, and they are not as unsightly to the public as the near-shore facilities. No matter which approach is used there are always some navigational hazards to shipping and other marine transportation.

Economical Concerns

Power output from a hydroelectric station increases with the flow rate of the water and its available head. For a given power output, less water is required for the higher head so the plant size would be smaller and the cost of construction and equipment would be lower. Unfortunately, taller waterfalls tend to be in mountainous regions and areas with lower population density. Therefore, transmission costs of electricity may be higher. Thus, before selecting the plant site, it is important to consider not only the total power but also the initial cost, operating cost, and cost of transmission. Typically the installed cost of hydro electrification projects is about $1,000-1,500 per kW for large, $2,500-3,000/kW for small (2.5-25 MW), and $10,000/kW for micro (below 500 kW) hydro projects.[18]

Environmental Concerns

Hydropower technologies have many of the same advantages of other renewable energy sources. They are clean, do not leave any waste, do not contribute to global warming, nor do they produce any of the air

[17] Thorpe, T. W,b"A Brief Review of Wave Energy," ETSU Report No. ETbSU-R-12, May 1999.
[18] See for example, Mayo, N., "Ocean Waves - Their Energy and Power," The Physics Teacher, 35, pp. 352-356, Sep. 1997.

pollutants that result from the combustion of fossil fuels.[19] The large reservoirs formed behind hydroelectric dams or tidal plants can provide a barrier against storm surges and benefit the public by providing beaches for swimming, fishing, and other recreational activities. Hydropower has an additional advantage over other renewable sources such as solar and wind in that it is a lot more reliable.

This does not mean that hydropower is without any environmental impact, however. Some of the concerns specific to hydroelectric and tidal projects are:

a. *Interrupting the natural flow of a river* – Change in river condition and adjacent lands may impact fish and aquatic population, threaten wildlife natural habitats, affect vegetation, and degrade shorelines.
b. *Affecting water quality* – Impounding water flow raises the surface water temperature which in turn affects the concentration of dissolved oxygen and other nutrients, and allows certain bacteria to grow in hydro rivers and reservoirs, which may potentially pose a health hazard.
c. *Causing floods* – Although dams can act as a flood control measure, there are also concerns for the flooding of lakes and rivers upstream of the plant. In addition, a dam failure in the event of an earthquake or other disaster could cause immediate flooding of the area below the dam, costing society countless lives and great economical loss.
d. *Impeding the natural transport of sediments* – Restricting the amount of water flow results in a reduction in sedimentation in the river downstream of the dam and an accumulation of silt in lakes behind the dam. As the storage volume becomes filled with silt, the generation capacity of the plant is reduced and the stored water may require periodic dredging and flushing of sediments at a considerable cost.
5. *Displacing Population* – Large hydropower projects will have a huge impact on the surrounding population. Many people have to be relocated which in turn may have considerable impacts on the local economy.

The major concerns raised against wave power are their vulnerability to violent storms, aesthetics, noise, and the additional costs if electricity is brought to land.

Summary

The ocean alone can fulfill all the energy needs of humans for a long time. Water is a clean source of energy, and environmental impacts are relatively limited. Among the drawbacks are the potential disturbances or destruction of some marine life and the interference with marine transport and shipping. For many, the installation of large wave-energy

[19] What differentiates hydroelectric and wind plants from thermal plants is that in hydroelectric and wind plants water or air flow directly through the turbine, whereas in thermal power plants, fossil or nuclear fuels heat water to steam before it flows through the steam turbine.

devices, overhead transmission lines, and their supporting facilities may not be aesthetically pleasing.

Except for the wave technologies, much of the hydropower potential in the developing countries has already been exploited. It appears that in the future, much of the hydropower development will be in developing countries especially in Asia and Africa where much of the world's small scale and low head hydro capacity exists.

Additional Information

Books

1. Bose, N. and Brooke, J., Wave Energy Conversion, Elsevier, 2003.
2. Ross, D., Energy from the Waves, Oxford University Press, 1995.
3. Cruz, J., Ocean Wave Energy: Current Status and Future Perspectives, Springer Series in Green Energy and Technology, Springer-Verlag, Berlin, 2008

Periodicals

1. *International Journal of Wave Motion,* Elsevier Science Publishing Company.
2. *International Journal of Renewable Energy,* Elsevier Science Publishing Company.

Government Agencies and Websites

1. National Oceanic and Atmospheric Administration (NOAA) Coastal Services Center *(http://www.csc.noaa.gov).*
2. European Commission on Tidal Energy *(http://europa.eu.int/comm/energy_transport/atlas/htmlu/tidal.html).*
3. OTEC, U.S. DoE, Energy Efficiency and Renewable Energy *(http://www.eere.energy.gov/RE/ocean.html).*

Non-Government Organizations and Websites

1. Wave Energy Council: Survey of Energy Resources *(http://www.worldenergy.org/wec-geis/publications/reports/ser/wave/wave.asp).*

Exercises

I. Problems

1. What is the gravitational force of attraction between Lisa Carr and a Ferrari parked in a showroom 100-m away? Assume Lisa and the Ferrari have masses of 50 and 2000 kilograms, respectively.

2. What is the force of attraction between an astronaut and a 500 kg console in a spacecraft? Is it zero? The astronaut has a mass of 75 kg and is standing 80 cm from the console.

3. Use data given in Appendix A to calculate the gravitational force of attraction between
   a. The earth and the sun
   b. The electron and nuclei of a hydrogen atom

4. An underwater turbine with rotor diameter of 2 meters is facing a strong wave at 10 m/s. Assuming rotor spins at 300 rpm, and for the water density of 1030 kg/m^3, calculate the maximum power output (kW) and torque delivered.

5. Calculate power delivered by a hydroelectric generating station, when water at a rate of 15 cubic meters per second is falling on a turbine located at a bottom of a 70-m waterfall. Assume water density of 1,000 kg/m^3.

6. What is the maximum power that can be exploited from a deep-water wave with a wavelength of 15 m and height of 5 m travelling at a speed of 4 m/s. Assume the combined turbine-generator efficiency of 85%.

II. Essay Questions

1. What are different methods of utilizing water as an energy source? Describe the advantages and disadvantages of each.

2. What were the earliest uses of hydropower? How did hydropower applications change throughout the centuries?

3. Which country houses the largest hydroelectric power station? Which waterfall is the tallest in the world?

4. What are the different types of hydro plants? What are their applications?

5. What is the difference between an impulse turbine and a reaction turbine?

6. Describe the principal of operation of an oscillating water column, a tapered channel, an Archimedes Wave Swing, and a Salter's Duck.

7. Which country is the largest producer of hydroelectric power? Which countries have the highest hydroelectric potentials that have not yet been exploited? Which countries have the highest potential for producing power from surface waves? From underwater currents?

8. How do tides form? What causes spring and neap tides? How can tides be predicted? How can we use tides to produce power?

9. Name three technologies which are best suited for near shore power production. Name three ways of producing power for offshore applications.

10. How much hydropower is generated for your country of birth? Is there a potential for exploiting waves, tides, and underwater currents for generating electricity? Discuss their environmental and economical impacts.

III. Multiple Choice Questions

1. The force of attraction between all masses is
   a. Electrical
   b. Magnetic
   c. Chemical
   d. Mechanical
   e. Gravitational

2. The best location for placing a water turbine is
   a. Close to the top of a waterfall
   b. Halfway down the waterfall
   c. At the bottom of the waterfall
   d. Does not make any difference
   e. Depends on the topography of the area

3. Energy can be extracted from water by
   a. Exploiting its potential energy as is done in hydroelectric and tidal plants
   b. Exploiting the temperature gradients in oceans as is done in OTEC plants
   c. Exploiting its mechanical energy as is done in ocean currents and surface waves
   d. Exploiting the energy in its nucleus as is done in fusion power plants
   e. All of the above

4. A pumped storage plant is
   a. The location where all pumps are stored
   b. A special plant for storing hydropower during the peak demand
   c. Used during emergencies, when quick delivery of power is demanded
   d. The auxiliary pump used to deliver water downstream of the hydro plant
   e. Used to act as a turbine when turbines are being serviced

5. In addition to using hydropower to produce electricity, hydro projects can also be used to support
   a. Recreation
   b. Irrigation
   c. Flood control
   d. Municipal water supply
   e. All of the above

6. The best way to control the amount of electricity generation is by
   a. Changing the head by moving the turbine
   b. Controlling the speed of the turbine
   c. Controlling the speed of the generator
   d. Controlling the flow of water
   e. Controlling the frequency of the transformer

7. Hydro energy accounts for about what percentage of global energy consumption?
   a. 1-2%
   b. 6-7%
   c. 10-15%
   d. 20-25%
   e. More than half

8. Which of the following is often cited as a major potential concern regarding the use of hydropower as a source of electricity generation?
   a. It contributes to global warming.
   b. It is a source of water and possibly air pollutions.
   c. Hydroelectric facilities can block fish passage.
   d. Reduces amount of water available for drinking and agriculture.
   e. All of the above.

9. What is (are) the main advantage(s) of hydropower over other fuel sources?
   a. It is economical to produce.
   b. It is a "clean" energy resource.
   c. It is a renewable fuel source.
   d. It provides additional benefits such as flood control and irrigation.
   e. All of the above.

10. A hydropower project that utilizes the river's natural flow for generating electricity during peak demand is called a
    a. Run-of-river project
    b. Storage project
    c. Waterfall project
    d. Overflow project
    e. None of the above

11. The electric generation capacity from a hydropower plant will depend on
    a. The flow rate of water
    b. The turbine's efficiency
    c. The head of the turbine
    d. The generator's efficiency
    e. All of the above

12. The largest hydroelectric plant in the world is in
    a. The United States
    b. Russia
    c. Brazil
    d. Tajikistan
    e. Canada

13. The leading producer of hydropower in the world is
    a. The United States

b. Russia
c. Brazil
d. France
e. Canada

14. Which of the following statements is correct?
    a. There are two high tides and two low tides every 24 hours.
    b. When the sun is along the line extending from the earth to the moon, we have higher-than normal high tides and lower-than-normal low tides (spring tides).
    c. The effect of the moon's gravity on the earth is the main cause of tide formation.
    d. The gravitational pull of the sun is much greater than that of the moon.
    e. All of the above.

15. High spring tides and low neap tides occur
    a. Twice a year following the solar cycle
    b. Twice a month following the lunar cycle
    c. Twice a day, roughly every 12 hours
    d. Once a month
    e. Once a day

16. Tidal energy has its origin in
    a. The phase of the moon
    b. The gravitational potential energy of the earth-moon-sun system
    c. Storms
    d. The rotation of the earth
    e. The change in surface temperature of the ocean

17. Spring tides occur
    a. During a full or a new moon
    b. When the moon is in the first and last quarter
    c. Mostly in spring
    d. When the moon is closest to the earth (at its perigee)
    e. When the gravitational forces of the moon and the sun are perpendicular to one another

18. Neap tides occur
    a. During a full or new moon
    b. When the moon is in the first and last quarter
    c. Mostly during falls and winters
    d. When the moon is farthest away from the earth (at its apogee)
    e. When the moon, the earth, and the sun are along a line

19. The maximum power generated from tidal energy
    a. Increases with the area of the basin and the square of average tidal range
    b. Is independent of the average tidal range
    c. Is independent of the basin surface area
    d. Occurs during the high tides
    e. Occurs during the low tides

20. Tidal power operates by building a barrier across:
    a. River estuaries
    b. Oceans
    c. Creeks
    d. Lakes
    e. Ponds

21. What percentage of the US electrical power comes from tidal plants?
    a. None
    b. Roughly 1%
    c. About 2-3%
    d. About 5-7%
    e. None of the above

22. The tidal interval is
    a. Time between two high tides
    b. Time between high tides and low tides
    c. Vertical difference between high and low tides
    d. Vertical difference between the highest high and lowest low tides
    e. The period when strongest tides occur during the year.

23. Tidal electricity generation does not require:
    a. Maintenance
    b. Storage system
    c. Waste disposal
    d. High capital cost
    e. None of the above

24. Power produced by hydro energy is often expressed in
    a. Gigavolts

b. Megawatts
c. Terajoules
d. Kilowatt-hours
e. None of the above.

25. Today, which renewable energy source provides the US with the most energy?
    a. Fossil
    b. Wind
    c. Solar
    d. Hydroelectric
    e. Nuclear

26. Which of the following statements is <u>not</u> correct?
    a. A major environmental concern regarding hydro and tidal plants is that the change in river condition and surrounding beaches may impact fish and marine life.
    b. Dams act mostly to control floods, but may also be the cause for flooding the lakes and rivers upstream of the plant.
    c. To prevent accumulation of silt, lakes behind dams must be occasionally cleaned at great costs.
    d. Unlike hydroelectric and tidal plants, wave plants have no environmental impact.
    e. Impounding water flow may cause certain bacteria grow in hydro rivers which could potentially pose a health hazard.

27. Which of the following statements is <u>not</u> correct?
    a. Since gravitational force decreases with square of the distance, the side facing the moon experiences the strongest pull.
    b. Tidal effects cause the earth to stretch along the earth-moon line.
    c. The oceans rise relative to land at both sides of the earth-moon line.
    d. The oceans rise relative to land at the point closest to the moon and falls relative to land at the point farthest away from the moon.
    e. Although the sun exerts a higher gravitational pull on earth than the moon, its tidal effect is smaller than the moon.

28. Which of the following statements is correct?
    a. A tidal day is slightly shorter than a solar day.
    b. A tidal day is exactly equal to a solar day.
    c. A tidal day is slightly longer than a solar day.
    d. A tidal day is the same as a lunar day.
    e. None of the above.

29. The major drawback to tidal power plants is that they
    a. Pollute the water
    b. Cause ozone depletion
    c. Heat the water
    d. Are too expensive
    e. Alter the tidal currents

30. The highest high tides and lowest low tides occur
    a. When the moon and the sun are nearly at right angles to each other
    b. When the moon and the sun are nearly in line with earth
    c. During summers when earth is closest to the sun
    d. During winters when earth is closest to the sun
    e. When sky is clear and moon is clearly visible

III. True or False?

1. Ocean waves have the potential to satisfy all of our energy needs.

2. The primary mover of surface waters is the wind.

3. The main application of the pumped storage plant is when demand is high and quick delivery of power (such as during emergencies) is needed.

4. There are no environmental concerns associated with hydroelectric plants because they are clean and no fuel is burned.

5. Tidal energy exploits the available head between the flood and ebb tides.

6. The tidal energy from the sun is negligible compared to that from the moon because the large distances between the sun and the earth make the pull of gravity by the sun very small.

7. Power outputs of most hydro plants are usually

expressed in megawatts.

8. Two common types of impulse turbines are Kaplan and Francis turbines.

9. The power delivered by waves is inversely proportional to the period between two successive crests.

10. Ocean waves are the result of the non-uniform heating of the ocean's surface.

11. During the night, water can be pumped from a lower lake to a higher lake in a hydroelectric power station. The water can be reused during peak power when additional energy is needed.

12. Wave size is affected by wind velocity, wind duration, and distance from the shore.

13. The potential energy of tidal power increases with the basin area and the square of the average range between the low and high tides.

14. Horizontal movement of the water accompanying tides is called a tidal current.

15. Each day, the high tides occur about fifty minutes later than the day before.

16. The speeds of currents in deep waters vary significantly from one location to another, and therefore an energy storage system is necessary.

17. The onshore wave plants are advantageous because they cost less and waves are stronger at the beaches.

18. There are currently no working tidal power plants in operation.

19. The power generated from a waterfall increases with the height of waterfall and flow rate of the water.

20. Waves have potential energy but no kinetic energy.

21. Power output from run-of-river plants vary widely with seasons and weather.

22. Underwater turbines are similar to wind turbines.

23. There exists a tremendous potential for exploiting the hydroelectric power in the United States.

24. Hydroelectric, tidal, waves, and other technologies that exploit mechanical energy of water do not have any adverse environmental impacts.

25. The impact of hydropower on global warming is minimal and not of major concern.

IV. Fill-in the Blanks

1. The best means of achieving the quick delivery of power during emergencies and power surges is through ________________ plants.

2. The largest hydroelectric project is being built in ___________.

3. ___________ turbines are most suitable for small hydro applications and heads of a few tenths to hundredths of meters.

4. The most suitable place for installing wave power stations is off the coasts of the ___________________ in the Atlantic Ocean.

5. ______________ devices have the advantage of being more efficient, and they are not as unsightly to the public as the near-shore facilities would surely be.

6. _________________________ between islands and around the coastal edges of oceans is best suited for exploiting wave's energy.

7. __________ tides form when the sun is at the right angle to the earth-moon line.

8. The leading producer of electricity by tidal power is _______________.

9. The power produced by underwater currents varies with the ___________ of turbine diameter and cube of water _______________.

10. Wave power is usually measured in ________ of wave front.

11. The cost per kilowatt of electricity generation is ____________ for smaller projects.

12. Two promising technologies for exploiting wave energy for onshore applications are oscillating water column and ______________.

13. The time between two successive waves is called the wave ____________.

14. Tides are primarily caused by ______________________.

15. When water moves in toward shore, it's called ___________ tide.

16. One way to harness wave energy is to channel the waves into a narrow funnel into a catch basin. A demonstration project called _____________ has been installed off a remote Norwegian island since 1985.

17. __________________________ is a device located onshore, installed on a floating buoy or fixed to the seabed; as wave surges into or recedes from it, air is pushed into or sucked out of a housing driving a turbine that is coupled to a generator that produces electricity.

18. An example of a floating wave energy convertor is a _______________.

19. We can extract power from moving water by installing _______________ anchored to the bottom of ocean floors; power can be extracted from water in the same way that it is extracted from wind.

20. In addition to the potential and kinetic energies contained in water waves, the temperature difference between the surface and deep ocean waters can be exploited in a process called _________________.

Thermal Energy

CHAPTER 5

Every great advance in science has issued from a new audacity of imagination ~ John Dewey (1859-1952)

Thermal energy is the energy associated with heat. Although the concept of heat is not new, its formal understanding only became clear during the nineteenth century when it was shown that heat is the result of the motion of molecules—a concept that laid the foundation for the laws of thermodynamics.

In this chapter, we will give formal definitions of heat and temperature and explain how heat can be transferred as a result of a temperature difference. Then we will discuss various modes of heat transfer and the laws of thermodynamics. Finally, we will show how these laws can be used to design practical devices and more efficient machines.

Heat and Temperature

Two of the earliest words that children learn are "hot" and "cold". They quickly learn to avoid touching hot stoves and getting too close to fires. They also learn to protect themselves with warm clothes in wintry weather and to stay in the shade on hot summer days. In spite of this common perception, precise definitions of heat and temperature are no less elusive than most abstract concepts in physics. As we shall soon learn, heat is a form of energy that flows from a hot object to a cold object as a result of a temperature difference that exists between them.

Heat

Heat is perceived as something which produces a sensation of warmth. The sensation will, of course, be stronger in higher temperatures. There cannot be any heat transfer between two objects that are at the same temperature. The amount of heat transferred depends on the temperature difference and the conductivity of the path between the two objects; the direction of heat flow is always toward the cooler object.

Earlier philosophers considered heat as a substance that was passed from hot objects to colder ones. This heat fluid was known as *phlogiston* (meaning flammable in Greek), which was thought to have a mass and the ability to flow in and out of objects during burning; it was considered to be the soul of matter. The concept was refined after Lavoisier an eighteenth century French chemist, showed that mass is a conserved quantity and does not change when a substance undergoes a chemical reaction. The new substance was called *caloric*, a mass-less fluid thought to flow from hot to cold objects.

It was not until the middle of the nineteenth century that Thompson and Joule showed that this theory is also wrong; heat is not a substance, but rather a manifestation of motion at the molecular level (kinetic theory). For example, when we rub our hands against each other both hands get warmer, even though initially they were at the same cooler temperatures. If the cause of the heat were a fluid, then it would have flowed from a (hotter) body with more energy to another with less energy (colder). Instead, the hands are heated because the kinetic energy of motion (rubbing) has been converted to heat in a process called "friction".

> **Question:** Why was the concept of heat as a fluid so well received for so many centuries?
> **Answer:** Because it met all the human notions of how a fluid should behave. It could pass from one body to another, similarly to the flow of water in a river or a pipe. When a hot object comes into contact with a colder one, the hot object cools and the cold one gets hotter as if some substance were passing from one to the other. Heat is also capable of running a steam engine in the same manner that water could turn a waterwheel.

Temperature

The word "temperature" seems so familiar to most of us that we often take it for granted. While our built-in senses do provide us with a qualitative characterization of temperature, our senses can often be unreliable and misleading. On a cold winter day, for example, an iron railing seems much colder to the touch than a wooden fence post, yet both are the same temperature. This "deception" by our senses arises because iron conducts heat away from our fingers much more readily than wood does. Similarly, if you place one hand in a bowl of warm water and one hand in a bowl of ice water, then remove both hands from the bowls, the hand from the hot water feels cold, while the hand from the cold water feels warm, though both hands are really feeling the same air temperature.

Until Scottish scientist Joseph Black (1728-1799), a founder of modern chemistry and a pioneer in the study of heat, no one established a distinction between heat and temperature. Black distinguished between the quantity (caloric) and the intensity (temperature) of heat.

Temperature Scales

Thermometers are commonly used to measure temperatures. To be useful, thermometers have to be calibrated—which may be done by measuring the rise of a column of fluid (such as water, alcohol, or mercury) as it expands when exposed to an environment. Referring to the fluid analogy, temperature as a measure of coldness or hotness of an object could easily be interpreted as the "height" of the caloric fluid within the thermometer.

Using two reference points makes it possible to divide the distance between them into a number of equal intervals. The first thermometers were divided into 360 parts, like degrees of a circle (thus the term "degree"). In 1708, Gabriel Fahrenheit used spirits as the fluid and a mixture of ice, water, and salt, the lowest attainable temperature in a laboratory setting at the time as the zero mark. The second reference point he used was the temperature of the human body, which was assigned the value of 96°. The choice of 96° is believed to be due to the fact that it is divisible by 2, 3, 4, 8, 12, 16, and 32. Today, the Fahrenheit scale uses the freezing and boiling points of water as 32° and 212°. According to this scale, body temperature is 98.6°F.

Swedish astronomer Anders Celsius, who assigned the numbers 0 and 100 to the freezing and boiling points of water, used what is known as the centigrade scale. In 1948, the International Committee on Weights and Measures adopted this scale as the temperature standard, calling it Celsius in honor of its inventor. As we will see when we introduce the laws of thermodynamics, there is a lower limit to the temperature that can be reached. This temperature, called the Zero Absolute Temperature (-273.15° Celsius or zero kelvin), is understood to be the temperature of a perfect crystalline lattice. There is no upper boundary to temperature, although the highest predicted temperature is 10^{38} K which was the suspected temperature at the very beginning of the creation of the universe, the Big Bang. Sustained fusion reactions require a temperature of 10^8 K, and thermal plasmas occur at temperatures of 10^4 to 10^6 K. Table 5-1 gives formulas for converting temperatures between Celsius, Fahrenheit, and Kelvin scales.

Table 5-1. Temperature Conversion Formulas

Equations for converting between Celsius (C), Fahrenheit (F), and Kelvin (K) temperature scales

°F = 1.8 °C + 32	(i)
°C = (°F – 32) / 1.8	(ii)
K = °C + 273	(iii)

Thermal Properties

Different substances have different abilities to retain or transport heat. For example, some materials, like Styrofoam and cork, are better insulators than aluminum, and some materials, like sand and brass, warm more quickly than grass. Thermal properties of materials are best understood by three quantities: heat capacity, specific heat, and heat conductivity.

The **heat capacity** of a substance measures its ability to retain heat. It is defined as the quantity of heat required to raise a substance by one degree in temperature. Water has a higher heat capacity than sand because it takes more heat to warm up by the same amount than sand does. In other words, for the same amount of heat, sand temperature raises more than water temperature. Closely related to heat capacity is **specific heat**, which is the heat capacity per gram of substance. Thus, a bigger object will have a greater heat capacity, but the same specific heat. The more loosely the components of a solid are held together, the higher the substance's specific heat. Graphite has a higher heat capacity than diamond because diamond's lattice structure is more tightly bound than the graphitic structure of carbon. In SI units, specific heat is expressed in kJ/kg.K.

> **Question:** Which do you think has the higher specific heat, lead or water?
> **Answer:** Lead atoms are over three times heavier than water molecules. Thus, a given quantity (by mass) of water has more molecules than the same quantity (by mass) of lead. Consequently, a given quantity of heat is distributed among fewer lead atoms than water molecules. This means that lead will experience a greater temperature increase and thus has a lower specific heat.

> **Question:** Licking a silver spoon that's been sitting in a very hot cup of coffee probably won't burn your tongue, but a spoonful of the same hot coffee dropped on your tongue could leave a blister. Why?
> **Answer:** The heat capacities of water and silver are 1 and 0.06 cal/g.°C, respectively. That is, silver contains much less heat energy than water at a given temperature.

Thermal conductivity is a measure of a body's ability to conduct heat. Table 5-2 gives values of thermal conductivity for several materials in order of their conductivity. As can be seen from this table, silver is a very good conductor of heat, with copper not too far behind. Air appears towards the end of the list, meaning it is a very poor conductor, but an effective insulator.

Table 5-2. Thermal Conductivities of Selected Materials, W/m.K (values at 20°C, unless otherwise stated)

Good Conductors		*Average conductors*		*Poor conductors (good insulators)*	
Diamond	2,000	Ice (0°C)	2.20	Brick, insulating	0.150
Silver	429	Concrete	1.70	Asbestos	0.090
Copper	400	Soil	1.50	Fiberglass	0.040
Aluminum	220	Glass	1.00	Glass wool	0.040
Iron	80	Water	0.60	Styrofoam	0.033
Lead	35	Epoxy	0.59	Air (dry)	0.026
Stainless steel	14	Body fat	0.20	Silica aerogel	0.004
Granite	3	Snow	0.16	Vacuum	0

Heat Transfer

Heat transfer is a result of temperature variation within a medium and occurs through one or more mechanisms of conduction, convection, and radiation.

Conduction

Conduction is the transfer of heat from molecule to molecule through a substance. If a steel bar is temporarily heated at one end, its molecules become agitated and move faster than neighboring molecules. When fast-moving molecules collide with slower molecules, energy is transferred from faster to slower molecules. The chain reaction moves along the bar until its temperature is uniform. While conduction does take place in gases and liquids, its effects are most pronounced in solids. The closer the molecules are packed, the easier conduction is. Heat transfer via conduction increases in materials with a greater heat conductivity and larger temperature gradients.

Question: Why do animals living in freezing climates often burrow into the snow to sleep?
Answer: In addition to furs and thick skin, air spaces in the snow help animals protect themselves in such harsh weather. Snow, a poor conductor, slows the loss of body heat. In freezing weather, an igloo would provide a warmer shelter than would a wooden shack because the snow and ice of the igloo are better insulators than wood.

Question: How do gloves protect our hands in the cold?
Answer: The temperature difference between hands (37°C) and outdoor air (say 0°C) is the same whether we wear gloves or not. However, with gloves heat follows a path of greater resistance, and the rate of heat loss from our hands reduces.

Convection

Convection is energy transfer by bulk motion. Unlike conduction, which is a microscopic phenomenon, convection involves the macroscopic interchange of energy between two mediums. Convection is most prominent in gases and liquids, where the distances between molecules are too large for conduction to be effective. Convection can be classified as natural or forced. In *natural convection,* fluid motion is a result of a density gradient, whereas *forced convection* is mainly due to a pressure gradient. Hot air balloons rise as a result of natural convection because they contain hotter, lower-density air. Helicopters are lifted by forced convection resulting from the pressure difference across their propeller blades. Dogs often pant to get rid of excess heat by forced convection.

It should be noted that heat transfer by convection and conduction are closely linked. Consider for example, the cooling of a hot plate by blowing air over it. As the cold fluid replaces the warmer fluid (convection), more heat is conducted away from the plate; the process continues until the surface temperature approaches that of the fluid, in this case the surrounding air. A fan placed in front of a car radiator functions in a similar manner.

In buildings, convection losses are due to air infiltration through the cracks, windows, and other openings in walls. Additional losses occur due to air movement inside and wind motion outside the exterior glasses and windows. In a typical building infiltration losses are the most significant and are comparable to losses by conduction. Common insulation materials such as fiberglass, rigid boards, cotton, and feathers work by creating tiny air pockets that slow down the convection flow of heat.

Convection is not only important at the local level, but also plays a role in large-scale movements of the atmosphere. The major winds are convection currents driven by temperature differences caused by non-uniform heating of the earth by solar radiation. The winds, in turn, drive the ocean currents.

> **Question:** To keep a body warm in a cold winter, would it be more beneficial to wear two layers of light clothing or one layer of clothing twice as thick?
> **Answer:** Two layers will work best because there is always some air trapped between the layers, providing additional insulation.

> **Question:** Is it best to add creamer to coffee immediately after the coffee is poured or to add it right before drinking the coffee? Assume that coffee is most desirable when it is hot.
> **Answer:** Creamer should be added as soon as the coffee is poured. Two effects are of importance: First, creamer makes coffee lighter in color, reducing its emissivity and heat loss from radiation. Secondly, adding creamer sooner decreases the temperature difference between coffee and the environment, reducing the rate of conductive and convective losses.

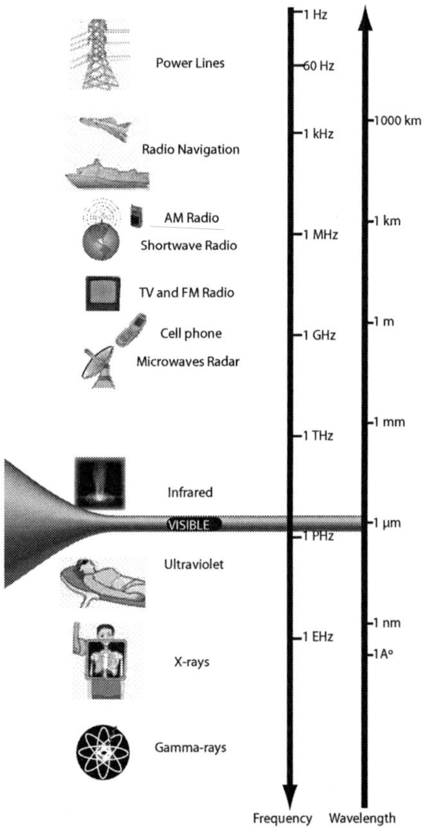

Figure 5-1
Electromagnetic wave spectrum

Radiation

Radiation is the transport of energy by photons of light migrating from a hotter to a colder surface. Unlike conduction and convection, both of which require a material medium to transport heat energy, radiation transports energy via electromagnetic waves of different wavelengths , even in a vacuum (Figure 5-1). Wavelength is the distance between two successive crests or troughs in a wave; the shorter the wavelength, the higher its energy is. No matter what the wavelength is, they all travel at 300,000 kilometers per second, the speed of light.

The most energetic radiations are gamma rays and x-rays. Gamma rays come from the nuclei of certain atoms with wavelengths smaller than the size of the atom; x-rays come from the innermost orbits of electrons. The least energetic radiations are from radars and radio waves, with wavelengths that can exceed several meters or kilometers.

Radiation at wavelengths shorter than 0.3 microns (high intensity ultraviolet) is dangerous to humans. Its effects can range from simple sunburn to cancer and death. Photons at these short wavelengths are sufficiently energetic to break bonds in molecules of living matter. Fortunately, most are filtered out by earth's atmosphere. Wavelengths between 0.3 and 0.4 microns (near ultraviolet) are weakly absorbed by clouds and dust in the atmosphere, while the rest reach the earth's surface. Microwaves have frequencies close to the resonance frequency of water molecules, and therefore, readily absorbed by water molecules, a feature exploited in microwave ovens for the rapid heating of food. Radar, TV, and radio waves have very long wavelengths (from many meters to kilometers) and thus are of very low energy; they are mainly of interest in communications.

The atmosphere is largely transparent to rainbow colors, which make up visible light and cover wavelengths in a very small range from only 0.4 microns for deep violet to 0.7 microns for bright red. Although earth's atmosphere is transparent to visible radiation, it is virtually opaque to

Wind Chill Factor

FYI ...

Air circulation promotes cooling for two basic reasons. It removes the warmed layer of air blanketing our bodies, which could potentially (in still air) act as insulation against conductive heat loss, and it promotes cooling by evaporation.

While some cooling can increase our comfort level, a strong wind can also create an unbearably chilling environment. On a windy day, air currents cause greater heat loss, as the warmer insulating air layer next to the body is continuously replaced by cooler, ambient atmosphere. The effect of wind on how cold we feel is conveniently expressed in terms of the **wind chill factor** (WCF). Wind chill factor describes the rate of heat loss from exposed skin due to the combined effects of wind and cold. The higher the wind speed, the higher the rate at which heat is removed from the body and the lower the body temperature becomes. It must be emphasized that WCF expresses an actual cooling rate and not simply some illusory sensation, as anyone living in the "windy city" of Chicago will attest to. For instance, the figure below shows that for an air temperature of 10°F and a wind speed of 15 mph, the cooling power of the moving air is equivalent to that of still air at -7°F. The wind chill factor is a good way to determine the potential of frostbite or hypothermia.

Wind (mph)	Temperature (°F)																	
Calm	40	35	30	25	20	15	10	5	0	-5	-10	-15	-20	-25	-30	-35	-40	-45
5	36	31	25	19	13	7	1	-5	-11	-16	-22	-28	-34	-40	-46	-52	-57	-63
10	34	27	21	15	9	3	-4	-10	-16	-22	-28	-35	-41	-47	-53	-59	-66	-72
15	32	25	19	13	6	0	-7	-13	-19	-26	-32	-39	-45	-51	-58	-64	-71	-77
20	30	24	17	11	4	-2	-9	-15	-22	-29	-35	-42	-48	-55	-61	-68	-74	-81
25	29	23	16	9	3	-4	-11	-17	-24	-31	-37	-44	-51	-58	-64	-71	-78	-84
30	28	22	15	8	1	-5	-12	-19	-26	-33	-39	-46	-53	-60	-67	-73	-80	-87
35	28	21	14	7	0	-7	-14	-21	-27	-34	-41	-48	-55	-62	-69	-76	-82	-89
40	27	20	13	6	-1	-8	-15	-22	-29	-36	-43	-50	-57	-64	-71	-78	-84	-91
45	26	19	12	5	-2	-9	-16	-23	-30	-37	-44	-51	-58	-65	-72	-79	-86	-93
50	26	19	12	4	-3	-10	-17	-24	-31	-38	-45	-52	-60	-67	-74	-81	-88	-95
55	25	18	11	4	-3	-11	-18	-25	-32	-39	-46	-54	-61	-68	-75	-82	-89	-97
60	25	17	10	3	-4	-11	-19	-26	-33	-40	-48	-55	-62	-69	-76	-84	-91	-98

Frostbite Times: 30 minutes · 10 minutes · 5 minutes

Question: At wind speeds of 4 mph or lower, wind chill temperature turns out to be warmer than the actual temperature. How do you explain this?

Answer: When the wind blows at low speeds, our body warms the layer of air next to our skin. Since air is relatively still it acts as insulation, protecting us from colder air farther away from the warmth of our bodies.

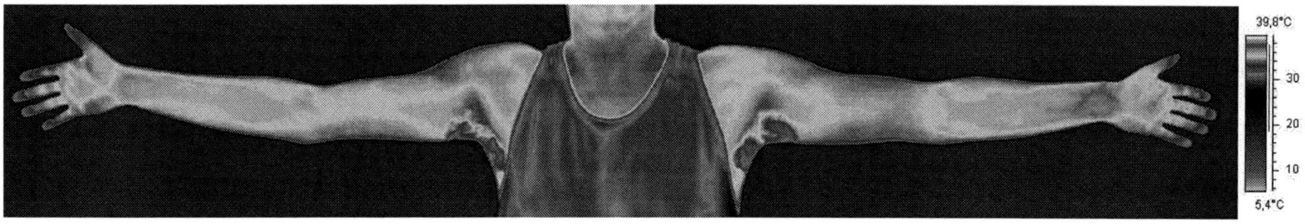

Figure 5-2
Thermal image of a man. The hotter regions (bare skin) are lighter in color, and cooler spots (tee-shirt, arm pits) are darker grey. The choice of shades of grey is arbitrary and does not imply a physical significance. *Photo courtesy of Thermotronics Inc., Brazil.*

infrared radiation, which has wavelengths longer than 0.7 microns. As we will see in the next chapter, this is the main property responsible for the atmosphere's behavior as a greenhouse; letting solar radiation in while trapping terrestrial infrared radiation results in the global warming phenomenon.

Whether a body emits energy at one wavelength or another depends on its temperature and surface properties. An object at room temperature, say 20°C, emits nearly all its energy in infrared. The human body at ordinary temperatures also emits in the infrared region. In fact, 98% of the radiation from a bare human body ranges from 4.78 to 75 microns in wavelength. Using a special infrared camera it is possible to "see" a human body or a passing car in total darkness. The image will not look like what we see with our single lens cameras, but will be a contour map of constant temperature regions. An infrared (thermal) image of a man wearing a tee-shirt is shown in Figure 5-2.

> **Question:** If human beings emit infrared, why do we see them in "visible" light?
> **Answer:** What we see is not emitted light, but the reflection of light from other sources (sunlight, fluorescent light, etc.). This is why in the absence of a light source (darkness) there is no light reflected to the eye and a person cannot see or be seen.

Unlike conduction and convection losses that increase with temperature differences between cold and hot objects, radiation losses increase as the differences of the temperatures to the fourth power ($T^4_{hot} - T^4_{cold}$), and become dominant only for very high and very low temperatures. In buildings, radiative losses are most significant when the surrounding terrain is either much colder or warmer than inside. Roofs can radiate a substantial amount of energy to the cold night sky. They also provide a low-resistance path to solar heat during the summer months. Window glass is much colder than adjacent walls during the winter, causing internal heat, such as heat released by the occupants or heaters, to migrate toward windows. This results in a larger temperature difference across the glass layer and causes more heat to escape through windows. Double-glazing the windows, closing the curtains, and adding additional insulation in the walls and attic can significantly reduce these losses.

Question: You have probably experienced colder room temperatures in the winter when drapes are not closed, even though the thermostat records air temperature that should be comfortable. Why?
Answer: Glass is transparent to radiation in the visible range, but is opaque to infrared. The radiation from cold glass windows to you is much less than yours to the window.

Question: You can easily feel the heat from the sun through a glass window, but behind a sheet of glass you do not feel much heat from a fireplace. Why?
Answer: Common window glass is transparent to the wavelengths of radiation between 0.3-2.5 μm. A large portion of solar radiation falls in this range, allowing both sunlight and solar heat to pass through. Flames, however, emit in wavelengths in excess of 2.5 μm, the region where window glasses are practically opaque.

Thermodynamics

Thermodynamics is made up of two Greek words: *therme* (heat) and *dynamis* (power). It is the science that describes the dynamics of heat and how it can be converted to power. Thermodynamics is a phenomenological theory derived from four very simple observations: 1) heat cannot flow between bodies of the same temperature; 2) heat and work are just two different forms of energy; 3) heat always flows from a hot body to a cold body; and 4) there is a temperature (called zero absolute temperature) that can never be reached. These observations have been refined and reformulated as the zeroth, first, second, and third laws of thermodynamics. These laws are important because they provide the basis for designing many machines and modern devices that change heat into work (such as an automobile engine or a power plant) or turn work into heat or cold (such as an electric heater or a refrigerator).

Equilibrium (The Zeroth Law of Thermodynamics)

Two objects at the same temperature are at equilibrium and remain at equilibrium until the temperature of one of the objects changes. On the other hand, if we put two objects of different temperatures next to each other, one object heats and the other cools until both bodies reach the same temperature. The zeroth law is considered obvious and will not be discussed any further.

The Zeroth Law of Thermodynamics: **Objects at equilibrium must have the same temperatures.**

Conservation of Energy (The First Law of Thermodynamics)

The principal of conservation of energy, or the first law of thermodynamics, implies that the energy of a system does not change as it goes from one state to another; only its form changes.

> ***The First Law of Thermodynamics:***
> **Energy can be neither created out of nothing nor destroyed into nothing, but it can be changed from one form to another.**

For example, a glass resting on the edge of a table has a certain potential energy. If the glass is knocked off the edge, its potential energy is converted to kinetic energy as it accelerates towards the ground. When the glass hits the ground the kinetic energy is converted to light energy (sparks), sound energy (a bang), thermal energy (heat), and chemical energy (the glass breaks). The first law of thermodynamics is the basis of all energy conversions from one form to another. Many of our biological activities are geared to perform these energy conversions. For example, during digestion, food molecules are broken down into progressively simpler molecules (chemical-chemical conversion); in the process some of the chemical energy stored in those molecules is converted to the thermal energy necessary to maintain our body temperature. Numerous practical devices have been designed that accomplish useful tasks through the conversion of energy from one form to another. Table 5-3 gives examples of different kinds of energy conversions.

Question: If energy can be neither created nor destroyed, how can people claim that there is an "energy shortage"?
Answer: Terms such as "energy shortage" and "energy waste" are misnomers. While total energy must remain constant, useful energy — that which can be used as fuel or perform work— may be in short supply. According to the first law, energy can never be wasted; it may only be converted to a form not readily usable to us. This is explained more when we talk about the second law of thermodynamics.

Question: If energy can be transformed, what is it transformed into?

Table 5-3. Examples of Different Kinds of Energy Conversions

FROM/TO	*MECHANICAL*	*THERMAL*	*CHEMICAL*	*ELECTRICAL*	*LIGHT*
MECHANICAL	Bicycle, Gearbox	Friction	Cigarette lighter	Wind generator, Microphone	Sparks
THERMAL	Gas turbines	Heat exchanger	Pyrometer	Thermocouple	Luminescence
CHEMICAL	Rockets, thermal engines	Food, Fires	Metabolism	Battery, Fuel cell	Candle
ELECTRICAL	Electric motor, Loudspeaker	Resistor heater	Electrolysis	Transformer	Light bulb
LIGHT	Galvanometer	Solar collector	Photosynthesis	Solar cell	Fluorescence

First Law and Nuclear Reactions

FYI ...

In the century and a half since the first law was formulated, the law has come into question on occasion. For example, nuclear reactions seemed to occur in violation of this law. Thanks to the pioneering work of Einstein and others, the first law continues to be upheld-- at least so far!

Answer: Although we often talk about transformation of energy, we should note that the nature of energy has not changed. We are only talking about its manifestation from one form to another.

Chaos and Disorder (The Second Law of Thermodynamics)

Countless processes have a preferred or "natural" direction. Heat flows spontaneously from hot objects to cold ones. Water will flow from high mountains toward rivers. Air rushes out of a punctured rubber balloon. Smells tend to diffuse outward to span greater distances. Humans grow older by the minute. When a house is left unattended, it quickly becomes disorganized. All these processes have one thing in common; they all tend to become more dispersed (chaotic). This is one statement of the second law of thermodynamics, which can be generally stated as:

> ***The Second Law of Thermodynamics:***
> **All natural processes tend to go from order (concentrated) to disorder (dispersed).**

More simply put — events happen in a certain direction. Although many people think the second law of thermodynamics must only be of interest to physicists, engineers, and those who deal directly with heat and energy, the second law has far-reaching consequences not only in predicting the fate of the universe, but also in all aspects of our daily lives. The second law gives us guidelines for what we can and cannot do. It tells us how to design better machines and provides us with a blueprint for using our resources more efficiently. In short, it gives us a sense of direction.

Question: It is often said that it is impossible to obtain negative absolute temperatures. Why?
Answer: Because absolute zero is the state where all molecular motion ceases and the complete order is achieved. Negative temperatures imply better than perfect order!

Question: What does the second law tell us about the ultimate fate of the universe? What should we expect after a severe earthquake? After dropping a glass cup from the edge of a table?
Answer: According to the second law, the universe will continue toward complete disorder (thermodynamic equilibrium), where all non-uniformities in temperature, electric potential, pressure, etcetera, vanish and eventually reach a heat death. Accordingly, it predicts that earthquakes flatten buildings – decreasing order. Similarly, dropping

a glass cup off a table edge will likely break it into many pieces, increasing the disorder. Remember that an unbroken glass cup, by the virtue of the careful positioning of the atoms in a lattice structure, is highly ordered.

Entropy

Figure 5-3
With the aid of a few hundred pounds of dynamite, demolition experts caused this hotel-casino in Las Vegas to go from an ordered state (lower entropy) to a disordered state (higher entropy).

Measuring the temperature of a room or the pressure of the tires of a car are probably easy and routine tasks. All that is needed is a thermometer or a pressure gauge. But how can we measure chaos? Is there a way to quantify how much dirtier my room is than yours? Ludwig Boltzmann (1844-1906), an Austrian scientist, attempted to do exactly that. The result of his studies is summarized in a single equation that defines **entropy** (from the Greek root meaning transformation) as a measure of randomness, or the total number of configurations that a system can assume. The concept of entropy is closely intertwined with the second law of thermodynamics in the sense that processes happen such that the total entropy of the universe is constantly increasing. When dealing with energy, the second law implies that it must flow from the one with the least amount of disorder (lower entropy) to the one with a higher amount of disorder (higher entropy).

Question: It is an easy task to mix two tablespoons of salt and pepper. It takes quite a bit of work, however, to separate a mixture of salt and pepper into its constituents. Why?
Answer: Salt and pepper by themselves are relatively orderly, but when mixed, disorder will increase considerably. Separating salt and pepper requires work because we need to create order by reducing the mixture's entropy.

Question: Visualize a deck of cards where all the cards are arranged according to suit and rank. How does the entropy change if the deck is shuffled a few times?
Answer: When suits are to be arranged in some orderly manner, rearrangement is limited. There are a great number of ways that a deck can be rearranged if a particular order is not demanded; the entropy is thus increases.

Question: An electric heater dissipates electrical energy into heat (and some light). What are the implications of the first and second laws of thermodynamics?
Answer: The first law assures that electrical energy is completely converted to heat (and light). The second law implies that the energy flow is from order (flow of electrons in the electric coil) to disorder (random motion of heated air molecules in the room).

Question: When a gas balloon is heated, it expands. What happens to the entropy?

Living Organisms and the Second Law of Thermodynamics

Digging Deeper ...

According to the second law of thermodynamics, nature always prefers chaos to order. Living organisms which formed from simple cells to complex beings seem to violate the second law in a big way. It should be noted that the second law applies only to isolated systems. Isolated systems cannot exchange energy or matter with their environments. The only truly isolated system is the Universe itself. Earth is not an isolated system because it exchanges energy with the sun and the surrounding atmosphere. Living organisms exchange both energy (heating or cooling by the environment) and mass (eating, breathing, and sweating) and therefore are not isolated systems. The sanctity of the second law is thus preserved!

A plant is an example of a heat engine which transforms light energy into chemical.

Answer: As the balloon expands, it opens more room for gas molecules to occupy. In other words, the possibility that gas molecules take different configurations increases. Thus the balloon's entropy increases.

Absolute Temperature (The Third Law of Thermodynamics)

As substances cool, their molecules move with less speed and their kinetic energy decreases. Actually, it can be shown that the kinetic energy of particles is directly proportional to the temperature. To rephrase, temperature can be seen as a measure of the kinetic energy of particles. It is therefore easy to understand that at temperatures of absolute zero, all molecular motions stop, and entropy approaches zero.

The Third Law of Thermodynamics:
There is a temperature so low that it cannot be reached. This temperature is called absolute zero kelvin (-273.15°C)

Thermal Devices

The first law of thermodynamics showed that we could convert energy from one form to another. When its brakes are applied, a car will slow down because of friction. As a result, both the tires and the road become a bit warmer. In this example, work has been turned to thermal energy or heat; therefore, work and heat are quantitatively the same. Experience tells us the reverse, cooling the tires and road to move the car backward, is not possible. This shows that thermal energy (heat) and mechanical energy (work) are qualitatively different - work can be turned entirely to heat, while the reverse is not true. This simple observation is a direct consequence of the second law of thermodynamics, which states that processes can occur naturally in one direction and not the other, although energy expenditure is exactly the same in both cases.

One of the applications of thermodynamics is in designing devices that transform one form of energy to a more useful form. Of course we wish to design these devices using the least amount of energy and with the highest

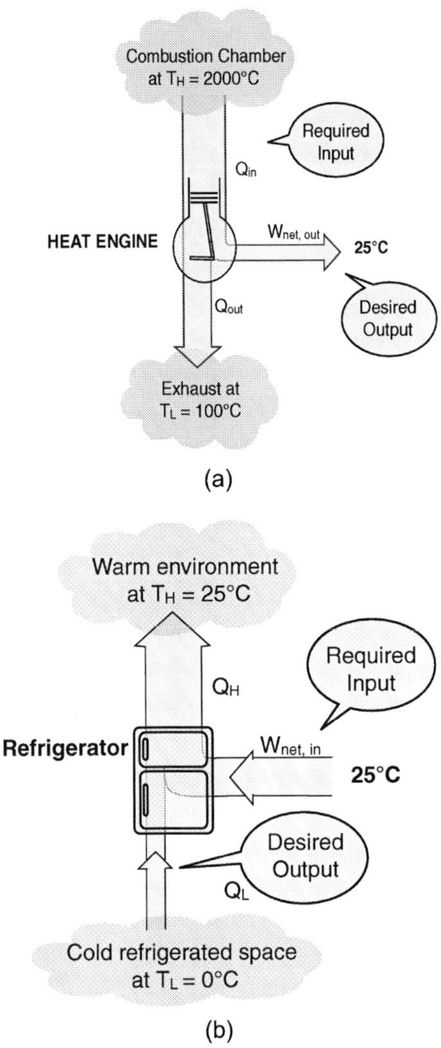

Figure 5-4
Heat engines and refrigerators

Table 5-4. Typical Efficiencies of Some Complex Machines	
Machine	***Efficiency***
Automobile	12-15%
Steam engine	50-75%
Steam locomotive	5-10%
Steam turbine	30-40%
Gas turbine	up to 40%

efficiencies possible. The first law of thermodynamics requires a minimum amount of energy to achieve a task. The second law of thermodynamics puts a limit on how efficient a device can be. From a theoretical standpoint, a device is most efficient (ideal) when it operates with no frictional losses; in reality, most systems have much lower efficiencies.

Some examples of thermal devices we use or are impacted by in everyday life are engines, power plants, refrigerators, heat pumps, and air conditioners. In all these devices, some form of energy (fuel) is consumed. Whether used in automobiles or jet aircrafts, heat engines convert part of this energy to shaft work that eventually runs the vehicle (Figure 5-4a). Power plants work in a similar fashion, but their work output is mainly in the form of electricity. Refrigerators, air conditioners, and heat pumps work in essentially the reverse direction; they use fuel energy to "pump" heat away from the space we want to cool or "pump" heat into the space that we want to heat (Figure 5-4b). No matter what the application, part of the energy is always discarded as waste energy into the surrounding atmosphere. In other words, it is impossible to build devices that convert 100% of the input energy into useful forms. Furthermore, because there are always some frictional losses, actual efficiency is always less than the maximum theoretical efficiency dictated by the laws of thermodynamics. Table 5-4 shows the typical efficiencies of several machines.

Internal Combustion Engines

Internal combustion engines are one example of thermal heat engines with widespread applications in power generation as well as in transportation systems. Basically, these devices burn a mixture of hydrocarbon fuels with the ambient air inside a combustion chamber (cylinder) to produce power. The power is transmitted through a shaft which can run a generator, drive the crankshaft of an automobile, or provide thrust to propel a jet aircraft. Among the most common internal combustion engines are gasoline and diesel engines and gas turbines. These devices are discussed in detail when we cover transportation systems in Chapter 14.

Efficiency

Wouldn't it be great if we could build a machine which puts out more work than we put into it? This machine would constantly create new energy and we would never have an energy shortage. If that were possible, we could build machines that, once started, did not need an additional expenditure of energy. This would be ideal, and many people have dreamt of making such machines. These machines, commonly known as "perpetual motion machines" (PMM), could operate forever. Sound too good to be true? It is. In the real world, all machines produce less work than the energy that goes into them. Useful energy is always lost as heat. That's the reality but, as shown by those who attempt to design PMMs, not everybody chooses to accept it. Unfortunately, the first law of thermodynamics precludes constructing such machines.

Cash vs. Gift Certificate

FYI ...

The difference between work and heat can be compared to that between cash and gift certificates from a major retail store. Cash and gift certificates have the same nominal value, in the same way that work and heat are quantitatively the same (1st law). The cash is of a higher quality because it has more diverse applications. Similarly, work is of a superior quality because it can be used in more ways. When purchasing an item from the retailer, a gift certificate has the same utility as cash. The same is true when work is used to perform a task that can be accomplished equally well by heat. For example, a resistance heater uses electricity (a form of pure work) to heat a house — a task that could easily be done using a gas heater. This does not mean that gift certificate is of no value outside the retail store, as it can be traded (for example, paying for the gas, but at a smaller amount than its face value). The same can be said for heat; we can use heat to produce work (heat engine), but at less than 100% efficiency.

Question: Why is it more advantageous to use a gas heater over an electric heater for heating a room?
Answer: Remember that most electricity is generated in power plants by burning some form of fossil fuels. In the process, 70% of energy is lost to the atmosphere (see "Heat Engines"). Additional losses occur during transmission, before the electricity is converted back to heat - a task that could be accomplished directly by burning gas in a gas heater. Using electricity for heating purposes is similar to using cash when you could have used the gift certificate instead!

Not only is the construction of such engines impossible, the second law of thermodynamics prevents the construction of engines that can convert "all" or "nearly all" of the heat input into useful work. This can be understood by noting that heat is a less ordered form of energy than work and complete conversion of heat to work would accompany a reduction of entropy (in violation of the second law), which is impossible.

Many inventors have proposed (and still propose) devices that violate the first and second laws of thermodynamics. Some, like Maxwell,[1] do so from a purely philosophical standpoint, but many have actually dared to propose such machines as practical devices.[2] Many have even received patents and accumulated great wealth from their "inventions". As you may have guessed, nobody was ever able to produce a working prototype.

Question: One of the consequences of the second law is that whenever energy is transformed from one form to another, at least some of the energy changes to a more dispersed form such as heat (recall that heat is a less ordered form of energy). Is it possible to design devices that transform energy in the opposite direction, i.e. take it from a dispersed form to a more concentrated (ordered) form?
Answer: The second law only precludes going from disorder to order for isolated systems. Many practical devices are not isolated and thus do not have to follow this restriction. For example, an air conditioner works by removing heat from a room and dumping it outside at a higher temperature. In this case we are re-concentrating energy. To do so, however, we need to spend even more energy in the form of electricity.

[1] James Clerk Maxwell (1831-1879) was a Scottish mathematician and physicist, most known for his theoretical formulation of laws of electromagnetism.
[2] Angrist, S. W, "Perpetual Motion Machines," *Scientific American*, 218:114-122, January 1968.

Question: Our solar system has been moving around the sun for a very long time and does not seem to slow down. Can the motion of the solar system be considered an exception to the second law?
Answer: If we look at the grand scale, motion of the Universe indeed conforms to the rules set out by the laws of thermodynamics. Energy is continuously poured in, externally and on a massive scale, to power the motion of all the galaxies and the motion of all that is within them. Even the solar system has to adhere to the laws put forth by nature!

The Ceiling on Efficiency

In an attempt to improve the efficiency of the early steam engine, the French scientist Sadi Carnot proposed an ideal engine that works between two reservoirs at different temperatures. Heat is removed from a source at temperature T_{hot}, part of which is discarded to a sink at temperature T_{cold}; the remainder is supplied as work. The best this engine can do is to achieve a loss of entropy from the source that just balances the gain in entropy by the sink. The net entropy production will then be zero. Under such ideal conditions, the efficiency of the engine is the maximum attainable and is equal to

$$\eta_{ideal} = \eta_{max} = 1 - \frac{T_{cold}}{T_{hot}} \tag{5-1}$$

T_{cold} and T_{hot} are temperatures of cold and hot reservoirs and must be expressed in kelvin. Kelvin temperatures are calculated by adding 273 to temperatures expressed in degrees Celsius.

Question: An inventor claims to have designed a cyclic engine that takes 1,000 Joules of heat and converts it entirely to work. How would you rate this claim?
Answer: His claim is impossible. The engine removes some heat from a reservoir, resulting in a drop in its entropy. Since work is perfectly ordered, no additional entropy is created. The net result is a decrease in the total entropy of the universe, in opposition to the principal of increasing entropy as stated by the second law.

Example 5-1: The same inventor comes back a few months later after having worked on a second engine. His new claim is that his engine can remove 1,000 J of heat from a source at 500 K, dump 400 J to a sink at a temperature of 250 K, and use the remaining 600 J as work to run a shaft. Would you invest in his company?
Solution: The ideal efficiency of an engine working between these two temperatures is 1-250/500 = 0.50; at best only 50% of the input energy can be converted to work. The proposed engine has an efficiency of 600/1000 = 0.60, higher than the ideal efficiency of 50%, which is an impossibility. Because of frictional losses, we can expect actual efficiency to be even lower than 50%. Once again, his claims cannot be realized.

Hurricanes

Digging Deeper ...

A hurricane (called "cyclone" in the southern hemisphere) can be viewed as an elegant example of nature's Carnot heat engine. It draws heat from the ocean by evaporating the water and turning it into steam, releases some of it to the atmosphere via radiative cooling and condensation, and does work during this process.[i] The Coriolis force converts the upwards suction of the air to spiral motion.

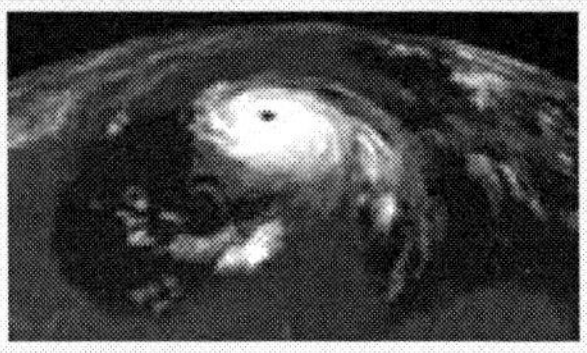

[i] *Emanuel, K. A. 1987. "The Dependence of Hurricane Intensity on Climate." Nature. 326: 483-485.*

First-Law and Second-Law Efficiencies

Real engines can never achieve the ideal efficiencies given by Carnot. First, real engines have always frictional and conductive losses that cannot be completely eliminated, thus limiting their maximum performance. Second, depending on its intended end use, a device may not be able to utilize the full potential of energy used to operate.[3] To quantify these limitations, two types of efficiency are defined.

The **first-law efficiency** is defined based on the first law principle of conversion of one form of energy to another, without any consideration to the quality of the energy resource. First-law efficiency is input/output efficiency, i.e. the ratio of energy delivered in a desired form and the energy that must be expended to achieve the desired effect. It does not differentiate between the qualities of the energy sources. The first-law efficiency can be less than (heat engines), equal to (friction), or greater than (refrigerator) one. In the latter case, it is commonly referred to as "the coefficient of performance" (see below).

> **Question:** What is the first-law efficiency for a heat engine? For an electric heater? For an electric motor? For an electric light bulb?
> **Answer:** A heat engine is a device that converts chemical energy (fuel) to mechanical work. Therefore, the first-law efficiency is the ratio of shaft work to heat input. Typically, first-law efficiencies of 25-30% can be achieved for automobile engines. An electric heater is a device for converting electrical energy to heat; in this case the first law efficiency is the ratio of heat given off to electric work needed to operate the heater (compare this to the first law efficiency of the heat engine!). Aside from small radiant losses through the heating coil, electric heaters are very efficient devices for their stated purpose and efficiencies close to 100% are achievable. Electric motors lose some energy through friction and coil heating and are generally around 90% efficient. Incandescent light bulbs are designed to convert electricity to light. A major fraction of the energy is, however, lost through heating the filament. The efficiency is rather low at around 5%.

Unlike the first-law efficiency that ignores the qualities of the energy, the

[3] The students who take a more advanced course in thermodynamics will learn of a new term, "*exergy*" or "*availability*," that distinguishes the part of energy in a system that can be converted to work from the part that cannot (is unavailable). Unlike energy, exergy is not conserved and destroyed in all real processes.

Gas vs. Electric (Revisited!)

Digging Deeper ...

At a department store you inquire about a gas heater for your house. The salesperson offers you two heaters, a gas heater with 70% efficiency and an electric heater with an efficiency of 97%. Which one would you choose?

The first inclination might be to purchase the electric heater. It is more efficient, it is cleaner, and might even be cheaper. A slightly more careful analysis will reveal that the salesman is actually giving you first-law efficiencies. The electricity used to run the electric heater is coming from a gas-powered generating station with 33% efficiency. The electricity is transmitted through a power line from the power station to your home. An additional 5-10% (let's say 10%) is lost through heating the wire (Joule heating). The overall efficiency (system efficiency) of the electric heater is then (0.33)(0.90)(0.97) = 28.8%. Assuming transmission losses of 5% through gas pipes, the first-law efficiency of the gas heater is (0.95)(0.70) = 66.5%.

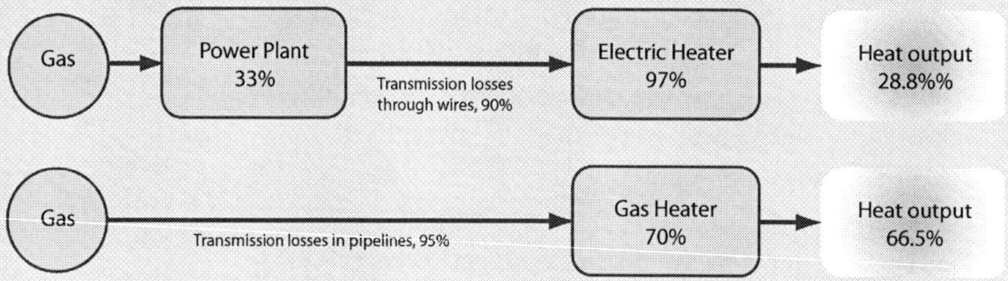

second-law efficiency compares the efficiency of an actual device (1st law efficiency) to that of the same or a similar device operated under ideal conditions. It measures the actual energy used as compared to the minimum amount of energy needed to accomplish the same task, i.e. the second-law efficiency is the ratio of the actual efficiency to that of an ideal device. By definition, the second law efficiency of all ideal devices is equal to one, and for all real devices is smaller than 1.

Question: What is the second-law efficiency of a heat engine?
Answer: To find the second-law efficiency, we must compare the heat engine to an ideal heat engine. An ideal heat engine is an engine that does not experience any frictional losses or other irreversibilities. The second law states that this engine must be a Carnot engine with the maximum theoretical efficiency given by equation 5-1. For example, an engine operating between a combustion temperature of 1,600 K and an exhaust temperature of 400 K has an ideal efficiency of 1-400/1600 = 0.75. For an automobile with an actual efficiency of 30% (first-law efficiency), the second-law efficiency is 0.3/0.75 = 0.40. That is, the engine operates at only 40% of its full potential.

Question: Air pollution is considered by many to be a direct result of incomplete (inefficient) combustion. Comment!
Answer: The common assumption that air pollution results from inefficient burning of fossil fuels is not correct. When a hydrocarbon fuel such as natural gas (methane) is burned in air, the product of the reaction is a mixture of water, carbon dioxide, carbon monoxide, oxides of nitrogen, and a number of other gases in such concentrations that maximize the overall entropy. Cleaner fuels, better burners, and more

exotic catalysts, although they increase combustion efficiency and reduce the pollutant emissions, cannot eliminate the air pollutants all together. In other words, complete combustion (of hydrocarbons), defined as burning with carbon dioxide, water, and molecular nitrogen as their only end products, is not possible and is directly in violation of the second law of thermodynamics.

Thermal Power Plants

As is the case for any thermal engine, thermal power plants are devices that convert heat to rotational shaft work, which can be coupled with an electric generator to produce electricity.[4] Unlike internal combustion engines, where the heat of combustion comes from fuel burned inside the engine, power plants use external heat – most often from fossil or nuclear fuels.

Most thermal power plants follow a thermodynamic cycle which consists of a boiler, a high-pressure steam turbine, a condenser, and a pump. Water is commonly used as the working fluid, although there are instances where ammonia and other refrigerants serve the purpose best. The cycle operates using a feed water pump to introduce water into a boiler or steam generator, where it boils and turns into high-pressure superheated steam. The superheated steam subsequently enters a steam turbine, where it is expanded and cooled to a saturated mixture of water vapor and liquid water. To close the cycle, the mixture must be turned into liquid and compressed to the boiler's operating pressure. This is accomplished by cooling the mixture in a condenser and circulating it back into the boiler using a feed water pump. To boil the water, energy is required. The source of this energy could be fossil fuel, nuclear fuel, solar heat, or a geothermal reservoir (see Figure 5-5).

Question: What is the advantage of recycling water (as is done in closed cycles) over discarding steam and pumping fresh water (as is done in open cycles)?
Answer: Recycling the same water not only conserves water, but also saves on the cost of filtering. This is essential to avoid corrosion and prevent the buildup of mineral deposits in the system.

To condense steam, it must be cooled. This is done by pumping cold water from a nearby ocean, lake, or river and diverting it to the condenser. A simple analysis shows that for operation of even a moderately sized power plant, a tremendous amount of cooling water is needed. This is the main reason that most power plants are built near a large body of water such as a river or a lake. In most instances, the water out of the condenser is hotter by 10°C or more, too hot to be returned back to its source. The thermal shock of hot water can prove especially harmful to aquatic organisms

[4] The major difference between thermal, hydroelectric and wind power plants are that in hydroelectric and wind plants, water or air flows directly through water or wind turbines, whereas with thermal power plants a working fluid is heated to its boiling temperature before it is passed through a steam turbine.

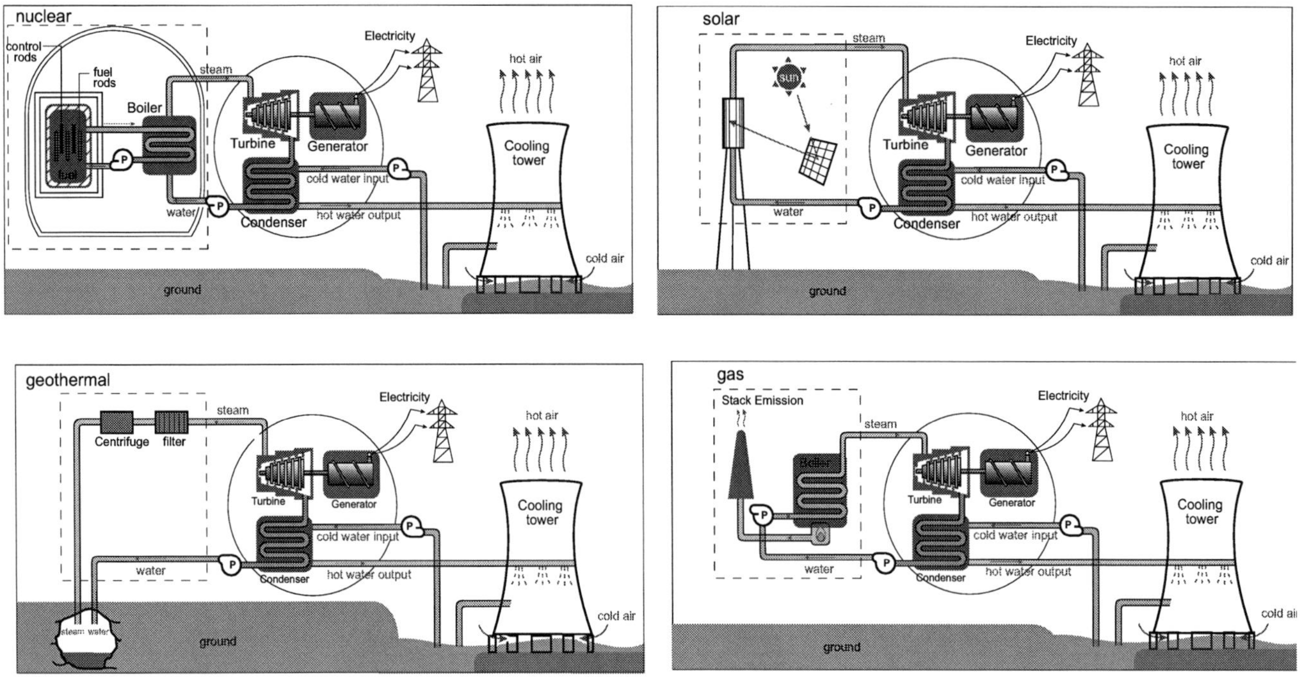

Figure 5-5
Thermal power plants consist of a boiler (or steam generator), a steam turbine, a condenser and a recirculating pump. Water boils in a boiler and turns into superheated steam. Steam will then expand in a steam turbine and deliver shaft work. The steam is condensed back to liquid water before being pumped into the boiler, beginning the cycle again.

whose survival depends on a narrow range of temperature fluctuations.[5] To safeguard fish and other marine habitats, there are regulations that require the condenser water to be cooled in a spray pond or a cooling tower before being returned to the lake or river from which it came.

Spray ponds are large, shallow bodies of water. Water from the condenser is sprayed across the large surface of the pond, where it is cooled by evaporation. Spray ponds are normally used in areas of low humidity and where land is available.

Figure 5-6
Cooling towers in Didcot Power Station in England.

Cooling towers can be either wet or dry. *Wet cooling towers* are similar to spray ponds except that water is passed through coils which look like a shower head. As water is dripped down, it is cooled by the ambient air and is collected in a basin. Heat is transferred from water to the ambient air, where it stays (Figure 5-6). The air is usually either sucked out by giant blower fans placed on top of the cooling tower (forced draft) or flows naturally upward to replace the less-dense warm air, which is rising out of the tower (natural draft). Wet towers are large and use two to three times as much water as cooling ponds. Furthermore, many people find the towers aesthetically displeasing. *Dry cooling towers* are cooled by conduction and convection. Warm water is passed through a heat exchanger where air is passed over the coils containing water, cooling it. This is very much like the

[5] The optimal temperature for plankton, the major source of food for many aquatic ecosystems, varies by only a few degrees. For example, the optimum temperature for growth of green algae is 30°C, whereas blue-green algae thrive at about 30-35°C. Thermal discharges therefore favor production of blue-green algae over green algae. Blue-green algae are a poorer source of food and can be toxic to fish.

radiators in automobiles. No matter what method of cooling is used, the heat always ends up in the atmosphere. As we shall see, this heat cannot be disposed of and therefore directly contributes to global warming.

Example 5-2: Estimate the thermal efficiency of modern power plants using fossil fuels.
Solution: A typical, modern steam power plant operates between temperatures of 600°C at the exit of the boiler and 35°C inside the condenser. An ideal (Carnot) plant operating between these two temperatures would have an efficiency of:

$$\eta_{ideal} = 1 - \frac{35 + 273}{600 + 273} = 65\%$$

Depending on the type of plant, practical plants have efficiencies far lower than this at around 30-40%. This is because much of the energy is used to raise water to high-temperature steam, or lost to friction in turbines and generators. In the case of fossil plants, if losses due to extraction, processing, and transport are included, the overall efficiencies will be much lower in the order of 10-20%. Figure 5-7 shows various losses at different stages of power generation from fossil resources.

Cogeneration

Cogeneration or CHP (combined heat and power) is the simultaneous production of electricity and heat. In this system, a primary power plant produces electricity, but unlike the simple power plants where exhaust is cooled to atmospheric temperatures, cogeneration utilizes the thermal energy left in the exhaust to drive a second thermodynamic device, which either produces additional power or is used directly for heating applications. Furthermore, some of the steam can be extracted at different pressures for industrial applications or home use.

Recent advances in gas turbine technology allow high-temperature combustion of natural gas, which can be used as the primary method of power generation. The exhausts of these turbines are sufficiently hot to be subsequently used in producing additional electricity or heat using conventional steam power plants, resulting in greater conversion efficiencies and lower pollution than the traditional generation methods. The addition of the second turbine boosts combined efficiency to about 60%, which is superior to conventional coal and nuclear power plants with efficiencies of around 33%. When the waste heat is used directly for industrial processes, efficiencies as high as 85% have been found to be possible.[6] This technology, generally referred to as the **combined cycle combustion turbine** (CCCT), has the additional advantage of releasing less heat into the atmosphere, limiting global warming and other environmental damage. The gas turbine operates by using gas (derived

[6] Manufacturing Energy Consumption Survey, Energy Information Administration, DoE, 1996 (*http://www.eia.doe.gov/emeu/mecs*).

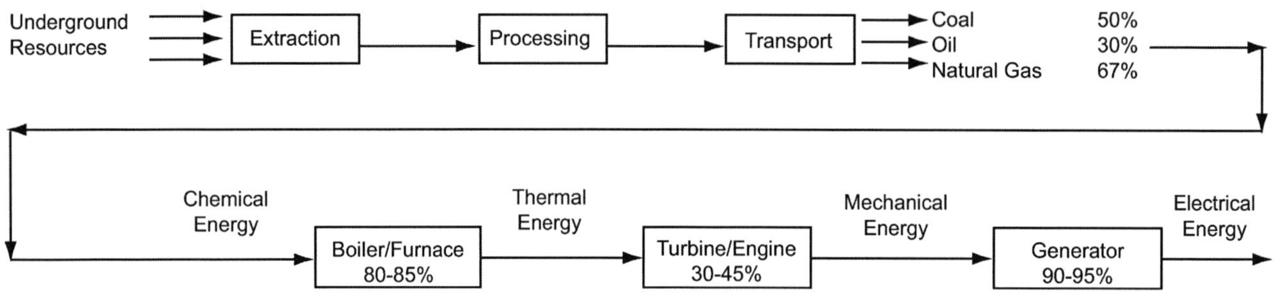

Figure 5-7
Fossil plant energy losses at various points from extraction to power generation.

from the gasification of coal or natural gas) as its primary fuel. Natural gas turbines are particularly advantageous over conventional oil and coal plants because there is no emission of sulfur and negligible emission of particulates. Also, NOx emission is cut by 90% and carbon dioxide emission by 60%. Another advantage of this technology is that power can be distributed. This means that small-scale power generation facilities can be constructed using hybrid systems consisting of solar panels, micro-turbines, and wind turbines that can produce enough electricity for small communities such as shopping malls, large office buildings, etc. When power is not needed, the excess electricity could be sold to utility companies, reducing their peak loads. We will discuss this issue in greater detail in Chapter 13.[7]

Question: Explain the limitations inherent in the operation of steam and gas turbine cycles. What is the advantage of a combined cycle over steam and gas turbine cycles operating alone?
Answer: Because of material limitations, mainly the hydrogen embrittlement used in piping, the peak temperature of the steam cycles is limited to below 1100°C. The gas turbines, on the other hand, have exhaust at temperatures as high as 450°C, and much of the energy is lost to the atmosphere. The combined cycle takes advantage of the high T_L of the gas turbine and the low T_H of the steam cycle to achieve a greater efficiency than can be achieved by either when used alone. In this technique, heat is added at a high combustion temperature (topping cycle) and rejected at the relatively low temperature of the condenser or smoke stack (bottoming cycle).

Example 5-3: A cogeneration plant uses combined gas turbines and steam power plants. Assuming a total of 200 kg of natural gas is burned every second, what is the plant's overall efficiency? The gas turbine has a thermal efficiency of 40%, whereas the efficiency of the steam power plant is only 33%. Assume methane has a heating value of 50,000 kJ/kg.

[7]A challenge problem for the more mathematically inclined: A cogeneration plant combines a gas turbine with efficiency η_{GT} with a steam turbine with efficiency η_{ST}. Show the cogeneration plant has a combined efficiency of: $\eta_{CC} = \eta_{GT} + \eta_{ST} (1 - \eta_{GT})$. What is the overall efficiency of a combined plant if gas turbine and steam turbine efficiencies are 40% and 33% respectively? Compare your answer with the example given in this chapter.

Solution: The total heat input into gas turbines is equal to the mass flow rate of the fuel multiplied by its heating value 200x50,000 = 10 MW_t (megawatts thermal). The power output from the primary cycle (gas turbine) is equal to heat input times the efficiency, 10x0.40 = 4 MW_e (megawatts electric), and heat rejected in the exhaust is 10-4 = 6 MW_t.

In simple cycles, this heat is normally rejected to the atmosphere. In cogeneration cycles, however, we can generate additional electricity by using this heat to drive a second steam turbine. Since the thermal efficiency of steam turbines is only 33%, we have an additional 6x0.33 = 2 MW_e (megawatts electric) from this turbine, for a total of 4+2 = 6 MW_e from both the primary and secondary systems, and a production of 50% more power by cogeneration. As a result, the overall efficiency of the cogeneration plant has increased to 6 MW_e/10 MW_t = 60%.

Air Conditioners and Refrigerators

The second law explains how the Universe is continuously slipping into chaos. This does not mean that there cannot be (on a local scale) a transformation from chaos to order. In fact, this is the main mechanism by which we can defy nature. When we build houses, we change the randomly distributed bricks, lumber, and clay into structured walls of living rooms, bedrooms, and kitchens. We bring order into our house by cleaning it. What the second law precludes us from doing is not creating order, but creating order without causing even more disorder somewhere else (usually in the neighborhood).

Air conditioners and refrigerators are devices that move heat away from a space in defiance of the common perception that heat moves from a higher to a lower temperature. The price is of course the expenditure of additional energy in terms of electricity and creation of disorder outside the immediate neighborhood of these devices.

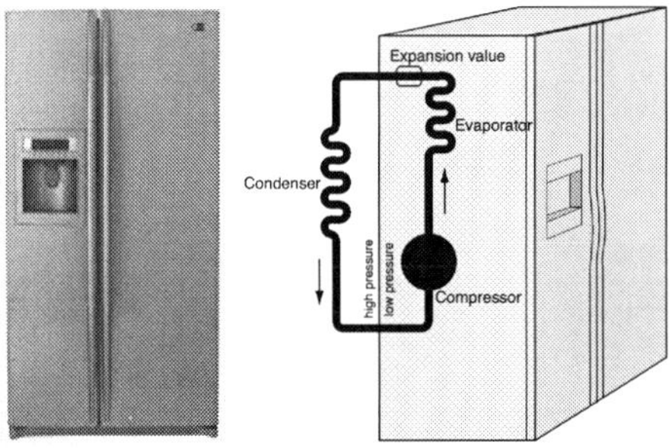

Figure 5-8
Refrigerator

The operation of these devices can be best described as the heat engines operating in reverse. Heat is removed from the space (inside the refrigerator or air-conditioned room) and dumped into a second reservoir at a higher temperature. In the case of the household refrigerator, the second reservoir is the kitchen. In the case of the air-conditioner, it is the outside air. Energy input in the form of electricity or heat is needed to make the uphill transfer of heat possible (Figure 5-8).

a Heat Pumps

As explained in the second law, the thermal energy of any substance vanishes only at absolute zero temperature (-273°C). This means even cool outdoor winter air has some thermal energy. The heat pump is a device that utilizes this thermal energy by bringing it into the house and heating its interior. Of course, by removing this heat, we make the outside temperature even cooler! In order to move this energy inside, a heat pump circulates a fluid called refrigerant which absorbs heat from outdoor air and releases it inside. Heat pumps can be used for cooling as well. This process is the reverse of the heating process; it removes the heat from the space to be cooled and dumps it into the already warm outdoor environment.

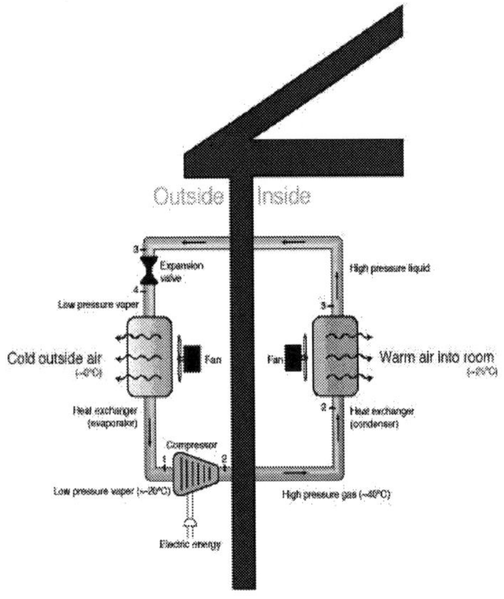

(a)

The operation of a heat pump as a heating device is shown in Figure 5-9a. The refrigerant flows inside a closed loop. At point 1, the refrigerant vapor enters a compressor where it is compressed to a temperature warmer than the indoor air. The compressor needs some energy, usually in the form of electricity, in order to function. At point 2, the heated refrigerant enters an indoor heat exchanger where it gives off its thermal energy and, with the aid of a fan, uniformly heats the room. In the process, some of the vapor is condensed back to liquid (point 3) inside the heat exchanger (thus the name condenser). The liquid is then expanded through a valve or capillary tube at point 4, where it rapidly expands and cools. The mixture then enters a heat exchanger, where the heat from outside air causes it to vaporize (thus the name evaporator). An outside fan facilitates this heat transfer. The refrigerant vapor leaves the evaporator at point 1 and the cycle repeats.

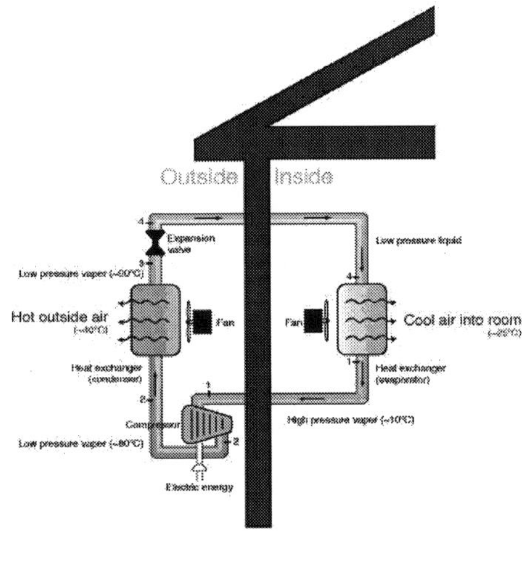

(b)

Figure 5-9
Heat pumps
a) heating cycle; b) cooling cycle

During summer, when cooling instead of heating is desired, the direction of flow is reversed (See Figure 5-9b) and the heat exchangers reverse functions. In this mode, the outside heat exchanger is a condenser where heat is extracted, whereas the inside heat exchanger acts as an evaporator, removing heat from the space where cooling is desired.

The performance of heat pumps is defined in terms of the amount of heating or cooling that is achieved per amount of work energy (electricity) that is needed to run the compressor. This is called the **coefficient of performance** (COP). The COP for practical heat pumps is around 4-5, meaning heat pumps can provide four to five times more heating or cooling than electricity consumed. In comparison, electric heaters have efficiencies close to 100%, oil and gas heaters have efficiencies of 50% to 90%, and wood stoves' heating efficiencies range from 20% to 60%. The COP decreases markedly once outside temperatures fall below around −5 or −10 °C.

In the United States, it is customary to specify the performance of a heat pump by two numbers, the COP as defined above for heating, and the **Energy Efficiency Ratio** (EER) for cooling. The EER is expressed in Btu/Wh, and is the ratio of cooling capacity given in Btu/h to required electrical power in watts.[8] For large commercial units, chillers are rated in kilowatts per ton. Because each ton of refrigeration is equal to 12,000 Btu, we have

[8] Remember that Btu is the US unit of energy, while Wh is the unit of energy in SI systems of units. So much for our creativity!

Energy Boosters

Mathematical Interlude ...

In line with the definition proposed for thermal efficiency of heat engines for refrigerators and heat pumps, it is most convenient to define efficiency in terms of the desired effect (cooling or heating) as compared to the work that is necessary to achieve this effect. For a Carnot heat pump and refrigerator, the following results are obtained:

$$COP_{ideal,\,HP} = \frac{T_H}{T_H - T_L} \qquad \text{(i)}$$

$$COP_{ideal,\,R} = \frac{T_L}{T_H - T_L} \qquad \text{(ii)}$$

T_L and T_H are temperatures of cold and hot reservoirs and must be expressed in the Kelvin scale.

Of course, construction of such devices is not practical because there are always friction and conduction losses that cannot be completely eliminated.

Example: A heat pump is to be used to heat a house during the winter to 20°C (293 K). The house loses heat at a rate of 100,000 kJ/h when the outside temperature drops to -8°C (265 K). What is the minimum power required to drive this heat pump?
Solution: Minimum power usage can be achieved when the heat pump works reversibly as a Carnot heat pump. The COP of this heat pump would be: COP = 293/(293-265) = 10.5. That means for each unit of power consumed, 10.5 units of heat can be supplied to the room. The power input to the compressor is then 100,000/10.5 = 9556 kJ/h (or 2.7 kW). If the house was going to be heated by an electric resistance heater instead, we would have needed 10.5x2.7 = 28.35 kW of electric power. This example shows that by investing a small amount of work in a heat pump we can supply a great amount of heat. For this reason, heat pumps can be appropriately called "*energy boosters*".

$$EER = 3.413 \times COP \qquad (5\text{-}2)$$

$$kW/ton = \frac{12}{EER} \qquad (5\text{-}3)$$

Summary

Understanding laws of thermodynamics is fundamental to understanding the workings of the universe. In this chapter we covered these laws and their relevance in designing useful thermal devices. In describing the working principle of such devices (a thermal power plant, a heat engine, a refrigerator, or any other thermodynamic system for that matter) we concluded that:

1) At least two reservoirs, one at a high temperature and the other at a low temperature, are needed. This is one of the important consequences of the second law of thermodynamics, which states that we cannot convert heat to work with a single heat source. In the case of a steam power plant, the steam generator and condenser are the two reservoirs. In internal combustion engines, the two heat reservoirs are hot combustion gases and cold exhaust gases.
2) Some form of energy is always needed to provide heat to the high-temperature heat reservoir. For power plants, energy needed to heat water comes from an external source

which could be fossil, nuclear, or less commonly solar or geothermal.

3) To increase efficiency, we must either increase the temperature of the source or lower the temperature of the sink. We are limited to atmospheric temperatures for the sink, but we can increase the source temperature. In power plants, we can raise steam temperature by increasing the boiler pressure (water boils at a higher temperature if the pressure is greater). Raising the compression ratio, minimizing heat losses, and using better fuels can similarly increase internal combustion efficiency.
4) Higher efficiencies not only reduce fuel consumption, but also protect our environment by producing fewer pollutants. In addition, higher efficiency means a smaller fraction of energy must be disposed of as waste heat, commonly termed "thermal pollution," which is a major source of global warming.

Additional Information

Books

1. El-Sayed, Y., The Thermodynamics of Energy Conversions, Elsevier Direct Science, 2003.
2. Cengel, Y. A., Heat Transfer: A Practical Approach, McGraw-Hill, Inc., 1998.
3. Rifkin, J., Entropy, The Viking Press, 1980.
4. El-Wakil, M/ M., Power Plant Technology, McGraw-Hill, Inc., 1984.

Periodicals

1. *Energy and Buildings,* Science Direct Elsevier Publishing Company. An international journal publishing articles about energy use in buildings and indoor environment quality.
2. *Energy Conversion and Management,* Science Direct Elsevier Publishing Company. This journal focuses on energy efficiency and management; heat pipes; space and terrestrial power systems; hydrogen production and storage; renewable energy; nuclear power; fuel cells and advanced batteries.
3. *Energy and Buildings,* Science Direct Elsevier Publishing Company, An international journal dedicated to investigations of energy use and efficiency in buildings.

Government Agencies and Websites

1. How Things Work *(http://howthingswork.virginia.edu).*
2. How Stuff Works *(http://www.howstuffworks.com).*
3. California Energy Commission Consumer Energy Center (*http://www.consumerenergycenter.org*).

Exercises

I. Essay Questions:

1. What is heat and its relation to temperature? How did the concept of heat change throughout the centuries?

2. The temperature of the water at the bottom of Niagara Falls is slightly warmer than the water at the top of the waterfall. To what do you attribute this temperature difference?

3. The highest temperature ever recorded in Death Valley, CA is 54.7 °C. What is the temperature in degrees Fahrenheit?

4. Define heat capacity, specific heat, and thermal conductivity. Does a substance with a high specific heat necessarily have a high heat capacity and thermal conductivity?

5. Which one is colder during the winter, an iron railing or a wooden fence post?

6. A wooden pan heated to 80°C can be picked up comfortably, but a similar pan made of copper can burn your hand. Why is this?

7. What are the three modes of heat transfer? Under what conditions do any of the modes become more significant?

8. Why do many frying pans have wooden handles?

9. Name the main components of a steam power plant and their functions.

10. Name the main components of a gas turbine, a heat engine, and a power plant. Explain their functions.

11. Name the main components of an air conditioning system and their functions.

12. What are the differences between an air conditioner and a heat pump?

13. As a car is coming to a stop, its kinetic energy continuously decreases. What happened to this energy?

14. Why would one use heat pumps when a heating coil will also do the job?

15. An inventor claims to have designed an engine that has eliminated nearly all frictional losses and uses insulating material with relatively little heat losses; the efficiency of this engine is therefore very close to unity. How do you evaluate this inventor's claims?

16. For each of the following devices, describe basic operation, typical efficiency, and the working fluid. In each case, explain what constitutes the source and the sink:
    a. Internal combustion engine
    b. Steam power plant
    c. Jet aircraft engine
    d. Refrigerator

17. Which of the following situations is allowed by the second law of thermodynamics?
    a. Designing a heat engine that exploits the temperature differences in the top and bottom of the oceans
    b. Converting 100 joules of heat energy into 100 joules of work
    c. Converting 100 joules of work into 100 joules of heat energy
    d. Removing heat from cold outside air to heat inside of a house in winter
    e. Reducing the waste heat from a power plant to zero with super insulating materials

18. Which of the following violate the second law of thermodynamics? Explain.
    a. You visit a friend's house and find it to be tidier than you found it in your previous visit.
    b. Spraying water on the pavement cools the air.
    c. A glass is dropped on a hard surface but does not break.
    d. A glass is dropped on a hard surface and shatters into many small pieces.
    e. A coin lands head up ten times in a row.
    f. You run a film backward to see a scene you missed.

19. The energy conversion efficiencies in a 3 step process are 60% for the first step, 40% for the second, and 80% for the third. What is the overall efficiency for this three-step process?

20. Thermal images of an airplane and an automobile are shown in pictures below. Explain what the different colors mean. Can you identify the hottest regions of these objects in these images? (*Images courtesy of Thermotronics Inc, Brazil*)

II. Multiple Choice Questions

1. Thermal energy results from the motion of
    a. Electrons
    b. Nuclei
    c. Molecules
    d. Mass
    e. All of the above

2. Heat is
    a. A substance with mass that flows from a high temperature toward a low temperature
    b. A mass-less substance that flows from a high temperature toward a low temperature
    c. The result of kinetic motion of particles
    d. Equal to the work required to run a heat engine
    e. The same as temperature

3. As a substance is heated
    a. Mass decreases because some of it goes into heat
    b. Mass increases because heat is added
    c. Potential energy of the molecules of the substance increases, but their kinetic energy does not
    d. Both potential and kinetic energies of the molecules of the substance increase
    e. None of the above

4. A BTU is a unit of
    a. Energy
    b. Power
    c. Heat flow rate
    d. Specific energy
    e. None of the above

5. Heat transfer in solids is mostly by
    a. Conduction
    b. Convection
    c. Radiation
    d. Conduction and convection, but not radiation
    e. Conduction, convection and radiation

6. Heat transfer in liquids is mostly by
    a. Conduction
    b. Convection
    c. Radiation
    d. Conduction and convection, but not radiation
    e. Conduction, convection and radiation

7. Heat transfer in flames is mostly by
    a. Conduction
    b. Convection
    c. Radiation
    d. Conduction and convection
    e. Convection and radiation

8. The heat transferred by conduction increases with
    a. The temperature gradient across the object
    b. The thermal resistance across the object
    c. The thickness of the object
    d. The volume of the object
    e. All of the above

9. Which of these statements is correct?
    a. Conduction is the transfer of heat by molecular interaction.
    b. Convection is the transfer of heat by bulk motion.
    c. Radiation is the transfer of heat by electro-

magnetic radiation.
d. All of the above.
e. None of the above.

10. The materials listed in order of increasing thermal conductivity are
a. Diamond, water, and air
b. Air, water, and copper
c. Aluminum, hydrogen, and vacuum
d. Copper, water, and air
e. Ice, Styrofoam, and silver

11. Radiant energy (light) travels
a. In a medium, but not in vacuum
b. In gases, but not in liquids and solids
c. In both a vacuum and a medium
d. Faster in a medium than in a vacuum
e. With infinite speed in a vacuum

12. Radiant energy (light) travels
a. A straight line
b. In waves
c. In either a straight line or in waves
d. Faster in a medium than in a vacuum
e. With infinite speed in a vacuum

13. Night vision cameras operate in the
a. UV range
b. X-ray range
c. Visible range
d. Infrared range
e. TV range

14. The reason sand gets warmer than water at sunny beaches is that
a. Sand is denser than water
b. Sand can trap heat much better than water
c. Sand has a higher specific heat than water
d. Sand has a lower specific heat than water
e. Sand is less transparent than water to solar energy

15. Why do most apples appear red to us?
a. Red apples absorb the red portion of visible light.
b. Red apples transmit the red portion of visible light.
c. Red apples absorb all colors except red.
d. Red apples absorb the infrared portion of visible light.
e. Red apples reflect the ultraviolet portion of visible light.

16. A lamp which has an efficiency of 15%
a. Converts 15% of thermal energy to light
b. Converts 15% of electrical power to light
c. Converts 15% of electrical power to light and heat
d. Converts 85% of electrical power to light
e. Uses only 15% of the incoming electrical power to work

17. A glass cup falls from edge of a table. The potential energy of the glass cup
a. Converts to heat only
b. Converts to heat and sound
c. Converts to sound only
d. Converts to static charge
e. Converts to sound and electricity

18. When you stir a hot cup of coffee with a metal spoon, the coffee
a. Warms mainly by conduction through the spoon
b. Warms mainly by convection
c. Cools mainly by conduction through the spoon
d. Cools mainly by convection
e. Stirring does not affect its temperature

19. A battery is a convenient device to convert
a. Electrical energy to mechanical energy
b. Chemical energy to thermal energy
c. Thermal energy to electrical energy
d. Chemical energy to electrical energy
e. Thermal energy to chemical energy

20. During photosynthesis
a. Light is converted to chemical energy
b. Light is converted to electrical energy
c. Light is converted to thermal energy
d. Thermal energy is converted to light
e. Thermal energy is converted to mechanical energy

21. According to the first law of thermo- dynamics,
    a. Energy cannot be created, nor can it be destroyed
    b. The available energy of a system in all real processes decreases
    c. Work can be converted entirely to heat
    d. Heat cannot be converted entirely to work
    e. All of the above

22 Entropy
    a. Increases for all real processes
    b. Decreases for all real processes
    c. Is a measure of disorder in the system
    d. Remains constant if energy remains constant
    e. Both a and c

23. Matter resources are depleted by being __________, whereas energy resources are depleted by being ___________.
    a. Dispersed, destroyed
    b. Consumed, lost
    c. Dispersed, transformed
    d. Destroyed, consumed
    e. Lost, dispersed

24. According to the second law of thermodynamics,
    a. One day the universe will reach a heat death
    b. One day we can invent devices to go back in time
    c. One day we can convert frictional losses completely back to work
    d. One day we can design refrigerators that do not require any work input
    e. One day all the processes cited above will be possible

25. According to the second law of thermodynamics,
    a. Some energy is lost when it transforms from one form to another
    b. Energy can be neither created nor destroyed
    c. Energy becomes more dispersed during any real process
    d. Entropy increases for all real processes
    e. All of the above

26. Efficiency of real heat engines can be shown to be equal to
    a. $1 - \frac{T_c}{T_h}$
    b. $1 - \frac{T_h}{T_c}$
    c. $1 - \frac{Q_c}{Q_h}$
    d. $1 - \frac{Q_h}{Q_c}$
    e. $1 - \frac{Q_h}{W_{net}}$

27. Thermal efficiency of a heat engine is a number between
    a. 0 and 1
    b. 0 and 100
    c. 0 and ∞
    d. - 1 and +1
    e. - ∞ and +∞

28. Coefficient of performance of a refrigerator is a number between
    a. 0 and 1
    b. 0 and 100
    c. 0 and ∞
    d. - 1 and +1
    e. - ∞ and + ∞

29. The efficiency of car engines can be improved if
    a. We design a heat sink that operates at a lower temperature
    b. We use better fuels
    c. We reduce friction
    d. We increase the compression ratios
    e. All of the above

30. The temperature of a healthy body can be expressed in different systems of units as
    a. 37 °C, 98.6 °F, and 310 K
    b. 100 °C, 212 °F, and 373 K
    c. 37 °C, 98.6 °F, and 37 K
    d. 37 °C, 98.6 °F, and 497 K
    e. 37 °F, 98.6 °F, and 310 K

31. To make fusion reactions possible, the gaseous mixture of deuterium and tritium must be heated

to 100 million degrees Celsius. This temperature is roughly equivalent to ______________ degrees Fahrenheit.

a. 55 million
b. 100 million
c. 180 million
d. 212 million
d. None of the above

32. According to the zeroth law of thermodynamics,
 a. Temperatures lower than zero absolute temperature can never be achieved
 b. The total energy of a substance will always remain the same
 c. Objects at equilibrium must have the same temperatures
 d. Heat always flows in the direction of decreasing temperature
 e. There is no such law

33. The "efficiency" of an electric heater is defined as the ratio of the
 a. Heat energy produced to the electrical input
 b. Output voltage of the heating element to the voltage of power supply
 c. Electrical energy into the heater to the thermal energy out of the heater
 d. Light output to the electrical energy input
 e. Heat energy produced to heat energy used to produce the electricity

34. Which of the following observations are consistent with the second law of thermodynamics?
 a. Aging is a natural process.
 b. Earthquakes tend to cause destruction.
 c. Heat flows from higher to lower temperatures.
 d. Some information is always lost through transmission.
 e. All of the above.

35. Which of the following observations are consistent with the second law of thermodynamics?
 a. If you puncture a hole in an inflated helium balloon, helium will flow out.
 b. The ultimate fate of the universe is a heat death where temperature, electric potential, pressure, etc. are uniform.
 c. We can never build a thermal engine with 100% efficiency.
 d. Even the cooling of a room is associated with the entropy increase of the entire universe.
 e. All of the above.

36. Which of the following observations is/are consistent with the first law of thermodynamics?
 a. Aging is a natural process.
 b. Energy is neither created nor destroyed.
 c. Heat engines can only reach efficiency of unity if there are no frictional losses.
 d. Two objects at the same temperature as a third object are themselves at the same temperature.
 e. All of the above.

37. Which of the following observations are consistent with the first law of thermodynamics?
 a. When we turn on a light bulb, some energy is lost as heat.
 b. A light bulb is a device that converts electrical energy into light and heat.
 c. In a bicycle, some mechanical energy is transformed into another form of mechanical energy.
 d. In a nuclear reaction, some mass is converted to energy.
 e. All of the above.

38. The four basic components of a thermal power plant cycle are
 a. Pump, compressor, turbine, and condenser
 b. Pump, turbine, boiler, and cooling tower
 c. Pump, boiler, turbine, and condenser
 d. Steam, water, river, and fossil fuels
 e. Heat, work, energy, and entropy

39. Saving one BTU of electric energy
 a. Conserves zero BTU of total energy
 b. Conserves less than one BTU of total energy
 c. Conserves one BTU of total energy
 d. Conserves about 3-4 BTUs of total energy
 e. Has no relationship to saving other forms of energy

40. Refrigerators
 a. Operate in violation of the zeroth law of

thermodynamics
b. Operate in violation of the first law of thermodynamics
c. Operate in violation of the second law of thermodynamics
d. Operate in violation of the third law of thermodynamics
e. Do not violate any laws of thermodynamics

41. If you leave your refrigerator door open, the average temperature in the room will be
    a. Lower than it was before
    b. Higher than it was before
    c. The same as it was before with uniform temperature throughout the room
    d. The same as it was before, but some regions will be hotter and others will be colder

42. Roughly, what percentage of energy from burning coal is converted to electricity?
    a. 10-20%
    b. 30-40%
    c. 50-60%
    d. 70-80%
    e. 90-100%

43. Jet engines
    a. Are reaction-type heat engines
    b. Require oxygen or air to operate
    c. Cannot operate in space because there is little or no oxygen
    d. Are cleaner to operate than gasoline engines, because they function under leaner conditions
    e. All of the above

44. Which of the following is <u>not</u> considered an internal combustion engine?
    a. Diesels using synthetic biofuels
    b. Gasoline engines
    c. Gas turbines
    d. Steam turbines
    e. All are internal combustion engines

45. Which of the following is <u>not</u> considered an external combustion engine?
    a. Stirling engines
    b. Steam engines
    c. Jet engines
    d. Solid-fuel rockets
    e. Both c and d

III. True or False?

1. Heat is a mass-less substance that resides inside an object.

2. Temperature is a property of the material from which an object is made and represents the amount of heat that is contained in that object.

3. Different objects will eventually reach the same temperature when they remain in contact for a long time.

4. Heat is energy transferred from an object at high temperature to another at low temperature.

5. Heat is the total amount of energy in an object.

6. Energy is conserved, but only if no friction is present.

7. Energy can be converted from one form to another.

8. Heat can be entirely converted to work and vice versa.

9. Heat can never flow from a low to a high temperature.

10. Specific heat is a measure of the amount of energy a substance can store.

11. The absolute best insulator is a vacuum as there are no molecules to pass on energy.

12. There is no limit to how much an object can cool.

13. There is no limit to how much an object can warm.

14. An ideal heat engine that works with no friction is 100% efficient.

15. The only known phenomenon that seems to violate

the second law of thermodynamics is a hurricane.

16. A transformer is a device to convert one form of electrical energy to another.
17. In a bicycle, one form of mechanical energy converts to another form of mechanical energy.
18. According to the second law of thermo-dynamics, nature always prefers order over disorder.
19. Since time always proceeds from past to present to future, running a film backward is in violation of the second law of thermodynamics.
20. Work, heat, and energy have the same units.
21. Work, heat and energy are identical in every sense.
22. Work is a higher quality form of energy than heat.
23. Solar energy is of a higher quality form of energy than fossil fuels.
24. Heat engines can be made close to 100% efficient if we find better materials and eliminate all frictional losses.
25. The construction of a time machine, although theoretically possible, will not be technically feasible for a very long time.

IV. Fill-in the Blanks

1. Based on the assumption of caloric theory, heat is a __________ called "caloric".
2. When burned, approximately ________ percent of gasoline energy ends up in the exhaust pipe.
3. Radiation coming from the nuclei of atoms is called _____________ rays.
4. Heat engines are devices in which thermal energy is converted to ___________.
5. No heat engine can operate with only one heat ______________.
6. It is impossible to construct a heat engine which has the efficiency of unity, unless it operates in an atmosphere at ________________ temperature.
7. The ratio of energy delivered to energy used is called the _______________ efficiency.
8. Combined cycle combustion turbine uses the high temperature exhaust of the gas turbines to achieve a greater ____________.
9. Heat pumps usually have __________ greater than one.
10. According to the ______ law of thermodynamics, temperature of zero kelvin can never be attained.

V. Project I - Energy balance

In this project you are asked to estimate your total daily energy consumption.

1. Make a table showing your daily food intake and the total food calories you consume. Convert this energy to joules.
2. Make a table showing all the electrical appliances you routinely use (electric oven, refrigerator, heater, stereo, lights, hair dryer, etc.) Estimate the number of hours you use each device. Multiply time by the power consumption (wattage) to calculate the total energy consumption in kilowatt-hours. If there is more than one person in your household, correct the result by dividing this number by the number of people sharing each device.
3. Assuming electricity was produced in a power plant with efficiency of 33%, how much thermal energy was needed to meet your electric power requirement.
4. Estimate total energy requirement for all the gas appliances in your house.
5. Estimate the total energy required for your transportation need. This can be done by dividing your daily travel in miles by the gas mileage your car gives (in mpg) and multiplying the results by energy content in one gallon of fuel.
6. Add items 1-5 to estimate your total daily energy consumption.

Project II - Gas vs. electricity

In this project you are asked to examine the electric and gas bills for your home for a month. Find a copy of your gas and electricity bills (if you don't get one, use those from your parents or a friend) and write down:

1. The amount of electricity used in kilowatt-hours.
2. The amount of gas used (in the United States this is usually given in therms, which is equal to 100,000 BTU).
3. The lowest, highest, and average prices charged during that month. Note that prices may vary depending on the month, time of the day and during peak hours.
4. Total amount of energy consumption (thermal and electrical) in megajoules and kilowatt-hours.
5. The average power consumed by you in kilowatts (Divide the total energy used by the number of hours in the month).
6. Determine average cost of electricity in \$/kWh and \$/MJ (1 kWh = $3.6x10^6$ J).
7. Determine average cost of gas in \$/therm and \$/MJ.
8. Which one (electricity or natural gas) costs more and why? Is the price differential justified? Explain.
9. How much do you save (lose) if you switch to all-electric appliances?

CHAPTER 6

Biomass Energy

All flesh is grass ~ Isaiah 40.6-8

Biomass energy refers to the energy of all products derived from living organisms. Just like fossil fuels, biomass is formed by the Sun which, through photosynthesis, converts carbon dioxide and water into organic matter. However, unlike fossil fuels that take millions of years to form, biomass can be produced in a short period of time. In fact, this is how the human body converts food into the energy it needs to perform daily tasks. Many of the materials we use and much of the by-products of our activities are biomass; they can be burned directly or converted into liquid and gaseous fuels such as methanol, ethanol, biogas and synfuel.

Although only a tiny fraction of the solar energy that reaches the earth is converted into biomass, it can fulfill all the world's energy needs. Currently, biomass contribution to overall energy consumption is small. However, if energy contained in food is included, biomass provides about 15% of all the energy consumed in the world and, after oil, coal, and natural gas, is our fourth largest energy resource.[1] Until the mid nineteenth century, wood provided 90% of energy used in the United States, before better coal technology and the discovery of petroleum reduced the demand for biomass. Today, about 3% of the US energy demand is supplied by biomass, mostly for residential heating and cooking in the form of firewood, but also indirectly as feedstock for the production of liquid and gaseous biofuels.[2] Some developing countries, mainly China and India, satisfy as much as one third of their total energy needs from biomass.

Tires, pulp, paper, wooden products, foodstuff, rice husks, peanut shells, fruit pits and animal waste are all examples of biomass. Municipal solid wastes are, for the most part, derived from plants and other organic matter. Unfortunately, they contain a variety of other substances that are highly toxic and could pose a health hazard if burned; due to these toxic components many people do not consider them to be biomass.

Although biomass energy is renewable, it is not clean, and when burned, like other conventional fuels, it produces pollution. The emissions are lower, however. Furthermore, the carbon dioxide produced from the combustion of biomass offsets that used up during its formation, maintaining zero net production of this dangerous greenhouse gas.

In this chapter, we will discuss the mechanism for the formation of biomass, its resources, and its many applications.

---

[1] Excluding food energy (See Chapter 1), wind energy is the fourth largest energy resource.
[2] EIA International Energy Annual Report, 1998.

Overview

When the earth was formed about 5 billion years ago, it was nothing more than a collection of extremely hot cosmic dust with no atmosphere. With time, these particles lumped together by the force of gravity forming an early atmosphere of mostly hydrogen and helium. This was followed by volcanic eruptions which produced water vapor, nitrogen, methane, sulfur, and a few other gases present in today's volcanoes. No free oxygen was yet present. As earth cooled, water vapor turned into liquid, filling up oceans. Small amounts of oxygen gas appeared only after sun's ultraviolet radiation broke up the water molecules. The condition was now just right for carbon dioxide and water to react through a process called photosynthesis, making the simpler organic compounds. As by-products, oxygen and ozone were also formed. As the atmosphere thickened, dinosaurs, birds, more complex animals, and eventually the primary form of man appeared.

Photosynthesis

Photosynthesis (photo=light; synthesis=build) is the crucial link between the sun and the chemical energy stored in all living organisms. It involves the removal of carbon dioxide from the atmosphere by plants and the combination of it with water and other nutrients from the soil to form carbohydrates. Carbohydrate molecules contain carbon (*carbo-*), hydrogen (*hydr-*), and oxygen (*ate*). In the process, oxygen is released into the atmosphere. For photosynthesis to occur, sunlight is needed. The overall reaction can be written as:

$$6\,CO_2 + 6\,H_2O \xrightarrow{Sunlight} C_6H_{12}O_6 + 6\,O_2 \qquad (6\text{-}1)$$

$$12\,CO_2 + 11\,H_2O \xrightarrow{Sunlight} C_{12}H_{22}O_{11} + 12\,O_2 \qquad (6\text{-}2)$$

or more generally:

carbon dioxide + water + sunlight ⟶ carbohydrate + oxygen

In this reaction, two important tasks are accomplished. Firstly, carbon is "fixed", that is, converted from its inorganic form (carbon dioxide) to its organic form (carbohydrates in the form of glucose or sugar). Secondly, the sun's dispersed light-energy is transformed into concentrated chemical energy. Once sugar is formed, it can be converted to starch for storage or combined with other nutrients such as nitrogen, phosphorous and sulfur to create more complex molecules such as proteins. Proteins are the building blocks of life, essential to the growth of the body as well as the brain. The best sources of proteins are meat, fish, eggs and milk.

But does all of the energy in the sunlight participate in photosynthesis? The light required for photosynthesis must be of the right frequency, around the red end of the visible light spectrum. To capture light for

Sulfur-Based Life

Digging Deeper ...

Until only a few decades ago, biologists thought that organisms in the deep ocean lived only on the debris of marine plants and dead animals falling from the surface to the bottom. Because of the high concentration of deep-sea plants near hydrothermal (volcanic) vents, it is presumed that another mechanism called **chemosynthesis** is responsible for producing this form of life.[i] According to this hypothesis, these organisms utilize chemicals rather than photosynthesis to metabolize carbon dioxide and water to form carbohydrate. Examples of vent animals include snails, crabs, limpets, mussels and tube worms.

Tube worms
Image: NASA Ocean Planet, Smithsonian

[i] *Little, C.S., and Vrijenhoek, "Are hydrothermal vent animals living fossils?" TRENDS in Ecology and Evolution, V. 18, No. 11, November 2003.*

photosynthesis, plants have special pigments called chlorophyll; these absorb only the red and blue portions of sunlight, reflecting the green. This is why many leaves appear green. Lower frequency light does not have sufficient energy so plants do not absorb it. Higher frequency light is too strong and the excess energy is wasted. The result is that photosynthetic efficiency is at a maximum for red light,[3] drops abruptly to zero for lower energies (infrared light), and falls slowly for photons of higher energies (blue and green lights). The net effect is that only 32% of solar energy participates in photosynthesis.[4] In addition to chlorophyll, plants have two other pigments to trap light in regions that chlorophyll misses. Carotenes absorb blue and blue-green light and anthocyanins absorb green and yellow light and are responsible for the leaves' yellow, orange, brown or red colors common during the fall season.[5]

Photosynthesis in Water

Photosynthesis is not only responsible for the production of food on land, but also for manufacturing microscopic single-celled plants called *phytoplankton* in the oceans. Phytoplankton is responsible for the nourishment of the entire marine food web. The lives of all animals that live in the sea -- with the exception of hydrothermal vent organisms (See box "Sulfur-Based Life") -- depend on phytoplankton for energy and minerals. The rate of photosynthesis is highest, not at the surface, but at a depth of about 10-20 meters where light intensity is not too high. As we go deeper, the rate of photosynthesis slows and eventually stops.[6]

Food Chain and Food Web

Plants use the energy stored by photosynthesis for their own growth, to acquire nutrients, to fight off insects and to create the oxygen required for respiration. Respiration is the reverse of photosynthesis; carbohydrate and oxygen are converted to carbon dioxide and water. In addition, a relatively large amount of energy is released and captured throughout the

---

[3] Near 6700 angstroms. One angstrom is equal to 10^{-4} microns or 10^{-10} meters (1 °A=10^{-4} μm=10^{-10} m)

[4] Thorndike, E. H., Energy and Environment, Addison-Wesley Publishing, 1976. p.33.

[5] Sunlight not only provides the energy for photosynthesis, but also causes chlorophyll to break down. In fact, plants continuously have to work to produce new chlorophyll to replace that which was destroyed. During autumn, in preparation for winter, trees absorb as much nutrients as they can before the leaves fall. As chlorophyll disappears another pigment called carotene, which holds up better in the sunlight, remains. Since carotene absorbs blue and blue-green, the leaves appear yellow.

[6] The study of algae growth in the euphotic zone shows that maximum photosynthetic activity occurs at a depth of only 10-20 m beneath the surface of water. The warmer surface temperatures limit the amount of carbon dioxide that can be dissolved. Below this point, the photosynthesis becomes light-limited, and so decreases with depth.

Living with Carbon Dioxide*

Digging Deeper ...

Carbon dioxide (CO_2) is a colorless, odorless gas that is 1.5 times as heavy as air. It is a component of all carbonated beverages. With water it forms carbonic acid, which has a mildly acid taste. As a by-product of fermentation in yeast, it is CO_2 that makes dough rise. The gas is easily liquefied by compression because of its relatively high critical temperature. When liquid CO_2 is allowed to evaporate, the vapor freezes to a snow-like solid at -56.2°C. The solid vaporizes without melting (sublimates) because its vapor pressure is one atmosphere at -78.5°C. This property makes solid CO_2, or dry ice, valuable as a refrigerant that is always free of messy liquids.

Carbon dioxide is also an efficient fire extinguisher because most substances will not burn in it. It is also easily generated and, best of all, it is cheap. Air containing as little as 2.5 percent CO_2 will extinguish a flame. Although the gas is not toxic, high concentrations will cause suffocation. The atmosphere contains only 0.04 percent CO_2 by volume, but still serves as a huge reservoir of the gas, which supports the photosynthetic activities of plants. Because a liter of water at 20°C can dissolve 0.9 liters of CO_2, oceans and lakes serve as rather sizable reservoirs of the compound as well. Although the oceans keep the CO_2 levels of the atmosphere relatively constant, the levels have increased detectably in the last few years due to the increase in the rate of burning of fossil fuels.

The early atmosphere contained no free oxygen. Almost all the earth's oxygen was bound in water and metal oxides. Carbon dioxide was also absent from the early atmosphere. Organic compounds would remain stable for a very long time under these conditions. It is therefore reasonable to assume that sea was the suitable laboratory to synthesize the first living organisms. In an environment devoid of oxygen, molecules could come together freely and experiment. Occasionally particles reacted to form new species. Colloidal aggregates grew to form molecules of larger size and higher complexity. For life to exist, a continuous influx and outflow of matter and energy must persist. In the absence of oxygen, only fermentation could provide such a mechanism. Energy is all that was needed for a cell to survive. Alcohol and carbon dioxide were waste products that had to be discarded. Gradually, the concentration of carbon dioxide dissolved in oceans increased until the water became saturated and CO_2 escaped into the atmosphere, making photosynthesis processes possible.

* Excerpts from a physics text currently under preparation jointly by this author and Professor Igor Glozman, Department of Physics, Highline Community College, Des Moines, Washington 98198.

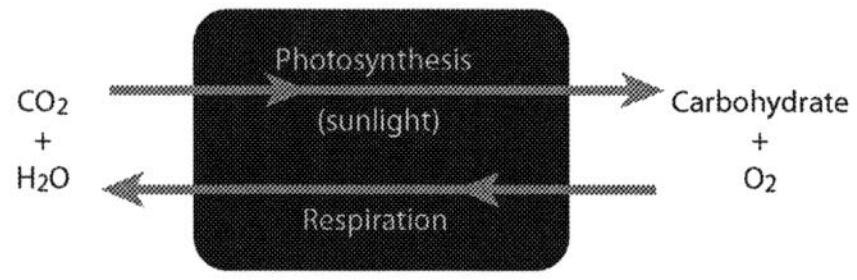

Figure 6-1
Photosynthesis and respiration. Carbon dioxide and water, through the process of photosynthesis, are turned into organic carbon. During respiration, the reverse happens and plants' and animals' organic carbon is broken down into carbon dioxide and water.

process (Figure 6-1). One major difference between the two processes is that while photosynthesis is possible only in the presence of sunlight (daytime), respiration occurs continuously, regardless of whether or not there is light (day and night). Humans continuously utilize respiration as the source of energy for daily activities.

$$C_6H_{12}O_6 + 6\,O_2 \longrightarrow 6\,CO_2 + 6\,H_2O + Energy \qquad (6\text{-}3)$$

This is the energy that ultimately, through the food chain, passes on to other living organisms to sustain life. The **simple food chain** starts from plants (producers) and passes on to other animals (consumers) through three or more stages called *trophic (nutritional)* levels. At the base of the food pyramid are autotrophs, which are responsible for the primary food production. *Autotrophs* (self-feeders or producers) are those organisms, such as green plants, algae, and certain bacteria that convert inorganic compounds into energy-rich organic compounds, mostly by photosynthesis.[7] *Heterotrophs* (other-feeders) are those organisms that cannot produce their own food, but rather feed on other organisms. Trees and most other plants are autotrophs. All animals, including humans, are heterotrophs, as are many microscopic organisms. Heterotrophs can be

[7] Biomass may also be produced without sunlight through the utilization of certain inorganic compounds called chemo-autotrophs.

divided into primary consumers or grazers (herbivores), colloquially referred to as vegetarians, and secondary consumers (carnivores), or meat-eaters. A simple three-level food chain is grass→sheep→wolf. Plant→insect→ frog→man is an example of a simple four-level food chain. Things are not always that simple in nature, however, as there are multiple types of prey for a single predator. Each link in the food chain can itself be linked with many other food chains, making a complex set of feeding relations which is more accurately called a **food web**.

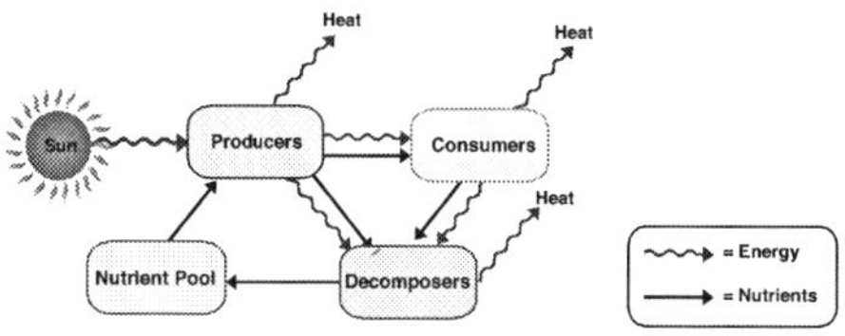

Figure 6-2
The flow of nutrients and energy through biological systems. Note that while nutrients are recycled over and over again, the sun must continuously supply energy.
[Adapted from Masters, G. M.,"Introduction to Environmental Science and Technology," John Wiley & Sons, 1974.]

Although nutrients undergo cyclic processes, energy is not cyclical (See Figure 6-2). Organisms use energy as fuel to help their movements and growth. In this fueling process, some energy is discarded as waste heat. To keep the system operating, as energy degrades, the sun must replenish it.

Energy Flow through the Biosphere

As was discussed above, photosynthesis is not only responsible for providing energy for a plant's own growth, but must also provide food for other consumers as they are passed through one trophic level to the next. The **photosynthetic capacity**, i.e., the rate at which solar energy is fixed by plant photosynthesis varies with temperature, rainfall and the biochemistry of the plant. The rate of **gross production** is therefore dependent not only on the amount of sunlight, the nutrient supply, and the stability of the surface water in an area, but also to geography. Overall, it is estimated that a mere 1% of the incident solar energy is responsible for the production of all plants, and only a portion of this energy is used for a plant's own respiration. The remainder is used by consumers and decomposers and is referred to as the **net production**. Additional losses cause efficiency to drop at each level of consumption beginning with plants; as a result, only ten percent of the energy is converted from one trophic level to the next (See box "Rule of 10"). For example, only 10% of the energy contained in plants is used to warm an herbivore and make it grow. The rest is excreted as waste. Similarly, our body uses only 10% of the energy we take in as food (plants or meat) for our various needs.

> **Question**: Plenty of sunshine is often treated as a necessary condition for growing plants. Does plant production increase with light intensity?
> **Answer**: Yes, to a certain extent. As light intensity increases, the chlorophyll traps more light, but as traps are filled, additional light will not be useful and may actually harm the plant.

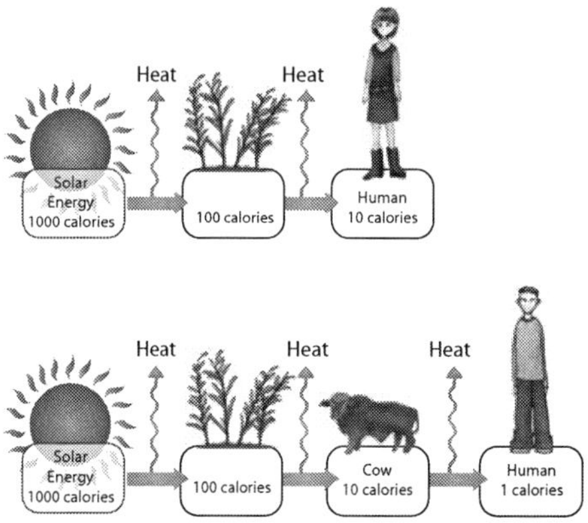

Figure 6-3
Energy flow in vegetarian and non-vegetarian diets.
[Adapted from http://Humboldt.earthsave.org.]

The demand for food production increases because of two factors: the increase in population and change to a more affluent diet. At the present rate of growth, it is estimated that by mid-century, the world population will increase to between 8 and 11 billion before gradually stabilizing. Affluent diets enriched with animal fat, meat, dairy, and eggs use up as much as three times more biomass than vegetarian diets (Figure 6-3).

Rule of 10*

Digging Deeper ...

Humans extract about 95 percent of the potential (chemical) energy available in food, but convert only about 10 percent of this energy into useful work, such as lifting weights and building muscle tissue. The rest is turned into heat and dispersed into the environment. Every time any organism eats another, some of the energy stored in the prey is released to the predator. Each of these storage steps along a food chain is called a *trophic level*. Estimates of energy loss between trophic levels vary widely but, on average, only about 10 percent of the energy fixed (captured and stored) by plants is ultimately stored by herbivores. Only about 10 percent of the energy that herbivores accumulate ends up being stored in the living tissues of the carnivores that eat them; and only 10 percent of that energy is successfully converted into living tissue by carnivores on the third trophic level. This pattern of energy loss is known as the ecological **rule of 10**.

Inefficient energy chains create the so-called *ecological pyramid*, in which each trophic level contains only one-tenth as much living tissue as the level beneath it. For instance, consider a person who decides to eat only red meat to gain one pound in weight. That person must then eat 10 pounds of beef. The cattle that produced the 10 pounds of flesh must have originally eaten about 100 pounds of fodder. That's why it makes good ecological sense that the larger animals, such as elephants, are vegetarians. By feeding at or near the base of the ecological pyramid, the larger animals make much more efficient use of energy, and the given ecosystem can support many more of them.

Of course, the ultimate source of energy on earth is the sun. Through photosynthesis, plants use most of energy they get from sunlight to grow roots, pump water, and so on. As a result, they convert only about one percent of the sun's energy into chemical energy, which is then stored in their molecules. Evidently, this is enough to provide for the whole planet, and even keeps gardeners busy.

* Excerpts from a physics text currently under preparation jointly by this author and Professor Igor Glozman, Department of Physics, Highline Community College, Des Moines, Washington 98198.

Raising domestic animals and livestock such as cows, sheep, goats, pigs, horses, and camels was a major factor that led to the development of ancient societies. These animals not only provided meat, milk, and milk products such as butter, cheese, and yogurt, but also fertilizer and plowing power needed to grow various plants. In fact, these mammals yielded several times more energy over their lifetime than if they were slaughtered and consumed as meat.[8]

Question: In India, cattle are considered sacred and cannot be slaughtered for food. How does this practice contribute to the food shortage in India?
Answer: Although on the surface this practice may appear unjustifiable, it is actually an effective way to combat hunger. Cattle are fed mostly with grass and crop wastes, but provide milk and other dairy products, which are sources of high-quality protein. Besides producing food, cow dung is used as fuel and a source of high-quality fertilizer.

Question: What are the consequences of vegetarian and non-vegetarian diets on total energy consumption?
Answer: Raising livestock for meat is a very inefficient way of generating food. For example, it takes far more resources (fuel, water,

[8] Diamond, J., Guns, Germs, and Steel: The Fate of Human Societies, W. W. Norton and Company, 1999. p. 88

etc.) to produce a kilogram of meat than a kilogram of corn. Adding in all the energy used for transportation, feedlot, and storage, it takes the equivalent of one gallon of gasoline to produce one pound of grain-fed beef. In other words, to provide the yearly average beef consumption in an American family, we require over 260 gallons of fossil fuel.[9]

The rate of food production varies widely among developed and developing countries. Modern agricultural machinery, better fertilizers, and more complex irrigation techniques have allowed developed countries to produce food at a rate exceeding the rate of their population increase. Unfortunately, the same cannot be said for developing and underdeveloped countries. In these countries, although yearly food production has remained relatively constant, the population has been steadily increasing. As a result, the amount of food per capita has been decreasing dramatically, and many countries have faced severe food shortages and even famine. Furthermore, the food surplus in the richer countries that was traditionally exported to the poorer countries is now being used either to feed the increasing appetite of the local population or is exported to other developing countries. At the same time, much of the high-quality food produced in less-developed countries is being exported to industrial countries. For example, for the last few years, the United States has been the world's largest importer of beef and fish while Latin America has been the major exporter of this same products.[10]

Units

Food energy is usually expressed in Calories. One Calorie (with a capital C) is equal to the heat energy that is required to raise the temperature of one kilogram of water by one degree centigrade. A smaller unit of thermal energy is a calorie (with a small c), which is 1/1000th of the food calorie:

$$1\ Calorie = 1\ kilocalorie = 1{,}000\ calories = 4{,}184\ joules$$

Unfortunately, the calorie notation can cause some confusion. What should be remembered is that when we talk about a food calorie, we are talking about kilocalories, whether capitalized or not. The same is true when we talk about weight; we usually mean mass, especially when we are using terms such as "weight loss."

Energy of Food Production

Food production has become less and less efficient as we have replaced the traditional pre-industrial non-mechanized practices with modern agricultural technologies and farming practices. Food production efficiency defined as the ratio of energy output (energy content of foods) to the energy input (energy used to produce, process, package, and

[9] Rifkin, J., Beyond Beef: The Rise and Fall of the Cattle Culture, Penguin Press, N.Y., 1992.

[10] United Nations Food and Agricultural Organization (*http://www.hsus.org/ace/352*).

transport food) has decreased from roughly 100 in the pre-industrial era to less than 1 today.[11]

The reason for this huge loss in efficiency is that much of the commercial production and delivery of our food on all stages (planting, irrigation, feeding and harvesting, processing, packaging and distribution) depends heavily on oil as the primary source of energy. Fossil fuel is important in cultivating the land, planting seeds, manufacturing fertilizers and pesticides, irrigating, and harvesting. Fossil fuel is also needed in the construction of roads and the transportation of farm workers and food. As a result, only one in every 7-11 calories of energy is available through food. Efficiencies of around 15% in Europe and 9% in the United States have been reported.[12]

Metabolism

It takes energy to stay alive. Whether we are eating or participating in rigorous exercise, we metabolize food as a source of energy. Even when we are not doing any work or are sleeping, we consume food energy. The minimum energy needed for survival and to maintain equilibrium of all vital functions (nervous, cardiovascular, respiratory, and digestive systems) when a body is at rest (sedentary) is called the *basal metabolic rate* (BMR). The brain and the liver, two organs which jointly make up only 4% of body weight, are responsible for half of all metabolic activity. As chemical energy in food converts to heat, body temperature tends to rise. To maintain a constant temperature, the body reacts by transferring the energy to the circulatory system and to the skin, where it eventually dissipates into the environment. BMR varies with sex, body size, general health, and age, and is generally higher in males and for heavier, healthier, and younger animals. BMR is about 1.45 watts for a rat, 80 watts for an average sized person, and 266 watts for a medium size cow.[13]

Table 6-1. Energy Expenditure (above basal)*

Activity	*Cal/h*	*W*
Sleeping	60	70
Eating, Reading	80	93
Sitting, standing , office work	100	116
Walking (average speed)	180	210
Walking downstairs	205	240
Walking upstairs	500	580
Golf	140	163
Jogging	400	465
Running	600	700
Tennis	400	465
Dancing (ballroom)	300	350
Dancing (disco)	430	500
Skiing (downhill)	300	350
Skiing (cross county)	600	700
Bicycling (20 km/hr)	450	520
Horseback riding, trotting	450	520
Horse racing, galloping	550	640
Boxing (sparring)	620	720

*For an average person (70 kg)

BMR does not account for any physical activity, so additional energy is needed to carry out daily tasks (See Table 6-1). Physical activities can be divided into aerobic and non-aerobic activities. During aerobic activities, oxygen breaks down carbohydrates, fat, and protein and converts them to energy. Examples of aerobic exercises are dancing, jogging, swimming, and biking. Anaerobic activities burn carbohydrates without oxygen with maximum bursts of energy of short duration. Examples are weight lifting, pushups, chin ups, and sprinting. The amount of daily energy needed is different for different people and can vary with gender, weight (mass), and levels of physical and mental activity. As a rule of thumb, it is usually assumed that an average man requires 2,500 Calories, whereas an average woman needs only 1,800 Calories per day in food intake.

Example 6-1: What is the minimum number of calories that an

[11] Green, B. M., Eating Oil – Energy Use in Food Production, Westview Press, Boulder, Colorado, 1978.
[12] Günther, F., "Fossil Energy and Food Security," *Energy & Environment*, Volume 12, Number 4, July 2001, pp. 253-272.
[13] Faughn, J., Life Science Applications for Physics, Harcourt, 1998.

Metabolism and Animal Kingdom

FYI ...

Unlike their cold-blooded neighbors, warm-blooded animals must always maintain a constant body temperature, irrespective of the temperature of the environment. In cold climates, they fight against losing too much heat; in hot climates, they try to stay cool by dissipating excess heat. The rate at which heat is generated in and lost from an animal's body depends very much on the size of the animal. Larger animals have more mass to "feed" and therefore generate more metabolic heat. Larger animals also have more surface area from which heat is lost. While it would seem that the larger animals have to contend with a greater heat loss, it is the specific heat loss (i.e., heat loss per unit body mass) that is more relevant to the animal's survival. Since mass is proportional to L^3 and surface area is proportional to L^2, the specific heat loss is actually proportional to 1/L (or $M^{-1/3}$). In other word, as the metabolic rate is lost through the body surface, it is logical to expect that an organism will adjust its BMR to overcome heat losses through the surface (skin) proportional to $M^{2/3}$. A similar relationship has also been found for plants.

Amazingly, we can actually predict (or at least rationalize) certain patterns of behavior of different animals on the basis of simple scaling arguments. A very small animal must compensate for its lost body heat by an almost continual intake of food. A mouse eats a food equivalent of about one-quarter of its body weight daily. The tiny shrew will die of starvation if forced to go without food for more than about 3 hours. A giant elephant, on the other hand, faces the opposite problem of getting rid of excess heat. It's no wonder that elephants take every opportunity to cool themselves off at waterholes. We can therefore understand why a mouse might want to be in a warm spot, while an elephant in a cool one. For insects, the surface to volume ratio is so much larger than that of any warm-blooded animal that they cannot possibly eat fast enough to maintain a constant body temperature. What to do then? Fortunately for the insects, the body temperature of a cold-blooded creature matches that of the surroundings. This provision of nature greatly reduces heat loss and consequently the insect's food requirements.

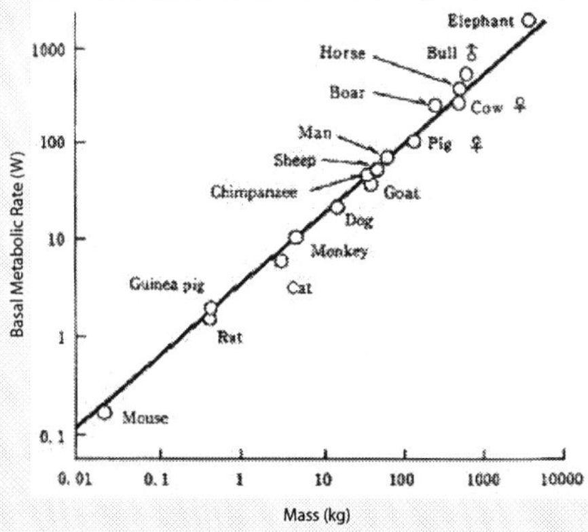

average person needs to barely stay alive?

Solution: Assuming BMR of 80 W, the daily caloric intake must be at least

$$80\ W = 80\ \frac{J}{s} x \frac{1\ kcal}{4.1868\ kJ} x \frac{1\ kJ}{1000\ J} x \frac{24 \times 3600\ s}{day} = 1650\ kcal/day$$

Question: How much weight does a starving male (zero food intake) lose in one week?

Answer: Assuming an average person requires 2,500 kilo-calories of food to meet his daily energy needs, and that each gram of fat metabolizes 9 Calories (37.6 MJ/kg), he burns 2,500/9=277 grams of his body fat each day (1.9 kilograms in one week). It is no surprise that people are known to survive many weeks without food.[14]

Question: Which one has a higher BMR, a tall and thin person or a short and fat person?

Answer: For the same mass, tall thin persons have a greater skin surface area and thus lose heat to the environment faster. They should have a higher BMR to maintain the equilibrium body temperature.

Question: Which organism has a higher basal metabolic rate, a

[14] They still need to drink water to prevent dehydration that may occur in a few days.

The Origin of Life

Digging Deeper ...

The simplest living organisms were developed about 3.5-4.0 billion years ago from organic molecules produced by chemical reactions in an atmosphere consisting principally of water vapor, carbon dioxide, ammonia, and methane, and devoid of oxygen (anaerobic), fueled by ultraviolet light from the sun. In addition, a substantial amount of organic molecules are believed to have been brought to earth by impacts from asteroids, comets, and meteors common during the early stages of earth's formation. There is strong evidence that the chemical reactions near deep-sea vents have also contributed. The minerals in the early primordial soup facilitated the coalescence of these molecules to form longer RNA (*ribonucleic acid*) strands that could replicate themselves. Some strands of the RNA molecules eventually combined to form DNA (*deoxyribonucleic acid*), the building block of what we know as life today.

The first microorganisms were chemoautotrophs that had access to very few enzymes and obtained their energy from inorganic molecules (such as those near hydrothermal vents and hot springs). As life evolved, other mechanisms, mainly photosynthesis, played more important roles. Early photosynthesis involved hydrogen sulfide, not water, and therefore produced no oxygen. Photosynthesis involving water evolved later, producing oxygen that was built up over the years in earth's atmosphere. In the beginning, oxygen reacted immediately with other organic and inorganic material and no molecules were released into the atmosphere; in the process, many early life forms became extinct. It took at least one billion years before oxygen concentrations rose to significant levels. As a result of the new chemistry, two types of molecules were formed: simpler single-organism prokaryotic molecules lacking a cellular structure (bacteria) and more complex *eukaryotic* cells (which ultimately led to the creation of a cell nucleus). These cells proved to be the key in the manufacturing of more complex molecules called adenosine *triphosphate* (ATP), which captured and transferred the energy that cells need for building all cellular components. Cells can be viewed as tiny chemical factories which provide processes needed to make complex organic molecules needed by all living organisms. With time, the cellular structures became more and more complex, eventually leading to multi-cellular species such as plants and, eventually, animals.

Early plants and animals lived in oceans where there were plenty of nutrients and they were protected from harmful ultraviolet radiation. As atmospheric oxygen converted to ozone, plants and animal could increasingly survive land environment and life as we know it today flourished.

hummingbird or an elephant?

Answer: Hummingbirds are very small birds, weighing approximately 2.5-4.5 grams. Because small birds have proportionately larger surfaces in relation to their body mass, they can lose heat faster and therefore have higher metabolic rates. Hummingbirds have the highest metabolic rate of any animal, roughly 12 times that of a human being and 100 times that of an elephant. Their BMR is around 29 W, which means they need to consume 600 Calories of food every day. No wonder each day they have to visit hundreds of flowers to gather enough nectar to survive.

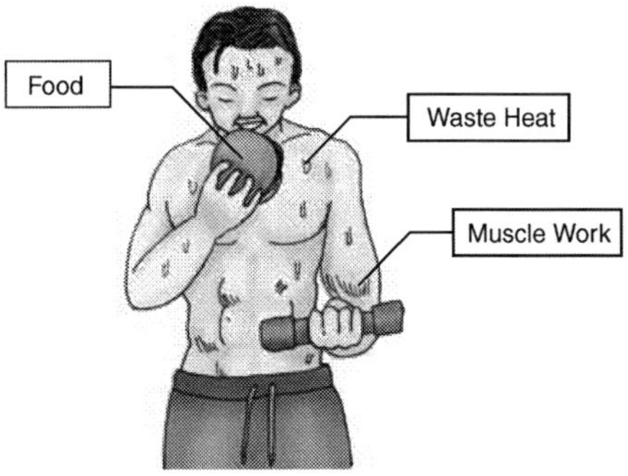

Figure 6-4
Human body is a heat engine, taking input energy in the form of food, converting a part of it into muscle work, and rejecting the waste heat as sweat and other excrements.

The Human Heat Engine

The human body functions like an engine, converting part of the food's chemical energy into useful mechanical energy when muscle cells carry out physical work, and dumping the rest into the environment (Figure 6-4). The conversion efficiency is about 17%, that is, only about one in every six units of energy in the food we eat is converted to muscle work. What remains is used as internal energy in our muscles, essential to maintaining our body temperature (through sweating and other forms of heat losses collectively called waste heat) or stored in body in form of fat . Metabolic efficiency is not the same for all people, and can vary between 15-25%

depending on gender, weight, condition of health, athletic abilities, and age. Although maximum power can be quite high, average power output is not and is limited to around 20 W of mechanical energy per kilogram of body mass. By combining the food production efficiency (~10-15%) and metabolic efficiency (~15-25%), a net figure for the human power efficiency is calculated at around 1.5-3.75%. In other words, on average, each unit of mechanical (muscle) energy comes at the expense of 17 to 66 units of primary (fossil) energy.

Example 6-2: Can we harness waste energy from the human body to perform work? What is the maximum power that can be utilized using this energy?
Solution: Assuming ambient temperature of 25°C, the Carnot efficiency is

$$\eta = 1-\frac{298}{310} = 3.87\ \%$$

where waste is taken to be at the body temperature of 37°C. While sitting, an average person can expend 116 W, of which 4.5 W is available in the form of useful power.

Example 6-3: Compare the cost of power generated by human muscles to that from nuclear or coal power plants.
Solution: An average human consumes 2,500 Calories of food every day. Each Calorie is equal to 1,000 calories or 4,180 joules, and each kilowatt-hour is $3.6\text{x}10^6$ joule, so 2,500 kcal = 10,450 kJ = 2.9 kWh

This amount of energy can be supplied by about 2 pounds of steak and 10 slices of bread for an average cost of $10 (or $300 a month for food cost). The energy cost is therefore estimated at around $3.45/kWh. The cost of electricity being charged to American consumers varies between 8-15 cents per kilowatt-hour, which is an order of magnitude cheaper than the cost of energy delivered by humans. No wonder machines are rapidly replacing people whenever possible.

Food, exercise, and dieting

The energy required by our body is provided by the food we eat. The major food categories are carbohydrates (mainly sugar, bread, and rice), proteins (primarily meat, milk, and eggs), and fats. In addition, most plant-based foods contain other nutrients like vitamins, minerals, and water which are drawn up from their roots during growth. Although they are small in quantity, these are essential ingredients in metabolizing the calories stored in food. The human body metabolizes these foods differently. While each gram of fat yields 9.5 kilocalories when burned, carbohydrates and proteins yield only 4.3 and 5.3 kcal of energy, respectively. Carbohydrates and proteins take some time to digest into simple sugars, whereas sugar is instantly metabolized. The energy contents (calorific values) of some

Food: The Facts
Did You Know That ...?
• It takes 127 calories of energy (aviation fuel) to transport one calorie of lettuce across Atlantic, 97 calories to transport 1 calorie of asparagus from Chile, and 66 units of energy to fly one calorie of carrot from South Africa. • Every year around 20 million people die as a result of hunger or malnutrition. • It takes 10,000 kilocalories of food to supply the roughly 3,000 kilocalories consumed per day by an average American. • It takes about 5.5 kilograms of protein in the feed to produce 1 kilogram of poultry and about 8 kilogram to produce one kilogram of pork.

common foods are given in Table 6-2.

Table 6-2. Food Calories (all values are averaged)

Food	*kcal*
Apple, large	100
Bacon, 2 strips	100
Banana, small	90
Beer, one glass	150
Bread and butter (one slice)	80
Butter (one teaspoon)	36
Cake, a slice	350
Chicken, fried (1/2 breast)	230
Cheeseburger	350
Cookie, chocolate chip	50
Coke, one glass	110
Doughnut	150
Egg (one)	80
French fries (regular)	250
Ice cream (one scoop)	300
Steak (1/4-lb)	200
Mayonnaise (1 tbsp)	92
Milk, one glass	166
Milk shake	420
Orange Juice (one glass)	120
Pizza, cheese, one slice	180
Potato chips, one serving	110
Hamburger	275
Spaghetti, one serving	400

Physiologists and physicians have studied, but differ in opinion on the root causes of weight gains by people. Some attribute the propensity to gain weight to a special gene and hormonal imbalance. Others believe bodies develop fat cells during childhood that become active as the body ages. Still others attribute obesity solely to overeating. No matter what the actual medical reason for weight gain, to maintain our weight, the total energy intake by food must balance the energy expenditure of normal basal metabolic rate and that expended by work and other physical activities. If the energy intake exceeds the energy expenditure, the excess energy is stored as fat. As a rule of thumb, we need to burn roughly 3,500 extra Calories to lose one pound of fat.

Example 6-4: According to one study[15], since the 1960s, the average individual living in the United States has increased caloric intake by about 250 Calories each day. How much weight would one gain by staying on the new diet?
Solution: The total excess intake of energy over one year is: 250*365 = 91,250 kcal. The added weight is 91,250/3,500 = 26 lbs (11.8 kg).

Example 6-5: A 50-kg woman is jumping rope at a rate of 50 times a minute for 15 minutes. Each jump on average raises the center of mass 0.5 m. How much energy does this woman consume?
Solution: The work performed per jump is W = mgh = 50x9.8x0.5 = 245 J. Total work performed is
(15 minutes)x(50 jumps/minute)x(245 J/jump) = 183,750 J = 44 kilocalories.

Example 6-6: A 110-lb (50 kg) woman pedals a stationary bike for 20 minutes at an average speed of 20 km/h (12.5 mph). Calculate power consumption rate, total Calories burned and the metabolic equivalent (METs). 1 MET is defined as the energy expenditure rate in kcal per hr per kilogram of body mass.
Solution: Referring to Table 6-1, energy expenditure rate (power) is given as 450 kcal/hr or 520 W. Total calories burned is E = P.t = 450 kcal/hr x 20 min = 150 kcal. The metabolic equivalent is calculated as

[15] Thompson, D. J., et al., "Lifetime health and economic consequences of obesity," *Archives of Internal Medicine*, 159:2177-83 (1999).

the power (450 kcal/hr) divided by her weight (50 kg), or 9 METs.

EX 6.6

Example 6-7: It is experimentally found that in the photosynthetic reaction given by Equation 6-1, for each mole of carbohydrate (CH_2O), 112 kcal of light energy is needed. What is the amount of energy supplied to our body through the consumption of one 12-oz cola containing 36 grams of sugar (glucose)?
Solution: The respiration reaction is the inverse of equation 5-1; for each mole of glucose (180 g) we consume 6x112 = 672 kcal (Calories) of energy is taken in. The soda contains 36x672/180 = 134 Calories.

Example 6-8: How many stairs should a 60-kg woman climb to burn off a chocolate chip cookie that has 100 Calories?
Solution: To maintain her weight, the woman needs to burn off 100 Calories. Assuming there are no other losses (friction from shoes and air currents, for example), this energy must be compensated for by the work that she does against gravity (i.e. gain in the potential energy), or 100 kcalx4.18 kJ/kcal = 418 kJ. The height she needs to climb is h = E/mg = 418,000/(60x9.81) = 710 m. Assuming each step is 20 cm in height, she must climb up 710/0.20 = 3,550 steps! In reality, the human body is only about 20% efficient at converting chemical energy into gravitational potential energy. In other words, 80% of the metabolic energy generated goes into heat. Consequently, the number of steps is really smaller by a factor of 5, but it still is a pretty big number, something to think about before you take another cookie.

Example 6-9: A 90-kg (200-lb) man has a cheeseburger, regular fries, and a glass of beer for lunch and a milk shake for dessert. How long would it take him to burn off the calories, if (a) he returns to continue his routine office work or (b) he goes to a gym and bicycles at a moderate speed?
Solution: The total caloric intake can be calculated using data in Table 6-2, as 350+150+250+420 = 1,170 Calories. An average person (70-kg) can burn off calories at the rate of 100 Calories/hr by sitting in his office or 450 Calories/hr by bicycling. Since the rate of energy consumption is higher for a heavier person, the 90-kg man burns calories faster at the rate of 100x90/70 = 129 Calories/hr doing office work, and 450*90/70 = 579 Calories/hr bicycling. The time

Solid Wastes in the United States: The Facts

Did You Know That ...?

- The United States generates 200 million tons of solid waste a year-- about 3 tons for every person, twice that of an average European.
- Every year we dispose of 6 billion plastic diapers and 200 million pounds of plastic soda bottles.
- The garbage from a typical American household consists of roughly 66% paper and other organics and 24% glass, metal, and plastic. The remaining 10% are rubber, textile, and other wastes.
- McDonald's produces over 34,000 tons of polystyrene cups, boxes and trays annually.

needed is 1170/129 = 9 hours of office work, or 1170/579 = 2 hours of bicycling. Clearly, the best way to lose the weight is by eating less. Exercising has many health benefits, but not necessarily the most effective for losing weight.

Example 6-10: In order to maintain constant weight, we must consume about 2,000 kilocalories of food energy per day of inactivity. How long could you live off your fat?
Solution: In two days of starvation and inactivity one would be able to lose about 4,000 kcal or about one pound of fat. On the average, 15% of an adult male's weight is in the form of body fat, and 22% for the adult female. Therefore, assuming about 30 pounds per person, we could, in principle, live for about two months (30x4100/2000 = 61.5 days) simply by burning off our own fat.

Example 6-11: Suppose a 65-kg person spends 8 hours sleeping, 1 hour performing moderate physical labor, 4 hours engaging in light activity, and 11 hours working at a desk or relaxing. Is this person likely to gain weight if he maintains a 3500-Cal daily diet?
Solution: Referring to table, we calculate the energy expenditure to be (8x70+1x460+4x230+11x115) x 3600 = 11.5 MJ = 2800 Cal/day. Since this person is taking in 3500 Cal and burning only 2,800 Cal, he will gain weight.

Among health care professionals, perhaps the best known method for assessing body size is the **Body Mass Index** (See box "Body Mass Index"). Other factors, such as the waist-to-hip ratio, may also be important in determining the ideal weight for a healthy person.

Power Generation by Biomass

Biomass is not only the source of our food, but also plays an important role in meeting our energy demands as a fuel. Wood can be burned directly in fireplaces or used in boilers to produce steam. Some developing countries,

Body Mass Index

FYI ...

Body mass index (BMI) is defined as body mass divided by height squared. It is an empirical formula defined by the National Institute of Health (NIH) to measure obesity and provide a guideline for weight control.

$$BMI = \frac{M}{h^2} \left[\frac{kg}{m^2}\right] \qquad \text{(i)}$$

According to this scale, adults with a BMI between 19 and 25 are considered to have a healthy weight, those below 19 are underweight, between 25 and 30 are overweight, and above 30 are considered obese. For easy reference, the recommended weights for adults of different heights are tabulated in metric and US units.

Height, cm	Healthy Mass, kg	Height, feet	Healthy Mass, lbs
150	43-56	5' 0"	97-128
155	46-60	5' 2"	104-136
160	49-64	5' 4"	110-145
165	52-68	5' 6"	117-155
170	55-72	5' 8"	125-164
175	58-77	5' 10"	132-174
180	62-81	6' 0"	140-184

such as India, use animal dung for heating and cooking purposes. In more developed countries, special plants are cultivated to be used specifically as biomass fuel. These include herbaceous plants (such as sugarcane, corn, cotton, sorghum, and bamboo), aquatic plants (such as algae and seaweeds), and crop residues (wheat and rice straws, and dead and dying trees, stalks, leaves, and cobs). The most suitable plants are perennial plants such as switch grass, poplar, willow, maple, and sycamore. These plants need little fertilizer or pesticide, and can help to reduce soil erosion.

Figure 6-5
Municipal solid wastes constitute a major source of energy.
Source: National Renewable Energy Laboratory, Photographic Information Exchange

Another source of biomass fuel is municipal solid waste (trash) consisting of everyday items such as product packaging, paint, yard waste, bottles, cans, newspaper, rubber, leather, textiles and other household items.[16] In 2001, the US produced 229 million tons of waste, divided equally among residential, commercial, and construction sectors (Figure 6-5). This represented a 2.6-fold increase since 1960, equivalent to 2.2 kg per person per day. Of all the garbage collected in the United States, roughly three-quarters is biodegradable and can be considered biomass. Over 72% of this is buried in landfills, about 15% is incinerated, and the rest is recycled or reused (Figure 6-6).

Direct Burning (Incineration)

Biomass can be burned in incinerators for heat, mixed with coal and used directly for heating, or converted to electricity in power plants. Conventional incinerators directly burn biomass to heat water and generate steam that runs a steam turbine. In modern incinerators, biomass is first converted into **producer gas**, which can then be burned inside gas turbines. In some instances, the waste heat from the gas turbine may be used to drive a secondary steam turbine, converting more of the fuel energy into electricity in what are commonly referred to as **combined heat and power (CHP)** units. Finally, the waste heat from these secondary plants is hot enough for hot-water and space heating that would otherwise be lost to the atmosphere. According to the US Department of Energy, domestic capacity for biomass generation utilizing combustion could reach 20-30 GW by the year 2020.

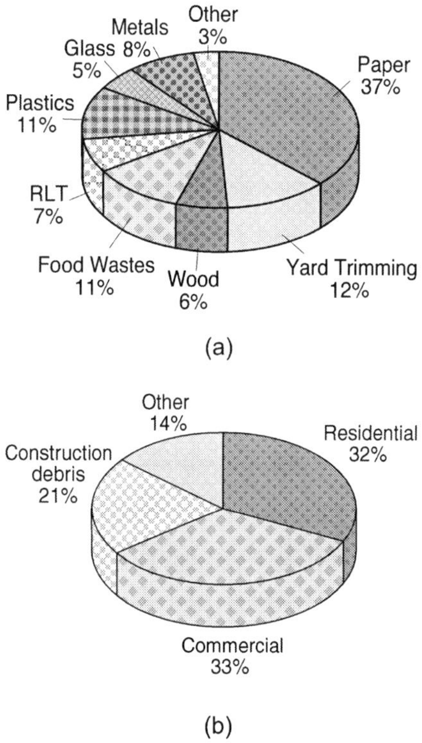

Figure 6-6
Composition of municipal solid waste in the U.S, (a) by type, and (b) by sector. RLT refers to rubber, leather, and textiles.
Source: US EPA 2001 Facts and Figures. http://www.epa.gov/garbage/pubs/msw2001.pdf.

About four-fifths of all biomass energy used today comes from burning wood and wood scraps. The remainder is from crops, garbage, landfill, and alcohol fuel. Materials used for direct incineration are usually scrap wood and sawdust produced by timber companies, solid wastes and garbage, and landfill gases. Most energy generated by biomass fuel is used locally.

Burning trash has the advantage of producing electricity and heat, and also reducing the volume of trash. The main problem with incinerators is that many of the pollutants associated with combustion remain and are released to the atmosphere. Some newer technologies burn these materials

[16] The word "municipal" implies anything that is operated and controlled by elected local city or county officials.

Biomass from Animal Waste

FYI ...

How much electricity can be utilized from the average daily droppings of the three most common farm animals?
Answer:

- Cow 3.0 kWh
- Pig 0.2 kWh
- Chicken 0.012 kWh

at much higher plasma temperatures. Plasma is gas heated (usually by passing strong electric currents through it) to temperatures around 10,000 °C, causing electrons to be stripped off the atoms. At these temperatures, hydrocarbons, PCBs, and other toxins break down, allowing them to be burned completely. The residues are mixed with soil and harden into inert and harmless rocks suitable for road gravel. Incineration accounts for 10-15% of all garbage generated in the US and for only a tiny one-tenth of one percent of all electricity supplies.[17] Biomass has lower calorific values as compared to fossil fuels. Table 6-3 compares the calorific (heating value) of several biomass fuels with fossil fuels.

Table 6-3. Calorific values for different biomass fuels

Source	*Heating Value* (MJ/kg)*
Sugar cane	21.5
Municipal solid waste	13.0
Garbage	19.7
Newspaper	19.7
Hardwood	20.4
Pinewood	21.3
Grass	18.5
Methane	49.8
Methanol	20.0
Ethanol	27.0

* Lower heating value

Thermochemical Conversion

In thermochemical conversion (also called *pyrolysis* or *gasification*) dry biomass is decomposed to simpler molecules (gaseous methane and liquid methanol) by high heat. No or little oxygen is used in the process, and no chemical reactions are involved. The methane can be collected and sold by natural gas utilities; methanol (also called wood alcohol) is of great interest in reformers to produce hydrogen for fuel cells or used directly in alternative fuel vehicles. Coal or other carbonaceous materials can be gasified to produce hydrogen, synfuel, and other valuable products (See Chapter 14).

Biochemical Conversion

Biochemical conversion could be either anaerobic or aerobic. In **anaerobic conversion**, biomass is put in anaerobic digesters, where it is exposed to microorganisms in the absence of oxygen and allowed to decay. The oldest form of biochemical conversion is the *fermentation* of grapes, corn, and barley by microscopic yeast to produce wine, beer, and other alcoholic beverages. The same process can be utilized to produce ethanol (also called grain alcohol) and other synthetic fuels. High-moisture herbaceous plants and marine crops are most suitable for biological digestion. Fuel alcohol is produced by cooking the biomass and converting the starch into sugar. The sugar is then allowed to ferment. Ethanol can be removed by distillation and can be used directly, or be blended with gasoline and used as fuel in internal combustion engines. New processes use enzymes to break down the cellulose part of the plants, allowing the entire plant (and not just the starch) to be utilized.

[17] Office of Technology Assessment, "From Pollution to Prevention: A Progress Report on Waste Reduction," Washington, D.C., 1987.

Ethanol Enigma

Digging Deeper ...

Ethanol is formed by fermenting sugar in grains such as corn, wheat, and rice, in much the same way that beer, wine, and liquors are made. The process is not new; in fact it has been around for many centuries. Before the Civil War, ethanol was used in the United States mainly as a lantern fuel and as a lubricant. In order to finance the war, US president Abraham Lincoln imposed a tax on liquors. Since ethanol was produced by fermentation of grains, it was taxed heavily and people turned to kerosene, which was poisonous and not suitable for drinking. The "Spirit Tax" was eventually repealed in 1906, and ethanol became an ideal fuel to run internal combustion engine once again. Fuel ethanol made a partial comeback, but the introduction of Prohibition in 1919 killed any chances of ethanol becoming a major fuel source. To make combustion more uniform, ethanol continued to be used as an additive, however. To increase market share, oil companies opted for lead additives, which soon proved to be a major health concern.

In 1933, Prohibition ended and ethanol was put to new uses in making synthetic rubber and the production of ethanol soared again. A decade later, as WWII came to an end, grains found new market and ethanol production was curtailed one more time. Lead continued to be the primary additive for many years until the increasing level of emission from automobile exhaust forced new regulations that resulted in development of catalytic reactors, which were poisoned by the lead additives in gasoline. Even after lead was removed from gasoline, oil companies chose aromatics like benzene, toluene, and MTBE. Benzene was eventually phased out in 1990s and many states are now passing legislations to ban MTBE.

Ethanol, as an alternative fuel or as an additive, has generated renewed interest. As we will show in Chapter 14, ethanol has a higher octane than gasoline and, because it is oxygenated, burns cleaner. The effect on production of the greenhouse gases is less certain. It was commonly presumed that, since ethanol is a biomass, no net greenhouse gases are produced. Recent studies, however, rebuke this claim on ground that to produce biofuels, virgin forests and grasslands must be cleared. About 2.7 times more carbon is stored in terrestrial soils and plant material than in the atmosphere, and this carbon is released when these areas are cleared (often by burning) and the soil is tilled. Cellulosic and sugar-based ethanol are considered best. The United States, under pressure from corn-producing states uses corn as source of its biofuel. Corn farming requires more land andonly yields 1/4 more energy out than the fossil energy you have to put in.[i]

In addition, serious environmental problems, such as higher food prices, soil erosion and nutrient leaching, and substantial demand on the world's land and water resources may not warrant production of biofuels in a large scale.

[i] *From Proceeding of the National Academy of Science of the USA, July 2006.*

The main source of bioethanol is sugar-containing plants, corn in the United States and sugar cane in Brazil. Sugar from these plants is fermented by yeast and bacteria to reduce carbohydrate to ethanol and carbon dioxide according to:

$$C_6H_{12}O_6 \longrightarrow 2\ C_2H_5OH + 2\ CO_2$$

Some energy of course is needed to produce the plant, to harvest it and ferment it into a biofuel. The ratio of the amount of energy produced when biofuel is burned to the amount of energy used to make the biofuel is the fossil energy replacement ratio (FER). FER is around 1.2-1.4 for corn ethanol, and about 8 for sugarcane. Therefore, it can be argued that not only can corn ethanol be economically viable substitute for fossil fuels, its impact on the production of greenhouse gases would be minimal.

Another important byproduct of anaerobic digestion is biogas. **Biogas** is a mixture of methane, carbon monoxide, and carbon dioxide gas which occurs naturally in the bottom of swamps, marshes and landfills, or is produced by the fermentation of human and animal wastes. Biogas is a well-known fuel for cooking and lighting in a number of countries. Biogas can be converted to liquid biofuels, which can potentially replace petroleum fuels.

Biochemical conversion is also possible in presence of air or oxygen *(aerobic conversion)*. These processes occur at much higher temperatures, and generally do not produce appreciable amount of useful fuel gases.

Summary

Biomass is a result of photosynthetic activities where carbon dioxide and water, in the presence of sunlight, combine to form carbohydrates. The conversion efficiency is rather low, however; only 0.3% of the incoming solar energy is converted to the chemical energy stored in plants. In contrast, as we will see in later chapters, solar cells can convert up to 40% of the solar energy directly into electricity.

Biomass is considered by many as a renewable source of energy. Others point out that in order for biomass to be truly renewable, the rate of cultivation of new trees must be equal to that of biomass consumption. That is to say, for every tree that is cut down and burned, another tree must be replanted.

Even if biomass were entirely renewable, it is not a clean source of energy. As is the case with fossil fuels, burning biomass creates major air pollutants such as carbon monoxide and particulates. In addition to these pollutants, biomass produces aldehydes which, although not carcinogenic, produce an unpleasant odor at high concentrations. Since there are low levels of sulfur and nitrogen in biomass, fewer products that cause acid rain are formed. Biomass, however, does not contribute to global warming and is relatively low in sulfur; it therefore contributes little to the acidity of the environment.

Besides pollution, biomass as a source of energy has been controversial for other reasons. Harvesting forests for fuel wood and lumber will ultimately lead to deforestation. Crop farms will not be able to produce such crops for a very long time, as soil erosion and massive use of herbicides and fertilizers will reduce their productivity. Furthermore, these farms are built on lands which could otherwise produce the food-crops necessary to feed many parts of the world. Burning trees, dung, and other animal wastes deprives the soil of its essential nutrients, reduces the yields, and makes the land less capable of supporting natural vegetation. It is therefore, important to use plants that are not suitable for food consumption, and are planted in marginal land unsuitable for food crops.

Trash and municipal solid wastes have become a major problem for many countries. Trash must be stored in landfills, shipped away to areas with lower population densities, or simply burned. *Landfills* are filling up fast, especially near towns. They are polluting, may leak hazardous chemicals into the ground and water, and are breeding grounds for rats, flies, and microbes that cause diseases, rot foods, and may react to produce other unwanted products. *Hauling trash* is not popular with its potential recipients. *Incinerators* are appealing to many because they reduce trash volume by 60-80% and produce energy at the same time. Incinerators, however, produce deadly smoke and ash, which require expensive scrubbers, filters, and other control equipment. Furthermore, many toxic metals such as mercury, lead, and cadmium cannot be effectively removed and will eventually find their way into the ground or be released into the atmosphere.

Recycling seems to be a sensitive solution to the problems of waste disposal. The problem with this approach is that it encourages waste, which ultimately increases energy use and generates a considerable amount of air pollutants. In addition, recycling demands consuming additional energy-- as much as 25% of the energy used in manufacturing the original material.

The only solution to our trash problem is to **create less trash**. This means we must produce only goods that are essential, made of materials that are harmless to the environment, last longer, and have been designed to be disposed with minimal waste when they are no longer needed.

Additional Information

Books

1. Sims, R., Bioenergy Options for a Cleaner Environment in Developed and Developing Countries, Elsevier, 2003.
2. Tillman, D., Combustion of Solid Fuels & Wastes, Academic Press, 1991.
3. Biofuels for Transport: Global Potential and Implications for Energy and Agriculture, The Worldwatch Institute, 2007.

Periodicals

1. *Biomass and Bioenergy,* Science Direct Elsevier Science Publishing Company.

Government Agencies and Websites

1. National Renewable Energy Laboratory: Biomass Research *(http://www.nrel.gov/biomass).*
2. US Department of Energy *(http://www1.eere.energy.gov/biomass).*

Non-Government Organizations and Websites

1. Biomass Energy Research Association *(http://www.bera1.org).*
2. American Bioenergy Association *(http://www.biomass.org).*

Exercises

I. Essay Questions

1. Why are most plant leaves green? Why do they turn yellow in the fall?

2. What does it mean to "fix" carbon? How can this be done?

3. What is a hydrothermal vent? Why do deep-sea organisms mostly grow near these vents?

4. Describe the processes of aerobic and anaerobic conversion. Give examples of physical activities that involve these conversions.

5. What are the various trophic levels? How does energy transfer from one trophic level to another?

6. What are the differences between thermochemical and biochemical conversions? Between the processes of aerobic and anaerobic conversion?

7. A basketball player is using an average 400 watts of power during a basketball match. How long does he have to play to burn off a Big Mac (750 Calories)? Assume a metabolic efficiency of 25%.

8. What is the BMR value for a 70 kg woman who is 155 cm tall? Is she overweight or underweight? If she is overweight, and if she intends to reduce her weight by aerobics only (no dieting) over 6 months, for how long does she have to exercise every day?

9. If the US were to meet its entire energy need (100 Quads) from sugarcane, how many tons of sugarcane would have to be planted?

10. Describe the difference between gross and net production. What are the factors that limit production efficiency from one trophic level to the next?

11. Which has a higher metabolic rate, a hummingbird or a human? How does size of an animal affect its metabolic rate?

II. Multiple Choice Questions

1. Biomass
   a. Is a form of solar energy
   b. Can be used to produce electricity
   c. Refers principally to wood, and animal and human waste
   d. Will likely increase in use because it is a free source of energy
   e. All of the above

2. Which of the following are considered to be carbon sequesters?
   a. Fossil fuels
   b. The oceans
   c. Thermafrost and tundra
   d. The atmosphere
   e. All of the above

3. During photosynthesis, a plant
   a. Creates energy
   b. Absorbs energy
   c. Gives off energy
   d. Transmits energy
   e. Reflects energy

4. Humans use chemical energy contained in food
   a. To maintain their body temperature
   b. To enable their bodies to move
   c. To perform mechanical work
   d. To feed neurons for transmitting electrical signals
   e. All of the above

5. Biomass energy refers generally to
   a. Wood and agricultural products
   b. Solid waste
   c. Landfill gases
   d. Alcohols
   e. All of the above

6. If 0.1% of solar energy that falls on earth is captured by plants, and 2% of that energy is involved in photosynthesis, what fraction of sunlight that hits the earth is converted to food?
   a. 0.2
   b. 0.02
   c. 0.002

d. 0.0002
e. 0.00002

7. Which of the following statements is incorrect?
   a. Autotrophs are mainly green plants and algae formed through photosynthesis.
   b. Heterotrophs cannot produce their own food.
   c. Heterotrophs can be divided into herbivores or vegetarians, and carnivores or meat-eaters.
   d. Nutrients are recycled over and over through biological systems.
   e. None of the above.

8. What kind of organisms can produce their own food?
   a. Autotrophs
   b. Herbivores
   c. Heterotrophs
   d. Carnivores
   e. All of the above

9. Which trophic level is represented by the greatest biomass in any ecosystem?
   a. Herbivores
   b. Primary carnivores
   c. Producers
   d. Secondary carnivores
   e. Top carnivores

10. What kind of organisms are at the top of the food web?
    a. Autotrophs
    b. Herbivores
    c. Heterotrophs
    d. Carnivores
    e. None of the above

11. If you eat a frog that eat insects, which in turn eat plants, you would be a
    a. Producer
    b. Primary consumer
    c. Secondary consumer
    d. Tertiary consumer
    e. Trophic consumer

12. Pyrolysis is
    a. Burning biomass at a very high temperature
    b. The process of hydrogenation of biomass by adding steam
    c. Breaking down the biomass matter by using heat in the absence of oxygen
    d. Thermochemical conversion at room temperature
    e. Disintegration by microscopic organisms in the absence of oxygen

13. Wood and other biomass account for about _____ percent of US energy consumption.
    a. 0.1
    b. 1
    c. 3
    d. 10
    e. 25

14. Respiration is
    a. An intermediate step in photosynthesis
    b. Another name for photosynthesis
    c. The reverse of photosynthesis
    d. Same as sweating
    e. Photosynthesis in water

15. Incineration
    a. Is becoming the most popular way to produce energy at very low costs
    b. Is considered as a good and inexpensive alternative to solid waste disposal
    c. Is not clean and produces harmful by-products that are highly toxic to humans and animals
    d. Is best for disposing of metals and plastics
    e. All of the above

16. "Rule of 10" refers to a hypothesis that states
    a. Roughly 10% of solar energy is involved in photosynthesis
    b. Biomass constitutes about 10% of US energy needs
    c. Humans use about 10% of their daily energy needs from food
    d. Only 10% of the energy contained in a given trophic level can be converted to the level below
    e. Humans convert roughly 10% of the energy contained in their food to useful muscle work

17. The human body works as a heat engine, converting _______ percent of the food energy to mechanical energy necessary to carry daily physical activities, while disposing the rest as waste heat.
    a. Less than 1%
    b. Between 1 to 5%
    c. About 15-25%
    d. About 60%
    e. Above 90%

18. Thermal treatment of biomass to methane is called
    a. Pyrolysis
    b. Incineration
    c. Fermentation
    d. Burning
    e. Decomposition

19. Which of the following statements is <u>not</u> correct?
    a. Body mass index is a measure of body's mass in proportion to height.
    b. Body mass index is calculated as body mass divided by height.
    c. Body mass index is calculated as body mass divided by height squared.
    d. According to the National Institute of Health studies, a healthy body should have a body mass index of between 19 and 25.
    e. None of the above.

20. The minimum energy required for a body to maintain all his vital organs function under rest is called
    a. Body mass index
    b. Basal metabolic rate
    c. Body's maximum respiration
    d. Survival index
    e. None of the above

III. True or False?

1. Biomass accounts for almost all the renewable energy used in the United States.

2. When burned, biomass does not contribute to any greenhouse gases.

3. Biodiversity refers to the number of species present in a given place at a given time.

4. Photosynthesis is possible only in air and in the presence of sunlight.

5. Photosynthesis was the first process to occur in nature after earth sufficiently cooled down, following its formation 5 billion years ago.

6. Heterotrophs are organisms that cannot produce their own food, but are fed by other organisms.

7. An organism which is fed by a producer is called a secondary consumer.

8. Autotrophs are placed at the base of food pyramid and are responsible for primary food production.

9. Net productivity is the gross productivity minus that which is metabolized by the producer.

10. BMR is generally higher for males and for heavier and healthier animals.

11. For two persons of a similar mass, the BMI is lower for the person who is taller.

12. To use 200 kilocalories in muscle energy, you need between 800-1000 kcal in food energy.

13. Biomass is a type of solar energy.

14. Biomass is stored chemical energy in plants.

15. Humans can convert about one-half of the energy contained in food they eat to muscle work.

IV. Fill in the Blanks

1. In addition to chlorophyll, plants trap light via two other pigments called _____________ and _________________.

2. The reverse process to photosynthesis where carbohydrate and oxygen react to produce carbon dioxide and water is called ________.

3. Half of all metabolic activities happen in two organs in the body: the _________ and the _________.

4. The three major food categories are __________, ____________, and ____________.

5. Roughly two-thirds of garbage from American households is made of __________.

V. Project I - Creationism and Evolution

How life was formed has been debated by scientists and theologians for a long time, without reaching consensus. Scientists strongly favor the theory of evolution as laid out by the noted British biologist, Charles Darwin (1809-1882). Theologians, however, believe that the odds of amino acids combining to form the necessary proteins by undirected means is so minute that the proteins needed for life could never have come into existence by chance or any natural processes. In this project you are asked to research arguments for and against each theory and answer the following questions:

a. What constitutes life? How can we distinguish living from non-living organisms?
b. How do different religions (Buddhism, Judaism, Christianity, Islam, etc.) view the origin of life? Are there any discrepancies among these religions in this regard?
c. Is there any contradiction between Christianity and the theory of evolution? Explain.
d. What are the main features distinguishing evolutionists and creationists in regard to the origin of life?
e. Is there a scientific basis for the creationists' point of view? What are they? Do these arguments withstand the accepted methods of scientific inquiry?
f. How does the doctrine of "Intelligent Design" differ from traditional creationists?
g. Does science rule out divine intervention?

Project II - Life in the Universe

Life as we know it is made of mainly four elements -- oxygen, carbon, hydrogen and nitrogen. A few other elements, notably calcium, phosphorous, potassium and sulfur, provide the bulk of nutrients we need to sustain life. Most oxygen is bound to hydrogen to form water, whereas carbon makes up cellular structure. In this project you are asked to research the web resources to find whether other forms of life can or cannot be sustained on Earth or any other place on the Universe. In particular try to answer the following questions:

1. What is unique about carbon that makes it so suitable for forming much of the internal structure of living organisms on Earth? (*Hint:* look at molecular structure of carbon and how it binds with other molecules).
2. Why do some scientists (and science-fiction writers) propose other forms of life, especially those based on silicon, as a possibility? Why is carbon-based life more prevalent over silicon-based life? What makes silicon less favorable?
3. What other elements have been suggested and why?
4. Why it is highly unlikely to have any other form of life such as nitrogen-based or iron-based life forms in the universe?

Project III - Diet and Exercise

In this project you are asked to keep a daily log of your daily food intake and tasks you perform. Then you are asked to estimate your caloric intake and the calories you burn as a result of daily routine, work, and exercise. For better estimates, it is recommended that you collect and average data for three 24-hour periods.

Basic Information:

Sex: ______ (male/female)
Mass: ______ kg
Height: ______ cm

1. Make a table listing all the foods you consumed. Categorize them into breakfast, lunch, dinner, and snacks.
2. Make a list the activities you performed and their durations.

<u>Now do the following calculations:</u>

1. Find the average daily total intake of food calories. Take the three-day average.
2. Estimate the average daily total calories from carbohydrates, protein, and fat, respectively. Assume 9 Calories per gram of fat and 4 Calories

per gram of protein and carbohydrates.

3. Calculate the average power you put out assuming you maintain a similar diet throughout the year.
4. Calculate your BMR. According to Harris and Benedict[19], BMR can be accurately calculated using the following equations. BMR is basal metabolic rate in kcal/day; M is body mass in kg, H is height in cm, and A is age in year.
   Men: BMR= 66 +13.7*M +5*H -6.8*A
   Women: BMR=655 +9.6*M +1.8*H -4.7*A
5. Estimate the total non-BMR calories burned through various activities. Which of the activities were aerobic and which ones were anaerobic?
6. Calculate your total daily energy expenditure by adding total BMR and non-BMR calories. What is power per mass (kW/kg)?
7. Carry the energy balance by taking the difference between your daily energy expenditure and energy from food intake.
8. Calculate your daily calorie needs by multiplying your BMR by:
   a. 1.200 if you are inactive (do little or no exercise)
   b. 1.375 if you are somewhat inactive (light exercise 1-3 times a week)
   c. 1.550 if you are moderately active (moderate exercise 3-5 times a week)
   d. 1.725 if you are active (extraneous exercise 5-7 times a week)
9. If you maintain a fairly similar routine for the entire year, how much mass should you expect to gain or to lose?

Work Sheet for Project III				
Name: ______________				
Mass: ______ kg	Height: _______ m	Sex: ___		
Food Intake:	*Breakfast:*	*Calories*	*Snack*	*Calories*
	__________	______	__________	______
	__________	______	__________	______
	__________	______	__________	______
	Lunch	*Calories*	*Dinner*	*Calories*
	__________	______	__________	______
	__________	______	__________	______
	__________	______	__________	______
Total: ______ **Calories**	__________	______	__________	______
Activities:	***Activity***	***Calories***	***Activity***	***Calories (a/n)****
	Sleeping	______	Light exercise	
	Sitting/Watching TV	______	________	______ (__)
	Walking	______	________	______ (__)
	Reading	______	Medium exercise	
Total BMR: __________ **Calories**	Driving	______	________	______ (__)
Total non-BMR: ________ **Calories**	Cooking	______	________	______ (__)
Total: ___________ **Calories**	Washing dishes	______	Strenuous exercise	
	Working: (describe)	______	________	______ (__)
	________	______	________	______ (__)
Power per kilogram: __________ **kW/kg**				
Net Energy Intake: __________ **Calories**				
Mass gain / loss: __________ **kg**				
Calorie needs __________ **kilocalories**				

* (a/n) aerobic/anaerobic

[19] Harris, J. A. and Benedict, F., "A Biometric Study of Basal Metabolism in Man," Washington, D.C. The Carnegie Institution, 1919.

CHAPTER 7

Fossil Fuels

My father rode a camel. I drive a car. My son rides in a jet airplane. His son will ride a camel. ~ A Saudi saying

Facts do not cease to exist because they are ignored.
~ Aldous Huxley (1894-1963)

Fossil fuels are the most widely used form of energy in the world. They power practically every device we use and are used in every manufacturing application that we can imagine. Even the electricity generated to power a television or light a house is most likely produced using fossil fuels. The United States is the largest consumer of fossil fuels, which account for about 86% of its energy needs.[1] Two-thirds of the electricity generated and almost all of the fuel used in transportation, comes from fossil fuels.

Among the fossil fuels, oil and natural gas are the most popular because of their ease of transport, handling, use, and relatively low emissions. As these supplies are exhausted, they must be substituted by other forms of fossils, such as coal, oil shale, and tar sands.

What is a Fossil Fuel?

Fossil refers to the remains of animals, plants, or other life forms that have been protected from decomposition and oxidation for a very long time. *Fuel* refers to anything that can be burned as a source of energy. Therefore, fossil fuels are the remains of animals and plants that have formed into materials that can be burned. Since the source of these materials is living matter, they have the same composition as living organisms. They consist of fats, oils, paraffin (waxes), carbohydrates (sugars, starch, and cellulose), and proteins. Sulfur, phosphorous, and metals –although they can be burned—are not considered fossils because they do not originate from organic matter.

Figure 7-1
Fossil fuel

Contrary to popular belief, fossil fuels are not the remains of dinosaurs; in fact, most fossil fuels were formed millions of years before the first dinosaurs lived. Since most lands were swamps, we can expect to find fossil fuels in areas that were once lush and housed many forms of plants and animals. The most likely sources are ancient trees, animals, fish, and tiny organisms that flourished in the oceans (Figure 7-1). Marshlands, subtropical and tropical swamps, lakes, lagoons, and river deltas are ideal sites.

[1] See Table 1-3, page 9.

Carbon Cycle

Since carbon is the major ingredient in fossil fuels, it is relevant to discuss the carbon balance in the atmosphere. Carbon exists primarily in the form of carbon dioxide, which is present in the air, dissolved in ocean water, or trapped in rocks and in plants through photosynthetic processes (See Chapter 7). Animals (including humans) consume plants as a source of nutrients and are part of the carbon cycle as well.

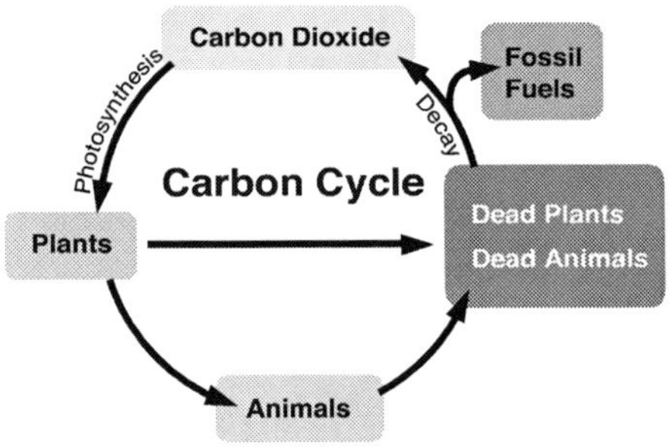

Figure 7-2
Carbon Cycle in the Atmosphere
[adapted from Energy and Fuel in Society, by Radovic and Schobert, McGrawHill 1992].

Depending on the ultimate fate of dead animals and plants, carbon may return to the atmosphere as carbon dioxide or turn into fossil fuel. If the environment dries up, dead plants and animals will become exposed to air and react with it to produce carbon dioxide, thus completing the carbon cycle (Figure 7-2). However, if the environment is flooded with mud, stagnant water, silt, or sand, oxygen is cut off and the proteins and carbohydrates decompose from bacteria into a wax-like, organic, insoluble material called ***kerogen***.[2] The time it takes for kerogen to transform into fossil fuel ranges from tens to hundreds of millions of years, and depends on its depth in the ground, temperature, and pressure.

Whether a fossil ultimately ends up as coal, oil, or gas depends on the original constituents of the kerogen and the conditions under which it is kept. If the origin of kerogen is from giant plants and stronger parts of plants such as lignin (material that strengthens the trunks and branches of trees), then carbon forms into very complicated structures and rings. These rings eventually connect together into a graphitic structure called coal. If the kerogen is formed from algae or plankton, they break into shorter chains that depending on their lengths, become a mixture of liquids (crude oil) and gases (natural gas). The deeper in the earth and more aged the kerogen, the greater the probability that the chains will break into lighter materials and a higher fraction of the kerogen will be in the form of natural gas. The carbon cycle is completed when fossil fuels are extracted and burned to form carbon dioxide and water.

Combustion of Fossil Fuels

Fossil fuel, whether in the form of coal, oil, or natural gas, reacts with oxygen in the air when burned, forming a number of products. In its simplest form, this chemical reaction can be written as:

$$\text{Fuel} + \text{Oxidizer} \longrightarrow \text{Product}$$

This reaction always produces heat. Chemical reactions that produce heat are called *exothermic reactions*. In addition to carbon and hydrogen, fossil fuels may contain oxygen and traces of other elements, such as sulfur and metals, which complicate the combustion. In the present analysis, we assume that all impurities have been removed and the reactions are such that all the fuels are burned to produce carbon dioxide and water vapor.

[2] *Keros* is the Greek word for wax, and *gen* means more.

For example, methane, gasoline, and coal burn in oxygen as:

Methane: $CH_4 + 2\,O_2 \longrightarrow CO_2 + 2\,H_2O$

Gasoline: $C_8H_{18} + 12.5\,O_2 \longrightarrow 8\,CO_2 + 9\,H_2O$

Coal: $C + O_2 \longrightarrow CO_2$

Such reactions, where the oxidizer is just enough to convert all fuels to carbon dioxide and water vapor, are called stoichiometric. When the reactant mixture has fuel or oxygen in excess of what is required for stoichiometric ratio, the mixture is fuel rich (oxygen lean) or oxygen rich (fuel lean), respectively. If air were used instead of oxygen, the reaction would remain the same except that the nitrogen in the air would appear unburned in the product. For example, the stoichiometric combustion of gasoline in air would be:

$$C_8H_{18} + 12.5\,(O_2 + 3.76\,N_2) \longrightarrow 8\,CO_2 + 9\,H_2O + 47.0\,N_2$$

(Air is a mixture of 1 part oxygen and 3.76 parts nitrogen).

On the mass basis:

114 kg gasoline + 1,716 kg air ⟶ 352 kg carbon dioxide + 162 kg water vapor + 1,316 kg nitrogen

or

1 kg gasoline + 15.1 kg air ⟶ 3.1 kg carbon dioxide + 1.42 kg water vapor + 11.54 kg nitrogen

Question: As a result of fossil fuel consumption, roughly 6 billion tons of carbon is produced each year. It is estimated from geological evidence that each year about 30 million tons of carbon goes into the formation of new fossil sediments. What is the significance of this data in estimating the lifetime of remaining fossil fuels?
Answer: The rate of consumption of fossil fuels is 200 times faster than the rate they are replenished by nature. This means fossil fuels are gradually being depleted and must be considered as a non-renewable source of energy.

Example: Calculate the stoichiometric air to fuel ratio necessary for burning the natural gas in air. Natural gas consists mainly of methane.
Solution: The stoichiometric reaction of methane is

$$CH_4 + 2(O_2 + 3.76\,N_2) \longrightarrow CO_2 + 2H_2O + 7.52\,N_2$$

16 kg methane+ 275 kg air ⟶ 44 kg carbon dioxide + 36 kg water vapor + 211 kg nitrogen

or

1 kg methane+ 17.2 kg air ⟶ 2.75 kg carbon dioxide + 2.25 kg water vapor + 13.2 kg nitrogen

Note that although methane and gasoline have very different structures, they have a similar air/fuel ratio and produce roughly the same amount of carbon dioxide per mass of fuel burned.

Incomplete Combustion

If chemical reactions are completed fully, there would be little pollution and hydrocarbons would be a relatively clean source of fuel. Unfortunately, these reactions do not reach completion and other gases such as nitric oxides (NO and NO_2) and carbon monoxide (CO) are always present. Because reactions take longer to reach completion than the time available in the combustion chamber, some hydrocarbons might also remain unburned in the product. Therefore, products will be exhausted before reaching equilibrium. The degree to which the reaction completes depends upon such variables as the combustion temperature and pressure, residence time, and mixture ratios. The details of the processes are outside the scope of this book and will not be discussed further. Those interested should consult more advanced texts on the subject.[3]

Heating Values

Heating value (also called calorific value or heat of combustion) refers to the amount of thermal energy that is released by burning a unit amount of fuel.[4] For gaseous and liquid fuels such as methane and gasoline, the heat of combustion is usually given per unit volume (37,000 kJ/m^3 for methane and 39,000 kJ/lit for gasoline), while for solids such as coal, it is expressed per unit mass (15,000 to 30,000 kJ/kg).[5] Roughly speaking, we can show the following relations between the energy releases of various forms of fossil fuels:

1 liter of petroleum ~ 1 kg of coal ~ 1 m^3 of natural gas
1 gal of petroleum ~ 14 pounds of coal ~ 150 ft^3 of natural gas

Table 7-1. Heating Values for Common Fuels*

Fuel	Calorific Value (kJ/kg)
Coal	15,000-30,000
Kerosene	43,000
Gasoline	44,000
Methane	55,600
Propane	50,500
Coal Gas	34,000
Methanol	19,800
Garbage**	19,800
Wood	20,000
Hydrogen	143,000

** Calorific values are approximate and can vary somewhat depending on fuel purity.*
*** Garbage from a typical American household*

It is interesting to note that, per unit mass, different fossil fuels have relatively similar calorific values (Table 7-1). This is logical due to the fact that all fossil fuels are hydrocarbons. Except for a slight variation in energy trapped between different bonds which form their molecules, these hydrocarbons are expected to release similar amounts of heat. For comparison, the heating values of hydrogen, firewood, and of garbage from a typical American household are also given. As we saw in Chapter 6, as much as 75% of the average American garbage content is paper, foodstuff, and yard waste; like fossil fuels, these derive their energy from living matter and thus can be burned in a manner similar to fossil fuels. Hydrogen is not a fossil fuel and does not contain carbon. Its high heating value is attributed to the much higher energy content trapped in H-H bonds relative to H-C bonds common in fossil fuels.

[3] See for example, Seinfeld, J. H. Air Pollution: Physical and Chemical Fundamentals, McGraw-Hill, Inc., 1975.
[4] Heating values are expressed either as higher (gross) or lower (net) heating values, depending on whether the products of combustion are cooled back to their initial reactant temperature, and whether water in the combustion product is condensed out. If water in the exhaust is in the form of vapor, the heating value is called net or lower heating value (LHV). If the water vapor is cooled so all the water is condensed out, then additional heat is available and the heating value is called gross or higher heating value (HHV). For most reactions the exhaust is relatively hot therefore little condensation takes place and LHV is a more realistic number to use. The HHV is most appropriate in certain devices such as natural gas furnaces used in residential units and air conditioners that use condensers.
[5] BTU equivalents of various energy sources are: 1 bbl crude oil = 5.8 million Btu; 1000 cu. ft gas = 1.03 million Btu; 1 kWh electricity = 3,400 Btu; 1 metric ton coal = ~30 million Btu.

US Fossil Fuels: The Facts

Did You Know That ...?

- Over 95% of all US fossil fuels are still intact in the form of coal.
- The majority of petroleum in the United States is in four states: Texas, Alaska, California, and Louisiana.
- Contrary to popular belief, the major supplier of crude oil to the United States is not Saudi Arabia. It is Canada!
- Without oil imports, the United States has enough oil to last 10-12 years.
- After Saudi Arabia, the United States is the largest oil-producing country.
- US import of petroleum has increased from 8.4% of its consumption in 1950, to 30% in 1973, and to 55% today. The percentage of import is likely to increase to 70% by 2020.
- Two-thirds of all US petroleum consumption is for transportation.

Fossil Resources

Major sources of fossil fuel are oil, coal, and natural gas. Tar sands and oil shale are also in large supply, but because of their high cost of production and their major adverse environmental impacts they have so far been largely ignored. As other resources deplete, these resources may prove to be a prominent source of fuel in the 21st century.

Petroleum

The term petroleum (nicknamed black gold) comes from the Latin roots *petra*, "rock," and *oleum*, "oil"; it literally means rock oil and refers to gaseous, liquid, or solid hydrocarbons found beneath the earth's surface in sedimentary rock formations. The two most common forms are natural gas and crude oil that, like coal, are formed from decaying animal and plant remains under very high pressures and temperatures underground. Unlike coal however, they can diffuse in porous sedimentary rocks and escape through the earth's surface. Large reservoirs are formed only if the migration is stopped and oil and gas are trapped by the impermeable cap rocks. Since there is always some water and salt present, the oil is usually found above a pool of brine. Natural gas accumulates above the oil (See Figure 7-3).

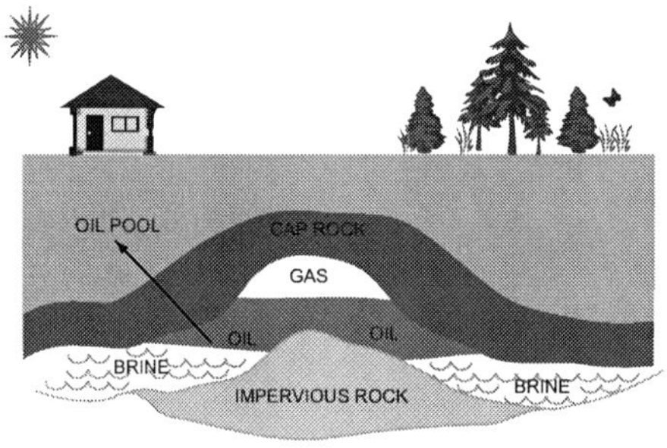

Figure 7-3
There is always some water and gas accompanying oil fields.

Petroleum is not only a convenient source of energy for generating electricity and powering our cars but also the main ingredients in much of the plastics, waxes, medicines, cosmetics, solvents, lubricants, feedstock, and a host of other petrochemical products used in our daily lives. In fact, because of its limited availability, many consider petroleum to be too valuable to be used in transportation, heating, and generating electricity.

History of Oil Exploration in the United States

Although oil has found its present place in the world for only the last century, it has been known to man as a useful product for a very long time. As early as 3000 B.C., oil was used by Mesopotamians (current Iraqis) as caulking to seal the cracks of buildings and joints in boats. The Egyptians used oil as a lubricant for their chariots, and the Romans used it to light their arrows before launching them at enemies. By 600 B.C., oil in Persia

Figure 7-4
"Colonel" Edwin Drake. The first commercial oil well in Titusville, Pennsylvania.

was routinely extracted from oil wells that were only a few meters deep and brought to the surface by buckets and pulleys to be used in heating, lighting, and other purposes.

In 1852, Abraham Gesner succeeded to distill kerosene from the crude which quickly replaced the whale oil which was more expensive, burned dirtier and was less luminous. The first commercial oil well in the United States was built by Edwin L. Drake in 1859 (Figure 7-4). It produced about 1000 gallons a day. Within one year, 2000 new oil wells were dug and a new industry was born. The oil boom in Pennsylvania did not last very long as wells dried up within a few years. Drake went bankrupt, dying a few years later.

In 1864, a young entrepreneur, John David Rockefeller, purchased his first oil refinery in Cleveland, Ohio. As the Civil War ended and the country was fast becoming industrialized, Rockefeller saw the opportunity and bought additional refineries. To assure control over market and protect himself from wide fluctuations in the oil market, he moved and eventually acquired control over every stage of oil production from extraction to retail; this included transport, research, marketing, and even the manufacturing of barrels. In 1870, he combined his smaller companies into the Standard Oil Company and offered shares to the public. Being the largest customer of the railway industry, he managed to negotiate a favorable rate for transporting his crude, manipulated the price ofkerosene, forcing many competitors to close their shops, file for bankruptcy, or sell assets to Rockefeller's Company. By 1879, Rockefeller controlled 90% of all refinery operation and a quarter of all petroleum in the US. At the time that the average American worker was making $500 a year, Rockefeller's wealth was estimated at several hundred millions.[6]

Threatened by his absolute control over the industry, several oil companies sued Rockefeller claiming the Federal Law prevents Rockefeller to operate any refineries other than in Ohio. Fearing loss of control, Rockefeller established the Standard Oil Trust.

On November 1, 1879, Thomas Edison was awarded a patent for his most famous invention the "electric lamp." The switch to electricity could adversely impact oil demand and price of the crude. The concerns did not last very long, however, when Henry Ford introduced his automobile (*the horseless carriage*), which was running on gasoline, a by-product of distillation of crude. Between 1900-1910, nearly half a million automobiles were sold, and with it petroleum found its prominent place in the American economy, and in doing so solidified Rockefeller's power to control the price of oil.

[6] Laughlin, R., John D. Rockefeller: Oil Baron and Philanthropist, Morgan Reynolds Publishing, Greensboro, North Carolina, 2004.

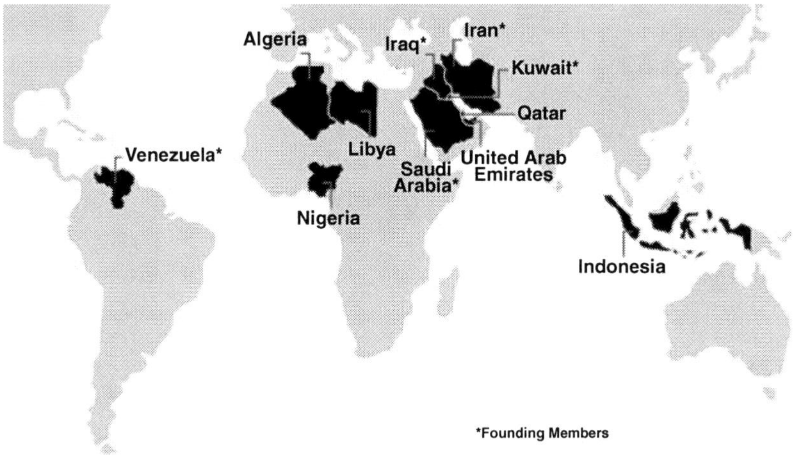

Figure 7-5
OPEC member countries.

Another threat soon was realized, when in 1901, the new oil discovery in Texas, attracted thousands of wildcatters and entrepreneurs.[7] Wildcatters refer to persons drilling oil wells in areas not known to have oil in advance. Overnight, the US production of oil doubled and new powerful oil companies such as Gulf and Texaco were formed that challenged the Standard Oil monopolistic power that eventually led to a law suit by the government which accused Standard Oil of artificially manipulating prices. In 1911, the Supreme Court ordered the company to break into several smaller companies.

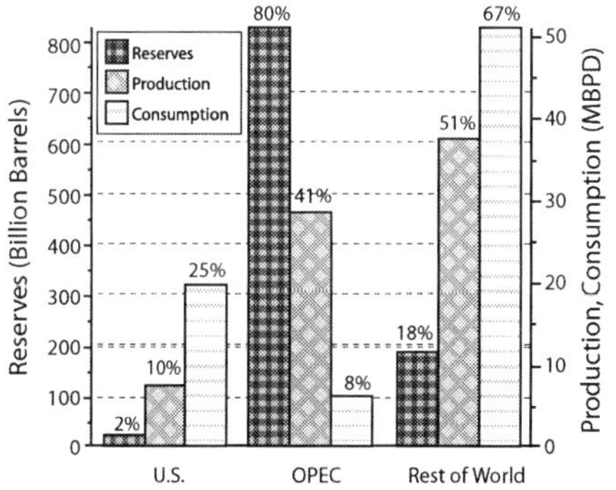

Figure 7-6
World Petroleum Reserves, Production, and Consumption.
Source: 2002 Transportation Energy Data Book, 2003.

Soon thereafter, World War I was in full swing, concluding only when the Allies successfully blockaded Germany's oil supply routes and caused much of the German industry, trains, and military machines to come to a halt. WWI ended in 1918 only to reemerge 20 years later as WWII; again, conflict over the control of major oil fields in Poland and Russia played a large part in starting the war.

With the discovery of huge reservoirs of oil in the Middle East in the years following WWII, there was a surplus of oil in the world market, and prices started to drop. To stabilize the sharply declining oil prices, as well as coordinate and unify petroleum policies, five countries (Iran, Iraq, Kuwait, Saudi Arabia, and Venezuela) established OPEC[8] in 1960. Since then, eight other nations (Algeria, Angola, Indonesia, Libya, Nigeria, Qatar, the United Arab Emirates, and recently Angola) have joined the group (Figure 7-5).[9] OPEC members collectively own 80% of the world's proven oil reserves and produce 41% of the world's oil, but consume only 8% of the oil (Figure 7-6).[10] In contrast, the United States – with a little more than 2% of the reserves – uses more than a quarter of all oil produced in the world (Table 7-2). At present, US imports roughly 60% of its oil needs; as US resources deplete, this percentage is expected to rise (Figure 7-7).

Table 7-2. Top Suppliers and Consumers of Petroleum*

Who has it?		Who uses it?	
Saudi Arabia	24%	US	25%
Iran	12%	China	8.7%
Iraq	10%	Japan	6.1%
Kuwait	9.1%	Russia	3.5%
UAE	8.8%	Germany	3.2%
Venezuela	7.2%	India	3.1%
Russia	5.4%	Canada	2.7%
Libya	3.5%	Brazil	2.7%
Nigeria	3.2%	S. Korea	2.6%
All Others	17%	All Others	42%

*Excluding canadian tar sands
Source: Oil & Gas Journal, Vol. 103, No. 47,Dec. 2005.

[7] Oilen, R., and Hinton, D., Wilcatters: Texas Independent Oilmen, Texas A&M University Press, 2007.
[8] Organization of Petroleum Exporting Countries (*http://www.opec.org*).
[9] In May 2008, Indonesia announced that it has become a net importer of petroleum and therefore is pulling out of the OPEC.
[10] Transportation Energy Data Book, Edition 23, 2003. Table 1.5.

A Brief History of the Contemporary Middle East: Oil and Blood

Digging Deeper ...

He who owns the oil, will own the world... ~ Henri Berenger, French Oil Minister in 1919.

European interest in the Middle East dates back to the 16th century when the Portuguese Navy seized the island of Hormuz in the Persian Gulf to control trade routes to India, China, and the African coast. In 1798, Napoleon crossed the Mediterranean Sea and overtook Egypt. The opening of the Suez Canal in 1869 provided Europeans with a shorter shipping route that expanded their trades with India, China, and the Middle East, and with it the rivalry between two major European powers, France and Britain, for control of Egypt and later the rest of the Middle East. In 1872, Persia granted the British Baron Julius Reuter, the founder of the Reuters News Agency, a seventy-year contract to operate the railway and exploit its minerals including oil. This followed by a second major concession in 1901, to another British financier William Knox D'Arcy for exclusive right to explore oil and construct pipelines from anywhere in Persia to the Persian Gulf. In 1909, the Anglo-Persian Oil Company (later to be renamed the British Petroleum) was established to oversee the exploration and production. Because of the strong Russian opposition, five northern provinces bordering Russia were excluded.[i] In 1908, major reservoirs of oil were discovered in Persia; shortly thereafter, the British fleet converted from coal to oil.

By the start of the 20th century, the British and French had seized much of the Ottoman's Arab territory, including Egypt, Sudan, Algeria, Morocco, and Tunisia. Kuwait, Qatar, Oman, and other Gulf states had become British protectorates. The Ottomans, however, had a firm grip on main Arabian (mainly Mesopotamia and Arabian Peninsula) oil fields. At the same time, internal combustion engines had been invented and oil had become a major source of contention in international affairs and as a major energy source driving the economy of the modern industrial world.

When World War I began in 1914, the Ottoman Empire sided with the Austrians and Germans against the Allies, principally Britain and France. The Arab masses were left to decide between 500 years of Ottoman domination and the European imperialism. As the Ottoman Empire weakened, the British sought the support of local Arab leaders, notably Hussein, the Sharif of Mecca, and his two sons Faisel and Abdullah (later to be installed as rulers of Saudi Arabia, Iraq, and Jordan) who fought along side the British against the Ottomans. WWI ended in 1918 only after Allied bombings of Romanian oil fields, which effectively cut much of the petroleum supply route to Germany.

In 1916, Britain and France secretly drafted the Sykes-Picot Agreement, in which the Arab world was divided into two spheres of influence; the Mediterranean coast was to be run by France, while Britain was to control Mesopotamia (current Iraq) and the Greater Syria. Palestine would be under international administration. By the time WWI had ended the Ottomans had lost control of the Middle East and the empire was reduced only to what is today's Turkey. The British had occupied Iraq and established bases in oil rich Kuwait Harbor. The Bolsheviks had succeeded in installing the first socialist government in Russia, threatening European Imperialism and giving rise to Arab nationalism and revolt.

The Paris Peace Conference (Conference of Versailles) in 1919 officially put an end to the war. As the absolute victors, Britain and France set to divide their war trophies and draw up a new map for the Middle East; Russia was largely bypassed. Britain was given the whole of the Arab Peninsula and a mandate to run Iraq, Palestine, and Egypt, while Syria and Lebanon became French protectorates. Except for the Russian oil, three companies - Standard Oil, British Petroleum and Royal Dutch Shell controlled almost all the petroleum production in the world.

After WWI, a bitter struggle for control of world oil reserves ensued. The British, French, and Dutch excluded US companies access to territories under their control. In 1920, Congress retaliated and passed the Mineral Leasing Act that punished those countries by denying them the right to produce oil in the United States. A settlement was eventually reached after all restrictions were removed by various parties to the conflict. Fearing that American oil reserves were being rapidly exhausted and to assure secure and continuous supply of oil for the foreseeable future, the US sought to establish ties with Middle Eastern oil exporters. In 1933, Standard Oil of California began exploration work in Saudi Arabia; it eventually merged with Esso (Exxon), Texaco, and Mobil and formed the California-Arabian Oil Company, later to be renamed the Arabian-American Oil Company (ARAMCO).

The Second World War (WWII) of 1939-1945 was the German attempt to reverse the defeat it suffered in WWI, and fulfill its imperial ambition over Europe, Russia, and the Middle East. Learning from its mistakes during WWI Germany had engaged in massive research and developmental work to extract synthetic oil from coal as well as stockpiling oil when needed.[ii] To cut off the oil supply routes of the enemy, in 1941, Italian forces stationed in Libya invaded Egypt; Germans followed suit and attacked the Suez Canal. Both attempts failed, however. In May 1941, the British took control of Iraq; Iranian oil fields were siezed later that year.

As the war dragged on and supplies depleted, Germany launched its ill-fated attack against Russia, at least in part, for the capture of oil fields in Grozny, Baku, and Caucasus. As the Russian Army retreated, they destroyed all fuel supplies, denying Germans their much needed fuel. Germany was eventually defeated and WWII ended in 1945. Ironically, war ended once again when Allied forces destroyed, sabotaged, or blockaded the Romanian oil fields and refineries in Ploesti.

The war not only ruined Germany, Japan, and Italy, but also crippled much of Europe, including the victorious France and England. The US had, however, belatedly entered the war and had the economic resources, manpower, and geographic means to help Europe and Japan rebuild, contain the spread of communism, and assume its political leadership over the Middle East and much of the world.

In the years that followed the WWII, the US, Japan, and other western powers underwent rapid economic expansion, and with it grew their need for more Middle Eastern oil. The disillusionment of many Middle Eastern Countries with the West over the establishment of a Jewish state in Palestine, and anger over what many considered as the exploit of their national resources, mandated reform. Reeling from the shameful concessions of the early 20th century to British Petroleum, Iran's nationalist Prime Minister nationalized its oil industry in 1951, a move that resulted in Shah's departure from Iran and instigated the subsequent 1953 CIA coup that brought back Shah to power. In 1956, Egyptian president Gamal Nasser nationalized the Suez Canal. In 1958, a military coup overthrew the monarchy in Iraq. Ben Bella led Algerian independence in 1962; the Palestine Liberation Organization was formed in 1964. In 1979, the Iranian Revolution overthrew the monarchy.

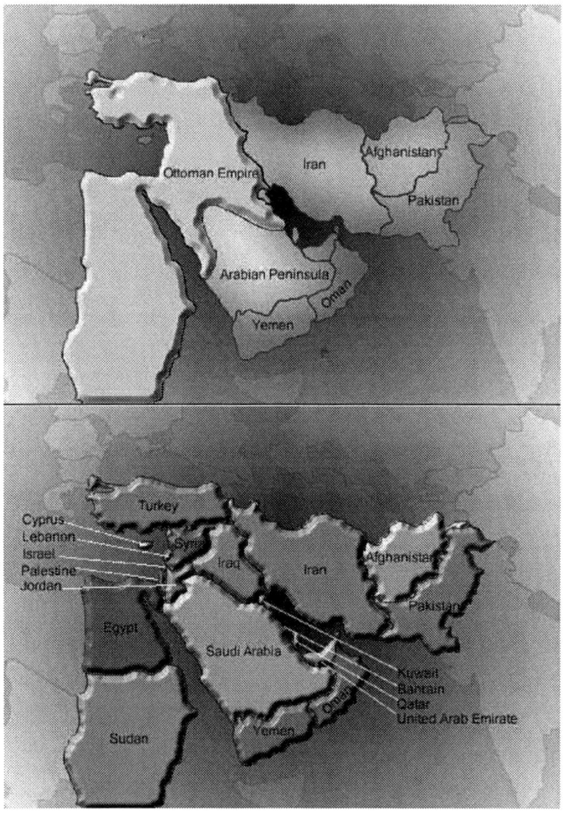

In September 1980, taking advantage of turmoil in the post-revolutionary Iran, Iraq invaded Iran, in part to seize control of the massive oil field on the Iran-Iraq border. Iraq's war effort was largely financed by Saudi Arabia and Kuwait and was politically supported by the United States, which provoked Iranians into taking American hostages. The war lasted eight years and left over a million estimated dead before the war was finally over.

Having accumulated a debt of over $17 billion, Kuwait demanded unrestricted access to the Ramallah oil fields that laid 95% in Iraq. Superior American technology in slanted oil drilling allowed Kuwait to extract oil and sell it under OPEC official prices in the oil market. In an effort to keep Kuwait from what he claimed as Iraqi oil reserves, Hussein invaded Kuwait in 1990, which prompted US retaliation and launched the First Gulf war in 1991 and eventual occupation of Iraq in 2003.

Arab disillusionment with the US role in defeating nationalist governments and its support of Israel during the 1967 six-day and 1973 Yom-Kippur wars only intensified their resolve in rejecting initiatives engineered by the US and its European allies. Lack of progress in ending the Palestinian/Israeili dispute and continuous support for Arab reactionary governments have provided the additional fuel to stir suspicion and anti-American hostilities throughout the Middle East and generated a rapid rise in terrorism and Muslim fundamentalism that led to the events of September 11, 2001, and recent wars in Lebanon, Iraq and Afghanistan and the loss of thousands of innocent lives.

[i] *Shwadran, B., The Middle East: Oil and the Great Powers, 3rd ed., New York, 1973.*
[ii] *Becker, P. W., "The Role of Synthetic Fuel in WWII Germany: Implication for Today?," Air & Space Power Journal, July-August 1981.*

With increasing dependence on foreign oil, developed countries will likely continue attempts to impose their will upon OPEC countries. Many of the oil producing countries (such as Iran and Venezuela) are, however, undergoing rapid economic expansion, and much of their oil production must be used to meet their domestic needs. This makes less oil available to export, with the result that oil revenues reduce for the exporter, and more expensive for the importer. This will necessarily put economic pressure on western and other developed countries, a prescription for regional and international conflicts and wars.

The 1991 Gulf War and the 2003 US invasion of Iraq are just two examples of many future conflicts that are bound to occur over control of the vast Middle Eastern oil reserves[11] (See box "A Brief History of the Middle East: Oil and Blood").

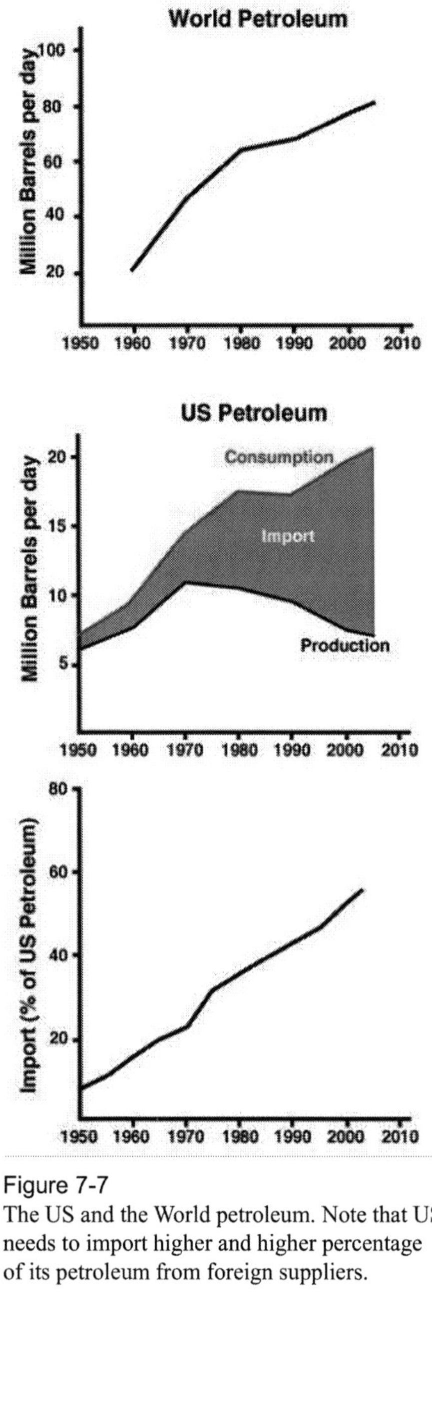

Figure 7-7
The US and the World petroleum. Note that US needs to import higher and higher percentage of its petroleum from foreign suppliers.

a From Oil Well to the Gas Pump

Finding Oil

In the early days of oil exploration, few technologies could locate oil reservoirs precisely; they were mainly limited to random drilling of exploratory oil wells (wildcats). Today, advances in geological science and the availability of new sensors make this process considerably more accurate. Although no one technology can be used to accurately verify the presence of oil or pinpoint its exact location, combinations of various technologies improve the probability of success. Two technologies frequently used in geological mapping are the gravimetric and seismic methods.

The gravimetric method uses variations in the local earth's density as an indication of the presence of local deposits of coal, petroleum, or other ores. If earth were a sphere, uniform everywhere, the gravitational forces would be the same everywhere. The force of gravity is not uniform however, and changes somewhat with earth's local density. On average, all objects are accelerated toward the center at the rate of $g = 9.8\ m/s^2$. Small variations of gravitational constant "g" can be accurately measured by a gravity meter and may indicate the presence of oil deposits. A gravity meter is essentially a mass-spring system in which the frequency of the oscillation is a measure of the gravitational constant.

The seismic method exploits changes in the velocity of sound as it travels from one medium to another. Waves can propagate through a medium indefinitely until they encounter a boundary or discontinuity and are reflected back. In the seismic exploration technique, a shock wave initiated by a surface explosion propagates through the ground. When it detects a change in density (such as when it encounters an oil or a coal deposit),

[11] According to Hall, D. "Oil and National Security," *Energy Policy*, Vol 20, No 11, 1992), Iraq was a threat to US interests in two ways: First it could use its vast oil revenue to develop weapons of mass destruction WMD), and second, the combined reserves and production capacity of Iraq and Kuwait equaled those of Saudi Arabia which gave Saddam the ability to influence the internal policies of the United States.

it reflects back. The reflected sound wave is measured by detectors (microphones) that are distributed at various locations around the detonation point. The time of arrival (velocity) and intensity (strength) of the reflected shocks, along with other data, are used to construct the 3-D geological map of the region and give information about the rock formation and potential presence of oil and coal deposits. Recent innovations in underground imaging and directional drilling allows development of 4-D geological maps (time is the fourth dimension) which track the motion of the fluid and increase oil discovery and recovery rates by 20%.

In addition to gravitational and seismic methods, other techniques have been used to measure the electrical resistance and other properties of the surrounding rocks and porous media.

Recovery

Once geologists pinpoint the location of a potential oil reserve, drilling begins. With luck, the well will reach the oil directly. If not, it may have to be closed and another one drilled nearby. The earlier oil wells were easy to find and only a few meters of drilling were needed. Today, we might have to dig many thousands of meters to find any oil at all.

Initially, oil is under sufficiently high pressure, which causes it to gush out naturally and no pumps are needed (flush production), but as more and more oil is extracted, a pumping facility can help maintain the flow of oil. The operation stops once the pressure has dropped below a point (*the settled production point*). This is the conventional method of production and is called *the primary recovery method*. About 30% of the oil can be extracted using the primary method.

Beyond this point, additional oil can be recovered only by **enhanced oil recovery** techniques. *Secondary recovery* involves injecting water, natural gas, steam, or carbon dioxide into dead wells to raise the pressure or push the oil that is otherwise unavailable into neighboring wells (Figure 7-8a). An additional 10-15% of the available oil can be recovered using the secondary technique. Unfortunately it raises the cost of oil production by 50 to 100 percent.

The *tertiary recovery* technique relies on reducing the oil viscosity. Viscosity is a measure of the ease with which a fluid flows (for example, water has a very low viscosity, but honey's is high). Viscosity can be reduced by raising the temperature, either by injecting superheated steam or by combustion. In the latter, a small underground detonation results in a shock wave that propagates across the oil deposit. The shock wave heats the oil and breaks it into smaller molecules, making it flow more easily (Figure 7-8b). Another 10-15% of the oil can be recovered using the tertiary method. The combined primary, secondary, and tertiary methods can recover about 50-60% of the oil deposit. Once these techniques are

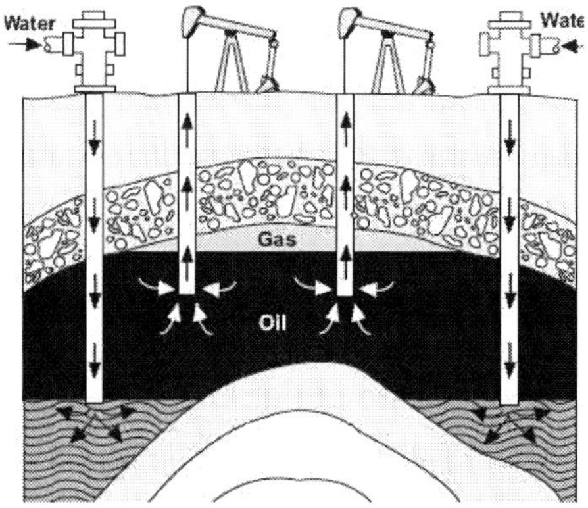

(a)

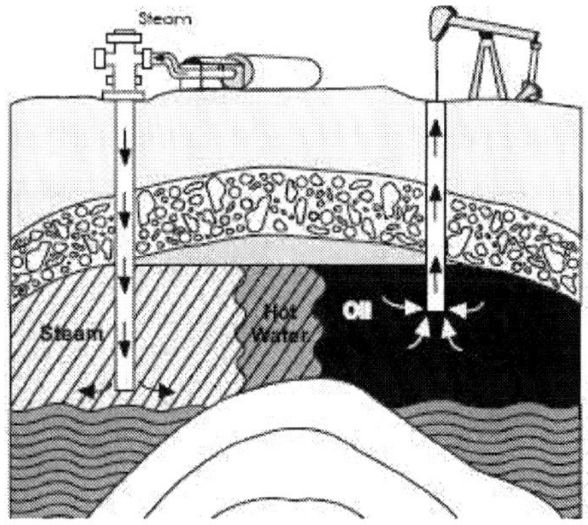

(b)

Figure 7-8
Enhanced Oil Recovery: (a) Secondary - water flooding, and (b) tertiary - steam injection.
Images Courtesy of Energy Information Agency, Office of Oil and Gas.

exhausted, the well is no longer usable and must be capped.

Occasionally, oil wells are discovered below ocean surfaces and drilling must be carried out offshore. The process is similar to onshore drilling except that derricks must be mounted on platforms built away from shorelines. Platforms can be permanently built and fixed to the ocean floor or may float – anchored with wire rope and chains to the sea floor.

Refining

When extracted from the ground, crude oil is not pure, but contains sand, water, and a number of salts. Sand settles to the bottom of storage tanks and is easiest to remove. Other contaminants are removed by electric or chemical means. The crude is then sent into a distillation column where, depending on their boiling points, they separate into heating oil, kerosene, gasoline, and various gaseous fuels such as butane, propane, and methane (See Table 7-3).

Table 7-3. Product of Petroleum Distillation

	Molecular Structure	Boiling Point (°C)	Use
Gaseous fuels	C1-C4	-164 to 30	Gas stoves, RVs
Ether	C5-C7	30 to 90	Solvent
Gasoline	C5-C12	30 to 200	Motor fuel
Kerosene	C12-C16	175 to 275	Diesel and jet fuel
Fuel oil	C15-C18	Up to 375	Furnace
Greases	C18 and up	Semisolid	Lubricant
Paraffins (waxes)	C20 and up	Solid	Candles
Asphalt and tar	---	Residue in column	Roofing and paving

Finite or Not?

Point/Counterpoint

P: Newly adjusted OPEC data proves that reserves may not be finite but are continously expanding.

CP: Much of the recent upward estimates of reserves are not warranted and were inflated by OPEC countries to increase their quotas. Furthermore, even with our advanced technologies, we have consistently found smaller deposits since the 1950s. In fact, today we are burning four barrels of oil for each barrel we discover.

P: Technology makes finding other resources easier.

CP: Newer technologies may increase the efficiency of oil recovery, which may extend the lifetime of petroleum a few years. It is doubtful however, that these technologies would meaningfully increase the total reserves.

P: There are always other substitutes.

CP: Other alternatives are of two kinds. Some, like solar and biomass, are more dispersed sources of energy and thus of lower quality. Others, like natural gas and coal, are themselves of limited supply and have niche applications.

P: The primary and secondary methods of recovery being used today can only extract a little more than half of the oil before wells are recapped. We have a lot of oil still underground that remains to be tapped.

CP: True, but there are available only if we are willing to pay much higher prices and if we find better methods of recovery. Then more oil resources are available and other resources such as tar sands and shale oils may become competitive.

Natural Resources: How Long Do They Last?

One of the most intriguing questions of our time is the remaining quantity of our natural resources (fossils, minerals, precious metals, etc.) and the length of time they will last. Many have tried to answer this question, often with greatly different outcomes. Depending on the extent to which reliable data are available and the assumptions made, the answer often varies by degrees of magnitude. Accurate data are rarely available and at best, are only good estimates. Even so, it is often manipulated to suit the geopolitical and economical interest of a particular country, politician, or corporation. For example, consider the proven reserves of petroleum. OPEC uses data on available reserves to set quotas on the volume of oil each member can sell. To increase income, member countries tend to overestimate their reserves. On the other hand, oil companies are taxed based on their known reserves; therefore, to pay lower taxes, these companies tend to underestimate their reserves.

Before delving further, we must distinguish between the reserves and resources that a certain region or country has. ***Resources*** are total quantities of a mineral in the crust of the earth -- the known and predicted deposits that may or may not eventually be sold for profit. The amount that has already been discovered or is believed to exist and can be exploited for profit (with today's technology) is commonly known as the ***reserve***. Resources continuously decline, whereas depending on its price and technology, reserves may or may not increase. Although we usually know the size of reserves to a good degree of certainty, we cannot accurately estimate what resources will be discovered in the future. The best we can do is to predict the probability of finding them. The uncertainty increases as we make our predictions farther into the future.[12]

Whether there is an upper boundary to a natural resource or not, that is, whether or not a resource is truly exhaustible, depends on whom we ask this question (see box "Finite or not?"). **Optimists** (also called cornucopians) argue that no natural resource is truly exhaustible. They point out that, as the reserves become scarcer, technological innovation will lead to the discovery of more alternative resources. In addition, production will become more efficient, which will reduce the cost and improve the reserves. **Pessimists** (some call them Cassandras or conservationists) argue that there will be a point at which all resources are depleted and the production process will eventually halt. While new technologies have enabled the discovery of new resources, the additional energy required may not justify further exploration. For example, we have to dig deeper and deeper to access oil as more and more is being extracted. That means the energy used per gallon of oil increases. Eventually we would reach a point where the cost of drilling and the energy needed

[12] The values reported for oil reserves can vary widely from different sources. Companies and countries are deliberately vague about the likelihood of reserves they report. The values reported can vary from a low P-10 (10% probability) to a high P-90 (90% probability) chance of finding the resource. Exaggerated estimates by oil companies will raise the value of their stocks, but may also increase their share of taxes. OPEC countries may have an additional incentive to inflate their estimates in order to boost their quotas for oil exports and obtain loans from international financial institutions. The median estimate P-50, known as proven reserves, is probably the best estimate as errors and biases tend to cancel each other out.

to recover the oil would outpace the economic benefits (the break-even point) and explorations would necessarily stop. As a case in point, since 1980, oil demand has continuously increased beyond any new amounts of oil discovered.[13]

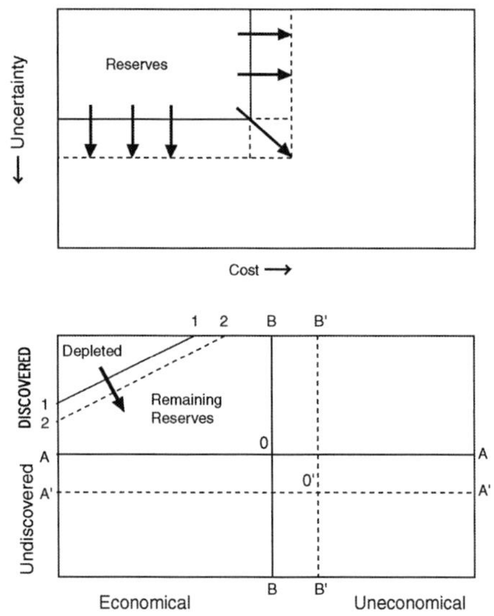

Figure 7-9
McKelvey's Box.

A model proposed by the US Geological Survey, called "McKelvey's Box," is often used to address the relationship between reserves and resources (Figure 7-9).[14] The size of this box represents the total amount of resources available. To the *optimist*, the box is not of a finite size, but varies with the cost of production and the probability that a resource can be found. If the resource is scarce, as the argument goes, its price continues to rise over time, the cost of production goes up, and there would be more effort in conservation and developing new technologies that can point to new discoveries and more efficient recovery methods.

The *pessimists* see the size of this box as finite, stressing that future discoveries and technological innovations cannot catch up with the demand, and that eventually the size of the reserves will shrink to zero. For example, data shows that US oil discoveries have decreased every decade since the 1950s (Figure 7-10) and that since 1962 (except in 1970 when Alaskan oil was discovered), the new oil reserves have never been enough to meet the rise in demand.

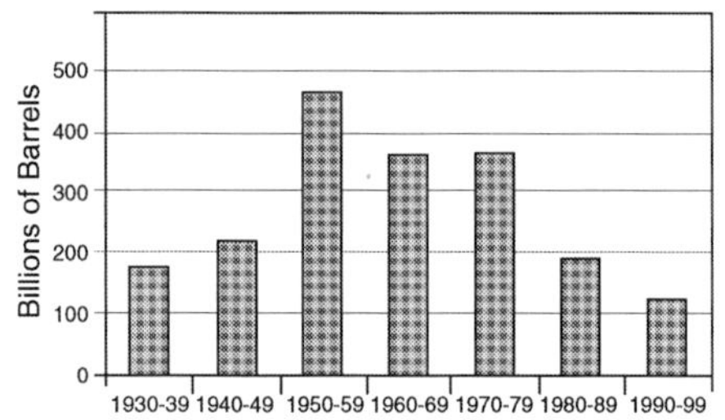

Figure 7-10
Oil discoveries by decade.
Source: Campbell, C. J., and Laherrere, J., "The End of Cheap Oil?" Scientific American, V.278, No. 3, March 1998.

Example 7-1: What is the upper limit to the quantity of oil that earth could contain? What is the exponential expiration time assuming that global oil consumption was 98.9 million barrels a day in 2000?
Solution: The earth could not contain any more oil than the volume of the earth itself. Earth has an average diameter of 6,400 km, so its volume is $Q = \pi d^3/6 = 1.37\text{x}10^{11}\ \text{km}^3 = 8.6\text{x}10^{20}$ barrels. If oil consumption continues to increase at a rate of 7% per year ($T_2 = 10$ years), the time it takes the oil to be expired is given by Equation (4) in Appendix C.

$$EET = (1/r)\ln(rQ/N_0 + 1) \qquad \text{(C-4)}$$

Substituting for r = 0.07, $N_0 = 98.9\text{x}10^6\text{x}365 = 3.61\text{x}10^{10}$ barrels/year, we get:

$$EET = \frac{1}{0.07}\ln\left(0.07.\ \frac{8.6\text{x}10^{20}}{3.6\text{x}10^{10}} + 1\right) = 303$$

Obviously the actual lifetime is much smaller than this as only a small fraction of the earth mass is composed of fossil fuels. Furthermore, this model assumes a constant rate of increase in oil consumption which is probably not a good assumption for many developing countries.

No matter which point of view is considered, as time passes technological

[13] Laherrere, J., "Forecast of Oil and Gas Supply to 2050." Petrotech 2003, New Delhi, 2003. Online at: *http://www.hubbertpeak.com/laherrere/Petrotech090103.pdf.*
[14] It must be noted again that this model applies equally well to other "nonrenewable" natural resources, although for the sake of clarity, we are using oil as an example.

innovation and better production techniques allow a larger portion of resources to become economically extractable. Furthermore, as resources are depleted and prices rise, recovery of additional resources makes more economic sense and line BB shifts to B'B' to the right. At the same time, expected discovery of more resources adds to the reserves and line AA moves down to A'A' (Figure 7-9 bottom). As the total reserves (Figure 7-9 top left box) expand, so does the amount consumed, and the diagonal line 1-1 moves to 2-2. Putting it all together, we can see that the quantity of the remaining reserves can change as a result of two factors:

1. It increases due to discovery of additional resources, higher prices, newer technologies, and more efficient extraction and processing techniques.
2. It decreases because of additional consumption.

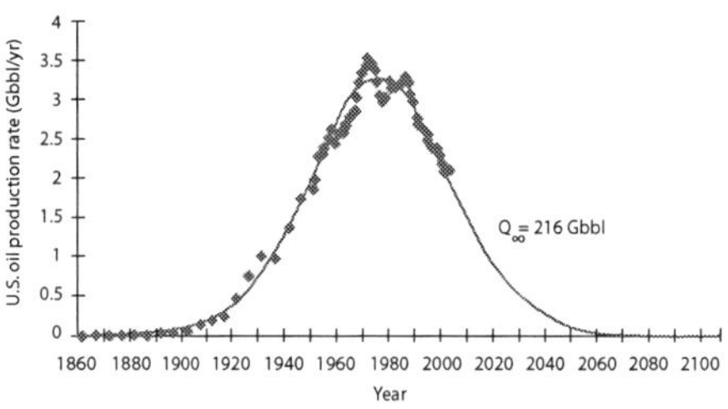

Figure 7-11
Hubbert's prediction of the US oil reserves. The area under the curve is the total accumulative resource.
Source: 1860 to 1965; M.K. Hubbert (Ref. 10); 1965 to present: BP, BP Statistical Review of World Energy (London: Pauffley Ltd, June 2004).

The remaining reserve is the difference between the two. As long as this difference remains unchanged or increases in size, the resource is not depleting and is therefore sustainable. Unfortunately, for fossil fuels the trend is in the opposite direction, and the reserves continue to shrink.

The most famous model pertaining to finite natural resources is that of geologist King Hubbert who, in 1965, predicted that US oil production would peak at about 1970 and decline thereafter (Figures 7-11). He also estimated that the world's oil production would peak at around the year 2000. Furthermore, Hubbert predicted that 80% of the reserves would be depleted in about 60 years, and that even finding an additional 50% in new petroleum resources would not extend the petroleum life more than a few years. Although the predictions were made some 40 years ago, they seem remarkably accurate. For example, Hubbert's predictions about US oil reserves, as well as those of the former Soviet Union and non-OPEC countries, have proven correct with only minor deviations. According to the latest worldwide estimates – even after including later discoveries of Alaskan, Siberian, Middle Eastern, and North Sea oils – the global production peak is expected to occur sometime around now (2007).[15] Note that even if the future discoveries extend the proven reserves to twice the current estimates, the peak in oil production will only shift by a few years (Figure 7-12). Major OPEC countries have not reached their peak productions (mid-points) and cannot yet be tested against Hubbert's hypothesis.

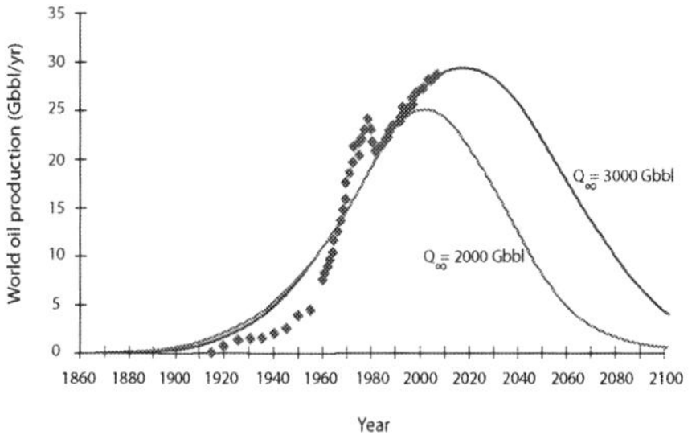

Figure 7-12
Effect of total reserve on lifetime.
Source: Energy Information Administration, Internal petroleum Monthly, July 2006, Tables 4.1a-4.1e and 4.3.

Question: Hubbert based his model on the assumption that resources have a finite life and that production follows a bell-shaped curve symmetric about the peak, where half of all resources have been depleted.[16] What was the basis of these assumptions?
Answer: Hubbert made the assertion that oil discovery, like shooting

[15] World Oil Production Estimates, Jean Laherrère at IIASA International Energy Workshop in Luxemburg, June 2001.
[16] Deffeyes, K. S., "The Hubbert's Peak: The Impending World Oil Shortage," Princeton University Press, 2001.

a target, is a statistical event.[17] There is a certain chance that a sharp shooter will hit the bull's eye, but there is also a chance he misses and lands some distance away. The probability drops as the shooter aims less and less accurately. Oil discovery is similar in the sense that the chance of drilling success is highly problematic. At the early stages, when a natural resource is discovered, the exploitation is slow and the chance of success in finding oil is small. Production is likely to rise exponentially as the most easily found reserves are exploited, technology becomes more mature, and additional uses for the resource are found. As reserves are used up and resources become scarcer, the probability of finding new resources declines; oil companies must dig deeper and production costs increase. Eventually, a point is reached where the rate of discovery reverses its course and starts to slow down. The rate continues to decrease until all resources are found and depleted.

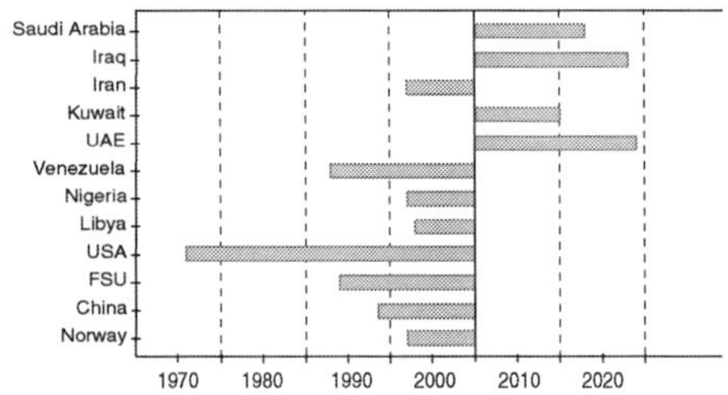

Figure 7-13
Production Midpoints (adjusted) for Major Oil Producing Countries from 2005.

In Figure 7-13, the midpoints, when peak production occurs, are shown for various countries taking 2005 as a reference. As the data show, the United States, China, the former Soviet Union, and most other countries have already passed their midpoints. Only Saudi Arabia, Iraq, Kuwait, and the United Arab Emirates will be reaching their peak productions sometime within the next two decades.

Table 7-4. Estimates of Proven Fossil Reserves, 2006

	Oil *(billion barrels)*	**Natural Gas** *(trillion cubic meters)*	**Coal** *(billion metric tons)*
US	30	6	247
Middle East	785	73	0.4
The World	1,371*	181	909

* Including Canadian tar sands
Statistical Review of World Energy 2007, British petroleum (http://www.bp/com)

Example 7-2: Assuming Hubbert's predictions continue to hold true for the foreseeable future and that all the remaining reserves in the world are uniformly distributed between every man, woman and child, what is the length of time until all petroleum is consumed? Redo the problem assuming US reserves are divided equally among its citizens.

Solution: According to the data in Table 7-4, the estimated total petroleum remaining (as of end of 2006) are 1371 and 30 billion barrels (Gbo) for the world and the US, respectively. During the same period, the world and the US's annual consumptions are estimated at 83.7 and 20.6 million barrels a day. The population of the world was estimated at 6.5 billion and that of the US at 300 million.[18]

Therefore:

[17] It is worth noting that the Hubbert model does not apply only to oil, but to other nonrenewable resources such as coal, natural gas, uranium, and minerals as well.
[18] US Census Bureau (*http://www.census.gov/ipc/www/idb/worldpopinfo.html*).

World:

Per capita total share: $\frac{1371 \times 10^9}{6.5 \times 10^9} = 163.8$ *barrels/person*

Per capta consumption: $\frac{83.7 \times 10^6 \times 365}{6.3 \times 10^9} = 5.73$ *barrels/person/year*

Years remaining: $\frac{211}{4.7} = 28.6$ *years*

US:

Per capita total share: $\frac{30 \times 10^9}{300 \times 10^6} = 100$ *barrels/person*

Per capita consumption: $\frac{20.6 \times 10^6 \times 365}{300 \times 10^6} = 25$ *barrels/person/year*

Years remaining: $\frac{100}{25} = 4$ *years*

There are, however, several flaws in this analysis. Among various important factors we can name:

1. *The total estimate of reserves is probably based on faulty data, inflated by the oil exporting countries.* For example, during 1988-1989, to increase their quota of oil imports (set by OPEC based on the total reserves of each member country), many Middle Eastern countries suddenly increased their estimated reserves by 246 billion barrels to a total of 660 billion barrels.
2. *The rate of consumption is not going to remain constant and will most likely increase with time.* The world consumption of energy is expected to continue to increase in the foreseeable future. This is a result of many factors; the population is expected to increase well into middle of the century, exports by OPEC countries will shrink, and as more migrants move to developed countries in search of better life and higher standard of living, they adopt the more energy intensive life styles.[19]
3. *Not all the oil in the reserves will be pumped as quickly as the oil extracted today.*[20] Today, most easy-to-reach reserves have already been depleted; it is required to dig deeper into the ground, under ocean floors, and remote areas to access the remaining petroleum resources.

On the other hand, there are undiscovered and inferred reserves that are not included in this analysis which could potentially extend the life of petroleum reserves for few additional years (See Table 7-5). The prospect is rather limited, however.

[19] Diamonds, J., Collapse: How Societies Choose to Fail or Succeed, Viking Penguin, 2005.
[20] A similar argument is proposed by C. J. Campbell in "The End of Cheap Oil," *Scientific American*, March 1998. The reserves at the time were 1020 Gbo, and the annual production was 23.6, giving the world 43 years of supply.

Table 7-5. Total Petroleum Endowment: 2000 Assessment *(billion barrels)*		
	World	**US**
Cumulative Production	708	169
Known Reserves	883	24
Reserve Growth (Inferred Reserves)	682	70
Undiscovered Resources	1,290	183
Remaining Resources	2,855	277
Total Endowment	**3,563**	**446**

Cumulative Production –The total amount produced as of 2000.
Known (proven) Reserves – Identified reserves that have not been exploited yet.
Undiscovered Resources – Resources postulated to exist from geologic information and theory outside of known fields.
Reserve Growth –The increase in a known resource that can be inferred from the past historical data.
Remaining Resources – Amount in discovered field that has not yet been produced. They are the sum of Known Reserves, Reserve Growth, and Undiscovered Resources.
Total Endowment – Total resources that ever existed (cumulative production and remaining reserves).

Source: USGS World Production Assessment 2000, http://energy.cr.usgs.gov.

Coal

Coal has been used longer than any other form of fossil fuel. Even early cavemen used coal (known as black rock) for heating purposes. In the United States and many other industrial countries, fuel wood was the dominant energy source until the sixteenth century. In the early seventeenth century, after many forests had been cleared and wood became scarce, coal substituted wood as the main source of fuel. The coal extracted from shallow streams was high in sulfur and burned with an irritating smell but also had some advantages. It burned at higher temperatures than wood and was more desirable for the smelting of irons and other metals. The coal byproduct, coke, was widely used in glass blowing, brick and tile manufacturing, and production of high purity metals that transformed the field of metallurgy.[21] Another byproduct of coal, the coal gas, revolutionized the field of lighting, first on the streets of London and soon thereafter in other large European cities (See box "Coal Gas"). As more and more coal was extracted, the need for better drilling equipment, safety devices, and hauling carts increased. This led to important innovations ranging from reciprocating pumps to atmospheric engines and wagon tramways.

In the early twentieth century, coal accounted for 90% of all energy supplies in the world, before it lost its dominance as an energy source to petroleum, a liquid fossil fuel that could easily be used in internal combustion engines. Today, coal accounts for a quarter of the world's primary energy consumption, and generates 40% of all electricity. Similar figures for the US are 25% and 50%.[22]

[21] Coke is produced by carbonization of coal in an oxygen-deficient environment and high temperatures.
[22] 2004 estimates; a. World Coal Institute *(http://www.worldcoal.org); b. US Geological Survey (http://energy.usgs.gov/coal.html).*

Coal Gas

FYI ...

Coal gas (also called **town gas** or **synthetic gas**) is produced by heating the coal or exposing it to hot steam in the absence of air to carbon monoxide and hydrogen. During the 19th century coal gas was used primarily in cooking and heating and later in the gas-lanterns used by most major cities. Because of its low caloric value and problems with tars, oil, and other pollutants, it was replaced by petroleum when the source was discovered.

Coal gas can be burned directly or converted to methane, gasoline, or other petroleum products such as plastics and photographic film.

Types of Coal

The quality of coal is typically categorized by its rank and grade. The **rank** of the coal represents its morphological development from peat to lignite (or brown coal), sub-bituminous, bituminous (soft coal), and anthracite (hard coal). The higher the coal is ranked, the greater is its carbon content, resulting in more energy being liberated when it is burned (Table 7-6). The **grade** of the coal determines its purity. Coal is of a better grade if its sulfur content is less and burns with lower emissions; coal is classified into low, medium, and high grades.

Table 7-6. Coal Ranks and their Properties.

Rank	Age *(million years)*	Carbon Content	Heating Value *(kJ/kg)**
Anthracite	350	85-95%	34,000 and up
Bituminous	300	45-85%	25,000-35,000
Sub-bituminous	100	35-45%	20,000-25,000
Lignite	60	25-35%	10,000-20,000

*Higher Heating Values.

Whether coal ranks higher or lower depends largely on how it was formed. The first stage of coal formation involves the compression of vegetation under the heavy weights of water and ground materials. These materials gradually turn into a dark-brown, compact organic material known as *peat*. Over time, peat is compressed and heated to form *lignite*. Lignite is a soft, brownish-black coal, containing about 30% carbon. It is also the lowest quality and the most abundant type of coal in the world. Traces of the texture of the original wood may even be found in pieces of lignite. At greater depths, lignite is transformed into *sub-bituminous, bituminous,* and ultimately *anthracite* coal. Therefore, it is expected to find anthracite in the very old deposits that reside in the deep layers and to find lignite in the younger deposits and in those closer to the surface.

Table 7-7. Major Recoverable Coal Resources Worldwide (2002 data)*

Country	Total Recoverable Reserves *(Billion metric tons)*	Percentage of the total
US	243	25.0
Russia	157	16.2
China	103	10.6
India	90	9.3
Australia	72	7.4
Rest of the World	306	31.5
Total	**971**	**100**

**EIA,2003 International Energy Annual Report, DOE/EIA-0219 (2003)*

Reserves and Resources

As Table 7-7 indicates, just five countries, the US, Russia, China, India, and Australia, own about 70% of the total world coal reserves – estimated at 976 billion metric tons. Despite what one might expect, with the exception of Iran, there are no coal mines in the Middle East.

Large coal deposits are found in 38 US states, with more than half of these in only three states: Wyoming, West Virginia, and Kentucky. Anthracite is most commonly found in Pennsylvania and accounts for only 1% of the US's coal reserves. About 70% of coal is bituminous; the rest is essentially lignite. Eastern and Mid-continent coalfields contain mostly bituminous coal, while sub-bituminous coal is predominantly found in the Western states. Most lignite is mined in Texas, Montana, and North Dakota.

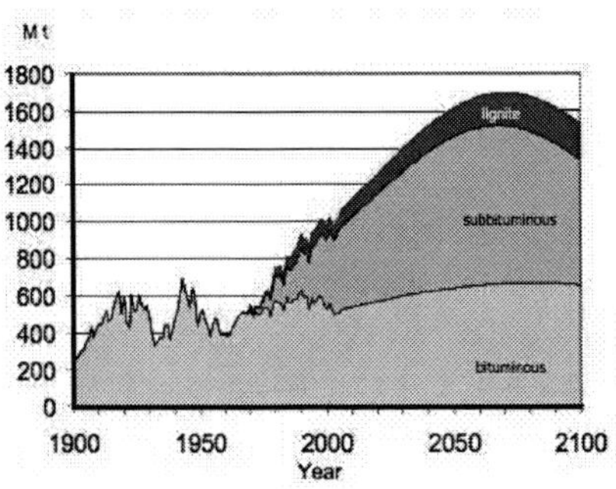

Figure 7-14
US actual and projected coal production.
Source: Historical data, EIA 2006.

Worldwide coal production, which reached 4.8 billion metric tons in 2004, accounted for more than a quarter of all energy sources and about 90% of all electricity generated. In the United States, coal production grew exponentially from 1860 to 1910. As we switched our source of primary consumption to petroleum and natural gas, coal production (and consumption) leveled off until 1972, whereupon the rate started to rise again.[23] The latest figures indicate coal consumption at one billion metric tons in 2004. The US's proven, identified, and ultimate (identified and undiscovered) coal reserves are estimated at 247 billion, 1557 billion, and 3968 billion metric tons, respectively.[24] The actual and projected US coal productions are shown in Figure 7-14.

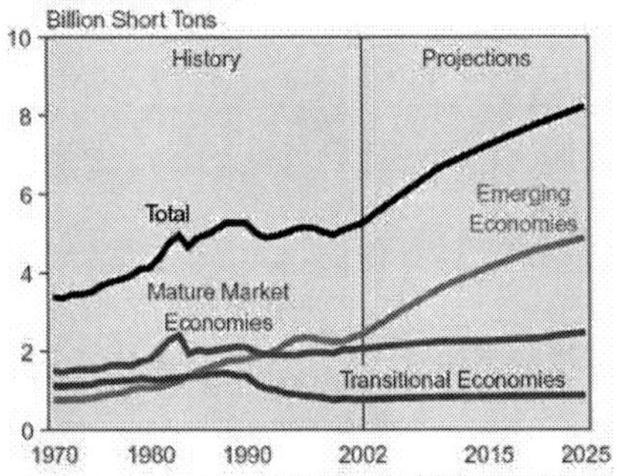

Figure 7-15
World coal consumption outlook by region (1990-2025).
Source: Energy Information Agency, 2003.

Although the US has the most coal reserves in the world, it is China that produces (and consumes) the most coal. The trend is expected to continue, and in the next two decades, coal use is projected to increase by another 36%, with the largest increases projected for China and India (Figure 7-15). In Europe, however, substantial declines in coal use are projected in favor of cleaner natural gas.[25]

Environmental Concerns

Coal is probably the dirtiest of all fossil resources; it produces more carbon per unit energy content, has the highest percentage of sulfur, and produces more nitric oxides. The relative impact on global warming and overall air quality is considered to be the greatest, producing twice carbon dioxide as oil for the same amount of energy (See Figure 7-16).

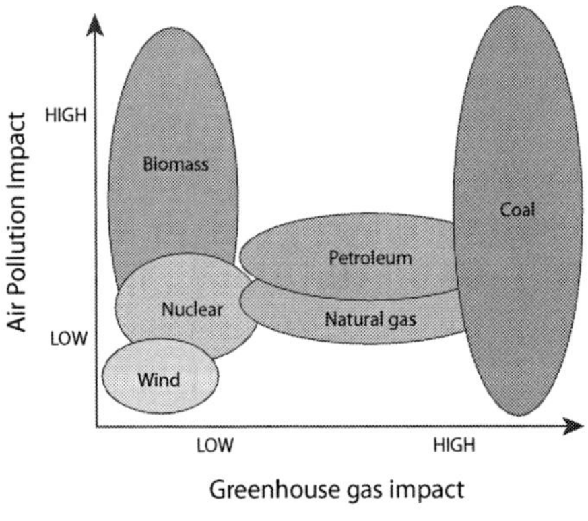

Figure 7-16
Impact of various energy technologies on global warming and overall air quality.

Depending on how deep the coal is buried, coal can be extracted by digging tunnels or strip-mining. Hazards associated with tunnel operation include cave-ins, explosions from dust buildup, carbon monoxide poisoning, and lack of sufficient ventilation. Although tunneling operations harm miners the most, strip-mining has the greatest environmental impacts. These include the destruction of fertile surface-soil, permanent changes in the landscape, and the possibility of acidic or alkaline drainage to the surface (Figure 7-17). In many industrial countries, public and environmentalist pressure has forced local governments to require coal mining companies to reclaim the land and restore it to its original form after strip-mining.

[23] Barlett, A., "Fundamentals of the Energy Crisis," *The American Journal of Physics*, 40, September 1978. pp. 876-888.
[24] EIA Website (*http://www.eia.doe.gov/cneaf/coal/reserves/chapter1.html*).
[25] Enery Information Agency Fact Sheet (*http://www.eia.doe.gov*).

Since coal is formed primarily underneath swamp beds, it contains a large amount of sulfur. To remove the sulfur, coal is crushed into small chunks and washed inside large water tanks. Since sulfur is heavier than coal, most sink to the bottom and is removed. Some of the sulfur is chemically bonded to carbon (organic sulfur) and cannot be washed off this way; it must be removed by adding chemicals that react with sulfur and break it loose.

Figure 7-17
Shale oil can potentially meet a great portion of the future energy demand.

Although most sulfur is removed by physical and chemical processes, the coal burned in power plants still contains considerable amounts of sulfur which, when burned, produce sulfur dioxide. In addition to sulfur dioxide, nitrogen oxides and particulates are produced. Modern power plants use wet or dry scrubbers to remove the remaining sulfur in their smokestacks before it is released into the atmosphere. In wet scrubbers, limestone slurry is sprayed onto the flue gases where it combines with sulfur to form a paste that is left behind. Dry scrubbers consist of a fixed sorbent bed of activated carbon, char, and alumina impregnated with copper. Large bag houses remove larger particles while electrostatic precipitators filter smaller particulates.

In addition to ecological concerns, coal extraction can be quite dangerous. It is estimated that several thousand miners are killed each year as a result of mine explosions, cave-ins, and carbon monoxide poisoning. As the energy crunch becomes tighter and the use of coal becomes more widespread, the number of mine accidents and fatalities are expected to increase.

Of course, coal has a number of advantages. It is relatively cheap; it can be transported by truck, ship, and rail; it is easy to store and burn, and can be liquefied to produce synthetic oil. Every ton of coal will yield about 5.5 barrels of liquid fuel.[26] It should be noted, however, that although there exists a large quantity of coal reserves to be exploited, if coal is used to substitute the current demand for petroleum, the coal supply would serve as only a very temporary solution to the problem.[27]

Oil Shale and Tar Sands

Two potentially rich sources of fossil fuel are oil shale (or shale oil) and tar sands (or oil sands). **Oil shale** is actually a misnomer. It is not shale, but a rock, and it doesn't contain oil, but rather a solid organic compound *kerogen*, which is tightly packed in clay, mud, and silt. **Tar sands** are grains of sand containing thick, viscous, soluble organic liquid called *bitumen* (Figure 7-18). The United States has two-thirds of the entire world's oil shale (along shores of Green River in Wyoming, Colorado, and Utah), and Canada has the largest percentage of tar sands in the world (in Alberta). In fact, if tar sand reserves are included, Canada would have the second largest

[26] Lumpkin, R.E., "Recent Progress in the Direct Liquefaction of Coal," *Science*, 239, p. 873, 1988.
[27] Hatfield, C. B., "How long oil supply grow?" M. King Hubbet Center for Petroleum Supply Studies, Newsletter #97/4-1-6, October 1997.

amount of petroleum reserves in the world after Saudi Arabia.[28] Russia and Brazil are also rich in oil shale, whereas Venezuela has the second largest deposits of tar sands (after Canada). Worldwide resources of tar sands are estimated at around two trillion barrels.[29] An additional 2.8-3.3 trillion barrels of oil can be recovered from oil shale.[30] Mining operations are very energy intensive; oil shale resides deeply underground and must be heated to high temperatures (around 400-500°C) before it releases any oil. In fact, the oil extract is not petroleum, but liquid kerogen. Steam is needed to hydrogenate kerogen into hydrocarbons that may then be refined into gasoline and other petroleum products. Tar sands must also undergo a similar process. Heating them to high temperatures causes the viscosity of the bitumen to drop, making it flow more easily out of the sand. Various petroleum products, such as kerosene can be manufactured by distilling the kerogen and bitumen oils.

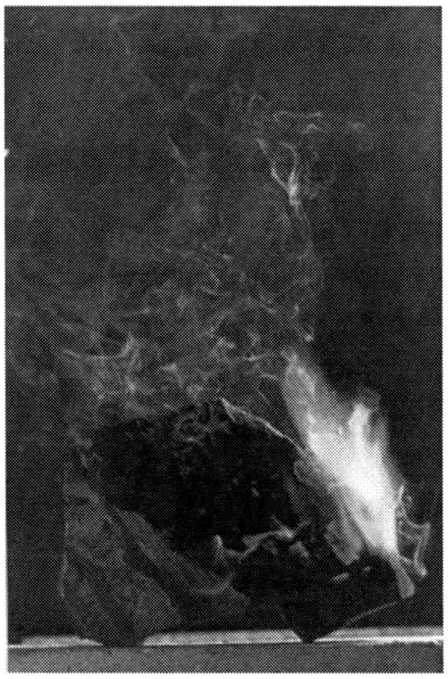

Figure 7-18
The huge equipment used in strip-mining operations illustrates the dramatic impact that mining has on neighboring communities.

In addition to their low yield (50-100 liters per ton of rock), oil shale and tar sands are not clean. These resources are rich in sulfur and nitrogen which can contribute significantly to the acid rain problem. Another environmental concern is the disposal of residues, called tailing, which occupy many times the volume of crude they produce. These problems have hindered widespread use of oil shale and tar sands. Canada is pursuing mining tar sand deposits aggressively; Estonia, Brazil, and China use oil shale to produce electricity and in production of cements. As the price of petroleum rise, these resources will find a more prominent role to meet our energy needs.

Natural Gas

The same factors that promote the formation of petroleum also help in the formation of natural gas. It is therefore expected that natural gas is found at locations near oil fields, often in association with petroleum (thus its name *associated* or *dissolved gas*). It is liberated when oil is brought to the surface in the same manner that carbon dioxide gas is liberated when someone opens a carbonated drink. If the gas diffuses through the pores of sedimentary rocks and accumulates in a reservoir other than the oil reservoir, the gas is called *non-associated gas*. Up to only a few decades ago, most natural gas discovered was flared off or reinjected into the ground, as it was considered to be of little use. Its use increased during the latter half of the twentieth century, as it proved to be a convenient and relatively clean fuel to be burned in gaseous burners for space heating and cooking (Figure 7-19).

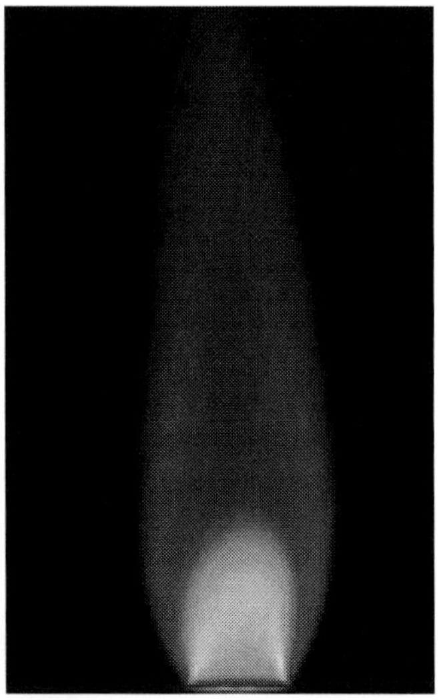

Figure 7-19
Natural gas, the cleanest of all fossil fuels.

Natural gas is primarily methane (CH_4), but a certain percentage of heavier hydrocarbons such as ethane (C_2H_6), propane (C_3H_8), butane (C_4H_{10}), and a small amount of pentane (C_5H_{12}) and hexane (C_6H_{14}) are

[28] George, R L, "Mining for Oil," *Scientific American*, March 1998.
[29] Duncan, D. C., and Swanson, V. E., "Organic-rich shale of the United States and world land areas," US Geological Survey Circular 523, 1965.
[30] "*Annual Energy Outlook 2006*," Energy Information Energy, February 2006.

also present. At normal temperature and pressure, pentane and hexane are liquid, but in elevated temperatures and pressures underground, they become gas and flow out with other gases. Commercial natural gas is primarily a mixture of methane and ethane. The propane and butane are liquefied and sold separately as *liquefied petroleum gas* (LPG) or *"bottled"* gas. For a long time, many producer countries used to burn natural gas at the well. This was very wasteful and caused huge environmental problems. Because of its lower density, it costs about four times as much to transport natural gas through pipelines as it does for crude oil; these gases are liquefied and shipped to their destinations as liquefied natural gas (LNG) in huge cryogenic containers aboard large tankers and ocean liners.

Table 7-8. Total Natural Gas Endowment *(bbo equivalent)**

	World	US
Cumulative production	292	142
Undiscovered reserves	866	88
Reserve growth (inferred)	610	59
Remaining (proven)reserves	799	29
Total Endowment	**2567**	**318**

* For definitions refer to Table 7-5.
Source: USGS World Production Assessment 2000 (http://energy.cr.usgs.gov)

Reserves, and Resources

The total amount of available natural gas is shown in Table 7-8. Roughly two-third of all natural gas reserves lie in Russia and the Middle East (Table 7-9). The United States with 193 trillion cubic feet has a little more than 3% of the world's natural gas reserves.[31]

Table 7-9. World Proven Natural Gas Reserves as of Jan. 1, 2006

Country	Total Recoverable Reserves *(trillion cubic feet)*	Percentage of the Total
Russia	1680	30.5
Iran	971	15.9
Qatar	911	14.9
Saudi Arabia	241	3.9
UAE	214	3.5
Other Mid.East	223	4.1
US	193	3.1
Rest of the World	1,679	24.1
Total	**6112**	**100**

Source: Energy Information Agency, 2006.

Because natural gas is relatively clean, its rate of consumption is increasing by an average of 2.8% annually, faster than other sources of fossil fuels (Figure 7-20). If consumption continues to grow at this rate, natural gas reserves remain available for another 75 years. However, this number is believed to be optimistic, as natural gas substitutes coal and petroleum in an effort to reduce the effect of greenhouse gases.[32]

Besides wells, however, there are other sources of natural gas such as hydrates that if they could be extracted, would provide energy for many hundreds of years. Hydrates are icy deposits of crystallized natural gas and water, buried under the extreme pressures and cold temperatures of the deep oceans and Arctic permafrost that have been formed by the disintegration of certain bacteria. It is estimated that gigantic hydrate fields around the world contain twice the energy of all other forms of fossil fuels combined. These reserves are virtually untapped, as there are major obstacles that have to be overcome before the reserves can be accessed. Any major changes in temperature or pressure may be capable of breaking down the material, releasing huge amounts of methane – a very potent greenhouse gas (See box "The Deadly Lake").[33] The cost of extraction is prohibitively high and the natural gas, even if it is extracted, has to be transported thousands of miles from Arctic regions before it can be used. One solution is to convert it to liquid synthetic fuel, such as methanol, before it is sent through pipelines.

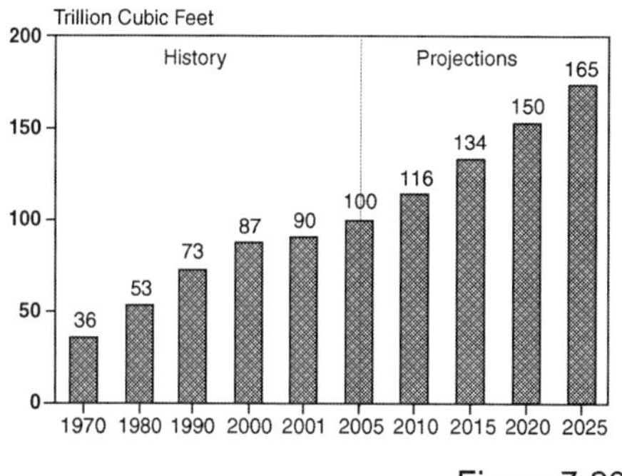

Figure 7-20
The outlook for worldwide natural gas consumption
Source: Energy Information Agency, 2005.

[31] Energy Information Agency Fact Sheet, *http://www.eia.doe.gov.*
[32] Meadows, D., Limits to Growth: 30-Year Update, Chelsea Green Publishing Company, 2002, p. 94.
[33] A few petroleum geologists have gone so far as to blame the loss of ships and airplanes in the Bermuda Triangle on sudden pulses of methane gas released from a hydrate layer below the Atlantic Ocean. As the ocean boils with a sudden squirt of methane bubbles, ships can be swallowed before having time to make distress calls, and plane engines might be choked by the rising plume of methane gas.

The Deadly Lake

FYI ...

Lake Kivu on the border between Congo and Rwanda is unique in the world in one important way: it conceals an enormous quantity of water -- 65 billion cubic meters -- lying dormant at the bottom of the cold lake.[i] Furthermore, the lake is continually being recharged with the gas, giving Rwanda an almost inexhaustible source of relatively clean energy. The source of this gas is believed to be tiny single-celled microorganisms called archaea that consume dissolved carbon dioxide from two nearby active volcanoes to produce methane. Some methane is also formed through bacterial fermentation of acetate in the lake bed sediments. Deeper waters can trap methane to higher concentrations. If water is brought to the surface, either by a pump or naturally through mixing or siphoning, it will lose its ability to hold the methane, releasing it in form of gas bubbles that can be collected and piped away. Methane, being 25 times less soluble in water than carbon dioxide, is bubbled first leaving carbon dioxide dissolved in the water. If water is saturated and the buildup of methane gas is allowed, it can form an explosive danger with catastrophic consequences. A fiery disaster or be a blessing in disguise for the 2 million people who live along the shore [ii]

[i] *Halbwachs, M., et al. "Investigation in Lake Kivu after the Nyiragongo Eruption of January 2002," European Community Humanitarian Office, Communauté européenne – ECHO 1 rue de Genève B- 1049 Bruxelles.*

[ii] *Bavier, J., "Deadly Mix of Gases Lurks in Congo's Lake Kivu; Explosive Methane Could Be Tapped To Generate Power," The Washington Post, June 17, 2007.*

Converting natural gas to liquid first requires breaking its chemical bonds using steam, heat and a nickel-based catalyst to produce a mixture of carbon monoxide and hydrogen known as *syngas*. This process is called steam reforming. Syngas is then blown over various catalysts to transform it into gasoline, diesel, and other liquid hydrocarbons.

Summary

No doubt fossil fuels have been and will remain as the most important source of energy. Much of today's technology has come about because of the availability of cheap and abundant fossil reserves. Unfortunately, in a relatively short time, we have managed to consume nearly half of our estimated conventional oil and much of our coal and natural gas resources. It is particularly noteworthy that by 1970, over half of the globe -- Africa, ME, Asia (except Japan)-- did not essentially use any oil. This is becoming particularly difficult as the demand for oil is increasing in much of the world. Developed countries, at least in foreseeable times, will consume more. As developing nations strive for better economic conditions, they will demand more share of the energy resources. Similarly, as Middle Eastern countries become industrialized, they will use more of their resources locally, leaving less available for export. Furthermore, with a majority of OPEC members in the Middle East, the political instability in this region can easily disrupt the flow of oil, increasing global conflicts and the potential for even more military interventions and war. This will have a spiraling effects of higher prices of petroleum, economic downturn, and even more conflicts. Unless we make drastic adjustments in our patterns of use, we may not be able to enjoy these resources for much longer.

Fortunately, there is hope that new sources of energy will become available at reasonable prices. Nuclear energy can offset rising demand for a short period. Tar sands and oil shale can extend the life of petroleum for some times. Similarly, methane hydrate is a huge potential source of natural gas. Alternative and clean sources of energy are also being developed at prices that are continuously being lowered. Until then, all we can do is to use the available reserves sparingly.

Additional Information

Books

1. Berkowitz, N., Fossil Hydrocarbons: Chemistry and Technology, Elsevier Academic Press, 1997.
2. Deffeyes, K. S., Hubbert's Peak: The Impending World Oil Shortage, Princeton University Press, Princeton, N. J., 2001.
3. Campbell, C. J., The Coming Oil Crisis, Multi-Science Publishing Company, 2004.
4. Tariq Ali, The Clash of Fundamentalisms: Crusades, Jihads and Modernity, Verso, 2002.
5. Pelletiere, S., Iraq and the International Oil System: Why America Went to War in the Gulf, Praeger Publishing, 2001.

Periodicals

1. *Oil and Gas Journal*, Technology, news, statistics, special reports, and analysis *(http://ogj.pennnet.com)*.
2. *Journal of Petroleum Technology*, The official journal of Society of Petroleum Engineers, Dallas.
3. *The Petroleum Engineer,* Petroleum Engineer Pub. Co.
4. *Journal of Petroleum Science and Engineering,* Elsevier, covers the fields of petroleum (and natural gas) exploration, production and flow.

Government Agencies and Websites

1. National Energy Technology Laboratory: The Strategic Center for Coal *(http://www.netl.doe.gov/coal)*.
2. National Petroleum Technology Office *(http://www.npto.doe.gov)*.
3. US Geological Survey *(http://www.usgs.gov)*.

Non-Government Organizations and Websites

1. Organization of Petroleum Exporting Countries (OPEC) *(http://www.opec.org)*.
2. Society of Petroleum Engineers *(http://sae.org)*.

Exercises

I. Essay Questions

1. Who are the top five producers and top five consumers of petroleum?

2. When was OPEC established and why? Name eight of OPEC's member states.

3. How much carbon dioxide is put into the air for every gallon of gasoline burned in an average car?

4. Where are the main sources of tar sands and shale oil? What are major impediments in using these resources on a large commercial scale?

5. How is coal classified? What are the characteristics of the best kind of coal?

6. What are the underlying assumptions in Hubbert's model? According to his model, when will half of all total reserves (past and future) be used up?

7. What is the difference between reserves and resources? In your opinion, is there a limit to the amount of reserves in the world?

8. What are some of the techniques for discovering oil? Explain the principle of operation for two of them.

9. What are the advantages and disadvantages associated with tar sands and oil shale? What is the role that these resources will play in meeting the future energy needs?

10. Explain the sources of conflict in the Middle East. How are they related to oil? Is there a sensible solution to problems gripping this region?

II. Multiple Choice Questions

1. Which energy source is used most by the United States?
   a. Coal
   b. Petroleum
   c. Natural Gas
   d. Solar
   e. Hydroelectric

2. After Saudi Arabia, the US
   a. Has the largest oil reserve
   b. Has the largest oil resources
   c. Is the largest oil producer
   d. Has the largest coal deposits
   e. All of the above

3. Gasoline is a product of refining
   a. Coal
   b. Petroleum crude
   c. Natural gas
   d. Propane
   e. Ethanol

4. What percentage of the world's proven oil reserves is in the United States?
   a. 2-3%
   b. 10-12%
   c. 15-20%
   d. 25-30%
   e. More than 30%

5. Natural gas is transported mainly by
   a. Pipelines
   b. Trucks
   c. Barges
   d. Trains
   e. All four, roughly equally

6. The three countries with the largest petroleum reserves in the world are
   a. Saudi Arabia, United States, and Russia
   b. Saudi Arabia, Venezuela, and the United Arab Emirate
   c. Saudi Arabia, Iran, and Iraq
   d. Iran, Iraq, and Kuwait
   e. United States, China, and Russia

7. The top three oil consuming countries in the world are
   a. Saudi Arabia, United States, and Russia
   b. United States, India, and China
   c. Saudi Arabia, Iraq, and Kuwait
   d. United States, China, and Russia
   e. United States, Japan, and China

8. Among the OPEC members that border the Persian Gulf are
   a. Iran, Libya, Kuwait, and Saudi Arabia
   b. Iran, Iraq, Algeria, and United Arab Emirates
   c. Iran, Iraq, Kuwait, and Saudi Arabia
   d. Iraq, Venezuela, Qatar, and Saudi Arabia
   e. Pakistan, Iran, Iraq, and Kuwait

9. The largest reserves of natural gas are in
   a. Saudi Arabia
   b. Iraq
   c. The United States
   d. The Russian Federation
   e. China

10. The largest reserves of coal are in
   a. Saudi Arabia
   b. Iraq
   c. The United States
   d. The Russian Federation
   e. China

11. Petroleum is primarily used in the United States in
   a. Transportation
   b. Generating electricity
   c. Heating and cooling buildings
   d. The petrochemical industry
   e. The pharmaceutical industry

12. In pockets containing petroleum,
   a. Oil is on top, water in the middle, and gas at the bottom
   b. Oil is on top, gas in the middle, and water at the bottom
   c. Water is on top, oil in the middle, and gas at the bottom
   d. Gas is on top, water in the middle, and oil at the bottom
   e. Gas is on top, oil in the middle, and water at the bottom

13. Most of the United States' oil imports are from
   a. Canada
   b. Mexico
   c. Saudi Arabia
   d. Alaska
   e. Europe

14. Over _______ percent of all petroleum reserves lie in the Persian Gulf region.
   a. 99
   b. 90
   c. 80
   d. 60
   e. 20

15. Natural gas combined-cycle combustion turbines
   a. Have greater combined efficiency
   b. Have lower rates of pollutant emissions
   c. Can be used as central or distributed
   d. Reject heat at lower temperatures
   e. All of the above

16. Propane is the fuel of choice on farms and in rural areas, mainly because it is
   a. Safer
   b. Portable
   c. Cleaner
   d. Cheaper
   e. More plentiful

17. Which country is the largest coal producer in the world?
   a. The United States
   b. Russia
   c. China
   d. India
   e. Germany

18. Which of the following statements is correct?
   a. Natural gas is formed by heating petroleum until it is turned to a gas.
   b. The largest reserves of natural gas are in Russia and Iran.
   c. Natural gas reserves can be found only where there is oil.
   d. After the Middle East, The United States is the largest producer of natural gas.
   e. Natural gas is composed of methane, ethane, and propane in roughly equal proportions.

19. Which US state is the largest producer of petroleum?
   a. California
   b. Louisiana

c. Alaska
d. Texas
e. Pennsylvania

20. Which fossil fuel creates the lowest amount of carbon dioxide per kilogram of fuel burned?
    a. Methane
    b. Methanol
    c. Coal
    d. Petroleum
    e. All four produce about the same amount

21. For each kilogram of gasoline we burn in our cars, we pollute about _______ kilogram of air.
    a. 1
    b. 15
    c. 100
    d. Over 1,000
    e. Cannot tell

22. The amount of a natural resource that may become available for use is called
    a. Reserves
    b. Reservoirs
    c. Resources
    d. Supplies
    e. Capacity

23. Proven reserve refers to a _________ percent probability of finding new reserves.
    a. 10%
    b. 50%
    c. 90%
    d. 99%
    e. 100%

24. Peat
    a. is the highest quality form of fossil fuel
    b. is the first stage in the formation of coal
    c. is embedded in tar sands
    d. is made by compressing animal dung
    e. is the nickname for Peter

25. Which of the following statements is true about tar sand?
    a. Tar sand is expected to play a significant role in energy production in the 21st century.
    b. About 2/3 of all tar sand deposits in the world resides in the US.
    c. Tar sand is a clean source of energy. Unfortunately it is expensive to produce.
    d. Per mass basis, tar sand yields a relatively large amount of energy.
    e. All of the above.

III. True or False?

1. Coal, petroleum, natural gas, and propane are called fossil fuels because they are formed from the buried remains of plants and animals that lived millions of years ago.

2. It is possible to find natural gas even if no petroleum is found.

3. It is estimated that over 70% of all of the world's recoverable natural gas has been used already.

4. Hydrogen is the cleanest form of fossil fuel.

5. Chemical reactions that release heat are called exothermic.

6. The secondary recovery technique relies on reducing the oil viscosity.

7. At the point at which oil production reaches its peak, half of all oil has been already consumed.

8. Peak production represents the mid-time that a particular resource is available to us.

9. The P-90 reserve figures generally over- estimates the amount of proven reserves.

10. With the existing technologies, over 90% of all petroleum reserves can be extracted.

11. During the next doubling time for coal production, we will use as much coal as we have used up to this point.

12. According to many experts, global oil production will peak sometime during this decade.

13. A great portion of coal reserves resides in the Middle East.

14. At the current rate of consumption, the remaining US coal resources should last an additional 300 years.

15. Oil shale is an oily substance found usually in seashells.

IV. Fill-in the Blanks

1. Coal reserves are typically categorized by ______ and __________.

2. Although __________ has the most coal reserves in the world, it is _________ which produces (and consumes) the most coal.

3. Tar sands are grains of sands containing a viscous carbonaceous substance called _____________.

4. The combustible fuel in the oil shale is a waxy solid called ______________.

5. The substance remaining in the bottom of the distillation column is __________.

6. The propane and butane removed from natural gas are usually liquefied under pressure and sold as __________________.

7. Natural gas is the cleanest form of fossil fuel because it has the highest __________ ratio.

8. The other name for liquefied petroleum gas is the ___________ gas.

9. The United States reached its peak oil production sometime during __________.

10. The total amount of a resource produced as of today is called _______________ production.

V. PROJECT I - Hubbert's Curve

Describe the major assumptions implicit in King Hubert's prediction concerning oil production. Use his estimates on the world petroleum production (shown on the cover of this book) to calculate the growth rate during decades spanning from 1900 to the present.

1. Plot the cumulative total production as a function of time. What does this graph look like? Why?
2. What is the size of total endowment?
3. What is the doubling time at around the turn of the century?
4. Based on his prediction, and assume the pattern of production follows his estimates, how much longer, do our petroleum resources last?

PROJECT II – Status of Energy Resources

In this project you are asked to investigate the status of energy resources, reserves, and use in the United States and compare it with those of the world as a whole. The main source of data is the Energy Information Agency of the US Department of Energy (*http://eia.doe.gov*), but other sources may need to be considered. Please answer the following questions:

1. What is the total annual energy consumption (in quads and billions of barrels of "oil equivalent") for various energy sources: oil, gas, coal, renewable, and nuclear? for various energy sectors: transportation, industrialized, residential and commercial?
2. What fraction of the primary energy consumption is used to generate electricity? Give the percentage from each source.
3. What are the ultimate sizes of the US and the world reserves?
4. What is the energy consumption per capita? How fast does total energy consumption increase with a rise in the gross domestic product (GDP)?
5. What is the energy consumption per every $1,000 of income? Compare US data with China, India, and the European Union.

Work Sheet For Project II				
	The United States		The World	
	Quads	Percent	Quads	Percent
1. Consumption				
a. Consumption by Source				
Oil				
Gas				
Coal				
Nuclear				
Renewable				
b. Consumption by Sector				
Transportation				
Industrial				
Residential & Commercial				
2. Electricity by Source				
Oil				
Gas				
Coal				
Nuclear				
Renewable				
3. Ultimate Size of Reserves		X		X
4. Energy Consumption Per Capita		X		X
a. Annual Rate of Energy Growth	X		X	
b. Annual Rate of Growth in GDP	X		X	

CHAPTER 8

Air Pollution from Combustion Sources

All substances are poisons. There is none, which is not a poison. The right dose differentiates a poison from a remedy. ~ Paracelsus (1493-1541)

Concentrate on what cannot lie. The evidence... ~ Gil Grissom (1956-)

Two of the main consequences of the second law of thermodynamics that must be taken into account when burning fossil fuels are the release of large amounts of waste heat and the emission of toxic pollutants into the surrounding air and water. Thermal pollution arises because the full energy content of the fuel cannot be converted to useful work and electricity. Chemical contaminants are produced because the combustion process cannot go entirely to completion.

Environmental pollution does not only affect human health, but also endangers plant and animal life, damages materials, affects visibility, and contaminates air, water, and soil. These effects cost billions of dollars in cleaning, repair, replacement, prevention, and lost productivity. According to the World Health Organization (WHO), an estimated 1.5-6 million people die of air pollution related diseases annually, mostly in developing countries where health care facilities are often inadequate, environmental laws are seldom enforced, and the high costs of air pollution control equipment are largely prohibitive.

In Chapter 5, we discussed the laws of thermodynamics and how they can be used to design power plants and other useful devices. In these devices, either heat is used to produce shaft work (automobile) and electricity (power plants), or work is used to move heat from a colder to a hotter space (refrigerators and heat pumps). In all cases, some heat must be discarded into the atmosphere in a process known as *thermal pollution*. Thermal pollution is not limited to mechanical devices, however. Any human activity, even the simple act of breathing, involves the dissipation of a substantial amount of heat to the surrounding air. Other phenomena attributed to emissions from combustion sources are smog formation, ozone depletion, and acid rain.

Global Warming

The earth's climate has undergone great changes over time. Until recently, these changes were solely a result of natural variation in the atmosphere. Human activities in the last century, particularly the burning of vast amounts of fossil fuels, have been responsible for alterations in atmospheric composition, changes in weather patterns, and an increase in global temperature of about 0.6°C. During the past 25 years, however, the rate of temperature increase has been even greater and totals to about 2°C if extrapolated over 100 years. In fact, the ten warmest years since 1860 (at which point measurements started) have occurred after 1998. According to data recently released by NASA, the earth is becoming hotter than it has been in the last 12,000 years, and if the current global warming pattern continues, by middle of this century, temperatures will exceed the highest level ever in the last million years.

Burning fossil fuels releases large amounts of carbon dioxide, nitric oxides, methane, and other pollutants acting as the culprits for global warming. These are collectively termed "*greenhouse gases*".

Greenhouse Gases

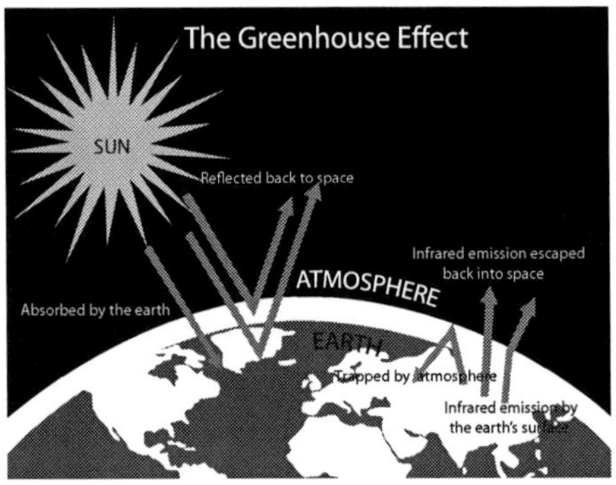

Figure 8-1
Global Warming

The link between carbon dioxide and atmospheric temperatures was presumed as early as the 19th century, but it remained merely speculation until the 1980s when enough data had been collected to show strong correlation between the two. In order to understand the global warming phenomenon, we must understand the greenhouse effect. Greenhouses heat to temperatures exceeding their surroundings because window glass (and some plastics) is transparent to incoming solar radiation but blocks most of the reflected infrared radiation from the plants and surroundings. Certain gases such as water vapor, carbon dioxide, methane, nitrous oxides, and CFCs behave in a similar fashion; they allow solar radiation to pass through to reach and heat up the ground and air above it but absorb and reradiates the reflected terrestrial (IR) radiation, blanketing the earth. Because no air is present outside the atmosphere, heat cannot escape by conduction and cooling can only occur by radiation. Carbon dioxide is the main product of the combustion of fossil fuel. Methane is leaked into the air from coal mines and pipelines, but is mostly produced by cattle and termites, or in wetlands, municipal waste dumps, and rice patties. Nitric oxides come primarily from fertilizers and animal waste. CFCs are non-toxic, inexpensive, and energy-efficient substitutes for highly toxic, potentially explosive ammonia and, until recently, were marketed under the trade name Freon, being widely used in fire extinguishers, refrigeration systems, and aerosol spray cans.

The degree to which greenhouse gases contribute to global warming depends not only on the nature of the gases and their concentrations, but also on the length of time that they remain in the atmosphere. Some gases remain for short periods of time while others, such as CFCs, may remain there for several hundred years.[1] Contrary to what some may suspect carbon dioxide is not the most damaging greenhouse gas. Methane traps 20 times more heat per molecule than carbon dioxide, nitric oxide 200

[1] For example, CFC-12, common in many refrigeration systems, has a half-life of 100-150 years. CFC-115 can stay in the atmosphere for up to 800 years.

times more, and CFCs could be 10,000 times more damaging than carbon dioxide. Carbon dioxide has been cited as the most important greenhouse gas merely because it is the most abundant; it is responsible for roughly one-half of measurable global warming.

Question: Water vapor is an effective greenhouse gas. Why isn't it cited as the main cause of global warming?
Answer: Indeed, water vapor is the largest contributor to greenhouse gases in the atmosphere. When we burn fossils, we are producing about twice as much water vapor as carbon dioxide; however, its emission is directly linked to the emission of carbon dioxide. Unfortunately, some politicians and lobbyists cite the role of water vapor as proof that global warming is a natural phenomenon having nothing to do with human activity.

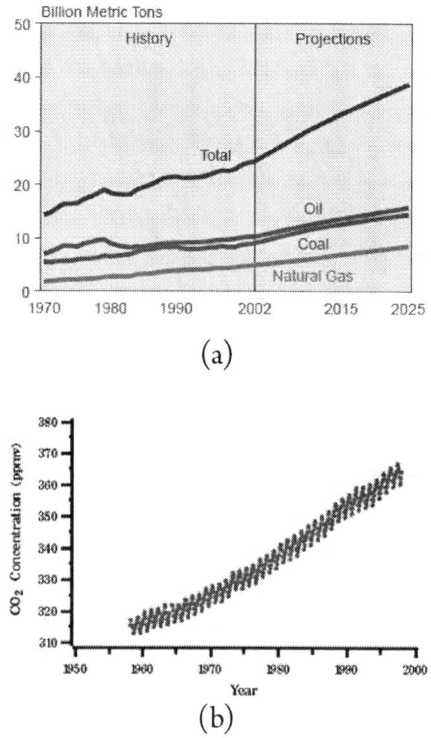

Figure 8-2
World carbon dioxide emission 1970-2020. (a) Total emission, and (b) Background concentration

An analysis of air bubbles trapped in Antarctic ice caps provides evidence that, for thousands of years, carbon dioxide concentrations in the atmosphere has been essentially constant at around 280 parts per million (ppm). Only in the last two hundred years, following industrialization and the rapid increase in the rate of consumption of fossil fuels, has man-made carbon dioxide caused these levels to increase.[2] The concentration is now roughly 370 ppm and is expected to rise to over 500 ppm by the middle of this century. As Figure 8-2 shows, the increase in carbon dioxide emission in the last few decades closely correlates with the consumption of fossil fuels. The seasonal variation of the background carbon dioxide concentration can be explained by noting that in spring and summer (in the northern hemisphere), when photosynthesis is more predominant, plants use some atmospheric carbon dioxide as food. It is returned to the atmosphere in fall and winter as plants die.

Until now, most greenhouse gases have been released by industrial countries, mainly the United States, and distributed roughly equally among the transportation, residential and commercial heating and cooling, and industry sectors (Figure 8-3). As developing countries become more industrialized and their consumptions of fossil fuels increase, it is expected that the percentage of greenhouse gases contributed by these countries will increase, causing global warming to accelerate in the foreseeable future.[3]

Cars and light trucks	20.2%
Trucks and Buses	5.4%
Aircrafts	3.6%
Boats and Vessels	1.3%
Rail	0.7%
Motorcycles	1.8%
Total Transportation	**33%**

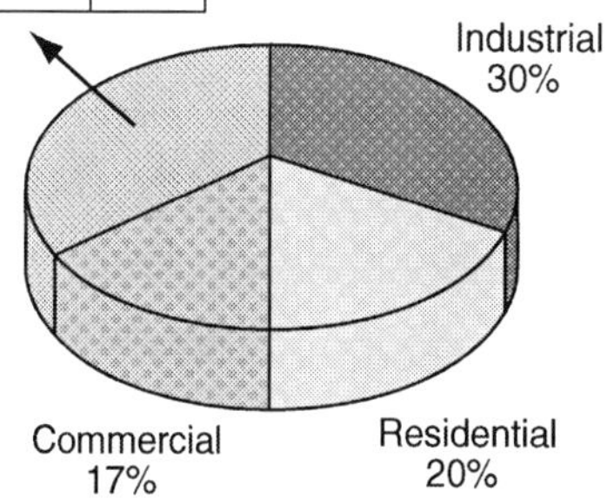

Figure 8-3
US carbon dioxide emissions from energy sources by sector in 2000.
Source: DoE/EIA-0573, 2000. http://www.eia.doe.gov/oiaf/1605/ggrpt/tbl5.html.

Question: Estimate the volume of carbon dioxide released into the atmosphere as a result of fossil combustion in the United States.
Answer: The latest data indicates that the US consumes about 112 quads[4] of energy annually. A simple calculation can be made to convert this to the total mass of fossil fuel consumed. Referring to the

[2] Neftel, A., E.,et al., "Evidence From Polar Ice Cores for the Increase in Atmospheric CO_2 in the past Two Centuries". (*Nature, 315, 45. 1985)*
[3] Recent data suggests that for the first time China surpassed the United States in total emission of greenhouse gases. Per capita production remains well below US, however. *(International Herald Tribune, June 20, 2007)*
[4] Each quad is one quadrillion BTU, each BTU is $2.51x10^{-7}$ barrels of oil, each barrel of oil is 42 gallons, and each gallon is 3.785 liters.

energy conversion tables given in Appendix A, we have:

112 quads = 1.12×10^{17} BTU = 2.81×10^{10} barrels of oil = 1.18×10^{12} gallons of oil = 4.47×10^{12} liters of oil

The United States satisfies about 86% of its energy needs with fossil fuels. Assuming a fuel specific gravity of 0.8, the total mass of fossil fuels consumed will be $4.16 \times 10^{12} \times 0.86 \times 0.8 = 2.9 \times 10^{12}$ kg fossil. Considering that, on average, hydrocarbon fuels produce about three times their mass in carbon dioxide, the total amount of carbon dioxide emissions is 9×10^{12} kilograms (9 billion tons). This figure accounts for about one quarter of all the carbon dioxide produced in the world.

Factors Affecting Global Warming

Various mechanisms affect the severity of global warming to different degrees. Among the mechanisms that increase the rate of global warming are:

1. Rising temperatures cause acceleration in the rate of decay of organic matter, producing methane, a major greenhouse gas.
2. Rising temperatures make forests drier and more vulnerable to pests, diseases, and fire. For example, it is estimated that by doubling of the carbon dioxide levels in the atmosphere, will raise earth's surface temperature by about 3°C; the US alone will lose up to 40% of its forests.[5]
3. With rising temperatures, the Arctic permafrost and tundra will melt and release an enormous amount of methane into the atmosphere, accelerating global warming. Tundra and permafrost cover about one-fifth of all land.
4. Oceans are huge sinks of carbon dioxide. The solubility of water is highly temperature dependent -- the colder its temperature, the higher the capability of water to dissolve carbon dioxide. As ocean temperatures rise, some of the stored carbon dioxide is released back into the atmosphere.

While most mechanisms act to amplify global warming, there are some that reduce the effect. Among the mechanisms responsible for reducing the rate of global warming are:

1. The burning of fossil fuel releases particles that act as condensation nuclei and increase cloud-cover, blocking sunlight from reaching the earth.
2. Higher temperatures increase rates of ocean water evaporation, thereby expanding low altitude cloud-cover.[6] The role of clouds in

[5] Cline, William R. 1992. "The Economics of Global Warming," Washington D.C. Institute for International Economics.

[6] The decrease in solubility of carbon dioxide with temperature is also the main reason that soda, beer, and other carbonated beverages become flat shortly after they are opened. Soda cans are filled under pressure. When opened, their pressure drops to atmospheric levels. Carbon dioxide is less soluble and leaves the beverage. Similarly, when the beverage warms up, its ability to dissolve carbon dioxide reduces and the beverage goes flat.

Global Warming: The Facts

Did You Know That ...?

According to the Intergovernmental Panel on Climate Change (IPCC),[i] the UN's scientific advisory body, as a result of global warming, it is expected that by the end of the 21st century:

- Every year, six billion tons of carbon dioxide is released into the atmosphere, over one quarter of which is produced by the United States. This number is proportionate to the U.S. share of total global petroleum consumption. China, Germany, and Japan are the distant second, third, and fourth largest carbon dioxide producers.[ii]
- The average sea level will eventually rise by half a meter or more, causing many countries like the Maldives and some of the Caribbean and Pacific Islands to disappear completely underwater. In the US, much of the coastline along the Gulf of Mexico and the entire Florida Everglades will be devastated. Many acres of public beaches, millions of birds, and many arctic animals like seals and polar bears will become extinct.[iii]
- Millions of people will be displaced. Tropical diseases such as malaria, once believed to be eradicated, will again become common. Up to 3.5 billion people could be affected.

[i] IPCC, Summary for Policy Makers: Economic and Social Dimensions of Climate Change - IPCC Working Group III (2001).
[ii] EIA International Energy Outlook 2001, Tables A-10 and A-11.
[iii] "Race to Save the Planet," A documentary narrated by Meryl Streep and Roy Scheider, WGBH/Boston Production.

global warming is particularly complicated as clouds differ in thickness, composition; they can travel and may or may not result in rain. More water vapor contributes to increased blanketing effects, both to keep the reradiated terrestrial radiation in (positive effect), and incoming solar radiation out (negative effect). Clouds net contribution in heating or cooling the atmosphere is not certain.

Consequences of Global Warming

There is considerable disagreement as to the severity of global warming and its long-term effects on the environment. Although environmentalists and the scientific community blame human greed and non-restricted use of fossil fuels, many politicians, oil companies, and even some researchers claim that the natural variations of solar activity have a much larger influence on climate change.[7] However, this point of view is the clear minority, and many discard it purely as a political ploy to support the hidden agendas of special interest groups.

One thing that scientists agree on-- and data confirm-- is that global warming, whatever the cause, can lead to major changes in the regional climate, or local weather, increased frequency of extreme events, unprecedented hot summers, more intense heat waves, heavy rainfalls, and severe droughts.[8] Among the potential consequences of global warming are:

1. *The air temperature close to the earth's surface increases, whereas the stratosphere cools.* It is expected that, as a result of global warming,

[7] See for example the article by Richard Lindzen in the *Wall Street Journal*, June 11, 2001.

[8] NOAA web site: see <*http://www.noaanews.noaa.gov/stories/s412.htm*>, April 18, 2000.

the earth will warm somewhere between 1.4-5.8°C by the end of the twenty-first century.[9] The severity of the consequences can only be appreciated if we compare this to the last ice age, during which time the earth's average surface temperature was only 5°C colder than it is today.

2. *The ocean water gets warmer.* Global warming also results in the warming of ocean waters and possibly even the collapse of the conveyor belt. The conveyor belt is a body of water that circulates between the Atlantic and Pacific oceans. It is shallowest in the Indian Ocean where the water is warmest. As water moves north it cools, becomes denser, and sinks to the bottom of the Atlantic near Greenland. The Gulf Stream is part of a conveyor belt in the south Atlantic that moves north, bringing mild weather to Europe; if weakened, it will cause the temperature in much of Europe to drop.[10]

   Warmer oceans also affect the diversity of marine life. As oceans warm, fish body temperatures, and with it metabolism and oxygen consumption rates, will rise. Unfortunately, the solubility of gases is reduced at higher temperatures and less oxygen is available, causing many species of fish to disappear. Higher temperatures also increase the rate at which organic waste decomposes, a process that competes for the dissolved oxygen and consequently, oxygen levels drop even further. Furthermore, water viscosity and density are reduced, causing increases in the rate of settlement of suspended particles in the water – possibly affecting aquatic food supplies.[11] Finally, warmer oceans mean fewer nitrates, an important ingredient for the production of carbon-absorbing phytoplankton.
3. *Ice caps in northern latitudes and Antarctica break up and melt.* Rising temperatures cause the melting of glaciers, thus further reducing the fraction of sunlight that is reflected to the sky. A rise in sea level of only a few feet will bury many of the smaller Pacific islands underwater. In low lying countries, such as Bangladesh and the Netherlands, millions of people will be displaced. Many coastlines and beaches including some in the United States, Australia, and Europe, will also be lost. As a result, atmospheric temperature rises even further, causing more ice to melt.[12]
4. *The climate becomes even more unpredictable.* Global warming does not mean <u>all</u> regions will experience warmer weather. As the upper layers of seawater warm, the intensity of hurricanes, typhoons, and tornadoes increases.[13] The change in rainfall pattern will result in droughts in some regions and floods in others. Forests and other natural resources will suffer from changes in the climate.
5. *Weather related diseases will increase.* Global warming can

[9] Intergovernmental Panel on Climate Change, IPCC Working Group ,2001.
[10] "Scientists see threat of abrupt world climate," Environmental News Network, December 13, 2001. Reuters article by Andrew Quinn, *(http://www.enn.com/news/wire-stories/2001/12/12132001/reu_45873.asp).*
[11] Hodges, L., <u>Environmental Pollution</u>, 2nd Ed., Holt, Rinehart and Winston, 1977. p.312.
[12] IPCC, Climate Change 1995: The Science of Climate Change, Oxford, Cambridge University Press, 1996.
[13] Recent studies suggest that much of the violent weather behaviors of the last decade, such as the 1991 cyclone in Bangladesh, 1992 floods in Pakistan, China, and Australia, and 1990-1995 El Nino and La Nina in the United States are related to global warming.

The Greenhouse Effect

FYI ...

It is interesting to note that life on our planet would not be possible if it were not for the greenhouse properties of the earth's atmosphere. Without an atmosphere, almost all solar energy incident on earth's surface would be reflected back to space, and earth would be a frozen planet with a global average temperature of -17°C (1°F). Various gases in the atmosphere, principally carbon dioxide and water vapor, are capable of absorbing the infrared radiation reflected from the surface, preventing heat from escaping. Nitrogen and oxygen do not absorb the infrared radiation and therefore are not greenhouse gases. As a result of presence of greenhouse gases, the actual global average temperature of our planet is around 15°C (59°F).

dramatically increase pests and outbreaks of infectious diseases such as tuberculosis and malaria.

6. *Agriculture will suffer.* Salt water will damage much of the low-lying coasts and reduce farmland, thus affecting the grazing of farm animals and production of food. Drinking water will also be in short supply.

Combating Global Warming

To slow down global warming, the concentration of greenhouse gases, carbon dioxide in particular, must be lowered. This can be best achieved by reducing fossil fuel combustion and converting to technologies that reduce the emission of harmful gases to the environment. In absence of adequate support for cutting emissions, we need to lower carbon dioxide levels in the atmosphere either by reducing the carbon in fuel (decarbonization) or by preventing carbon dioxide's release into the atmospheric air by storing it in other mediums (sequestration).

Decarbonization can be accomplished by switching to alternative fuels with higher H/C ratios. Coal has no hydrogen, and is therefore considered to be the dirtiest of all fossil fuels. Natural gas has four hydrogen atoms for each carbon and is considered cleaner than other types of fossil fuels. Of course, the best option is to remove carbon altogether and use pure hydrogen as fuel. Hydrogen is, however, highly explosive and has a low density. Additionally, it must be kept at high pressure, which makes it even more dangerous. A good way of utilizing hydrogen is through fuel cells. Fuel cells are devices that combine hydrogen and oxygen in the air to produce electricity (see Chapter 14). The only product is water vapor, which is not a pollutant.

Sequestration refers to the long-term storage of carbon underground or in oceans so as to prevent the buildup of carbon dioxide concentration in the atmosphere. Earth's oceans have the capacity to hold 45 trillion tons of carbon - 60 times as much as the atmosphere. 2.4 trillion tons are contained within organic matter, such as trees and soil. Untapped fossil fuel reserves could add another 10 trillion tons.[14] This means that the oceans alone can absorb all the carbon dioxide that will eventually be produced. However, there are problems with this scenario. As carbon

[14] Savage, Neil, "Greenhouse Effects, R.I.P.," *Discover, August 2001.*

dioxide dissolves in water, the water becomes more acidic. This could harm marine organisms, demonstrating that one environmental problem potentially leads to another equally important environmental problem.

> **Question:** What would have happened if oceans could not dissolve the carbon dioxide in the atmosphere?
> **Answer:** The amount of carbon dioxide in atmospheric air is a tiny fraction of that dissolved in ocean. Without water the atmospheric carbon dioxide concentration would have been much higher, making the atmosphere much hotter and unsuitable for most forms of life. In fact, the planet Venus apparently retains much of its carbon dioxide in its atmosphere, keeping its surface temperature at a searing 470°C (880°F), today (See box "Venus").

Figure 8-4
Large amounts of carbon dioxide are trapped in vast forests and large bodies of water.

Carbon dioxide can also be sequestered by planting trees (Figure 8-4). Photosynthesis helps by removing carbon dioxide and converting it to carbohydrates and food (Chapter 6). This also improves air quality and beautifies the landscape. Unfortunately, trends point in the opposite direction. In the last few decades, much of the existing rainforests have been lost to logging and used as wood fuel or cut down to pave the way for urbanization. Another option is to inject the carbon dioxide back into coal mines, oil fields, and saline aquifers. This is being done to some extent in enhanced oil recovery and the reclamation of abandoned mines.

Other less serious approaches have been proposed in which fleets of aircraft crisscross the skies releasing dust into the low atmosphere, millions of hydrogen-filled aluminized balloons are sent into the sky, or thousands of mirrors are put into orbit, all in order to reflect sunlight back to space.[15] We could adjust the numbers to balance cooling as a result of reduced insolation with increased heating from emissions of greenhouse gases.

The Kyoto Protocol

In response to growing concern and mounting evidence of global warming, the United Nations hosted the 1992 Earth Summit in Rio de Janeiro, Brazil, where leaders of most industrialized countries agreed to adopt policies aimed at protecting biodiversity. Gases specifically targeted were carbon dioxide, methane, sulfur hexafluoride, HFCs, and CFCs. Furthermore, the Rio agreement established a legal ground to protect a country's biological

Venus

FYI ...

Compared to the earth's atmosphere which except for some trace gases is composed entirely of nitrogen and oxygen, Venus' atmosphere is 96% carbon dioxide. Much of the carbon dioxide in earth is dissolved in ocean waters, sequestered in plants, or locked up in carbonated rocks, releasing only trace amounts of carbon dioxide to the atmosphere. Venus does not have any liquid water and almost all carbon dioxide has outgassed into its thick atmosphere. As a result, the runaway green house gases play a much more dominant role in Venus, maintaining its temperature in excess of 470°C.

[15] Begley, S., "On the Wings of Icarus," *Environment, May 20, 1991.*

assets by giving host countries the intellectual property rights on genetic information of their unique animals or plants. Unfortunately, because of the low price of petroleum and lack of enforcing authority, the use of fossil fuels and the emissions of carbon dioxide continued to increase.

The failure of the Rio Summit prompted many to organize a meeting in Kyoto, Japan in 1997. This meeting was to devise plans to fight climate changes and their adverse consequences on the world's weather, agriculture, and ecosystems. Many countries favored putting limits on fossil fuel consumption, phasing out CFCs and other chlorine and bromine-based products, banning future deforestation and planting new trees, increasing efficiency, and promoting renewable energy sources. The United States argued against the measure and considered the protocol to be biased, unfair, and politically motivated. If ratified, it puts unfair strain on the US economy, while giving unfair advantages to countries that are not covered under this treaty, like China and India. There will be more incentive for US companies to transfer their manufacturing to these and other less-industrialized countries that are inherently less efficient and thus the greenhouse gas problem will actually worsen.

Developing nations opposed the US view, stating that the cost of cleanup should be borne solely by rich nations that had done the most damage over the past years and arguing that their fledgling economies should not pay the price for the petroleum-driven economic growth of developed nations. Specifically, most criticized the United States for being the biggest contributor to global warming. The US, with 5% of the world population, is responsible for one fifth of all carbon dioxide emissions. A compromise was finally reached, stipulating that by 2012, the so-called Annex 1 countries (the developed countries and the Eastern European countries

Global Warming Controversy*

Point/Counter

A minority of scientists take the view that there are unproved and insufficient evidence that links the global warming to man-made carbon dioxide emission. Some cite the poor measurements, inaccurate instruments and crude computer models, preconceived ideas, or biased interpretation of the theoretical or analytical data; others attribute the phenomenon to natural and cyclic fluctuations in the solar activity, cosmic rays, ocean currents, or other unknown causes; yet others see it as a response to political (or even academic) pressures on scientists, and in the US at least, a partisan issue.

Even among those who believe that global warming is real and due to human impact, there are debates as to the extent of the problem, appropriate response and the time that it must be implemented.

The ultimate judgement on whether the global warming is a natural phenomenon or a result of human activity might have to wait sometime, hopefully before it is too late. One thing is true however that overwhelming majority of reputable scientists, various professional societies and academies of science, as well as US National Academy of Science, and the American Association for the Advancement of Science have sided with the United Nations' IPCC conclusions and consider climate warming as real and dangerous, and must be dealt with effectively and immediately.

**For an excellent review of scientific, political, and social aspect of the global warming controversy see the Wikipedia and references that follow (http://en.wikipedia.org/wiki/Global_warming_controversy).*

of the former Soviet block) were to reduce their collective emissions to 5.2% below the 1990 benchmark. The non-Annex 1 countries would be exempt from these restrictions for a ten-year grace period. However, to induce other nations (notably China and India) to join the agreement, it required rich countries to establish funds to finance costs of replacing CFCs in poor countries by benign refrigerants. In addition, the protocol allowed the signatories to meet emission reduction requirements by devising an emission trading market, which allowed Annex-1 countries to receive emission credits for helping the non-Annex-1 countries reduce emissions. In order to be binding under international law, the treaty would need ratification by countries responsible for at least 55% of the global greenhouse gas emissions of 1990.

In 2001, the United States officially pulled out of the treaty, effectively making the treaty enforceable only if Russia, responsible for 17% of the greenhouse gases, ratified the treaty. In February 2005, after some modifications, Russia agreed to sign the treaty and Kyoto Treaty officially went into effect. Thirty five industrialized countries have signed a binding commitment to curb their greenhouse gas emissions by 2012 to overall levels that are 5.2% below 1990 levels. The European Union has committed itself to cut emissions by 8%, Japan and Canada by 6%, while Russia has agreed to limit emissions to 1990 levels. Developing countries are exempt from this protocol, although many have expressed a desire to sign on and are planning to reduce emissions voluntarily. Australia was the last country who signed the agreement, leaving the United States the only developed country who has not ratified the treaty.

By 2006, China had already overtaken the United States as far as total carbon dioxide emission. With China's heavy reliance on coal for production of its massive amount of electricity, the situation is not getting any better soon. At the time of this writing China is producing 2.5 times greenhouse gases as it did in 1991 when Kyoto Protocol was initially drafted. India's contribution has risen from 2.9% of the world emission in

The Killer Lake[i]

FYI ...

On August 21, 1986, a sudden release of over one billion cubic meters of carbon dioxide gas suffocated 1,800 people and numerous herds of cattle living near Lake Nyos in the African nation of Cameroon. Lake Nyos is thermally stratified; colder water at the bottom warms up as it gets closer to the surface. Due to the cold temperatures and high pressures at the bottom, the concentration of dissolved carbon dioxide can reach near saturation levels. Under normal conditions, the stratification limits vertical mixing, allowing the gradual release of the trapped gases. It is believed that a heavy storm triggered a thermal instability in the lake that brought the deeper, gas-rich water to surface. As water warmed up, its capacity to dissolve carbon dioxide decreased and gas bubbles began to form. This process is identical to the sudden opening of a can of soda or a bottle of champagne. When the cap is removed, there is no more pressure to keep the gas dissolved in the soda and the gas fizzes. The cloud of gas is estimated to have reached one hundred meters above the lake surface and extended out to 10 miles into the valleys near the lake, killing villagers while they slept.

[i] *Fialka, J. J., "Do Bubbles of Champagne Die? Alas, No," Wall Street Journal, November 17, 1997, Section B.*

1991 to 4.4% in 2008.[16]

Carbon Abatement Strategies

The major strategies used to meet the Kyoto benchmark are:

1. *Emission Trading* -- The establishment of an emission market where emission credits are traded. This way, companies and countries could continue producing pollutants as long as they could offset their higher emissions by buying credits from companies or countries that opted for cleaner technologies or deployed emission control strategies. The rationale for this option is that as long as the total emissions are reduced on a global scale, it does not matter where these reductions are made. US is the prime proponent of this approach.
2. *Carbon Capture and Storage (CCS)* -- This is the process by which the carbon dioxide produced at the power plant is captured and stored (sequestered) instead of being released into the atmosphere.
3. *Carbon Tax* -- Taxes are based on the amount of carbon emission per unit energy production (kg/kWh, or lb/Btu). This scheme is used in most European countries.
4. *Renewable Resource Mix* -- Proposed in Britain, electrical suppliers are obligated to acquire a certain fraction of their energy from renewable sources. The fraction is gradually increased until their obligations under Kyoto Treaty is met.

Outdoor Air Pollution

The major sources of outdoor air pollution are transportation, stationary sources, and industrial processes. Motor vehicles account for about a third of all carbon dioxide, 98% of all carbon monoxide, and about half of all nitrogen oxides and hydrocarbons emissions (Figure 8-5).[17] Stationary sources such as large diesel engines, coal power plants, and refineries are major contributors to sulfur emission and particulates. Industrial pollution is mostly a result of manufacturing a huge number of new and non-biodegradable chemical compounds produced from petroleum products.

Figure 8-5
Pollutants emitted by fossil power plants and cars are responsible for 2/3 of all pollution. The rest is from industries.

The sources of pollution are many, and the detailed discussion of their harmful effects is outside the scope of this book. Air pollutants are divided into two categories: The so-called criteria air pollutants and hazardous (toxic) air pollutants. The **criteria air pollutants** are those common in mobile and stationary sources (tailpipe and smokestack emissions) which are thought to cause heart and respiratory problems and may cause cancer. They are particulates, sulfur dioxide, carbon monoxide, nitrogen oxide, ground-level ozone (smog) and lead. Carbon dioxide is not usually considered a pollutant, but because of its increased global background

[16] Energy Information Administration website, *www.eia.doe.gov/pub/international/iealf/tablehlco2.xls*.

[17] US Environmental Protection Agency, National Emission Inventory Air Pollutant Emission Trends website (*http://www.epa.gov/ttn/chief/trends*).

Table 8-1. Total US Emission of Carbon Dioxide and Criteria Air Pollutants, 2002										
	CO_2		CO		NOx		HC		Particulate	
	Tons*	%	Tons*	%	Tons*	%	Tons*	%	Tons*	%
Transportation	1875	32	86.6	77.3	11.5	54.3	7.2	43.7	1.0	3.5
Stationary Sources	2250	39	4.4	4.0	8.3	39.3	1.0	6.1	2.5	8.7
Industrial	1665	29	2.7	2.4	0.9	4.0	7.0	42.1	1.1	3.8
Others	0	0	18.3	16.3	0.4	11.0	1.3	8.1	24.3	84.0
Total	5790	100	112	100	21.1	100	16.5	100	28.9	100

* million short tons
(a) Includes transit and intercity; (b) Certificated routes and general aviation; -- Data not available.
Ref: Davis, S. C., Diegel, S. W., US Department of Energy, "Transportation Energy Data Book," ORNL-6978, 2007. Tables 12.1.

and its effect on global warming, it is sometimes treated as such. However, carbon dioxide emission is not regulated by the federal government. The **hazardous air pollutants** are less common air pollutants, but they are carcinogenic or may cause damage to the immune or neurological systems. These include volatile organic compounds (unburned hydrocarbons), ammonia, and a wide range of chemicals used as solvents, sterilizers, and components in industrial processes. Table 8-1 shows the total US emission of criteria air pollutants for various sectors in 1999. In addition to these pollutants, radio nuclides emitted from nuclear sources are also of great concern and will be discussed separately in Chapter 12.

Depending on the source of the pollution, air pollutants can be considered as primary or secondary.

Primary pollutants are those coming directly from the source and may be natural or anthropogenic (human caused). Natural sources include dust, pollen, sea salt nuclei, volcanic ash, smoke, and particulates from forest burning. The primary anthropogenic source of pollutants results from fossil fuel combustion and includes: carbon monoxide, the oxides of nitrogen (mainly NO and NO_2), the oxides of sulfur (mainly SO_2), unburned hydrocarbon, and particulates (mainly soot, ash, and metal traces). Because almost all air pollutants come from combustion sources, the amount of emission correlates closely with the amount of fossil fuel consumption.[18]

Volatile Organic Compounds (VOC)[19] are the carbon containing compounds derived from living things. Fossil fuels are the main source of VOCs released to the atmosphere and include gasoline, industrial chemicals such as benzene, solvents such as toluene and xylene, and many other chemicals used as dry cleaning solvents.

Carbon monoxide forms when there is not enough oxygen or time for the complete oxidation of carbon to carbon dioxide. When inhaled,

[18] US Environmental Protection Agency, National Air Quality and Emissions Trends Report, Annual 2000.
[19] The EPA does not list VOCs as criteria pollutants, but because of their impact on the formation of smog they are included here.

Dinosaurs and Air Pollution*

FYI ...

Dinosaurs dominated the earth for nearly 130 million years but became victims of an unthinkable mass extinction some 65 million years ago. In 1979, Luis and Walter Alvarez, a father-son scientist team, suggested that extreme pollution was in fact a major contributor to the dinosaurs' demise. The scientists proposed that an asteroid or comet had smashed into the earth and caused a series of catastrophic events that wiped out the dinosaurs within a blink of a geological eye. A compelling piece of evidence discovered by the two scientists was an excess of iridium and other rare elements in a rock layer determined to be about 65 million years old. In 1992, further support for the hypothesis came in the form of a giant crater found in Mexico spanning about 300 km.[i]

The asteroid that possibly produced the crater is estimated to have had a mass of approximately $2x10^{16}$ kg, which collided with the earth at a speed 10 km/s. This amounts to about 10^{24} J of kinetic energy, which is roughly equivalent to about 2 billion Hiroshima bombs. The debris from the impact might have risen as high as several kilometers into the atmosphere.

In addition to polluting the air with dust and deadly acid rain, the impact caused huge tidal waves and earthquakes — much larger than anything experienced in recent history. And if that were not enough, the combined effect of a darkened, wintry sky and the release of vast quantities of greenhouse-prone carbon dioxide (normally trapped in the soil and water) would undoubtedly produce calamitous extremes of cold and/or heat in the climate. Such abrupt environmental changes evidently exceeded the ability of the dinosaurs — and certainly of many other living organisms — to adapt and condemned them to remain forever as spectacles of extinction.

[i] *Alvarez, L., Alvarez, W., "Giant Crater Linked to Mass Extinction," Science News, Vol. 142, No. 7, p. 100, August 1992.*

** Excerpts from a physics text currently under preparation jointly by this author and Professor Igor Glozman, Department of Physics, Highline Community College, Des Moines, Washington 98198.*

it diffuses through the lungs into the blood stream where it reacts with the blood hemoglobin displacing the oxygen. Consequently, the body will be deprived of oxygen resulting in headaches, fatigue, impaired judgment, and in extreme cases, suffocation and death. The overwhelming majority of all carbon monoxide emissions are from cars. Other major contributors include stationary power plants and solid waste disposal.

Nitrogen dioxide is emitted from a reaction between oxygen and nitrogen in the air at combustion temperatures. The main sources of nitrogen dioxide are transportation and stationary power plants, but some are also released from fertilizers and other chemical plants. The most important health effect associated with nitrogen dioxide is the reduction of the body's resistance to infection. Nitrogen dioxide also limits plant growth, causes corrosion in metals, and is a major cause of global warming, acid deposition, and photochemical smog. In the presence of sunlight, nitrogen dioxide reacts with hydrocarbons to form ozone – a major lung irritant and contributor to bronchitis and asthma.

Sulfur oxides (SO_2) are generated by the combustion of sulfur-rich fuels in stationary power stations, refineries, and industrial processing plants. In the atmosphere, it can be further oxidized to form SO_3 and eventually reacts with water to form sulfuric acid. Some of the sulfuric acid will subsequently react with trace elements emitted alongside other pollutants to form submicron size sulfate particles – a major cause of acid fog, acid mist, and acid rain. The major impacts are reduced visibility and damage to vegetation, materials, and human health.

Particulates include soot, lead, dust, ocean salt, and metal debris. The main sources of particulates are power plants and diesel cars, although smaller sources such as wood stoves, agricultural burning, and dust also contribute. Particulates can vary in size from a fraction of a micron to many hundreds of microns. Larger particles are responsible for reduced visibility and will eventually settle, whereas smaller particles become airborne and may be inhaled, diffuse, and be adsorbed and deposited on the interiors of the lungs, potentially contributing to lung damage and cancer. Particles larger than 10 microns are entirely blocked by the nose. Those below 10 microns (PM-10) pass through the upper respiratory system. Smaller particles in the range of 0.5-5 microns (PM-2.5) can travel as far as the bronchioles before they are deposited onto the inner walls. Depending on their physical and chemical structure, particles can be toxic (trace of metals such as lead, cadmium, mercury, etc.), or they can interfere with normal breathing (dust). Some particles, such as ocean salt and fine soot, can act as nuclei sites for water condensation and fog formation.

Lead is a poisonous metal emitted from old automobile exhausts, paints, storage batteries, and pipes. The lead problem became particularly significant when it was routinely added to gasoline to reduce knocking and to allow for the construction of higher compression engines in the 1970s. Due to its adverse health effects, lead was eventually removed from gasoline. Depending on its concentration, lead can damage the lungs and may lead to convulsions, brain damage, and even death.

Secondary pollutants result from the further interaction of the primary pollutants with the atmosphere. Ozone is considered the main secondary pollutant. Ground-level ozone is produced by the combination of pollutants from smoke stacks, cars, paints, and solvents in the presence of sunlight.

Ozone is a pale blue gas with a pungent smell that allows it to be detected even in very low concentration. In fact, the word ozone is derived from Greek *ozein* meaning "to smell." Major health problems associated with elevated levels of ozone in the atmosphere are coughing, headaches, nausea, and chest pains. Ozone also causes shortness of breath, asthma, chronic bronchitis, and fatigue. Asthma epidemics, especially in children in industrialized countries, can be directly attributed to ozone.

Effects of the Air Pollutants

Depending on the nature of the contaminants, dose rate, and duration, air pollution can cause health complications in humans and animals, damage plants, deteriorate materials and property, and affect global climate.

The health effects associated with air pollution are numerous and can vary from burning eyes and nose, irritated throat, and breathing difficulties to more severe effects such as cancer, birth defects, brain and nerve damage, and death. Gaseous pollutants can enter the body through the eyes, skin, or respiratory tract. Depending on the size of the particulate, debris can then be deposited in different areas of the respiratory tract such as the nose, mouth, or lungs. The effects of pollution on animals are similar to those in humans and result in respiratory and eye damage. Water pollutants have a similar effect on aquatic life. Depending on their types and concentrations, toxins and other air pollutants can have significant health consequences over a relatively short duration or even during one-time exposure. A testament to this is the notorious London episode of 1952 which, as a result of thick fog and a high concentration of carbon and sulfate particles, caused many thousands of people to become sick, 4,000 of whom eventually died. The 1984 release of methyl isocyanate from a Union carbide pesticide plant in Bhopal, India killed 2,500 people and injured numerous others. Another example is the Exxon Valdez oil spill in 1989. In this incident, the huge tanker carrying 1.2 million barrels of Alaskan crude oil ran aground and emptied its cargo into the Prince William Sound. Although no human life was lost in the accident the cost to the environment and local economy was enormous (Figure 8-6).[20]

Figure 8-6
Oiled crow in the aftermath of the Exxon Valdez oil spill

Many of the emissions released into the atmosphere are directly or indirectly *corrosive and attack buildings and materials* at a great cost. Dust and soot particulates, especially in humid air, cause the erosion and soiling of buildings, paintings, and sculptures. Nitrogen and sulfur oxides cause deterioration of cottons and nylon, corrosion of metals, and fading of dyes.

Atmospheric pollution can affect the environment by changing the climate and reducing the visibility. It damages plants by destroying their chlorophyll, which causes yellowing and the dropping of their leaves. It makes lakes, rivers, and aquifers toxic and corrosive, which kills fish and other marine species and damages crops.

Smog

Smog is a result of physical and chemical activities between various pollutants released from *smoke* stacks in the presence of *fog* (thus the word *smog*). **Classical smog** refers mainly to plume of sulfur oxides and particulate matter generated from the combustion of coals and

[20] "Valdez Spill Toll is Now Called Far Worst," New York Times, April 18, 1992.

petroleum products. It is formed primarily in the early morning, during winter months, and in places of high humidity where condensation of water vapor over smoke particles is easier. The notorious London "fog" discussed above is a good example of this type of smog.

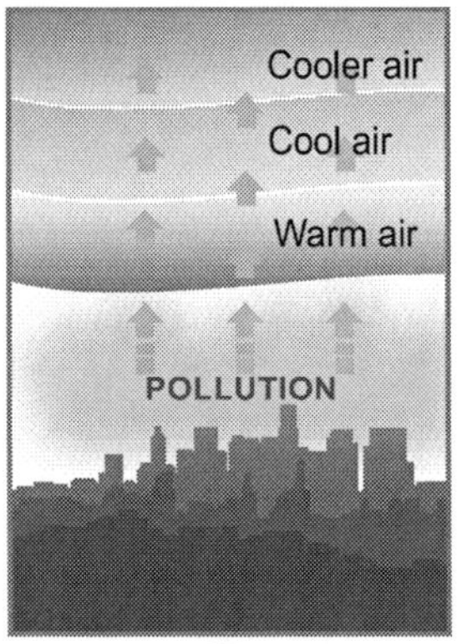

Normal Pattern

Photochemical smog is very different from classical smog because it is formed under specific meteorological conditions and only when a large amount of sunlight is available. It forms when nitric oxides and hydrocarbons react in the presence of sunlight to produce ozone; this ozone can be further oxidized to produce nitric dioxide and other photochemical oxidants.

$$\begin{cases} NO + 1/2\ O_2 \longrightarrow NO_2 \\ NO_2 + HC \xrightarrow{h\nu} NO + O \\ O + O_2 \longrightarrow O_3 \end{cases}$$

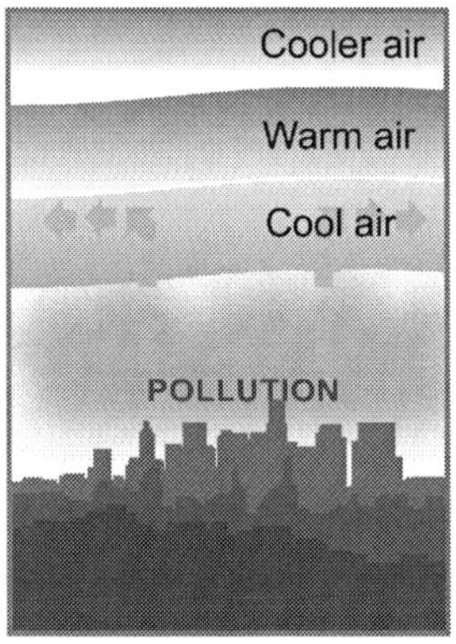

Thermal Inversion

Figure 8-7
Thermal inversion

The Los Angeles Basin is a prime location for photochemical activities because of the large number of cars and power stations, the availability of sunlight, and its proximity to water. Furthermore, the basin is surrounded on three sides by the San Fernando mountain range, which prevents the dispersion of the pollutants. To make the matter worse, Los Angeles has many days in which a thermal (or temperature) inversion layer blankets the basin and prevents the upward mixing of pollutants with atmospheric air. *Thermal inversion* refers to the condition where, as a result of a large amount of emissions from cars and smoke stacks, temperature increases rather than decreases with height (Figure 8-7). Under these conditions, a warm layer is sandwiched between a layer of cold air near the surface and another cold layer in the upper atmosphere. Thermal inversion is more severe when wind is calm and pollutants can stay in place for many hours or even days. The smaller the inversion's height, the less volume is available where pollutants can mix and the greater their concentration. The situation can be visualized as having a pot (mountains) filled with

Figure 8-8
The Los Angeles Basin

water (oceans) and other ingredients (pollutants) with the lid closed (the inversion layer) to prepare a stew (photochemical smog). All that is missing is the heat, which is provided by the abundant Southern California sun (Figure 8-8).

Acid Rain

Caused by air pollution, acid rain refers to any kind of acid precipitation such as acid fog, acid sleet, acid snow, acid mist, or any other acidic particles or aerosols that fall to the ground or remain suspended in the atmosphere. The main components responsible for the acidity of these particles are oxides of sulfur and nitrogen that are dissolved in water and deposited on the surfaces of suspended particles. Acid rain has a detrimental effect on human and animal health, endangers plant and marine life, and also degrades materials. The health problems associated with acid rain include lung diseases and respiratory ailments. Its primary effect on plants is the leaching and blocking of the uptake of nutrients and essential minerals, such as calcium and magnesium, from the plants' roots. Acids are also highly corrosive and attack metals and other construction materials.

The measure of the acidity of a solution is its pH value. The pH scale ranges between 0 and 14. Pure water has a pH of 7 and is considered to be neutral. Acids have pH values lower than 7, and alkalines have pH values higher than 7. The lower the number, the more acidic the substance is. The scale is logarithmic, which means a drop of one in the pH level represents a tenfold increase in the acidity. Thus a pH of 6 is 10 times more acidic than water; the pH of 5 is 100 times more acidic. Even unpolluted rainfall is normally slightly acidic, with a pH of around 5.6, because it includes dissolved carbon dioxides. Precipitation with a lower pH than this is considered acid rain.

Acid rain affects lakes and ponds in an adverse way. The severity of the problem depends on the nature of the lake. In some parts of the world, lakes are in limestone (calcium carbonate), which neutralizes acid. The tendency to neutralize acid provides a buffer against acid rain. In other parts, including the northern US, Canada, and most of Europe, lakes are in granite and have less buffering capacity. Therefore, acid rain poses a greater threat to fish, salamanders, and frogs in these areas. At higher levels of acidity, bacteria and water plants also die. Under these conditions, lake water looks pure, while a healthy lake might have cloudy water because of the natural plant life in it. To increase the alkalinity of the soil and water, some people have proposed liming them; however, this process has proven prohibitively expensive.

Many nations have enacted laws to limit their sulfur emission, mainly by installing scrubbers. Scrubbers remove sulfur dioxide by spraying powerplants' exhausts with a watery, limestone mist to produce sludge that can be dumped or used to make boards suitable for home building.[21]

[21] Moore, Curtis, "Green Revolution in the Marketing," January/February issue of *Sierra*.

Ozone

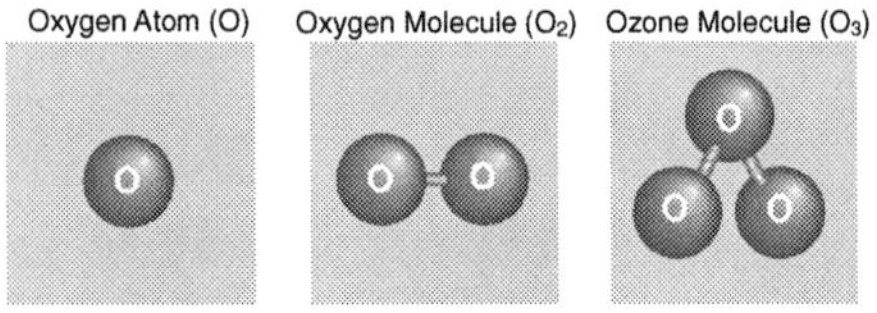

Figure 8-9
Molecular structures of oxygen and ozone

A molecule of ozone contains three atoms of oxygen bound together (Figure 8-9). In the lower atmosphere (troposphere), ozone is formed from the reaction between nitrogen oxides and volatile hydrocarbons in the presence of sunlight. In the upper atmosphere (stratosphere), ozone is produced from the reaction of oxygen with ultraviolet radiation from the sun. Ozone is highly reactive, attacking and oxidizing almost anything it contacts.

The Good Ozone and the Bad Ozone

The ozone layer may be regarded as the earth's sunglasses. It serves as a shield, protecting earth from the harmful effects of ultraviolet radiation. Most of the ozone is found at around 15-35 km above the ground (Figure 8-10). As a result of large-scale meteorological weather systems, the ozone layer somewhat varies in thickness from one day to the next and greatly fluctuates with the seasons. Depending on its frequency, ultraviolet light can be classified as UV-A (320-400 nm), UV-B (290-320 nm), or UV-C (100-290 nm).

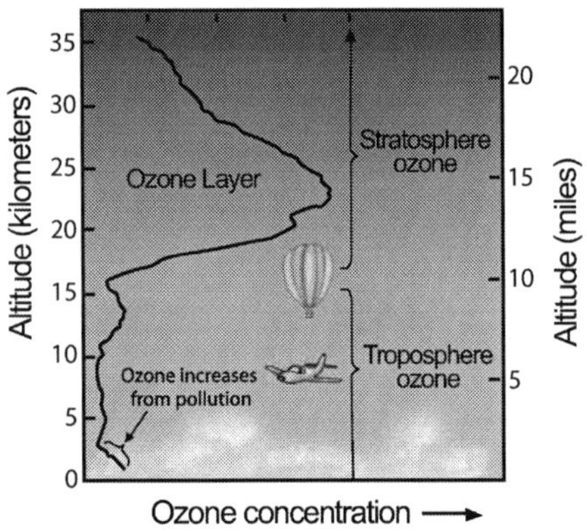

Figure 8-10
Atmospheric Ozone: 90% of all ozone resides

UV-A, also known as ***black light***, is responsible for skin tanning and has the lowest energy. It is generally considered to be harmless, although some new studies have linked some damage to the retina with long term exposure to this type of radiation. UV-C has the most energy, but is entirely absorbed by the outer layers of the earth's atmosphere and, under normal circumstances, is not of concern. Because it is highly effective in killing germs, bacteria, viruses, mold, fungi and spores, UV-C is used in hospitals and some industrial applications. UV-B has a lower energy than UV-C, and most is absorbed by the ozone layer in the upper atmosphere, however, even the small amount that reaches the ground results in long term cataracts, skin cancer (malignant melanoma), and breakdowns of the immune system in a large number of people (Figure 8-11). A 1% reduction in the thickness of the ozone layer will result in a similar increase in the intensity of UV-B radiation and a 2% increase in the probability of skin cancer. The effect is more severe at higher latitudes (such as in Scandinavian countries) and higher altitudes.

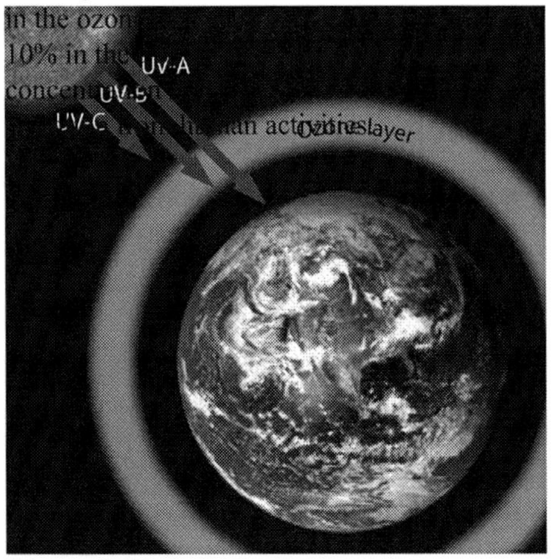

Figure 8-11
Ultraviolet radiation

In contrast to stratospheric ozone that is considered "good," the ozone in the lower atmosphere is "bad." As discussed in the previous section, it is formed mainly as a result of photochemical reactions between nitric oxides, hydrocarbons, and sunlight. Furthermore, ozone is a strong oxidant and easily attacks metals, rubber, and other materials. At high concentrations, ozone is harmful to plants. It damages crops and slows forest growth. In humans, ozone reduces immunity to various diseases, reduces lung capacity, and causes chest pain, coughing, and asthma.

Air Pollution: The Facts

Did You Know That ...?

- Almost 50% of smog-producing pollutants come from cars.
- Every year, as many as 200,000 people die in the United States as a result of diseases related to air pollution.
- Air pollution costs Americans at least $150 billion a year in health care and lost work productivity.
- Radon is the second leading cause of lung cancer.

Ozone Depletion

The reduction in the amount of ozone in the stratosphere is another environmental problem widely attributed to the increased use of fossil fuels. This reduction, commonly referred to as the "ozone hole," was first reported over Antarctica in 1985. Since then it has been recorded over the Arctic and elsewhere around the globe (Figure 8-12). Since 1978, in areas where ozone concentration has been routinely measured, there has been a 4-5% loss of stratospheric ozone over the continental United States.

Figure 8-12
The Antarctican one hundred dollar bill signifies the ozone depletion problem over the continent.

Among the chemicals most threatening to the ozone layer are chlorofluorocarbons (CFCs). CFCs are odorless, non-toxic, non-flammable and inexpensive chemicals originally used as a substitute to ammonia, which was the primary coolant in refrigerators. Marketed under the trade name Freon, it quickly found its way into a large number of other products such as paints, propellants, insulation, seat cushions, solvents, and a foam lowering agent for aerosol cans. In addition to CFCs, methyl bromide (a product mainly used in pest control and fire extinguishers) and nitrogen oxides (produced during combustion and a common component of fertilizers) are also shown to have similar destructive properties. Supersonic passenger aircrafts, rockets, and space shuttles release chlorine gases directly into the atmosphere and thus contribute to the depletion of the ozone layer.

The mechanism responsible for ozone destruction was first investigated by Rowland and Molina,[22] who showed that the depletion is a result of chlorine, fluorine, and bromine compounds transported from the lower to the upper atmosphere. Once in the stratosphere, these chemicals are broken down by high concentrations of ultraviolet rays which, in turn, decompose ozone molecules into molecular and monatomic oxygen. Although vehemently attacked by various manufacturers of CFCs, the experimental data became so convincing that the production and use of CFCs was eventually banned. Rowland and Molina received the 1995 Nobel Prize in chemistry for this pioneering work.

Although ozone depletion is widespread over the entire globe, it is of particular significance in the Antarctic. The severity of depletion in the Antarctic is due to its extreme cold temperature (as low as -90°C in winters) and the presence of ice crystals that hover over Antarctica and act as catalysts to breakup CFCs and release chlorine needed to destroy

[22] Rowland, F. S., and Molina, M. J., "Ozone Depletion: 20 years after the Alarm," Chemistry & Engineering News, August 15, 8-13, 1994.

Ozone: The Facts

Did You Know That ...?

- Ozone is thinnest at the equator and thickens toward the poles, so the farther from the equator you live the less radiation you'll receive.
- A single chlorine free radical reacts with as many as 100,000 molecules of ozone before it reacts with other molecules and becomes immobilized.
- Even if the use of ozone-depleting agents altogether stops today, it will still take 50-100 years before the stratospheric ozone returns to its original concentration.
- If we were to replace the ozone loss in the upper atmosphere, we would have to pump in 90 megatons of ozone. Even if we could produce this much ozone, the energy required for its delivery would be unthinkable.
- Like carbon dioxide, ozone is a greenhouse gas, though ozone's contribution to global warming is small compared to other greenhouse gases.
- As a result of ozone depletion, in the United States alone, the incidence of melanoma has gone up from 1 in 1500 in 1935, to 1 in 150 in 1991, to 1 in 75 in 2000.

ozone. There are no such ice clouds around the North Pole and so the ozone hole is relatively smaller.[23]

The Montreal Protocol

The rapid increase in the rate of ozone depletion led to an international agreement -- known universally as the 1987 United Nations Montreal Protocol-- obliging the signatories to gradually phase out the production of CFCs and other ozone depleting compounds. According to this agreement and the subsequent amendments (London, 1990; Copenhagen, 1992; Vienna, 1995; Montreal, 1997 and Beijing, 1999), a set of interim steps were devised that put a cap on the production of ozone depleting materials. Responsibility for the implementation of the Protocol and subsequent mandates remains within individual countries and no reinforcement authority is assigned. As a first step, it was recommended that CFCs, halon, and other ozone depleting emissions should be substituted with hydrofluorocarbons (HFC) and hydrochlorofluorocarbons (hydrogenated CFCs or HCFCs). HFCs do not have chlorine and are therefore not ozone-depleting, but are not as efficient as a coolant, either. HCFCs have a much shorter lifetime in the atmosphere and are much less harmful than CFCs to the ozone layer. These products are scheduled to be phased out by 2030 and substituted with harmless alternatives.

The preliminary results are encouraging, as the amount of CFCs released by major industrial countries has fallen by 99%.[24] It should be noted, however, that despite a substantial decrease in the release of ozone depleting materials into the atmosphere, because of their long lifetimes, these chemicals remain in the atmosphere for many decades so repair to the ozone layer is very slow. It is expected that by 2050, stratospheric ozone concentrations will return to its original levels.

[23] Fahey, D. W. et al., "Twenty Questions and Answers about the Ozone Layer," 2002 Scientific Assessment of Ozone Depletion, Les Diablerets, Switzerland, 24-28 June 2002.
[24] National Renewable Energy Laboratory Fact Sheet, DOE/GO-102000-1048, May 2000.

Ozone Myths
Point/Counter
P: There is no scientific evidence that the ozone hole actually exists. CP: Although there is no hole per se, there is enough data that indicates the thinning of the ozone layer.
P: The Ozone hole exists only over Antarctica CP: Although the ozone hole is larger over Antarctica, recent measurements show similar but smaller effects over the Arctic and other locations, except the tropics.
P: CFC molecules cannot reach the stratosphere because they are heavier than air. CP: Unlike the stratosphere, which is relatively calm, the troposphere is very turbulent. This turbulence ensures that all gases whether they are lighter or heavier than air, are well mixed in the troposphere.
P: Sea salt is a much bigger source of chlorine than CFCs. CP: Unlike CFCs, sea salt particles are soluble in water and are washed out in the lower atmosphere within days or weeks. In fact, no sodium can be detected in the stratosphere.
P: The chlorine from man-made CFCs is negligible compared to the chlorine gases released from occasional volcanic eruptions. CP: Although volcanic plumes are rich in hydrogen chloride, they also contain considerable amounts of water vapor. The chlorine is efficiently scavenged by rainwater and ice and removed from the lower atmosphere, and therefore is unlikely to be transported to the upper atmosphere.

Indoor Air Pollution

We spend over 90% of our lives indoors, so the quality of indoor air we breathe is particularly important. Unfortunately, over the past few decades indoor air quality has continuously deteriorated. In an effort to reduce heating and cooling loads, many newer buildings have been weather-proofed by insulating walls and ceilings and by sealing windows. Little did anyone realize that these cost and energy saving measures would cause a multitude of problems resulting in what is commonly known as the "sick building syndrome." It is estimated that 20-30% of people are exposed to toxins, and losses of as much as $400 billion dollars are attributed to sick buildings.

There are three basic sources of indoor contaminants:

1. *People:* People are carriers of many contaminants such as bacteria, germs, allergens, and not-so-obvious pollutants such as perfumes and fragrances. Even the simple act of breathing can emit many chemicals and shed millions of dead skin flakes.
2. *Building materials:* These include construction materials such as formaldehyde outgassing from particle boards, paints, glues, waxes, fiberglass, paper, dust, lint, carpet fibers, dirt, detergents and other cleaning materials, and, in the case of old buildings, asbestos from ceiling tiles.
3. *Processes:* Major contributing processes are tobacco smoking and indoor appliances, such as copying machines, computers, gas water heaters, gas stoves, and oven ranges.

The adverse effects of various gaseous pollutants were covered earlier when we discussed outdoor air pollution, so only those of other pollutants present mostly in indoor environments are given below:

Radon is a naturally occurring gas derived from the radioactive decay of uranium-238. It diffuses through rocks and soil and enters through cracks into concrete and clay basements. When ingested, a small amount diffuses through the lungs causing ionization of molecules and other unwanted chemical reactions along its path that may cause lung cancer and other ailments. According to EPA estimates about 20,000 persons are developing lung cancer as a result of exposure to radon in the home. Areas with naturally high concentration of radon are Brazil, China, India and Iran. Ionizing radiation will be discussed in greater detail in Chapter 12.

Tobacco smoking is a significant contributor to indoor air pollution. Worldwide, over 2-3 million people die annually of smoke-related diseases such as lung cancer, emphysema, and bronchitis. The number in the United States is around 400,000. It has been reported that even a single cigarette in an office can quickly raise the levels of particulate matter to over 30 times the EPA standards.[25] In addition to tar and nicotine, tobacco smoke contains large amounts of carbon dioxide, formaldehyde, hydrogen cyanide, and benzene.

The concern about smoking became significantly greater among the general populace when it was discovered that smoking was not only a hazard for the smokers themselves, but for people around them. As evidence on the dangers of secondhand smoking mounted, smoking was banned in public places such as office buildings, theatres, shopping malls, buses, and airplanes in many US and Canadian cities. Other countries are slowly following their leads.

Asbestos is one of the most dangerous contaminants still found in older buildings. Because of its fire resistancy asbestos has been widely used in woven fiber mats, fireproof suits, gloves, gaskets, brake shoes, and even gas masks. The most popular applications of asbestos have been its use as insulating material in roof panels and corrugated walls.

Reports of fibrotic lung damage (asbestosis) among textile workers and ship insulators, massive increases in the number of lawsuits, and pressure from labor unions have forced the EPA to ban the use of asbestos in most products. Today, asbestos is used in US Navy submarines, and space shuttles' solid fuel boosters carry asbestos-impregnated rubber liners to protect the steel casings from the heat of takeoff.

[25] Neumann, R. J., "Smoking and Air Pollution Standards," Science, 182:335-336, 1973.

Pesticide is a general term that refers not only to insecticides (for killing insects), but also herbicides (for controlling weeds), fungicides, and fumigants and constitutes as a major contributor to indoor air pollution.
Mold refers to the growth of microorganisms on damp surfaces. Much of the mold found indoors comes from outdoor sources.

Other harmful substances found in everyday household products include formaldehydes, benzene, and chlorinated compounds. Formaldehyde, a component of many glues, resins, paints, surface coatings, particleboards, insulation foam, and wood paneling, is present in most furniture, shelving, and cabinetry. The major health effects associated with formaldehydes are headaches, dizziness, burning of the eyes and throat, nausea, and dermatitis. Benzene is known to cause leukemia and is present in most petrochemical products, tobacco smoke, and also many household products (such as glues and paints). Chlorinated substances are found in many dry cleaners, spray aerosols, and paint removers. PCB or polychlorinated biphenyl is used in insulation, lubricants, paints, and fluorescent lights. PCB is carcinogenic and is known to damage the liver, the skin, and to cause birth defects.

There is no federal law to regulate or limit indoor air pollutants. Several states have enacted their own regulations in terms of building codes, and some industries have adopted standards and developed guidelines that are mostly voluntary. In the absence of federal laws, it is best that we use our common sense and implement strategies that safeguard our health and the health of our loved ones. To reduce complications associated with indoor air pollution and "sick building syndrome," it is best to remove their source by eliminating or reducing the use of toxins and substituting them with more environmentally friendly chemicals, proper maintenance, and increased ventilation. The costs are somewhat higher, but considering the cost of exposure to indoor pollutants, many feel it is worthwhile.

Control and Prevention

The best way to control pollutants is to prevent them from forming in the first place. The obvious choice is to reduce fossil fuel consumption and switch to cleaner sources of energy. Short of that, cleaning the fuel and subsequently modifying combustion processes is probably our next best option. As a last resort, we may opt to remove gaseous and particulate pollutants from the exhaust before they are released into the atmosphere. Fortunately, technologies exist that can help reduce environmental impact at every stage in the process. New technologies have made many alternative and renewable sources of energy cheaper and, in some instances, competitive with fossil fuels. Several new processes have been developed that show the promise of more efficient combustion and reduced air pollution. Cogeneration and combined cycle combustion turbines have helped to improve efficiency and reduce pollution considerably.

Coal combustion has become cleaner as well. For example, in fluidized bed combustion, coal is pulverized and suspended on a bed of air, thus reducing combustion temperature which reduces nitric oxide emission.

These technologies can reduce particulate and sulfur emissions to a great degree but still produce substantial amounts of gaseous emissions which must be removed before they are released into the atmosphere (end-of-the-pipe control technologies). Depending on the nature and size of the particulates, they can be removed by cyclones, bag houses, filters, scrubbers, or electrostatic precipitators. Gaseous pollutants can be removed by physical and chemical adsorption. Sulfur dioxide emission can be controlled by fuel desulphurization and limestone scrubbing. Nitric oxides and oxides of carbon are best removed by catalytic converters and other flue gas treatment techniques.[26] Automobiles are another source of emission. Vehicular emission is controlled by catalytic converters, where small pellets of palladium and platinum transform carbon monoxide and nitric oxides into carbon dioxide and nitrogen. Automotive emission will be discussed in greater detail in Chapter 14.

Air Pollution Standards

There are no air pollution standards that are universally adopted by all countries. The World Health Organization (WHO) has provided *Air Quality Guidelines* for key pollutants (SO_2, NO_2, CO, O_3, and lead) to enable countries to set their national or regional air quality standards within the context of existing environmental, social, economic, and cultural conditions.

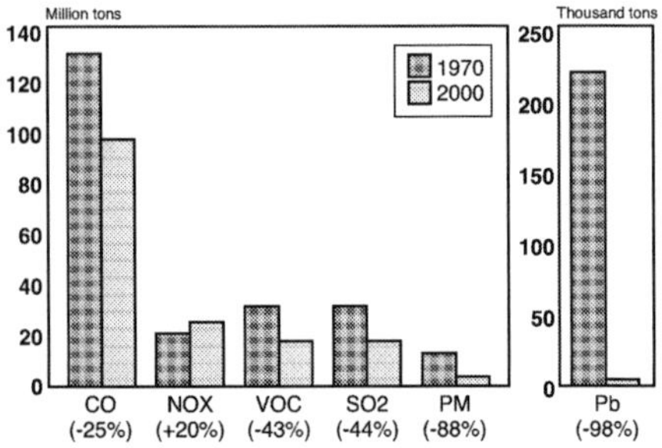

Figure 8-13
Comparison of the emissions for six criteria pollutants in the United States for the period 1970-2000.
Source: Environmental Protection Agency, http://www.epa.gov

Because air pollution does not recognize any boundaries, various states and governments must establish guidelines that regulate interstate and inter-continent pollution transport. In the United States, the agency responsible for setting, monitoring, and enforcing air quality standards is the Environmental Protection Agency (EPA). Congress has empowered the EPA to enact the Clean Air Act (CAA), passed in 1963 and amended in 1970, 1977, and 1990, to develop **National Ambient Air Quality Standards** (NAAQS) to set criteria for improving average air quality at the federal level. Individual states can set more stringent standards if they desire. Furthermore, it is each state's responsibility to develop *state implementation plans* that set the procedure to achieve the state and federal goals. As a result of more and more stringent regulations, US air quality has continuously improved since 1970 when the EPA was established (See Figure 8-13).

Each of the criteria pollutants is assigned a primary standard intended to set limits that protect health and a secondary standard intended to protect public welfare by preventing environmental and property damage. For areas already cleaner than NAAQS, a three-tiered system

[26] Cooper, C. D., Alley, F. C., "Air Pollution Control-A Design approach," Waveland Press, Inc., 1994.

has been established to prevent significant deterioration of air quality. In Tier 1 regions (mainly national parks), the emission levels were frozen at their 1997 level when the air quality standards were established. Tier 2 included most regions and allowed some reduction in air quality, whereas the quality of the Tier 3 regions could eventually deteriorate to the minimum air quality standards. Table 8-2 compares the current guidelines established by the WHO and the US EPA for various pollutant concentrations and exposure times.

Table 8-2. Guidelines for Common Pollutants

		WHO	US EPA
Pollutant	Exposure time	Concentration ($\mu g/m^3$)	
CO	15 min	100,000	---
	30 min	60,000	---
	1 hour	30,000	40,000
	8 hours	10,000	10,000
NO_2	1 hour	400	---
	1 year	40	100
SO_2	10 min	500	---
	24 hour	125	365
	1 year	50	80
O_3	1 hour	175	235
	8 hour	110	157
$PM_{2.5}$	24 hours	---	65
	1 year	---	15
PM_{10}	24 hours	---	50
	1 year	---	50
Lead	1 quarter	---	1.5
	1 year	0.5	---

Example 8-1: A monitoring station near a coal power plant measures sulfur dioxide concentrations of 200 $\mu g/m^3$ between the hours of 7:00 am and 5:00 pm. It drops to 60 $\mu g/m^3$ at all other times. Determine whether this power plant satisfies the WHO's guidelines.

Solution: Within any 24-hour period, the power plant emits 200 $\mu g/m^3$ for 10 hours of operation and 60 $\mu g/m^3$ for the remaining 14 hours. The average 24-hour emission is calculated as:

$$\text{Average Daily Concentration} = \frac{(10x200)+(14x60)}{24} = 118\ mg/m^3$$

Although the power plant does narrowly meet the WHO's guidelines in any 24-hr average, unless there is a considerable downtime, the power plant is likely to exceed the 1-year average emission guidelines. Geographical areas that do not meet the emission standards are called *non-attainment* areas.

Air Quality Index

The Air Quality Index (AQI) was established by the EPA as a simple indication of how clean the air is on a given day and what precautions, if any, need to be taken in carrying out daily activities.[27] The AQI can be viewed as a yardstick to measure the cumulative effects of the criteria air pollutants on human health. It is calculated based on a number of factors such as ambient air temperature, wind conditions, and concentrations of various pollutants. The value of 100 is arbitrarily assigned to air which barely meets the national air quality standards. A value lower than 100 means acceptable air quality and successively lower values mean comparatively cleaner air. Values above 100 signify unhealthy air. The higher the AQI number, the more health risks to individuals. Under these conditions, sensitive groups such as children, the elderly, and adults with heart disease, asthma and other respiratory complications are at higher risk and must limit their outdoor activities. When the AQI climbs above 200, everyone is affected. AQI above 300 is considered hazardous and everyone should avoid prolonged outdoor exposure (See Table 8-3). In the United States, the AQI is usually below 100 for most communities, but may exceed this value a few times a year. In recent years, the AQI has rarely gone beyond 200. Similar standards are set in other major cities around the world (Figure 8-14).

Table 8-3 Air Quality Index

Air Quality Index (AQI) Values	Levels of Health Concern	Colors
When the AQI is in this range:	*...air quality conditions are:*	*...as symbolized by this color:*
0 to 50	Good	Green
51 to 100	Moderate	Yellow
101 to 150	Unhealthy for Sensitive Groups	Orange
151 to 200	Unhealthy	Red
201 to 300	Very Unhealthy	Purple
301 to 500	Hazardous	Maroon

Figure 8-14
This billboard in a Tehran street, informs the drivers that all criteria pollutants are well within their safe allowable limits.

[27] Environmental Protection Agency, *http://www.epa.gov/airnow/aqibroch.*

Summary

Air pollution, both indoor and outdoor, is responsible for hundreds of thousands of deaths each year. Fossil fuels are responsible for thermal pollution and an overwhelming majority of the pollutants released into the atmosphere. Many of the products used indoors are petroleum-based and therefore can be eliminated only if they are replaced with more environmental-friendly materials. Even renewable energy technologies are not entirely "clean". For example, it has been suggested that to generate the same amount of electricity with wind and solar plants, we still need some fossil fuels for manufacturing these devices. Furthermore, solar cells and batteries use large quantities of highly toxic materials (such as arsenic, cadmium, selenium, boron, etc.) that may prove even more harmful than contaminants associated with fossil fuel use. Since most of these products have a limited life, they must be disposed of and occasionally retrofitted with new systems, so finding cleaner manufacturing techniques is essential.

The Kyoto Treaty is expected to have only a limited success unless it includes provisions that requires reducing greenhouse gas emissions by all countries. This will be possible, only if the industrialized countries help the less-developed countries with technologies that enable them to reduce their carbon footprint with no or little economic hardship.

Additional Information

Books

1. Gore, A., An Inconvenient Truth, Penguin Books, 2007.
2. Roleff, T., Pollution: Opposing viewpoints, Greenhaven Press, 2000.
3. Walsh, P. J., Dudney, C. S., Copenhave, E. D., Indoor Air Quality, CRC Press, 1984.

Periodicals

1. *Environmental Science and Technology*, published by the American Chemical Society.

Government Agencies and Websites

1. Environmental Protection Agency *(http://www.epa.gov)*.
2. Occupational Safety and Health Administration (OSHA) *(http://www.osha.gov)*.

Non-Government Organizations and Websites

1. Intergovernmental Panel on Climate Control (IPCC), *(http://www.ipcc..ch)*.
2. United Nations Environment Programme *(http://www.unep.org)*.
3. World Health Organization (WHO) *(http://www.who.ch)*.

Exercises

I. Essay Questions:

1. What is the difference between weather and climate? What are the major factors that determine weather and climate?

2. What factor(s) determine(s) global temperature? Explain how.

3. Describe what the greenhouse effect is and how it contributes to global warming.

4. What are the sources of pollution? Which sector contributes most?

5. What does sequestration of carbon dioxide mean? Describe a few methods by which atmospheric carbon dioxide can be sequestered.

6. How does acid rain form? Where is the acid rain problem more severe and why? What are its effects on health, the environment, and water supplies?

7. Why do some highly polluted lakes appear to be clean?

8. Is a higher concentration of ozone in the lower atmosphere good or bad? In the upper atmosphere? Explain.

9. What is the difference between primary and secondary pollutants? Name a few pollutants in each category.

10. What are the criteria pollutants and why are they important? Which pollutants are considered criteria air pollutants and why?

11. What do primary and secondary pollutants refer to? What are the main primary and secondary pollutants in our atmosphere?

12. What is the air quality standard and how is it determined? What constitutes healthy air quality? What about hazardous?

13. What was the main purpose of the Rio Summit? The Kyoto Treaty? The Montreal Protocol? How have they been implemented?

14. What are the indoor air pollutants and how do they affect our health?

15. Sea salt actually puts more chlorine into the Earth's atmosphere than anything else, but is not a contributor to ozone depletion. Why not?

16. How do you assess the outcome of Montreal and Kyoto Protocols? Name a few accomplishments and shortcomings.

II. Multiple Choice Questions:

1. The increase observed in atmospheric levels of carbon dioxide over the past 100 years is most likely due to
   a. Variations in the rate of lunar eclipses
   b. Melting of thermafrost
   c. The burning of fossil fuels
   d. Deforestation
   e. The advent of nuclear reactors

2. What factor(s) determine(s) global temperature?
   a. Concentration of greenhouse gases in the atmosphere
   b. Earth's albedo
   c. Seasonal variation
   d. Solar intensity
   e. All of the above

3. Ozone has the chemical formula ________, is generally ____________ in the lower atmosphere, and is ________________ in the stratosphere.
   a. O_3, beneficial, detrimental
   b. O_3, detrimental, beneficial
   c. O_3, detrimental, detrimental
   d. O_2, beneficial, detrimental
   e. O_2, detrimental, beneficial

4. The gas most responsible for the global warming at present is
   a. O_3
   b. CO_2
   c. CFC
   d. NOx
   e. SO_2

5. On a per molecule basis, which of these gases is most effective in causing global warming?
   a. Carbon dioxide
   b. Methane
   c. Ammonia
   d. Nitric Oxides
   e. CFCs

6. The so-called "greenhouse effect" is mainly due to
   a. The opacity of ozone and CO_2 to infrared radiation
   b. The transparency of CO_2 and H_2O to infrared radiation
   c. The opacity of CO_2 and H_2O to infrared radiation
   d. The transparency of ozone to ultraviolet radiation
   e. All of the above

7. Which of the following gases is <u>not</u> considered to be a greenhouse gas?
   a. Nitrogen
   b. Methane
   c. Nitric oxide
   d. Chlorofluorocarbon
   e. Water vapor

8. Carbon monoxide is considered a highly poisonous gas because
   a. CO can be oxidized to CO_2 which is a strong greenhouse gas
   b. CO can cause extensive damage to DNA
   c. CO can diffuse and prevent air from entering the lungs
   d. Hemoglobin has a higher affinity for O_2 than for CO
   e. Hemoglobin has a higher affinity for CO than for O_2

9. Carbon monoxide is dangerous to human health because
   a. It causes suffocation because it can react with hemoglobin in the blood
   b. The greenhouse effect will warm up the earth to a point where life will be affected
   c. It will reside in the lungs and produce cancer
   d. It is an important ingredient of photo-chemical smog
   e. CO can cause extensive damage to DNA

10. Which of the following statement is <u>not</u> correct?
   a. The severity of global warming problems depends on the nature, concentration, and their residence time in the atmosphere.
   b. We can use the concentration of air bubbles deep in the Antarctic to estimate the degree at which the earth has warmed up.
   c. In the worst case scenario, many Caribbean and Pacific Island countries will be buried underwater by the end of this century.
   d. Rising the global temperature results in increase in solubility of carbon dioxide in the ocean water.
   e. Rising temperature causes acceleration in rate of methane production by organic matters.

11. The primary source(s) of SO_2 in the atmosphere of a large city is/are
   a. Waste disposal
   b. Incineration
   c. Automobiles
   d. Volcanic ash
   e. Power plants and other stationary sources

12. What is the most abundant gas in the universe?
   a. Hydrogen
   b. Oxygen
   c. Nitrogen
   d. Carbon dioxide
   e. Water vapor

13. Which of the following statement(s) is/are correct?
   a. 90% of all ozone resides in the ozone layer in the stratosphere.
   b. Transportation is the largest source of carbon monoxide, volatile organic matters, and nitrogen oxides.
   c. Nitrogen is the most common gas in the atmosphere.
   d. The largest source of photochemical smog in Los Angeles is automobiles.
   e. All of the above.

14. The most destructive photochemical pollutant in smog is
    a. Particulates
    b. Carbon monoxide
    c. Carbon dioxide
    d. NOx
    e. Ozone

15. Ozone depletion is a result of the
    a. Reaction between nitric oxides and hydrocarbons in the presence of sunlight
    b. Reaction between carbon dioxide and ozone on the upper atmosphere
    c. Leaching mechanisms that remove ozone from the lower atmosphere
    d. Reaction between chlorine and bromine compounds with ozone
    e. Reduction in photochemical activities in the stratosphere

16. The seasonal variation in background carbon dioxide concentration is a result of
    a. The uncertainties inherent in the CO_2 monitoring devices
    b. The more energy use in winter than in summer
    c. The more energy use in summer than in winter
    d. The change in the level of photosynthetic activities
    e. None of the above

17. Photochemical smog has its name because
    a. The chemical industry in Los Angeles produces photons
    b. Sunlight converts O_3 to NO
    c. Sunlight converts NO_2 to sulfuric acid
    d. A combination of NO, HC, and sunlight leads to the formation of O_3
    e. A combination of SO_2 and sunlight leads to the formation of O_3

18. Thermal inversion refers to
    a. A condition in which cooler air sits on top of the warmer air
    b. A condition in which, as a result of a large amount of emissions from cars and smoke stacks, temperature increases rather than decreases with height
    c. Changes in temperature between summer and winter seasons
    d. Cooling of the atmosphere as we move to higher elevations
    e. Decrease in the concentration of pollutants in the upper regions of the atmosphere

19. The most environmentally harmful emissions from automobiles powered by gasoline engines are
    a. CO, SO_2, and NOx
    b. Hydrocarbon, CO, and NOx
    c. CO_2, SO_2, and NOx
    d. CO, O_3, and NOx
    e. Hydrocarbon, particulates, and CO

20. Which of the following statements is correct?
    a. The measure of the acidity of a solution is its pH value.
    b. The pH scale ranges between 0 and 14.
    c. Pure water has a pH of 7, and is considered to be neutral.
    d. Acids have pH values lower than 7, and alkalines have pH values higher than 7.
    e. All of the above.

21. The primary source of man-made SO_2 is ________; it harms people, animals, vegetation, and material through the formation of ___________.
    a. Gasoline burning, nitric acid
    b. Hot springs, sulfuric acid
    c. Coal burning, sulfuric acid
    d. Coal burning, hydrocarbons
    e. Automobiles, smog

22. Which of the following statements about sulfur dioxide emission is true ?
    a. More than two-thirds of all sulfur dioxide emissions in the US are from chemical smelting plants.
    b. The main consequences of large sulfur dioxide emissions are acid rain and ozone destruction.
    c. The primary source of sulfur dioxide emission is coal combustion.
    d. Due to the lack of environmental regulations,

the poorest countries are responsible for most sulfur dioxide emissions.

e. All of the above.

23. It is feared that, with the use of many SSTs (supersonic transports) and the release of Freon, the ________ in the upper atmosphere will decrease, which in turn will cause an increase in the level of ____________ at the surface of the earth.
   a. Ozone, ozone
   b. Ozone, ultraviolet
   c. Ultraviolet, ultraviolet
   d. Carbon dioxide, global warming
   e. Ozone, greenhouse effect

24. Which of the following statements about CFCs is (are) true?
   a. CFCs are more harmful than HCFCs.
   b. The most effective mechanism for removal of CFCs from the atmosphere is by rain.
   c. CFCs are the main cause of global warming.
   d. Stratospheric ozone is produced mainly by the incomplete combustion of fossil fuels.
   e. All of the above.

25. The reason skin cancer can be caused by an increase in ultraviolet radiation but not in visible light at earth's surface is
   a. That the wavelength of UV is longer; hence, it is more destructive
   b. That there is much more UV than visible light at earth's surface
   c. The greenhouse effect
   d. UV has more energy than visible light and can breakup (ionize) biological molecules more readily than light
   e. All of the above

26. Which of the following statements is correct?
   a. The most harmful type of UV radiation endangering our environment is UV-C.
   b. UV-C has been cited as the main cause of increase in skin cancer in humans.
   c. Almost all incoming UV-C radiation is blocked by the earth's atmosphere.
   d. For better results, most tanning machines use UV-B radiation.
   e. UV-A is completely blocked by the atmosphere and therefore is of no concern to humans.

27. You can easily feel the heat from the sun through a glass window, but behind a sheet of glass you do not feel much heat from a fireplace. This is because
   a. Glass is transparent to visible light and opaque to infrared
   b. The fireplace emits essentially pure ultraviolet radiation
   c. Glass is opaque to visible light and transparent to infrared
   d. The infrared component of solar radiation is more intense than that of fire
   e. None of the above

28. The 1987 United Nations' Montreal Protocol
   a. Was drafted to limit consumption of fossil fuels to pre-1990 levels
   b. Obliged signatories to cut carbon dioxide emissions
   c. Was designed to gradually phase out the production of CFCs and other ozone-depleting compounds
   d. Established guidelines for initiating an "emission market"
   e. Provided incentives for voluntary reduction of greenhouse gases

29. Which chemicals are most responsible for breakdown of ozone molecules?
   a. Carbon dioxide and carbon monoxide
   b. Sulfur and nitrogen oxides
   c. Sulfur and carbon oxides
   d. Compounds containing chlorine or phosphorous
   e. Compounds containing chlorine or bromine

30. In which layer of the atmosphere is most of the earth's ozone concentrated?
   a. Exosphere
   b. Mesosphere
   c. Thermosphere
   d. Stratosphere
   e. Troposphere

31. Which of the following chemicals have been

banned in most of the world because of their role in destroying the ozone layer?
   a. DDT
   b. Peroxide
   c. Mercury and heavy metals
   d. Chlorofluorocarbon
   e. All of the above

32. Radon is
   a. A poisonous gas produced in many industrial processes
   b. Found in a naturally occurring radioactive gas derived from the decay of uranium-238
   c. Mostly absorbed by the atmosphere and is not of particular interest except for certain regions in the United States.
   d. A strong rock mostly used in building foundations
   e. A reference to photons emitted in radio frequency range

III. True or False

1. A perfect engine produces no thermal pollution.

2. The best strategy to slow down the rate of ozone depletion is to ban or reduce the use of products that contain chlorine or bromine.

3. The so-called "greenhouse effect" is due, at least partially, to the opacity of CO_2 and H_2O to infrared radiation.

4. The reason that skin cancer can be caused by an increase in ultraviolet radiation but not visible light at earth's surface is that UV wavelength is longer; hence, it is more destructive.

5. The primary source of SO_2 in the atmosphere of a large city is automobile exhaust.

6. The increase observed in atmospheric levels of carbon dioxide over the past 100 years is most probably due to the SST flights in the stratosphere.

7. The brownish color of Los Angeles smog is due to particulates from the many oil-burning plants in the area.

8. Acid rain is a severe problem found mainly in areas with high traffic congestion.

9. Deforestation is of grave concern because it enhances pollution levels and reduces the capacity to remove carbon dioxide.

10. No methane is put into the atmosphere by natural processes.

11. Even if no more ozone depleting materials were to be released into the atmosphere, it would take many decades before ozone concentration in the stratosphere returns to its pre 1950s level.

12. The seasonal variation in carbon dioxide concentration can be traced to changes in levels of photosynthesis activity in plants.

IV. Fill-in the Blanks

1. Long-term storage of carbon underground or in oceans to prevent buildup of carbon dioxide concentration in the atmosphere is referred to as ______________.

2. The main ingredients of ________________ smog are sunshine, hydrocarbon, water, and nitric oxides.

3. It is widely believed that ozone destruction is a result of _____________ and _____________ compounds transported from the lower to the upper atmosphere.

4. There are three basic sources of indoor contaminants: _____________, ____________, and processes.

5. In the United States, the ________ is the agency responsible for developing, monitoring, and enforcing the air quality standards.

6. As of 2006, __________ has overtaken ________ as the largest contributor of greenhouse gases.

V. Project I - Status of international treaties

In this project you are asked to carry out a literature survey and comment on the status of various international treaties and what different countries have done to comply with the established guidelines and recommendations.

1. Highlight major milestones that are to be achieved by the Kyoto Treaty and Montreal Protocol. As of today, what should have been accomplished and by whom?
2. What were the responsibilities of the country of your birth in meeting these requirements? Have they been met? If not, why?
3. How do you rate the overall success of these treaties? Have they achieved their general goals? What were the shortcomings?

Project II - Status of the air quality around the world.

Much of the discussion in this chapter was centered around the air pollution problems in the United States and what the US is doing in terms of air quality standards. This project concentrates on air pollution problem in the countries other than the United States. Please conduct a literature search on the Internet and other resources and answer the following questions.

1. What country were you born in or resided in for a major part of your life?
2. How is the air pollution problem in the country different from that of the United States? Why?
3. What are the sources of pollution? Which sector contributes the most? The least?
4. How do the air quality standards differ in the country from those of the United States?

Project III - Global Warming and Climate

In the movie "An Inconvenient Truth," Al Gore suggested that as a consequence of global warming there has been an increase in the number of hurricanes, tornado, and other violent storms over the past few decades. Check the internet and compile the data on the frequency of these events for the last 50 years. Discuss the data and verify whether such correlations exist. If you find such correlations, does that prove this increase is as a result of global warming? Can you give an alternative explanation?

Project IV - Inventory of household chemicals

Take an inventory of the chemicals found in your house (kitchen, laundry, garage, etc). Classify them by their applications as solvents, detergents, paints, etc.

1. List their chemical compositions and identify them as toxic, non-toxic, volatile, organic, or biodegradable.
2. Which of these chemicals are carcinogenic, poisonous, or flammable?
3. Which of these products are causing global warming? Which ones are ozone depleting? Which ones are considered to be health hazard?
4. Can you substitute any of these products with more environmentally friendly (green) products? List.
5. What are the proper methods of disposal for each of these products?

CHAPTER 9

Geothermal Energy

All these reckonings of the history of underground heat, the details of which I am sure you do not wish me to put before you at present, are founded on the very sure assumption that the material of our present solid earth all round its surface was at one time a white-hot liquid. ~ Lord Kelvin (1824-1907)

Geothermal energy is heat energy from magma (molten rock) deep within the earth, brought to the surface naturally by geysers and springs, by drilling into hot reservoirs or magma, or by exploiting the thermal mass of soil and ground water to drive a heat pump. Geothermal energy has been used since ancient times when the Romans, Greeks, and Japanese - among others - used naturally heated mineral water for bathing, cooking, heating, and medical applications. Today, hot springs and spas[1] are still some of the most attractive spots for recreation and therapeutic bathing. In addition, geothermal energy is used for power generation, space heating, and in greenhouses, swimming pools, and steam processing for industrial applications. Geothermal energy can be considered renewable as long as the rate of heat extraction from a reservoir does not exceed the rate it is recharged by the earth's heat, which, depending on its temperature, may take tens to hundreds of years.

Although only 10% of the earth's surface are at temperatures hot enough to be used as a practical energy source, the earth's geothermal resources are large enough to provide many times the energy needs of the entire population.[2] There are areas where a substantial amount of geothermal energy diffuses through the crust and reaches the surface of the earth. Low- to moderate-temperature geothermal reservoirs are present on most continents; the largest resources in the world are believed to be in Iceland. High-temperature geothermal resources are predominately found in volcanic ranges and island chains. Other especially attractive regions for geothermal energy are in Central America, Indonesia, East Africa, and the Philippines.

Worldwide, geothermal energy provides about 8,000 megawatts of electricity and another 8,000 megawatts of direct-use thermal energy. The largest geothermal electricity generation plant in the world is located in the Geyser area in Northern California, generating about 1,000 MW – about 1/3 of the total US geothermal output and enough electricity to supply a city of over a million homes (Figure 9-1).[3]

In the past few decades, researchers have found new ways to trap natural

---

[1] Named after town of Spa in Belgium, famed for healing hot mineral springs since 14th century.
[2] Norton, G. A., and Goat, C. G., "Geothermal Energy – Clean Power from the Earth's Heat," US Geological Survey, USGS, Circular 1249, 2003.
[3] Kutscher, C.F., "The Status and Future of Geothermal Electric Power," National Renewable Energy laboratory *(http://www.nrel.gov)*.

Figure 9-1
Dry steam plants at the Geysers
Source: National Renewable Laboratory.

thermal gradients in dry rocks. When this technology matures, geothermal energy promises to become an important source of renewable energy available virtually anywhere on the planet.

The Earth

The earth is the third planet from the sun and has an average radius of 6,400 kilometers. Its internal structure includes an inner core, an outer core, lower and upper mantles, and a crust (Figure 9-2). The inner core – the center of the earth – is a sphere that is 2,400 kilometers in diameter. Because of the extreme pressures in the center, the earth's inner core remains solid and becomes molten only in the outer core, allowing rocks to "flow" and form the convection cells within the mantles. Near the top of the mantle, rock cools sufficiently by conduction, causing it to solidify and form the earth's crust. Heat transfer causes warming of the rocks and fluid, providing what we call the geothermal energy.

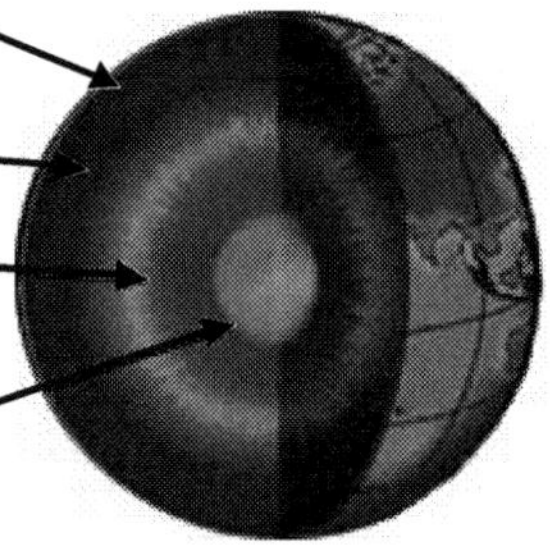

Figure 9-2
Earth's internal structure.

The inner core consists mainly of iron, nickel, uranium, and various heavy metals. The ultimate source of geothermal energy is the radioactive decay of the isotopes of these materials and the original heat produced from the formation of the earth due to gravitational collapse.

The outer core is 2,300 kilometers thick and composed of liquid molten metals, iron, nickel, and lighter elements. The outer core along with earth's rotation is believed to be responsible for the earth's magnetic field. The temperature of the outer core is relatively uniform and varies from 5000°C to about 4,000°C at the outer edges. Surrounding the core are the lower and upper mantles. The combined mantles constitute 80% of the earth's volume and about 67% of its mass. The lower mantle is 2,300 kilometers thick and is made of highly viscous fluids. The upper mantle is solid, has a thickness of about 600 kilometers, and may contain pockets of high-temperature volcanic reservoirs. The top layer is the earth's crust, which is not a continuous sheet of rock, but is broken into solid plates that float independently on top of the hot mantle. The crust is only about 40 kilometers thick and is comprised of all the earth's water and landmasses. The temperature drops rapidly in the upper mantle and crust at a rate of 20-30°C/km until it reaches surface temperatures. Magmas reside at the

Ring of Fire

FYI ...

The horseshoe region encircling the Pacific Ocean covering western coasts of southern, central, and northern Americas, the Alaskan southern coast, Japan, Philippine, Indonesia, and New Zealand houses many hundreds of active and dormant volcanos and thus is called the "**Ring of Fire**." This region is also the area most frequented by earthquakes and volcanic eruptions and is suitable for exploiting geothermal energy.

Volcanoes are formed when convection currents originating deep inside the earth cause lateral forces that push one oceanic plate beneath the other, causing it to melt in a process called *subduction*. Some of the molten rock or *magma* rise find their way toward the surface and flow as *lava*.

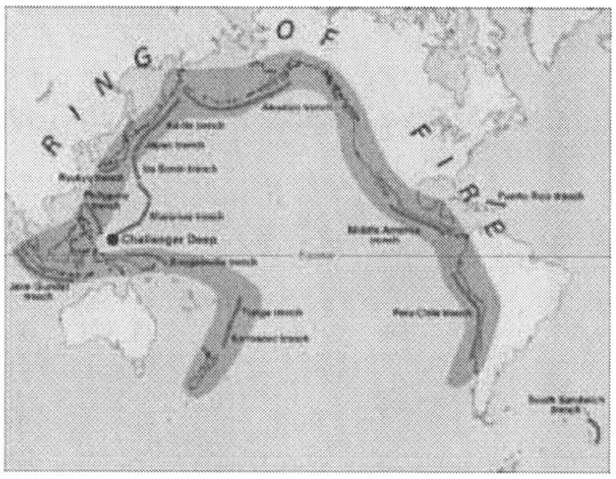

The "Ring of Fire." The solid lines are locations of deep-sea trenches formed when oceanic plates collided (*Image Courtesy of USGS*).

depths of 5-10 kilometers beneath the crust and range in temperature from about 650 to 1,300°C, depending on chemical composition. Moving toward the surface, the temperature decreases; heat is transferred to nonporous, hot, dry rocks, which become increasingly porous and permeable near the surface. The capacity of a porous rock to transmit water is called *permeability*. As the rock becomes saturated, it will form reservoirs of water, steam, or both, which will eventually find their way to the earth's surface through faults and cracks, making geothermal energy available for exploitation.[4]

> **Question:** Considering that our best drills can barely cut through a very small depth of the earth's surface, how can we know anything about the earth's interior structure?
> **Answer:** We know that the average density of the earth (as measured by dividing its volume by mass determined from its gravitational force field) is higher than the surface density, so it can be concluded that the earth's interior must be denser than its surface. Furthermore, detailed analyses of the seismic waves following an earthquake give us information about the internal structure. The speed of propagation of the waves is a function of the density of materials through which they travel.

Geothermal Resources

The earth is a geologically active planet with distinct features such as volcanic eruptions and earthquakes. The energy required to drive these activities is the internal heat stored during earth's formation. Because of its relatively smaller size and lower internal heat, the moon is devoid of these activities.

[4] Kious, W. J., and Tilling, R. I., "The Dynamic Earth: The Story of Plate Tectonics," US Geological Survey, 2001. The electronic version of this book can be downloaded from *http://pubs.usgs.gov/publications/text/dynamic.html*.

Geothermal resources are of four types: hydrothermal, pressurized, hot dry rock, and magma. Hydrothermal resources are the most common source of geothermal energy and are referred to as underground reservoirs containing hot water or steam. Pressurized resources are high-temperature, high-pressure brines trapped in porous rocks. Hot dry rocks (HDR) refer to solid slabs of rock that can be up to several kilometers thick. Finally, magma is the molten rock in volcanic formations rising up from deep within the mantle. Almost all geothermal sources currently being utilized are hydrothermal.

Depending on how the geothermal eruption appears, a hydrothermal source can be classified as a hot spring, a warm spring, a fumarole, or a geyser. Hot springs refer to upwelling of ground water with temperatures above that of the human body. It is called a warm spring when the temperature is lower body temperature but higher than the surrounding atmosphere. When a reservoir does not contain adequate water to seep through and is eventually converted to steam, the geothermal resource is referred to as a fumarole. Geysers are intermittent hot springs that, depending on geological conditions, erupt in regular intervals ranging from a few minutes to hours or months apart (See box "The Old Faithful"). Sixty percent of all geysers in the world are in Yellowstone National Park in Wyoming; others are scattered elsewhere, mainly in California, Italy, New Zealand, and Iceland.

Geothermal resources typically lie from between a few hundred to a few thousand meters below the earth's surface. Deeper reservoirs are generally higher in temperature and therefore yield higher efficiencies. The cost of the power plant itself is less, but drilling costs increase exponentially with depth, increasing the overall cost of extracting energy from deeper mines. With current technologies, only reservoirs within roughly four kilometers are considered to be economically viable sources of geothermal energy.

Question: Can geothermal energy be considered a renewable source of energy?

a Old Faithful

Digging Deeper ...

A geyser is like a periodically erupting pressure cooker. The one known as "Old Faithful" has been a popular attraction in Wyoming's Yellowstone National Park for many years. Geysers work the same way a coffee percolator works; a long narrow column of water is heated from below. In the case of geysers, the source of this heat is volcanic activity and magma. Because of its weight, water reaches temperatures well above its boiling temperature under atmospheric conditions. As the lower layer of water turns into steam, it expands, pushing out the upper layer lifting the column of water above it. As the eruption continues, the weight of the column of water, and with it the pressure at the bottom, decreases and water turns liquid again. After the water begins to reenter the channel, the whole process repeats in 50-70 minutes.

Answer: The earth, like the sun, has an immense amount of stored energy which will by all accounts last billions of years into the future. Unlike solar energy, which is readily available, we must tap geothermal resources. With the current technology, only hydrothermal resources are economically suitable and these have a limited lifetime before they are exhausted. (For example, the power-generating capacity of Geysers has dropped by 40% since only a decade ago). These resources can be recharged in the future as they are gradually reheated by the internal heat of the earth and, depending on their locations, can take several decades to hundreds of years to become operational again. It has therefore been suggested that geothermal energy be considered as a sustainable resource, one whose usefulness can be prolonged or sustained by optimum production strategies and methods.[5] In the case of the Geysers, wastewater from a nearby community has increased the efficiency; it is estimated that at the rate of 1000 megawatts, geysers will remain sustainable for a few decades.

Finding Resources

Short of drilling directly into the hydrothermal reservoir, *chemical geothermometry* is perhaps the best tool for surface exploration of deep geothermal resources. The concentrations of the many minerals dissolved in underground reservoirs are highly temperature dependent. The temperature of an underground geothermal reservoir can therefore be estimated by measuring of chemical composition of various minerals in the hot spring water. The ratios of concentration of sodium, potassium, calcium, magnesium, and lithium, as well as the isotopes of individual atoms are commonly used as indicators for viability of the subsurface reservoir for geothermal exploitation.

Applications

Depending on the source and temperature, geothermal heat can be used either directly or by converting it to electricity. Geothermal resources can be classified as low-temperature (less than 100°C), moderate-temperature (100-200°C), and high-temperature (greater than 200°C). In addition to hot water and steam reservoirs, hot dry rocks and magma have enormous potential. In principal, the natural gradients present in these rocks can be used to extract an unlimited amount of energy - if the technology to exploit them is developed. Magma cannot be economically developed in the near future and will not be discussed further.

Hot Water and Steam Geothermal Power Plants

The highest temperature resources are generally best for electric power generation. When dry (superheated) steam is available, it can be directly expanded through a turbine to generate electricity. For hot water reservoirs under high pressure, the geothermal fluid is brought to the surface and sprayed into a tank. As the pressure drops, some of the water flashes into steam. The steam is subsequently cleaned and piped directly into steam turbines, which drive electric generators. The remaining water is injected back into the reservoir to help maintain its high temperature and pressure. Dry steam reservoirs (fumaroles) are highly efficient, but rare. The largest plant of this type in the world is at the Geysers in northern California.

Moderate steam temperatures are not practical for use in turbine facilities because the steam condenses before it has appreciably expanded through the turbines. Under such circumstances, binary-cycle power plants are more efficient. In a binary plant, the geothermal water is pumped at high pressure through a heat exchanger where it passes its internal heat to a secondary fluid with a lower vapor pressure (alcohol, isopropane, or other refrigerants), causing it to boil and turn into high-pressure steam. The steam is then used in a closed-loop cycle to drive a turbine/generator assembly. Because binary plants do not use the steam or hot water directly, the fluids can be reinjected back into the reservoir. This maintains the pressure and prevents any toxic or noxious gases from entering the atmosphere. To optimize efficiency, flash and binary cycles are often combined in a hybrid design. One such plant is the binary power plant near Mammoth Lakes in California; this system transfers heat from steam at 170°C to isobutene, which vaporizes and drives the turbines for a net generating capacity of 37 megawatts.

Warm Water Systems: Direct Use

At temperatures below 100°C, geothermal sources are available in the form of liquid water. Hot water can be pumped through pipes and used directly for residential, office, and greenhouse heating, domestic water heaters, swimming pools and spas, or can be passed underground to boost agricultural and aqua-cultural production in colder climates. Worldwide, around 12,000 megawatts of thermal energy is available from warm water systems for direct use. Iceland boasts the title of the largest consumer of hot water thermals; 87% of all homes are heated with geothermal water.[5]

Hot Dry Rocks

Another method for using geothermal energy is to utilize the heat from aquifers and hot dry rocks (HDR). Hot dry rocks are impermeable solid slabs of hot rocks found a few kilometers below the surface. Granite usually contains trace amounts of radioactive uranium and thorium and therefore is substantially hotter than surrounding nonradioactive rocks.

Cold water is injected through a bore hole to a bed of hot dry underground rock under pressures high enough to fracture the surrounding rock. Water is heated through conduction as it moves toward one or more nearby production wells where water can be extracted and, depending on the temperature, used either directly or for the generation of electricity. Since

[5] Gawell, K., and Greenberg, G., "2007 Interim Report: Update on World Geothermal Development," US Geothermal Energy Association, May 2007.

the pressure is high, water is still liquid even at temperatures of 200°C or more. When electricity generation is of interest, the hot water gives off its heat to a secondary fluid such as a refrigerant and turns it into vapor before returning to the well to become reheated -- closing the loop. The refrigerant steam drives a turbine-generator to produce electricity. A plant operated in such a closed-loop fashion is virtually pollution free and sustainable.[6] An experimental plant of this type was constructed by the Los Alamos National Laboratory in New Mexico in early 1980s, which operated for 20 years before it was shut down.[7]

Many technological issues must be resolved before this type of plant is suitable for commercialization.

Enhanced Geothermal Systems (EGS)

Hydrothermal reservoirs are useful only when they are in the geographical areas where rocks are sufficiently permeable to allow easy flow of fluid. Enhanced Geothermal Systems essentially use the same technology developed for creating reservoirs in HDR. The fractures induced in surrounding rocks increase the permeability, which allows geothermal exploration in areas that otherwise were deemed inoperable. Many geothermal plants in Europe and Japan employ EGS techniques, thus increasing the productivity of existing hydrothermal reservoirs rather than creating new ones.

Magma

Magma is hot molten rock in the earth's crust, much of it resides within the 5-km of the surface. Thermal energy from magma can be recovered by drilling holes and injecting cold water through magma. The magma heats the water into steam which will rise buoyantly through a second pipe. Magma is, however, highly corrosive and when solidified, creates a layer of insulation that limits the operability of such systems. Finding materials that can withstand hot corrosive magma for an extended time is a major obstacle to commercial development of this technology.

Geothermal Heat Pumps

Heat pumps, as the name implies, are devices that move heat from one place to another up the temperature gradient, i.e. removing heat from cold outside air in the winter and deliver heat to hot outside air in the summer. Geothermal heat pumps (GHPs) take advantage of the relatively constant temperature of water underground from one season to another. Because rocks and soil are good insulators, they are rather insensitive to variations in the ambient air temperature. They are warmer than the ambient air temperature above ground in the winter, and colder during the summer. Below a depth of approximately two meters, the temperature of the soil in

[6] Hooper, G., and Duchane, D., "Hot Dry Rock: An Untapped Sustainable Energy Resource," US Department of Energy Website *(http://www.ees11.lanl.gov/EES11/Programs/HDR/documents/HDREnergy.pdf)*.

[7] *Dateline Los Alamos*, Monthly publication NO. W-7405-ENG-36, Los Alamos National Laboratory, 1995.

most of the world's regions remains stable between 7°C and 20°C.

Geothermal heat pumps exploit the earth or groundwater as heat sources by transferring heat from the soil to the house in winter and as a heat sink by transferring heat from the house to the soil in summer. In the simplest form, a mixture of water and ethylene glycol solution (antifreeze) circulating in underground pipes or loops is used as a medium for transferring heat into or out of a building. In the heating mode (during winters), the mixture is colder than the surrounding ground; thus it absorbs heat from the ground and warms up. The heated liquid is pumped into the building where it transfers its heat through a heat exchanger into the room. In the cooling mode (during summers), the process is reversed. Here, the water/glycol mixture is hotter than the surrounding soil, and thus it releases its energy to the ground and cools. The cold water exchanges heat with a refrigerant, circulating through a heat exchanger located inside the heat pump. A blower cools the air by forcing it across the refrigerant coil.

Figure 9-3
Different loop configurations used with geothermal heat pumps – (a) horizontal, (b) vertical, (c) coil.

Just like conventional heat pumps, geothermal heat pumps can operate in a closed- or an open-loop cycle. In closed-loop systems, a small pump is used to circulate the fluid. In open-loop systems, water from an underground aquifer is piped directly from a well to a building, where it transfers its heat to a heat pump. The water is then returned to the aquifer through a second well some distance away. Open-loop systems are simpler and less costly, but can be used only when there is an abundant supply of ground water.

Depending on the location and the availability of ground space, loops can be installed either horizontally or vertically and in linear or loop configurations (See Figure 9-3). *Horizontal* installation is the most cost-effective for residential sites and in new construction where sufficient land is available. It requires trenches between one and two meters deep. *Vertical* installations are used in large commercial buildings, in places where land area is limited, where land is too rocky to dig trenches, or where digging causes major disturbances to the existing landscape. To save space, the pipe may be coiled or looped into a spiral. When there is an adequate body of water, submerging the loop into the water may be the most cost effective strategy. In such instances, the supply line is run underground from the building to the water and coiled into circles a few meters below the surface to assure that it is not susceptible to winter freezing.

Geothermal heat pumps are rated in two ways: the *coefficient of performance* (COP)[8] and the *energy efficiency rating* (EER)[9]. Although COP is defined for either heating or cooling, EER is used to indicate the cooling efficiency. Geothermal heat pumps can be installed practically anywhere and, when compared to conventional cooling and heating systems, use 30-50% less electricity. If a geothermal system is planned before a building is

[8] COP is the ratio of heating or cooling achieved per unit of work supplied (See Chapter 5).
[9] EER = COP(cooling) x 3.413

constructed, installation is easy, and cost is relatively small. Furthermore, domestic hot water production is essentially free during summers. Unlike conventional rooftop models, geothermal heat pumps are small and can be installed indoors, typically in a basement or attic.

Worldwide, there are currently more than a half million geothermal heat pumps installed, totaling a thermal output of 7,000 megawatts; nearly 70% of them are in the United States.

Advantages and Disadvantages

Unlike wind, solar, or tidal plants, geothermal power plants can deliver power continuously and thus provide base-load electricity.[10] Furthermore, geothermal plants are not vulnerable to weather changes, no storage is needed, and distribution is not an issue. Geothermal reservoirs are, however, limited to specific geographical areas.

Depending on the resources and on power demand, geothermal plants can be constructed in any size-- as small as 100 kW (convenient for local grid applications and rural electrification) to many hundreds of megawatts for base-load and load-demand applications and for national grids. Modular plants can be built so that capacity can be added as the need for power increases.

Like most renewable energies, direct-use systems require a larger capital investment as compared to traditional systems, but the lack of fuel cost and lower operating expenses offset the initial investment in only a few years. The cost of power production varies greatly from 2.5 to 10 cents per kilowatt-hour, depending on such factors as the size, depth, location, and temperature of the reservoir. As the price of petroleum and other fossil fuels increase, geothermal sources become more competitively priced or even cheaper than fossil plants. The currently installed US capacity of direct-use systems totals to 470 MW, enough energy to heat 40,000 average-sized houses.

Environmental Concerns

Compared to fossil fuels, geothermal energy produces much less pollution. Current geothermal fields produce 60% less carbon dioxide that a natural gas fueled electricity-generating power plant produces,[11] with none of the nitric oxide (NOx) gases and much lower sulfur compounds than coal and oil power plants. Geothermal plants do, however, produce trace amounts of hydrogen sulfide, a gas that smells like rotten eggs-- usually associated with hot mineral springs-- and which can be harmful at high concentrations. Some arsenic, mercuric and other toxic gases are also released by the plant. Newer binary- and combined-cycle geothermal plants inject these gases back into the geothermal wells and produce

[10] Base-load refers to minimum electricity needs, independent of the time of the day.

[11] Combustion of bituminous coal emits about 900 kg of carbon dioxide per MWh; the corresponding numbers for natural gas and geothermal plants are 300 and 120 kg per MWh.

virtually no air emissions.

Geothermal production can adversely affect the environment by degrading of geothermal features, promoting ground subsidence, and increasing seismic activities, all of which occur as a result of declining reservoir pressures. Another concern is the release of hot waste water containing significant amounts of toxic substances such as lithium, boron, mercury, and arsenic into existing waterways. This impact can be mitigated by controlling reservoir pressure through adjusting the discharge rate and reinjecting the geothermal fluid back into the ground. Binary plants work in this fashion and typically have no releases.

Summary

Geothermal energy has been used for many thousands of years by people who enjoyed natural hot water springs for bathing and therapeutic reasons. Today, geothermal resources are not only used for direct applications, but also to generate electricity. Currently, most geothermal resources are limited to steam and hot water reservoirs. Innovative technologies are being developed that enable us to exploit the temperature gradients in hot dry rocks, allowing geothermal energy to be used for electricity generation anywhere in the world. This requires the ability to drill approximately 5 to 20 kilometers into the earth's surface and at reasonable cost, a technology that is not currently available. The exploitation of energy from volcanic eruptions and magma flow remains in the distant future.

Additional Information

Books

1. Dipippo, R., Geothermal Power Plants: Principals, Applications and Case Histories, Elsevier, 2005.
2. Dickson, M. H., Fanelli, M., Geothermal Energy: Utilization and Technology, Stylus Pub., 2005.
3. Ochsner, K, Geothermal Heat Pumps: A Guide for Planning and Installing, Earthscan Ltd, 2007.
4. Gupta, H. , and Roy, S., Geothermal Energy: An Alternative Resource for the 21st Century, Elsevier, 2007.

Periodicals

1. *Geothermics,* Direct Science Elsevier Publish. Company, publishes articles on geothermal energy resources and technologies.
2. *Geotimes,* Journal of the American Geological Institute.
3. *Geo-heat Center Quarterly Bulletin,* covers how-to articles on various geothermal applications and equipment, progress in research and development activities of direct heat utilization
4. *Journal of Volcanology and Geothermal Research,* an international journal on the geophysical, geochemical, petrological, economic, and environmental aspects of volcanology and geothermal research.

Government Agencies and Websites

1. National Renewable Energy Laboratory Geothermal Energy Program *(http://www.nrel.gov/geothermal)*.
2. Idaho National Laboratory Geothermal Program *(http://geothermal.id.doe.gov)*.
3. US Department of Energy Geothermal Technology Program *(http://www1.eere.energy.gov/geothermal)*.
4. California Energy Commission *((http://www.energy.ca.gov/geothermal)*.

Non-Government Agencies and Websites

1. Geothermal Resources Council *(http://www.geothermal.org)*.
2. Geothermal Energy Association *(http://www.geo-energy.org)*.

Exercises

I. Essay Questions

1. What countries have the most potential for using geothermal energy?

2. Describe the principal of operation of geothermal heat pumps. Where can they be used?

3. Can geothermal energy be considered a renewable energy? Explain.

4. What are the uses for low-temperature geothermal resources?

5. How can we use the earth's temperature gradient to drive a heat engine?

6. What are the factors that constitute a good geothermal location? Where are such sites located?

7. Explain how chemical geothermometry works and how it is used to explore the ideal sites for geothermal excavation?

8. With current technologies, which types of geothermal sources can be exploited?

9. Is geothermal renewable? Sustainable? Explain.

10. What are the environmental considerations of geothermal energy?

II. Multiple Choice Questions

1. How do temperature and pressure vary with depth inside the earth?
   a. Both temperature and pressure increase with depth.
   b. Temperature increases with depth, but pressure decreases.
   c. Pressure increases with depth, but temperature decreases.
   d. Temperature increases with depth, but pressure remains relatively constant.
   e. Both temperature and pressure decrease with depth.

2. What distinguishes earth's inner and outer cores?
   a. Because of its extreme temperatures, the inner core is a gas. The outer core is cooler and remains liquid.
   b. Both the inner and outer cores are liquid. The only difference is their temperatures.
   c. Both the inner and outer cores are solid. The only difference is their temperatures.
   d. The inner core is liquid. It solidifies gradually as temperature drops toward the surface.
   e. The inner core is solid. The outer core is liquid.

3. The inner core is composed mostly of
   a. Iron
   b. Silicon
   c. Sulfur
   d. Oxygen
   e. Steam

4. The average thickness of the crust is
   a. Less than 1 km
   b. 30-50 km
   c. 100-200 km
   d. 200-1000 km
   e. More than 1000 km

5. Where is the source of earth's magnetic field?
   a. Crust
   b. Mantle
   c. Inner core
   d. Outer core
   e. Electromagnetic radiation from the sun

6. Today, which renewable energy sources provide the US with the most energy?
   a. Solar
   b. Wind
   c. Hydropower
   d. Geothermal
   e. Waves and Tides

7. On average, the energy available from geothermal resources
   a. Is much greater than the solar influx
   b. Is about the same as the solar influx
   c. Is much less than the solar influx

8. The source of geothermal energy is
   a. Radioactive decay of elements below the earth's crust
   b. Heat still left from the time when the earth was formed
   c. Chemical reactions among gases trapped below the earth
   d. Underground nuclear explosions
   e. Both a and b

9. The temperature in the upper mantle and crust falls at the rate of
   a. 1°C/m
   b. 1°C/km
   c. 5-10°C/km
   d. 20-30°C/km
   e. More than 100 °C/km

10. Which layer of the earth's interior is liquid?
    a. Crust
    b. Mantle
    c. Inner core
    d. Outer core
    e. None is liquid

11. Which layer of the earth's interior has the greatest density?
    a. Crust
    b. Mantle
    c. Inner core
    d. Outer core
    e. All about equal

12. Which layer of the earth's interior has the greatest volume?
    a. Crust
    b. Mantle
    c. Inner core
    d. Outer core
    e. About the same

13. Which layer of the earth's interior has the least thickness?
    a. Crust
    b. Lower mantle
    c. Upper mantle
    d. Inner core
    e. Outer core

14. The major obstacle(s) associated with the use of geothermal energy is/are
    a. That many sites are difficult to access
    b. That they produce hydrogen sulfide which is an environmental pollutant
    c. That no technology exists for exploiting temperature gradients in hot solid rocks
    d. High initial cost
    e. All of the above

15. Most of the geothermal energy available today is from
    a. Volcanoes
    b. Hot water springs and geysers
    c. Hot dry rocks
    d. Desert surfaces
    e. All of the above

16. Binary-cycle power plants are most often used
    a. When the geothermal source is at a temperature below 100°C
    b. When the geothermal source is at a moderate temperature, between 100-200°C
    c. Only when steam is available at temperatures above 300°C
    d. Only when they are in association with hybrid design plants
    e. Only when underground sources contain large fractions of alcohol and other refrigerants

17. Depending on their size, geothermal plants can be
    a. Used for base load electricity production
    b. Designed to follow daily load demands
    c. Used for mini-grid and rural electrification
    d. Used for direct thermal use
    e. Used for all of the above

18. The largest geothermal power plant in the United States is located near
    a. San Francisco
    b. The Rocky Mountains
    c. Hawaii
    d. Los Angeles
    e. New York

19. Geothermal heat pumps can be used for
    a. Space heating
    b. Space cooling
    c. Both space heating and cooling
    d. Processing steam
    e. All of the above

20. Geothermal energy
    a. Is essentially a zero emission source of energy
    b. Produces some pollution, but not as much as fossil fuels
    c. Produces some radioactive waste, but with a short half-life
    d. Produces more pollution than fossil fuels, but with no radioactive waste
    e. None of the above

III. True or False?

1. Boron, mercury, and arsenic are all types of pollution that come from geothermal plants.

2. Earth receives most of its life-supporting energy from geothermal sources.

3. Temperatures within the earth's inner core could reach millions of degrees.

4. The outer core is composed of liquid molten metals, mostly iron and nickel.

5. Most geothermal resources lie only a few meters below the earth's surface.

6. Binary geothermal plants usually use geothermal water at high pressures and moderate temperatures.

7. HDR is an innovative technology that can possibly meet many of the energy needs of the 21st century.

8. Geysers are considered renewable only if they are used at a sustained rate.

9. Binary plants are advantageous over non-binary plants, because they can be designed virtually pollution-free.

10. It is best to decide on the use of geothermal plants before a building is constructed.

IV. Fill-in the Blanks

1. Geothermal resources can be categorized as hydrothermal, ____________, ____________, and ____________.

2. The largest concentration of hydrothermal activities lie in __________________.

3. __________ are vents in the earth's surface that release only steam.

4. The fluid often used in geothermal heat pumps is a mixture of water and __________ solution.

5. Geothermal plants produce trace amounts of ____________, a gas that smells like rotten egg.

6. _________ installation of heat pumps is most cost-effective for residential sites and new construction where sufficient land is available.

7. Geothermal plants produce little carbon dioxide and practically none of the ______________ commonly produced in power plants using fossil fuels.

8. Upwelling of ground water with temperatures above surrounding atmospheric temperature but lower than body temperature is called ___________.

9. The water spewing out of hot springs have temperatures _______ that of human body.

10. The performance of a geothermal heat pump is measured by its _______________.

Solar Energy

CHAPTER 10

Give me the splendid silent sun with all his beams full-dazzling.

~ Walt Whitman (1819 - 1892)

Since ancient times, humans have demonstrated a basic understanding of solar energy by designing shelters that protected them from the cold of winter and the heat of summer. Socrates suggested that houses should be built facing south in order to maximize the heat of the winter sun. Anasazi Indians designed their dwellings under cliff overhangs, maximizing the solar gain in winter and providing shade in summer. The sheltering overhangs also reduced radiant heat loss at night and protected against cold winds. Furthermore, some houses were built with thick stone and clay walls and had small windows to retain heat during the day and release it gradually at night.

Today, solar power has been proven to be highly effective not only in providing heat to buildings and swimming pools, but also in cooling applications and for the generation of electricity. As with many other sources of energy, there are limitations on where this energy source can be used. The amount and intensity of solar energy available varies depending on geographical location, the position of the sun in the sky, weather conditions, and the orientation of the collector and surrounding objects. Solar energy can be used worldwide, especially in regions with high insolation. This includes large areas of Africa, Australia, the Middle East, Central and South America, and the southwestern United States.

In this chapter, we start by looking at the sun itself and the amount of radiation that can potentially be intercepted by various collectors on earth. We will discuss passive and active systems, as well as thermal and photovoltaic systems for utilizing solar energy for heating, cooling, and producing electricity.

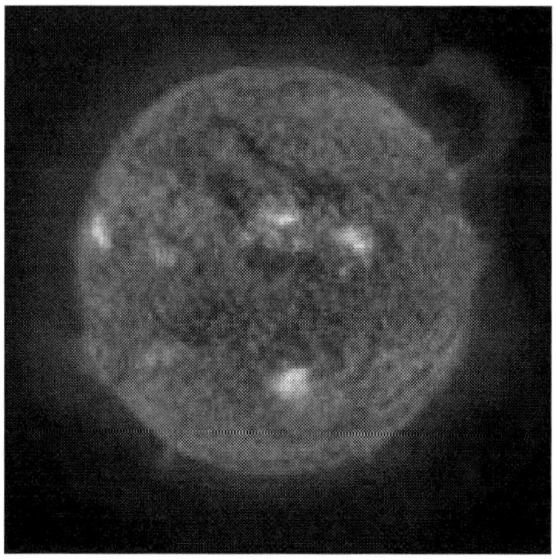

Figure 10-1
The sun
Source: NASA http://www.nasa.gov.

The Sun

The sun is a 1.4 million kilometer wide sphere of extremely dense, hot gas that resides approximately 150 million kilometers from the earth (Figure 10-1). Composed predominately of hydrogen and helium, its central temperature and pressure are estimated to reach 15-25 million kelvins and 300 billion atmospheres – conditions suitable for nuclear fusion to take place. As we will discuss later in Chapter 11, fusion involves combining atoms of hydrogen into a helium nucleus and heat. The sun loses mass at a rate of about 4.3 billion kilograms every second while producing an enormous amount of energy.[1] At its outer layer (called the photosphere),

[1] This loss of mass is nothing to worry about! By some estimates, the sun will have enough energy to support life on earth for another five billion years.

ionized gases both absorb and emit a continuous spectrum of radiation, causing this layer to be essentially opaque. Temperatures drop to about 5,800 K at the surface of the sun.

Solar Position

The geometric position of the sun as seen from a particular place on the surface of the earth varies from day to day and hour to hour. At any given instance, the sun's position can be fixed by two angles, altitude and azimuth (Figure 10-2).

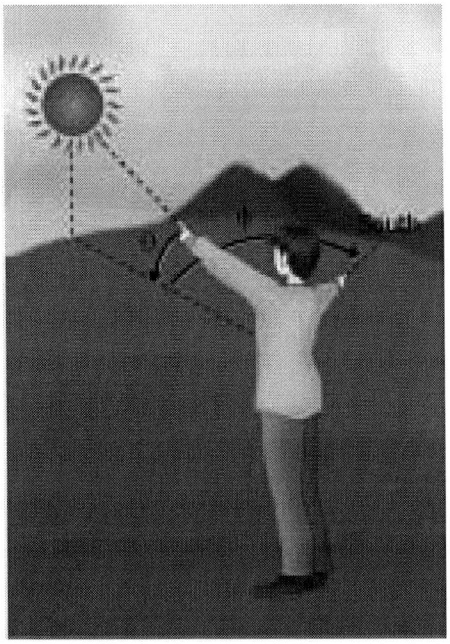

Figure 10-2
The position of the sun in the sky can be determined by knowing the solar altitude (θ) and azimuth (ϕ) angles.

Altitude is the angle between the sun and the horizon. When the sun is on the horizon (at sunrise and sunset), this angle is zero. Solar altitude is at the maximum at solar noon. The complement of solar altitude angle, or the angle of the sun from a vertical line directly overhead, is called the **zenith** angle.

Azimuth is the angle between a north-south line on the earth's surface and the horizontal projection of the sun's rays; it is measured from true south. By convention, solar azimuth is negative before noon and positive after noon.

To calculate the sun's angle we need to know not only the relative position of the sun, but also the geographical location of the observer on earth. Two angles define the position of any point on the surface of the earth.

Latitude is the angular distance measured along a meridian north or south of the equator. All points on the equator have a latitude of zero; the north and south poles have latitudes of +90° (90° N) and –90° (90° S). Los Angeles is located at a latitude of 33.9° N.

Longitude is the angular distance measured east or west from the Prime (International) Meridian -- an imaginary circle passing through Greenwich, England. Since the period of rotation of earth about its own axis is 24 hours, each hour covers 360/24 = 15° longitude. Los Angeles has a longitude of 118° W. Calculating the relative position of the sun in the sky from earth at different locations and times is outside the scope of this book, but readers are encouraged to refer to more advanced texts.[2]

Solar Insolation

Solar *insolation* (not to be confused with insulation) is the amount of radiation arriving from the sun that collides with a flat surface. This includes the direct (beamed) radiation from the sun, scattered (diffused) radiation from the sky, and reradiated (or reflected) radiation from the surroundings. Both direct and diffused radiation depend on a number of

[2] See for example, Duffie, J, and Beckman, William A., "Solar Engineering of Thermal Processes," 2nd. Ed., John Wiley and Sons, 1991.

factors such as the day of the month, the time of day, and the weather. The reflected radiation depends on the topography of the surrounding area, the amount of shade, and the reflectivity (also called *albedo*) of the earth. Albedo varies significantly with the type of material and whether the surface is exposed or covered by water, snow, or vegetation. Atmospheric conditions affect the intensity of the light as it reaches the collector. Factors affecting the light intensity are cloud cover, humidity, and atmospheric composition.

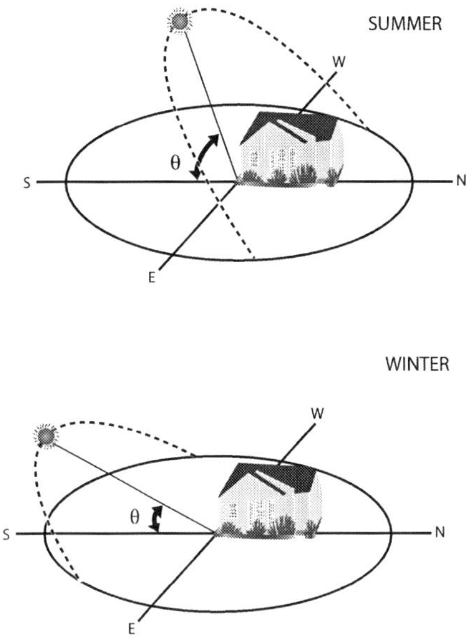

Figure 10-3
Summer and winter sun paths in the northern hemisphere.

It is a common experience that the sun rises higher in the sky in summer than winter. It may also be noted that, in the northern hemisphere, the sun rises south of due east in winter and north of due east in summer. As winter proceeds into spring and summer, the sun rises earlier in the morning, moves higher in the sky, reaches its highest point at noon, and sets later in the afternoon (See Figure 10-3). The opposite is true in the southern hemisphere.[3]

Unlike what some might suspect, the total solar energy incident on a south-facing window is greater during the winter than in the summer in spite of the fact that days are longer in the summer. Two factors contribute to this:

1) Although the sun rises earlier in the summer, it remains north of the window before it sets in the afternoon. In winter, the opposite is true -- sun rises and sets south of the window; therefore, south-facing windows are subject to incident sunlight for the entire length of the solar day. For example, a house in Long Beach, California located at 33° 48' latitude in the northern hemisphere experiences about fourteen hours of daylight (5:30 am to 7:30 pm) on Summer Solstice (June 21) and ten hours of daylight (7:30 am to 5:30 pm) on Winter Solstice (December 21). The south-facing windows in this house receive sunshine for only seven hours (8:30 am to 3:30 pm) during summer and receive the full ten hours of sunshine during the winter.
2) The summer sun is higher in the sky than the winter sun. In the winter, the lower sun strikes the windows more directly than in the summer when the sun is higher. For example, every square meter of the window of a house in Long Beach receives, on average, 220 watts of solar energy per hour of daylight in the winter, whereas the same window receives 40% less energy in the summer. Furthermore, because of the higher angle of incidence in summer, more of the sunlight is reflected off the glass than would be during the winter.

Of all the radiation emitted by the sun, only a very small fraction is intercepted by the earth's atmosphere. This amount is called the ***solar constant*** and represents the energy from the sun, per second, that falls on a surface perpendicular to the sun's rays and just outside the earth's

[3] From now on, we will limit our discussions to the northern hemisphere.

Solar Energy: The Facts

Did You Know That ...?

- Every second, about 4.3 million tons of solar mass is converted into massive amount of energy, totaling to equal to roughly one billion hydrogen bombs.
- Photovoltaic modules covering 0.3% of the land in the United States could supply all its electrical demand.
- The sun emits radiation at all possible wavelengths, from very short- wavelength, high-energy cosmic and gamma rays to very long-wavelength, low-energy radar and radio waves. Interestingly, a large fraction of solar radiation (48%) falls in the narrow band between 400-700 nm ($4x10^{-7}$ to $7x10^{-7}$ m) which is visible to our eyes. Another 46% of the radiation is in the form of heat and falls within the infrared region.

atmosphere. The solar constant is not truly "constant", however. Because the actual orbit of the earth around the sun is elliptic, the average distance varies by about 3.4% during the year and the solar constant varies between 1,300-1,400 W/m^2. Like the sun, the earth is also emitting radiation, but at a much lower rate. The earth's (terrestrial) emission is in the infrared region of the spectrum and averages at around 172 W/m^2.

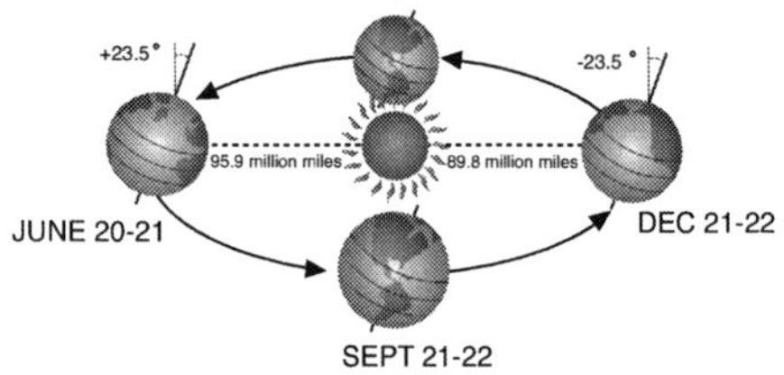

Figure 10-4
Earth travels around the sun along an elliptical orbit about once every 365 days.

Question: The data collected on the motion of the earth around the sun shows that the distance between the earth and the sun is slightly shorter in the winter than in summer (Figure 10-4). Why then are summers warmer than winters?

Answer: The earth's rotational axis is not perpendicular to its orbit around the sun, but tilts by 23.5 degrees. In the northern hemisphere, the angle of the earth's rotation tilts toward the sun in the summer and away from it in the winter. In southern hemispheres, the earth's tilt is in the other direction and the seasons are reversed.[4]

To calculate the amount of energy reaching a surface, we not only need to know the position of the collector plate, but also its tilt – the angle that the collector makes with a horizontal plane. It is obvious that the highest amount of solar energy will be received by a collector that is perpendicular to the sun's rays. The sun is, however, rarely directly overhead, and its position changes throughout the day, limiting the solar power averaged over a 24-hour period to only one quarter of the solar constant value, or 340 W/m^2. Due to the absorption and scattering by particles and clouds, only 18% of this amount actually reaches the earth. Therefore, for maximum collection efficiency, many designs employ tracking systems that move the collectors to directly face the sun at all times. Tracking systems, however, can be very costly and are used only in special applications.

Solar Heating and Cooling

Heating and cooling of a space is needed when the temperature falls below

[4] The seasonal variation of the lengths of days and nights can be explained by noting the relative position of the sun with respect to the earth. Twice a year on the vernal (first day of the spring) and autumnal (first day of the fall) equinoxes, day and night become equal in length. The vernal equinox has been celebrated throughout history as the time of rejuvenation and rebirth, thus marking the start of a New Year in the Zodiac calendar. The summer solstice occurs around June 22, and represents the longest day of the year. The shortest day of the year falls on winter solstice, at or around December 22 in the Northern Hemisphere.

Window Glass

FYI ...

Window glass is a simple and economical way of getting both light and heat conveniently into a living room or an office building. However, in many instances, such as during the hot summer days, we would like to block the sun's heat while still allowing the light into the room. This is traditionally done by laminating window glass with ultra-thin layers of silver, or by adding infrared-absorbing dyes that are broken down by strong sunlight and scatter light, thus giving the glass a smoky haze.

A new technique has recently been developed that blocks the infrared by an order of magnitude and, at the same time, allows visible light to pass through.[i] The key is to sandwich small particles of lanthanum hexaboride in the glass. These particles strongly absorb the near-infrared radiation, but are so small in size (smaller than the wavelength of the visible light) they cannot scatter sunlight, leaving the glass highly transparent.

[i] *Schelm, S. & Smith, G. B. Dilute LaB6 nanoparticles in polymer as optimized clear solar control glazing. Applied Physics Letters, 82, 4346 - 4348, (2003).*

or rises above a desired value. To determine how much heating or cooling is necessary, we must determine how much heat enters or leaks out of a building. A simple procedure for the rough calculation of thermal load is presented in the box below. The more detailed calculation is outside the scope of this book, but the interested reader is referred to texts and journals dedicated to the topic.[5]

Passive Heating

Passive heating refers to the collection of solar energy without use of electrical or mechanical power. The simplest form of passive heating is the direct gain of solar energy as it passes through window glass (Figure 10-5a). In a more complex system, a medium with a large thermal mass (such as a concrete wall, or a drum of oil or water) stores the solar energy before releasing it at a later time when it is needed. Because they have fewer (if any) moving parts, passive designs are simpler, more reliable, more durable, and cost less than active systems. Passive solar systems can be designed in a variety of forms.[6]

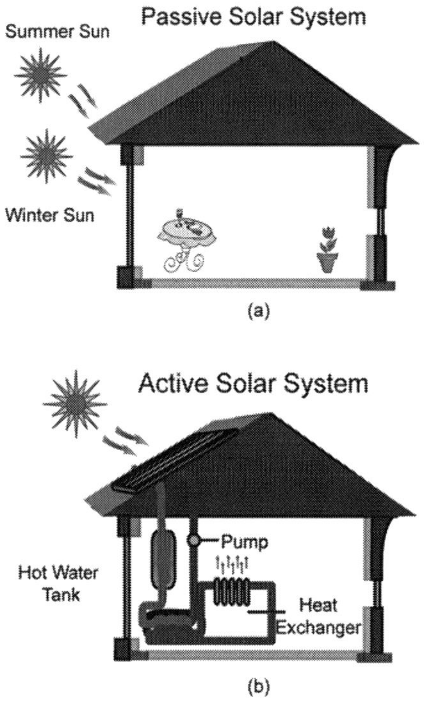

Figure 10-5
Passive and active solar systems

Solar rooms are south-facing rooms with large windows, thick walls, and well-insulated roofs. Large windows allow maximum gain of direct solar radiation. Double and triple glazing of windows can significantly reduce heat losses from the room, while only moderately reducing the solar gains. Thick walls have large masses and high thermal storage capacity.

Solar chimneys allow cool ambient air into the bottom of a glass collector where it is subsequently heated by incident sunlight and then rises by natural convection to the top before being released into a room. Solar chimneys are advantageous in that they can eliminate direct sunlight and glare and reduce heat losses during the night. More elaborate designs allow the amount of circulating air (and thus heat input) to be adjusted

[5] See for example ASHRAE Journal of Heating, Ventilating, and Air Conditioning.
[6] Anderson, B., and Wells, M., Passive Solar Energy, Brick House Publishing, New Hampshire, 1981.

as necessary.

Active Heating

The main component of most active systems is rooftop solar collectors, where solar energy heats a working fluid like water and either stores it in a hot water reservoir or distributes it directly to interior spaces through pipes or ducts. For most applications, temperatures of about 100°C are sufficient and flat plate collectors are the most convenient. For applications that require higher temperatures, energy must be concentrated and solar concentrators such as lenses and mirrors are necessary.

Flat plate collectors consist of a number of tubes through which a working fluid (such as water or air) is heated by solar energy. Tubes are arranged in parallel inside an airtight collector box that is covered by a sheet of glass or plastic and insulated on the back. To absorb the maximum amount of energy possible, the back-plate is painted black.[7] The collectors are placed on rooftops or in open areas and tilted at an angle that maximizes solar insolation. Ideally, collector plates should be perpendicular to the sun's rays at all times. This requires costly and complex tracking devices, however, so fixed flat plate collectors are often used instead. For optimal efficiency, collectors are faced south (north if in the southern hemisphere) and tilted at an angle equal to the latitude of their location. **Evacuated tube collectors** are similar to flat plate collectors except that the glass tubes are replaced with two concentrating tubes. Fluid flows inside the inner one while the outer tube is evacuated. This arrangement eliminates conduction and convection losses, which increases the collection efficiency.

The most common uses of flat plate collectors are for domestic hot water systems, pools, spas, and space heating.[8] Usually a gas or an electric-powered heater is added as a backup for periods when sufficient solar energy is not available. Evacuated-tube solar collectors can be used under cloudy conditions, so their year-round efficiency is higher. They are considerably more expensive and maintenance costs are higher. A typical domestic hot water system is shown in Figure 10-5b.

Solar Cooling

In Chapter 5, we reviewed the principle of operation of a refrigerator (or an air conditioning system). Cooling occurs as a result of a refrigerant removing heat from the refrigerated (air conditioned) space and becoming vaporized (step 1). The vapor is then heated to a high temperature and pressure by a compressor (step 2); it then enters a condenser where it is condensed and becomes liquid (step 3). Finally the liquid is expanded

[7] Hoagland, W., "Solar Energy," *Scientific American*, V. 273, pp 170-173, September 1995.

[8] Solar water heaters, sometimes called domestic hot water systems, can be either passive or active and open or closed loop. Passive systems rely on the principle that water in the collector becomes lighter and rises as it heats, while cooler water in the tank sinks, causing circulation by natural convection (*thermosiphon*). No pump is needed for water circulation through the collectors, making the system simpler and less expensive. For passive systems to work, the tank must be above the collector. Active systems use pumps to assist circulation. *Open-loop* systems are popular in mild climates where there is no danger of freezing. These systems require a circulation pump and are therefore inherently active. *Closed-loop* systems use a mixture of water-glycol antifreeze mixture and thus can be used in area where freezing is a possibility.

in an expansion valve, allowing the refrigerant to cool to the evaporator temperature (step 4). The refrigerant is now ready to remove additional heat and repeat the cycle. The point to remember is that some energy in the form of electricity or heat is needed to carry out step 2. The source of this energy can be natural gas, oil, or in this case, solar energy.

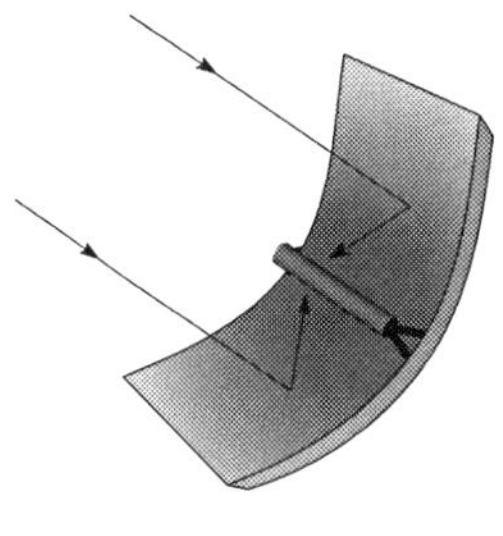

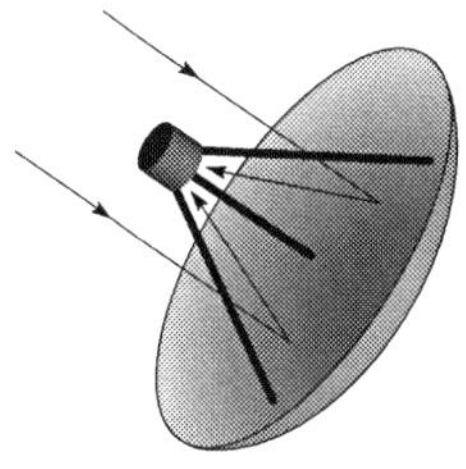

Figure 10-6
Parabolic trough and parabolic dish.

Solar Concentrator

When temperatures higher than 100°C are demanded, solar energy must be concentrated.[9,10] The degree to which solar energy is concentrated by a given collector is called the concentration ratio and is defined as:

$$C.R. = \frac{\text{Collector Aperture Area}}{\text{Receiver Surface Area}} \qquad (10\text{-}1)$$

Concentrators come in two basic configurations: parabolic troughs and parabolic dishes (Figure 10-6). *Parabolic troughs* are basically long sheets of metals bent along an axis to concentrate sunlight on a tube containing a liquid and placed along the focal line. *Parabolic dishes* are similar except that they are bent to form dishes. Since parabolic dishes distribute the energy over a smaller collector surface (around a focal point), they have a higher concentrating power than parabolic troughs. Furthermore, it is a matter of common sense that the bigger and more curved the collector area is, the higher the degree of concentration and collection efficiency will be. As Table 10-1 shows, parabolic dishes with concentration ratios as high as 1000:1 can be constructed; these can heat water to superheated steam at 1,200°C. Temperatures as high as 4,000 °C have been reached by a combination of various concentrator methods (Figure 10-7).

Table 10-1. Solar Concentrators

	CR	T_{max}	η_{max}
Flat Plate Collectors	1	100°C	20%
Parabolic Trough	10-50	400°C	56%
Parabolic Dishes	30-1000	1200°C	75%
Power Tower	---	2000°C	85%

Electrical Power Generation

In Chapter 5, we saw that whenever there is a temperature gradient, a heat engine can be designed that converts thermal energy to work. According to the second law of thermodynamics, the efficiency of any heat engine can be improved by increasing the temperature differences between the two heat reservoirs. In the case of fossil or nuclear power plants, the chemical energy in the fossil fuel or the nuclear energy in the uranium is used as a heat source to boil water into steam. Solar power plants work in a similar fashion except that the heat energy comes from the sun.

Figure 10-7
The largest solar furnace in the world is in Odeillo, France. A heliostat consisting of 63 mirrors (not shown here) automatically track the sun and reflect its rays into a fixed parabolic mirror reaching temperatures of 3800°C for a maximum of 1 MW of power.

Solar Thermal Power Systems

There are several ways that thermal energy from the sun can be used to produce electricity; these include solar chimneys, solar collectors, and solar power towers.

It was shown earlier that **solar chimneys** can be used to heat a room by

[9] Manchini, T., et al., Solar Thermal Power, in Advances in Solar Energy, Vol. 11, American Solar Energy Society, Boulder, CO, 1997.
[10] Readers interested in detail designs of solar concentrators should refer to Solar Energy Systems Design by W.B. Stine and R.W. Harrigan (John Wiley and Sons, Inc. 1985. An updated copy of text is available electronically at http://www.powerfromthesun.net).

drawing hot air through a duct. Alternatively, the rising current of air can drive a gas turbine to produce electricity; a conversion efficiency of 2-3% can be achieved.[11,12] The capital cost is relatively high, but operating costs are very low, fuel is free, and the power station has a long lifetime. To increase efficiency, a heat exchanger can be placed at the outlet of the turbine to capture the remaining thermal energy in the air column and preheat the incoming air.

Figure 10-8
SEGS Solar field in Mojave Desert, California.
Image courtesy of SunLabs, Dept. of Energy

The **solar trough system** consists of a trough with a parabolic cross section and a reflective surface on the inside. The sunlight impinges on the surface and is reflected to a receiver tube placed along the focal line of the parabola, where a working fluid is heated to about 400°C. The receiver is blackened to increase its absorption efficiency. The heat is transferred through a series of heat exchangers to convert water to superheated steam, which can be used directly or to power a turbine/generator system to generate electricity. The SEGS (Solar Electric Generating Station) plants are the world's largest parabolic trough facilities; they operate at three sites in California's Mojave Desert and range in size from 14 MW to 80 MW for a combined electric generating capacity of 354 MW. Each plant consists of a collector field of many troughs aligned on a north-south axis, which reflect sunlight onto pipes filled with oil. Each trough is mounted on a tracking device that can follow the sun from east to west during the day to ensure that the sunlight is continuously focused on the receiver pipes (Figure 10-8). Because oil loses its heat quickly, gas-fired generators are used to assist solar collectors in meeting demand during nights and cloudy days.

Figure 10-9
Solar Two.
Source: National Renewable Energy Laboratoty.

Solar power towers consist of the same main components as other power plants, except that the steam generator is replaced with a large array of mirrors, called a *heliostat*, arranged around a tower. A central processor controls each mirror individually in order to track and at all times focus the sun's rays to a receiver located at the top of the tower. A fluid that circulates through the receiver carries the heat and passes it through heat exchangers to heat water to steam. Steam subsequently expands in a steam turbine, which is coupled with a generator to produce electricity.

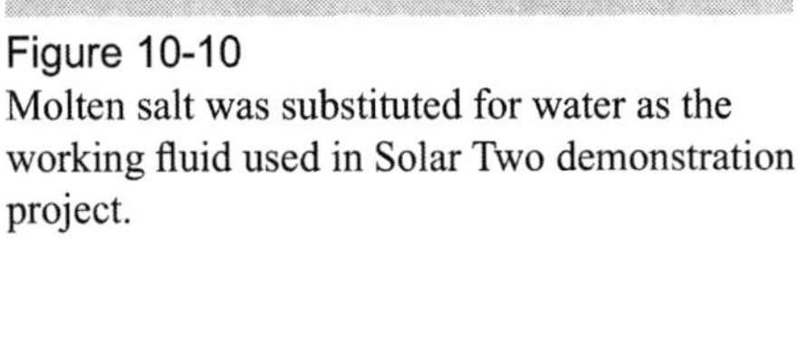

Figure 10-10
Molten salt was substituted for water as the working fluid used in Solar Two demonstration project.

Using this concept, two demonstration plants, called Solar One and Solar Two, were designed by the Sandia National Laboratory and operated by Southern California Edison in Daggett, California. Solar One was a 10 MW plant where water was heated directly in the solar receiver to generate steam. The plant operated from 1982 through 1988 and was eventually decommissioned and replaced by Solar Two, in which water was replaced with a mixture of molten nitrate salt as the working fluid (Figure 10-9). The advantage of molten salt is that it remains liquid to a higher temperature, and thus its thermal efficiency is increased (Figure

[11] Schlaich, J., The Solar Chimney: Electricity from the Sun, C. Maurer, Geislingen, Germany, 1995.
[12] Von Backström, T.W. and Gannon, A.J., "Compressible Flow through Solar Power Plant Chimneys". *ASME Journal of Solar Energy Engineering*, Vol. 122, No. 3, pp. 138-145, 2000.

10-10). Solar Two had an additional advantage in that it could generate 7 MW of electricity for three hours after sunset. The Solar Two plant operated from 1996 to 1999, after which point it was decommissioned due to a lack of funding. Based on lessons learned, BrightSource Energy is building the largest series of solar installations with capacity of 1,300 megawatts in Mojave Desert. When completed in 2013, the facility will provide electricity to 845,000 Southern California homes, doubling the total US solar output.[13] Similar plants are being built in Spain, Australia, and elsewhere.

Dish/engine systems are similar to trough systems, except they use parabolic mirrors. Due to the higher concentrating power of dishes and their ability to track the sun in all directions, higher temperatures (750°C) are achieved. Since dishes have a smaller aperture than trough reflectors, these systems are best suited for small-scale power production (10-50 kW) as a stand-alone unit in remote areas, away from power grids. The system can, however, be expanded by adding more modules as the load increases (Figure 10-11). Furthermore, these systems can be used in solar-only applications or hybridized with fossil fuels during periods of high-energy demand or at night. Unlike parabolic troughs and central tower systems that run steam turbines, dish systems use Stirling engines[14] or gas turbines.

Figure 10-11
25 kW dish Stirling system.
Source: University of Nevada, The Center of Energy Research (http://www.me.unlv.edu/research/cer/research.htm)

Hybrid systems integrate solar thermal technology with existing coal-fired plants or high efficiency combined-cycle plants. This allows distribution, higher thermal efficiency, and continuous operation of the plant even after the sun has set.

Which of the solar thermal power plant schemes is the best is not obvious. Each system has its advantages. Towers and troughs are best suited for a large grid-connected power system in the 30-200 MW range, whereas

Table 10-2. Characteristics of Concentrating Solar Power Systems

	Parabolic Trough	*Power Tower*	*Dish / Engine*
Status	Commercially Available	Demonstration Stage	Prototype Stage
Plant Size[1]	30-320 MW	10-200 MW	5-25 kW
Operating Temp (°C)	390	565	750
Concentration Ratio	10-0	----	30-1000
Peak Efficiency	21%	23%	29%
Annual Efficiency[2]	14-18%	14-19%	18-23%
Capacity Factor[3]	25-50%	20-77%	25%

[1] The upper range indicates 2030 projections
[2] Projected based on pilot-scale testing
[3] The % of the year the technology can deliver solar energy at the rated power

[13] BrightSource Energy Inc., http://www.brightsourceenergy.com.
[14] Stirling or "Hot Air Engine" refers to a class of engines in which heat is provided from a non-combusting external source such as nuclear or solar energy. Because the temperature of heat source and sink can be controlled and kept uniform throughout heat supply and heat rejection part of the cycle, thermal energy of Stirling cycle can reach that of the Carnot cycle.

dish/engine systems are modular and can be used singly in remote applications or be grouped together to power a small community and expanded as more power is needed. Table 10-2 summarizes characteristics of concentrating thermal power systems.[15]

Ocean Thermal Energy Conversion (OTEC)

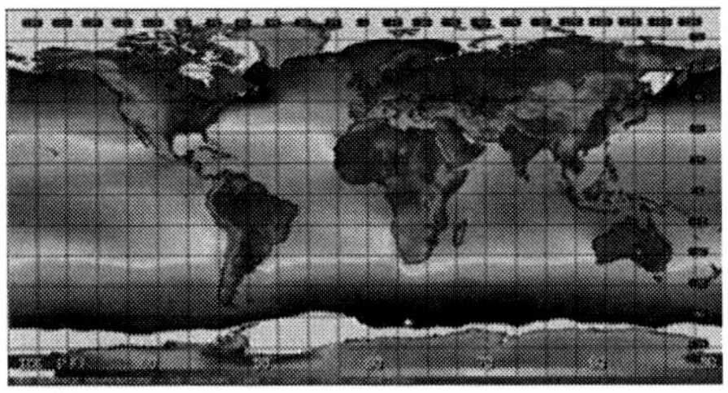

Figure 10-12
The map shows the contours of oceanic temperature differences between the surface and 1,000-m depth. Orange and red colors are regions with a ΔT of at least 20°C.

Oceans cover over 70% of the earth's surface area and are a huge source of thermal energy. Ocean Thermal Energy Conversion (OTEC) takes advantage of the temperature difference between the warm surface waters and cold deep waters of oceans to drive a steam turbine and produce electricity. Tropical waters are especially suitable because of the higher surface temperatures (Figure 10-12). Ocean waters between the tropics of Cancer and Capricorn (latitudes between 23°N and 23°S) typically have highest surface temperatures around 25°C-30°C (77°F-86°F), while the water below 900 meters is only 5°C (41°F). The temperature differences are very steady over time between days and nights and from season to season.

French engineer Jacques D'Arsonval was the first to note the huge potential for using thermal gradients in the ocean to produce power. His proposed system was not implemented until 1930 when his student, George Claude, built the first OTEC plant at Mantanzas Bay, off the coast of Cuba. Although the system functioned, the work needed to run the pumps and other devices (work input) was more than the work output. Furthermore, the difficulties in lubricating underwater pipes, salt corrosion, and growth of algae (biofouling) made the system inoperable and the project was abandoned. Many of the technical difficulties facing Claude have since been resolved. Recently, there has been a renewed interest in OTEC plants and demonstration plants have been constructed near Kailua-Kona on the island of Hawaii. In the United States, Florida and Puerto Rico could also utilize OTEC Technology. The Indian government has also actively sought to invest in this technology and has built a 1-MW floating OTEC plant.

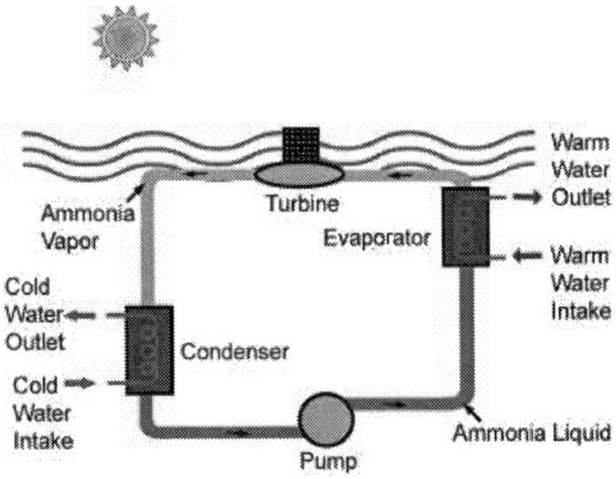

Figure 10-13
Closed-Cycle OTEC

OTEC plants can be constructed to operate in either closed or open cycles. In open-cycle systems, water itself is the working fluid. Open-cycle systems work by boiling the surface water at very low pressures (flash evaporation). After expansion through the turbine, steam is condensed back to liquid water via a condenser, which is then cooled by deep seawater before being discharged. The output of the condenser is desalinated water, one of the desirable byproducts of the OTEC system. Closed-cycle systems use the ocean's warm surface water to vaporize a working fluid, such as ammonia or another refrigerant. The vapor expands through a turbine, turning a generator to produce electricity. The vapor is eventually condensed by transferring its heat to the cold water of the deep ocean, before being pumped back to the evaporator to complete the cycle. A schematic of a

[15] Taylor, Craig, et al., "Concentrating Solar Power in 2001," SolarPaces Corp, 2001.

closed-cycle OTEC plant is shown in Figure 10-13. In addition to open and closed cycles, OTEC can also be designed as a hybrid. In a hybrid cycle, steam is generated by flash evaporation of the warm surface water and then used as a heat source for a closed cycle.

Example 10-1: What is the maximum theoretical efficiency of an Ocean Thermal Energy Conversion (OTEC) plant?
Solution: The maximum efficiency is limited to its Carnot efficiency, which can be calculated as:

$$\eta_{Ideal,\,OTEC} = 1-\frac{(273+5)}{(273+25)} = 6.71\%$$

Because of the power required for pumping cold temperature water from the bottom of the ocean and other frictional losses, the efficiency of a practical OTEC device is lower than this value at around 2-3%; this is far below the 30-35% thermal efficiencies commonly found in conventional oil- or coal-fired steam plants. Unlike conventional plants, however, there are no fuel costs.

Advantages and Disadvantages

OTEC could potentially fulfill a major part of our energy demand with few adverse environmental effects. Because of the huge water resources at our disposal, the technology potential could be significant, especially for those countries that have few fossil resources. Among features that distinguish OTEC from other renewable sources of energy is the wide range of side benefits it offers. Besides desalination, OTEC can provide industrial cooling, space air conditioning, and can allow chilled soil agriculture and aquaculture.[16]

Cooling – After steam is expanded in the turbine, it is condensed in the condenser. The condensate is relatively free of impurities and can be used in a number of applications such as air conditioning, desalination, and chilled soil agriculture.

Aquaculture – Deep water is not only clean, but also has a significantly higher amount of nutrients that allow enhanced growth of marine algae and many marine organisms, such as salmon and lobster, that could not otherwise grow in a tropical environment.

Sink - Deep seawater is a huge heat sink for many industrial applications that require dumping of waste heat.

The environmental impact associated with OTEC operation is relatively minimal and limited to some changes in local marine life and the release of CO_2 as the cold water from the deep oceans are brought to the warm

[16] Daniel, H. T., "Ocean Thermal Energy Conversion: An Extensive, Environmentally Benign Source of Energy for the Future," Natural Energy Laboratory of Hawaii Authority, Kailua-Kona, HI 96740.

water surface. This release is however negligible compared to the amount of CO_2 release that would result from the use of fossil fuel to generate a comparable amount of energy. The major obstacle to OTEC development is the initial cost and limit of suitable sites near shorelines. The OTEC cold water pipes make up almost all of the costs, so OTEC sites cannot be more than a few kilometers away from the shoreline. Most suitable sites are in areas away from population centers which make the power transmission costs even higher. It is likely that the best solution is to use the electricity generated by offshore OTEC platforms to produce alternative fuels such as methanol and hydrogen which can be loaded onto tankers or flow through pipes.[17]

It has also been suggested that the solar thermal energy stored in the reservoirs behind hydroelectric dams (thermocline) be used to increase the power output of conventional hydroelectric plants. The data suggests that, for typical thermoclines of 15°C, heat engines can effectively increase the gravitational potential (head) by tens to hundreds of meters and thus can double the hydropower efficiency for many hydroelectric power plants.[18]

Solar Ponds

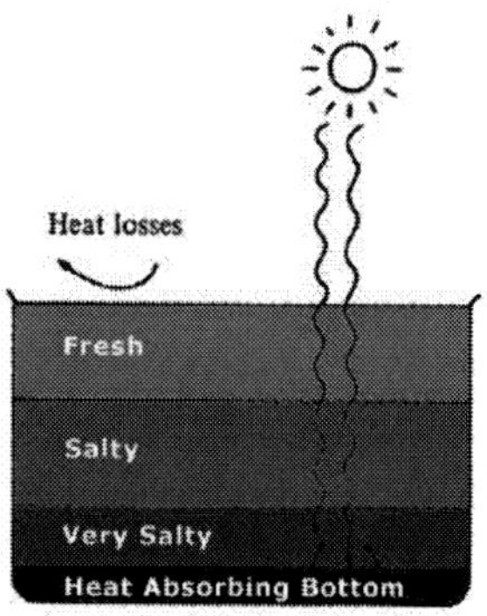

Figure 10-14
Solar pond

Unlike ocean waters where warmer water is at the surface, solar ponds are shallow lakes of salty water where higher temperatures are in the bottom layers and colder temperatures are near the surface. As solar radiation penetrates shallow waters, it is absorbed at the bottom, raising its temperature. If the water is fresh, the buoyancy causes mixing of the water and temperature will soon become uniform throughout. If water contains some salt, it becomes heavier than fresh water and sinks to the bottom, retaining the heat and temperature gradient (Figure 10-14). Temperatures as high as 95°C can be reached at the bottom layers. The temperature difference between the bottom and the surface layers can be used to design and construct large-scale power production and desalination plants. Heat is extracted from hot salty fluid at the bottom of the pond and passed through heat exchangers to heat a working fluid, causing it to evaporate. The vapor is used to drive a turbine, similarly to conventional steam power plants. Fresh water is produced as the byproduct of these processes.

To operate the plant continuously, the salt concentration gradient must be maintained. As a result of convection, there is always some diffusion of salt from the bottom to top layers. To maintain stability, salt must be added at the bottom and removed at the top. Larger ponds and calmer wind conditions are preferable. Larger ponds have a larger surface to perimeter ratio, and convective effects are less important. To prevent wind from disturbing surfaces and to reduce salt mixing, smaller ponds install

[17] Avery, W. H., and Dugger, R. D., "Hydrogen Generation by OTEC Electrolysis and Economical Energy Transfer to World Markets via Ammonia and Methanol," NREL Business/Technology Books: ISBN 0899343317, 1997.
[18] McNichols, J. L., et al., "Thermoclines: A Solar Thermal Energy Resource for Enhanced Hydroelectric Power Production", *Science*, 203, pp 167-168, January 1979.

Solar Crock Pot*

FYI ...

Indian workers have devised a cooking system that makes no demands on the critically short supply of wood, their traditional energy source. They dug a pond about 1 m deep and 8 m in diameter and lined it with impermeable plastic. The pond was filled with water and enough salt poured in to give a saline concentration of 20 percent at the bottom. The concentration of salt gradually decreased from the bottom to the top of the pond. The salt gradient prevents convection, as the salt water is too heavy to rise. Thus the water at the bottom absorbs solar heat energy throughout the day and reaches a temperature over 80°C at the bottom. Food is placed in sealed pots and left in the pond for a whole day. This kind of cooking is equivalent to an electric "crock pot" which takes 75 W for 8 hours.

* Excepts from a physics text currently under preparation jointly by this author and Professor Igor Glozman of Department of Physics, Highline Community College, Des Moines, Washington 98198.

suppression rings that cover pond surface. Additionally, many ponds contain a large amount of salt making them naturally suitable as solar ponds. The largest solar ponds are in Israel, where the hot, dry climate is ideal for their operations (Figure 10-15).

Figure 10-15
The largest solar pond operated to-date is a 210,000 square-meter pond at the Dead Sea, Israel, which generates 2.5 MW of peak electric power.

Solar ponds have found numerous applications in drying, desalination, process heat, refrigeration, and power generation. Their cost is considerably lower than that of flat plate and photovoltaic solar collectors. Besides the initial cost of construction, the only cost associated with solar ponds is that of maintenance. This includes preventing growth of algae in the upper convection layer and maintaining the salt gradient.

Photovoltaics

Photovoltaic (PV) systems convert solar energy directly into electricity using specially designed collectors known as solar cells (Figure 10-16). The photovoltaic effect was first observed by Alexander Becquerel in 1839, although it took another century before any serious attempt was made to design a practical device. In 1968, Glaser[19] proposed to place a satellite solar power system in a geosynchronous orbit that collected sunlight and beamed it to collectors that then generated electricity (See box " Solar Power Satellites").

Figure 10-16
Photovoltaic Cells
Source: National Renewable Energy Laboratory, Photographic Information Exchange (http://www.eere.energy.gov/pv).

By the mid-1970s, solar cells were primarily used in space for powering the instruments aboard space stations and in satellites. As the technology matured and the cost of production decreased, solar cells found new applications in everyday appliances, from pocket calculators and wristwatches to small remote applications such as highway call boxes, message boards, and traffic signals. Worldwide, one gigawatt of electricity is generated by photovoltaics. Japan currently leads in solar cell manufacturing and controls nearly 40% of the global solar cell market; Germany is second and the United States is third. Compared to fossil fuels, the cost of electricity generation by solar cells is expensive, but can quickly become competitive as manufacturing costs drop. Because they can be installed at any capacity, photovoltaic power is an ideal method for powering remote homes or villages too remote to connect to a grid.

[19] Glaser, P. E., "Power from the Sun: Its Future," *Science*, Vol. 162, pp. 957-961 (1968).

Principle of Operation

To understand the photovoltaic (or photoelectric) effect, it is best to visualize sunlight as small packets (quanta) of light called photons. The other equally valid view considers light as a wave. Photons are emitted at different frequencies and, therefore, have different wavelengths and energies. When photons strike a cell, some are absorbed, some are reflected, and some pass through. Only photons beyond a certain frequency have sufficient energy to dislodge electrons off a substrate and flow into an external circuit producing electricity.

To facilitate the movement of electrons, an electric potential must be constructed. This can be done by sandwiching two layers of silicon wafers together; one layer is doped with a material that has one less electron - is positively charged (p-layer) - and the other contains a material with an extra electron (n-layer) in their valence shell. To make a positively charged semiconductor, silicon is doped with impurities such as boron, aluminum, or indium. To make a negatively charged semiconductor, silicon is doped with impurities such as arsenic, antimony, or phosphorus. Commercial silicon cells are made by pressing two thin silicon wafers together, a very thin top layer doped with a small amount of phosphorous (n-type) and a thicker bottom layer doped with a small amount of boron (p-type).

When an n-type and a p-type semiconductor are pressed together, the imbalance of electrons across the p-n junction causes electrons to migrate from the n-type to the p-type material. If a wire connects two layers across a load (for example a light bulb or an electric motor), an electric field is created. As long as there is solar energy available and the circuit is closed,

Solar Power Satellites (SPS)

FYI ...

A solar panel facing the sun in near-earth space receives about 1,350 watts of sunlight per square meter. Because of the sun's angle and the attenuation by the atmosphere, the radiation received on the earth's surface is considerably reduced. Furthermore, solar cells can only work during the day, so they can only receive sunlight half of the time. If solar collectors are placed in a geostationary orbit* (36,000 kilometers high), all of the extraterrestrial radiation can be collected and converted to electricity. This electricity can then be beamed as microwaves, which can pass unimpeded through clouds and rain and be received on the earth before it is converted back into electricity. This electricity is in the form of direct current, which can then be converted to the 50- or 60-cycle alternating current electricity used in homes. Because the panel always faces the sun, electricity would be generated 24 hours a day and there will be no need for storage.

Such a system has indeed been proposed by the US Department of Energy and NASA. A preliminary design consisted of a 5 km x 10 km rectangular solar collector and a 1 km diameter circular transmitting antenna array. The SPS would weigh 30,000 to 50,000 tons and provide 5 billion watts of electricity, equivalent to ten 500 MW conventional coal or nuclear power plants. We are not likely to see such a huge orbiting solar collector beaming energy anytime soon. What probably makes more sense is to build a small proof-of-concept demonstration before extensive resources are dedicated toward fully-operational solar power satellites.

*Alternatively, some have proposed to place the collectors on the surface of the moon.

Source: Glaser, P.E., "Power from the Sun: Its Future," Science, Vol. 162, pp. 856-861, 1968.

Metals, Nonmetals, and Semiconductors

Digging Deeper ...

Every atom is comprised of a nucleus and a number of electrons orbiting the nucleus. Electrons can only move in certain orbits. To move an electron from a lower orbital to a higher orbital, it must absorb a photon of energy proportional to the difference between the two energy levels. Similarly, an electron jumping from a higher to lower orbit emits a photon of light with a frequency proportional to the energy difference. Electrons naturally occupy the lowest available energy levels (valence bands), so upper energy levels (conduction bands) remain vacant. The energy needed to move the electrons from the highest valence band to the lowest conduction band is called gap energy. The greater the gap energy, the more difficult it is to move electrons and the less conductive (insulator) a material is considered.

In non-metals, energy levels are separated by large gaps; therefore the transition from one energy level to the next is associated with frequencies that construct a discrete band. Metals, on the other hand, produce a continuous band because the electrons are free to move through the body of metal and pass through several energy levels. Metals have a continuous conduction band and are therefore good conductors of electricity. Semiconductors have intermediate gap energies.

Because of its abundance and low price, silicon is the most popular semiconductor material. Atoms of silicon have four electrons in the outer valence shell and are therefore electrically neutral. Silicon is an intrinsic semiconductor, i.e. at room temperatures, electrons do not have sufficient energy to jump the gap. By raising the temperatures, some electrons gain enough energy to overcome the gap and move into the conduction band. If the source of energy is removed, the electrons fall back into their previously occupied valence band.

Depending on the processing method, silicon can have different structures:

Monocrystalline silicon is grown from a single atom of silicon and has a perfectly uniform molecular structure, making it ideal for the efficient transfer of electrons through the material. Single crystalline silicon is difficult to manufacture and very expensive.

Polycrystalline silicon consists of, not one, but a number of crystals. It is much cheaper than single crystalline silicon to produce, but the different crystalline structures introduce boundaries that increase internal resistance and impede the flow of electrons. The greatest advantage of polycrystalline silicon is that it can be deposited monolithically as a thin film on a glass substrate. Unlike single crystalline silicon, which must be grown from a single atom (ingot), layer upon layer of polycrystalline thin-films of the required materials can be sequentially deposited to a desired thickness. Polycrystalline silicon cells are less expensive than monocrystalline cells, but their efficiencies are also lower.

Amorphous silicon does not have a crystalline structure and atoms are not arranged in any particular order. Amorphous silicon contains large numbers of structural and bonding defects, where electrons and holes recombine and limit the maximum current that can flow. It is the cheapest, however, making it suitable for use in low-power consumer devices such as wristwatches and calculators. Amorphous silicon cells can be sprayed onto glass plates and onto a variety of flexible substrates like metal foils and plastic foils, eliminating much of the production cost.

the current will continue to flow. Current is directly proportional to how much light strikes the module. To increase the current output, the top surface is coated with an anti-reflection material. Without the coating, an additional 30% of solar energy would be lost due to reflection. A typical commercially available cell, 10 cm x 10 cm, produces about 1.5 watts (0.5 volts and a current of about 3 amperes) of electric power when exposed to strong sunlight. A typical setup is shown in Figure 10-17.

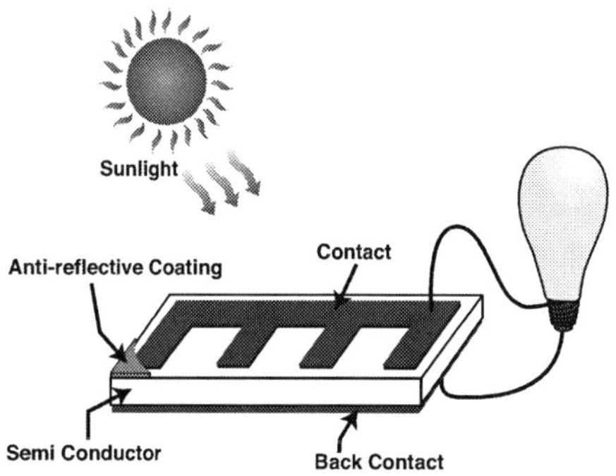

Figure 10-17
Schematic representation of a solar cell

To achieve the proper current and voltage, cells are connected in particular configurations. Normally, modules are designed to supply electricity at a certain voltage by connecting a sufficient number of cells in series. The current can be adjusted by connecting a number of such modules in parallel. PV panels can then be grouped together to form large solar farms

233

which generate electricity that can be used locally, stored in batteries, or fed into the commercial transmission grid. The output of photovoltaic cells is in the form of DC (direct current) that can be stored in batteries or used as a source for powering various devices. If it is to be used to generate electricity, the output must be converted to AC (alternating current) and amplified by transformers to high voltages before being transmitted through commercial electrical grids.

Collection Efficiency

Solar cells are often characterized by their conversion efficiency – defined as the percentage of incident power that is converted into electric power. The efficiency of solar cells is limited by the number of photons of a precise frequency that can knock out the electrons from the junction. Photons of lower frequency do not have sufficient energy, whereas higher frequency photons may cause overheating of the substrate and subsequent loss of efficiency. Because of its abundance, silicon remains the most common cell material used today. However, higher efficiency cells have been constructed of gallium arsenide, cadmium telluride, cadmium sulfide, and other materials (commonly called III-V materials because they are impregnated with material from groups III and V of the periodic table). Efficiencies in the range of 5% for amorphous silicon and 17% for monocrystalline cells can be expected for commercially available silicon solar cells. Efficiencies as high as 33% may be obtained under concentrated light conditions.

To increase the flux of photons, scientists have developed a spherical cavity treated with highly reflective coating and lined with multiple solar collectors. Different cells are made from different materials with different energy gaps. Each cell is covered by a filter that allows only the light with the appropriate frequency to pass through. The remaining photons are reflected back and forth within the cavity until the proper cells absorb them. Efficiencies as high as 48% have been reported.[20] Another innovation, the cascade or tandem cell, is constructed by stacking several single-junction cells with different band gaps on top of each other. The top layer has the highest band gap, capturing the high-energy photons and allowing the photons with lower frequencies to reach the lower cells with smaller band-gaps.

The major problem with flat panel cells is that, once they are installed with a particular orientation, they only collect maximum sunlight if the sun is directly overhead. As the sun's incident angle changes, the light intensity drops and less power is produced. The basic idea behind the spherical cells is that a spherical receiver can collect light from all directions. This allows the capture of direct beams of light as well as light diffused from clouds and reflected from buildings, thus spherical cells would have efficiencies up to 50% greater than those of flat cells. An attempt to improve efficiency

[20] Leine, J. D. et al., Proc. 15th IEEE Photovoltaic Specialist Conference, Las Vegas, IEEE, 141, 1991.

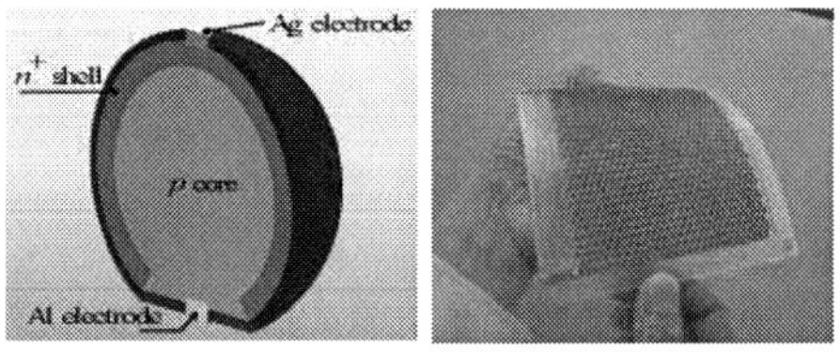

Figure 10-18
Spherical Solar Cells.
Image Courtesy of: Kyosemi Semiconductor Corp., Japan (http://www.kyosemi.co.jp/).

and reduce costs, a group of Japanese scientists have recently succeeded in developing spherical cells using single crystal silicon droplets (0.5-2 mm in diameter) to make the p-layer (Figure 10-18). A thin layer of n-type film is diffused and, except for a small opening for electrodes, covers the surface of the microsphere. The microspheres can be lined up along a string and connected in a series or parallel configuration with fine copper wire and are mounted on a white resin reflection plate covered with a transparent layer. The spherical solar cell module is highly flexible and can be made to match the contours of buildings or cars, thus providing integrated power sources in windows, roofing materials, canopies and other surfaces.[21]

Another promising technology is *the multi-junction solar cell.* Unlike conventional photovoltaic cells that absorb light only in the red part of the light spectrum, these cells consist of several microthin layers of light absorbing materials sandwiched together. Each layer absorbs a different color of light, increasing the overall efficiency. The technology has been implemented on the Mars Rover, but remains too expensive to be used in commercial products.[22] a

Thermophotovoltaics

Photovoltaic cells principally respond to light in the visible wavelengths. Thermophotovoltaic (TPV) cells work on the same principal, except that they are sensitive to infrared energy and use heat instead of light as their energy source.[23] The source of heat could be a burning gas, a radiant furnace, waste heat, or another source in the 700-1,700°C range. TPV can also work by focusing solar radiation on to an intermediate absorber, which then re-emits it as thermal radiation. These devices are capable of converting as much as 1-10 watts of power per square centimeter, a far better power density than solar cells, which can only deliver about 100 mW of power per square centimeter. Because thermophotovoltaics are solid state devices (no moving parts), they have found applications in heating, cooling, and as sensors for temperature stabilization.

Thermophotovoltaics (TPV) have been used for many years by NASA to power spacecrafts. The substantial drop in price of semiconductor ceramic material has prompted investigators to develop TPV systems that use waste heat from furnaces and burners to generate power, or make heat pumps that can provide cooling or heating. Bismuth-telluride materials are among the most favored because they can operate at temperatures as low as 150-200 degrees centigrade.

Electrically Conductive Plastics

Instead of conducting electricity and producing light, as is done in flat-panel

[21] Kyoto Semiconductor Corp., Japan *(http://www.kyosemi.co.jp).*
[22] Olson, J, and Kurtz, S., "Multi-Junction, Monolithic Solar Cell Using Low-Band-Gap Materials Lattice-Matched to GaAs or Ge," Patent no. 6,281,426, National Renewable Energy Press Release *(http://www.nrel.gov/news/press/2007/502.html).*
[23] In principle, these devices are made by linking together a large number of modern semi-conductor thermocouples in a series. A thermocouple consists of two electrical conductors of different materials joined together at both ends. If the junctions are kept at different temperatures, an electric current is generated and flows through the conductors.

Solar Sail

FYI ...

Its time to set sail for the stars ~ Carl Sagan

Did you know that solar light can exert a gentle pressure just like water squirting out of a nozzle onto a plate? A team of Russian and American scientists have pioneered Cosmos-1, the first solar sail spacecraft which uses the impact of light particles or photons off of a reflecting surface to propel a spacecraft without heavy and expensive onboard fuel sources. Sunlight pressure is enough to accelerate the spacecraft at the rate of 100 mph a day. That doesn't sound like much, but it can quickly add up over time. The effort opens a new chapter in space travel, offering a hope for future interplanetary missions.

The first such spacecraft was launched on June 21st, 2005 from a Russian submarine in the Arctic Barents Sea. Unfortunately, the spacecraft was lost during the launch and experiment had to be postponed.

Source: Matloff, G. L., Deep Space Probes: To the Outer Solar System and Beyond, Springer-Verlag New York, LLC, 2005. 2nd edition.

displays, electrically conductive plastics run in reverse - they absorb light and produce electricity. Conventional solar cells are relatively expensive. The low cost and inherent flexibility of the electrically conductive plastics mean that they can be deposited as coatings hundreds of times thinner than the silicon crystals used in conventional solar cells and can be used on a range of materials such as glass and roofing tiles. Efficiencies as high as 3-4.5% have been reported.[24]

Environmental Impacts

Photovoltaics are considered to be among some of the most environmental friendly energy alternatives. Their operation produces no pollution, does not contribute to global warming, and is noise free. Although PV operation is generally clean, its manufacturing (and disposal) is associated with the production of some of the most toxic materials such as cadmium and arsenic. Furthermore, some fossil or other non-renewable energy sources are normally used during their production. With proper safety precautions, the total emission is, however, small as compared to fossil burning.

Summary

Solar technology can be used for space heating and cooling, to generate electricity, or in various industrial and scientific applications such as drying, desalination, and manufacturing. It also can provide power to operate electrical appliances, and for telecommunication and lighting in remote areas.

For solar energy to capture a major utility market share, large power plants with capacities comparable to a coal or nuclear plant in the range of 100-200 MW must be developed. The major drawback of solar energy is the diffuse nature of sunlight that requires a large surface area for capturing

[24] Fairley, P, "Solar on the Cheap," *Technology Review*, January-February 2002.

enough light. Another problem is that solar energy is intermittent, available only during daylight and in favorable atmospheric conditions. One way to address this difficulty is to store the power in large, bulky, and expensive batteries. Another is to design a pumping storage water facility that stores water in an elevated reservoir during the day and releases it downhill to turn a water turbine and a generator as needed. Many homes use a backup system to supplement their primary solar systems when necessary.

Compared to current PV technologies with typical efficiencies of 5-17%, solar thermals are more efficient and can reach efficiencies of up to 30%. Unlike solar concentrating technologies that rely on direct radiation, photovoltaic cells rely mainly on indirect radiation and therefore can be used in areas with few hours of sunlight.

Currently, none of the solar electric generating stations can compete economically with the low cost of electricity generation using conventional coal and nuclear power plants. Both the capital and the operating costs are much higher, and the solar stations take up a much larger land area. Further research is needed to design larger heliostats and mirrors, as well as lenses with better optical properties and lower costs.

Additional Information

Books

1. Markvart, T., and Castanar, L., Solar Cells: Materials, Manufacture and Operation, Elsevier Publishing Company, 2005.
2. Galloway, T., Solar House, Elsevier Publishing Company, 2004.
3. Stine, W. B., and Harrington, R. W., Solar Energy Systems Design, John Wiley and Sons, Inc., 1985.

Periodicals

1. *Solar Energy*, Direct Science Elsevier Publishing Company, the official journal of the International Solar Energy Society, covers solar, wind and biomass energies.

Government Agencies and Websites

1. National Renewable Energy Laboratory: Solar Research (*http://www.nrel.gov/solar*).
2. Energy Efficiency and Renewable Energy: Solar Energy, US Department of Energy (*http://www.eere.energy.gov*).

Non-Government Agencies and Websites

1. American Solar Energy Society (*http://www.ases.org*).
2. Solar Electric Power Association (*http://www.solarelectricpower.org*).
3. California Solar Center (*http://www.californiasolarcenter.org*).

Exercises

I. Essay Questions

1. What does each of the angles of altitude, azimuth, latitude, and longitude represent? What angles are needed to locate the position of the sun from a point in space? From a point on the surface of the earth? What are factors that affect solar insolation?

2. How do solar chimneys work, and how can they be used for generating electricity?

3. What are advantages and disadvantages of passive and active solar energy systems?

4. What are the principles of operations of OTEC and solar ponds?

5. How do photovoltaics work? What are their advantages over solar thermal for power applications?

6. What are the advantages of electrically conductive plastics over solar cells?

II. Multiple Choice Questions

1. The sun is a sphere consisting mainly of
   a. A solid core covered by a layer of molten liquid and hot gases
   b. A liquid core covered by a mixture of hydrogen and helium gases
   c. A mixture of hot helium and hydrogen gases
   d. A mixture of noxious ammonia and sulfur gases
   e. The same material that makes the planet earth

2. Solar thermal energy has its origin in the heat released by the
   a. Various chemical reactions at the center of the sun
   b. Various chemical reactions at the surface of the sun
   c. Fission reactions occurring at the center of the sun
   d. Fusion reactions occurring at the center of the sun
   e. Latent heat as liquid hydrogen converts into gaseous form

3. Solar energy is ultimately responsible for which kind of energy?
   a. Fossil fuel
   b. Photovoltaic
   c. Wind
   d. Wave
   e. All of the above

4. The sun's interior and exterior temperature are roughly
   a. 5 million and one million degrees celsius
   b. 10 million and 10,000 degrees celsius
   c. 20 million and 5,500 degrees celsius
   d. 50 million and 10 million degrees celsius
   e. 5,000 and 300 degrees celsius

5. Seasons are primarily a result of
   a. Rotation of the earth
   b. Rotation of the sun
   c. Varying distance of earth from the sun
   d. Tilt of the earth
   e. Orbiting of the earth around the sun

6. Which of the following is <u>not</u> a form of solar energy?
   a. Biomass
   b. Wind
   c. Wave
   d. Hydropower
   e. Geothermal

7. The fraction of solar energy that falls in the visible range is
   a. Less than 1% of the sun's total radiant energy
   b. About 10% of the sun's total radiant energy
   c. About 50% of the sun's total radiant energy
   d. About 90% of the sun's total radiant energy
   e. More than 99% of the sun's total radiant energy

8. In what frequency range does the earth radiate the most?
   a. Cosmic ray
   b. Ultraviolet
   c. Visible

   d. Infrared
   e. Radio wave

9. The infrared portion of solar radiation consists of roughly
   a. 1% of total radiation
   b. 10% of total radiation
   c. 25% of total radiation
   d. 46% of total radiation
   e. 90% of total radiation

10. The angular location of a point on earth relative to the equator is called the
    a. Latitude
    b. Altitude
    c. Zenith
    d. Meridian
    e. Azimuth

11. The angle between a north-south line on the earth's surface and the horizontal projection of the sun's rays is
    a. Latitude
    b. Altitude
    c. Zenith
    d. Meridian
    e. Azimuth

12. The sun's location relative to an observer on the ground can be determined by
    a. Azimuth and altitude
    b. Latitude and altitude
    c. Azimuth and latitude
    d. Latitude and longitude
    e. All four angles

13. Albedo is
    a. The fraction of incident light that is absorbed by earth
    b. The fraction of incident light that is reflected by earth
    c. The fraction of incident light that is transmitted through the atmosphere and reaches earth.
    d. An organism exhibiting deficient pigmentation
    e. Unnaturally strong sexual drive in humans

14. Solar insolation at a point on the earth depends strongly on
    a. The day of the month, the time of the day, and the weather
    b. The topography of the area and the amount of shade
    c. The albedo of the earth
    d. The type of clothing we wear
    e. All of the above

15. The best orientation for installing fixed flat plate collectors is
    a. Parallel to the ground surface, because maximum radiation occurs when the sun is directly overhead
    b. Tilted at an angle roughly equal to the latitude of the location where they are going to be installed
    c. Tilted at an angle roughly equal to the complement of the latitude of the location where they are going to be installed
    d. Perpendicular to the ground surface to occupy the least space
    e. It does not make any difference, because the sun is traveling across the sky

16. Flat plate collectors
    a. Are large arrays of photovoltaic cells
    b. Work by using sunlight to electrolyze water to its components hydrogen and oxygen
    c. Are faced toward the wind to catch the most wind
    d. Are useful for heating domestic water heating systems
    e. None of the above

17. The majority of the solar radiation reaching the earth's surface is
    a. Ultraviolet and infrared
    b. Infrared and radio wave
    c. Visible and ultraviolet
    d. X-ray and gamma ray
    e. Visible and infrared

18. An overhang on a southern-facing window is useful because
    a. It blocks the summer sun while enhancing

heating in the winter
b. It prevents rain from hitting the side of the house
c. It prevents squirrels from looking in
d. It makes for more space in the ceiling for insulation
e. It allows rain to drain efficiently

19. For maximum efficiency, flat plate solar panels
a. Must be horizontal
b. Must face south and vertically
c. Must be painted white
d. Must face north
e. Must be perpendicular to the sun's rays at all times

20. The best solar system for producing process steam at temperatures of 300-400°C is
a. Flat plate collector
b. Parabolic trough
c. Parabolic dishes
d. Heliostat with flat mirrors
e. Photovoltaics

21. The primary drawback of electricity produced by solar energy is that it is
a. Intermittent
b. In the wrong frequency
c. Unpredictable
d. Dispersed
e. All of the above

22. Some of the byproducts of an OTEC plant include
a. Desalination
b. Space air conditioning
c. Chilled soil agriculture
d. Industrial cooling
e. All of the above

23. OTEC systems
a. Can be used anywhere in the northern hemisphere
b. Can be used only with areas with hot summers
c. Is best in latitudes of between 10° N and 10° S
d. Only in Hawaii
e. Oceans with a minimum of 50°C difference between surface and deep water temperatures

24. OTEC technology can be used to
a. Generate a large amount of electricity without the environmental impacts associated with fossil fuels
b. Produce a large amount of drinking water
c. Produce some fruits, vegetables, and marine organisms otherwise impossible to produce in a tropical environment
d. Act as a large heat sink for many industrial processes
e. All of the above

25. The main advantage of solar cells over other methods of electricity generation is that
a. They are cheaper than the power the utilities can produce by burning coal
b. Most utility plants are located in sunny areas
c. They are clean sources of energy with no known adverse environmental impact
d. Their power correlates with the utilities' daily load patterns, because the power is available when it is needed most – during daylight hours
e. Both c and d

26. Which one of the following ways of harnessing solar energy does not involve thermal energy?
a. Active heating
b. Passive heating
c. Indirect generation of electricity
d. Direct generation of electricity
e. Passive cooling

27. PV cells are most frequently used in
a. Consumer products, such as calculators, wristwatches, and solar-power radios
b. Remote power applications of 100 W or less
c. Areas close to or far from utility grids
d. Satellites
e. All of the above

28. The most common material used in PV modules today is
a. Single crystalline silicon

b. Polycrystalline silicon
c. Amorphous silicon
d. Cadmium telluride
e. Gallium arsenide

29. Which of the following statements is correct in regard to solar ponds?
    a. For a solar pond to be effective, surface water temperature must be about 50°C warmer than the water at the bottom of the pond.
    b. Temperature as high as 95°C can be reached at the bottom.
    c. Works best in areas of high winds.
    d. Works best in the same geographic areas that OTEC technologies are applicable.
    e. All of the above.

30. Thermophotovoltaics
    a. Is another name for solar cells
    b. Works best with visible lights
    c. Is most sensitive to ultraviolet lights
    d. Used mainly when waste heat is available
    e. Operates only when temperatures in excess of 2000°C are available

III. True or False?

1. The Sun is the ultimate source of all our energy.

2. Albedo is another name for reflectivity of the surfaces on the earth.

3. Factors affecting light intensity are cloud cover, humidity, and atmospheric conditions.

4. The total energy incident on a southward-facing window is greater during summer than winter.

5. OTEC technology is an economical way of producing power from a few kW to hundreds of MW.

6. The fraction of solar energy in the visible range is called solar insolation.

7. Since dishes have a smaller aperture than trough reflectors, these systems are best suited for small-scale power production or as a stand-alone unit in remote areas away from power grids.

8. Both solar ponds and OTEC plants exploit the difference between higher surface temperatures and colder temperatures in the bottom layers.

9. The higher the frequency of photons, the more energy and the higher efficiency solar cells will have.

10. Since the output of solar cells is in the form of direct current, they cannot be directly fed into electrical grids.

IV. Fill-in the Blanks

1. The angular distance measured east or west from the Prime (International) Meridian is called the ____________.

2. The amount of solar energy that reaches a surface on earth depends on the position of the surface and ___________.

3. The first OTEC plant was constructed off the coast of _____________.

4. For a solar pond to work, there must be a higher concentration of __________ at the bottom than at the top.

5. A typical 10 cm by 10 cm commercial cell produces as much as __________ watts of electrical power.

6. ____________ cells are advantageous over flat panel cells, because flat panel cells only collect maximum sunlight if the sun is directly overhead.

7. Current PV technologies have typical commercial module conversion efficiencies between _____________ percents.

8. OTEC systems using water as the working fluid are of the ____________ type.

9. ________________ plastics are special kinds of plastics that collect light and convert it to electricity.

10. The higher the ______________, the higher will be the collection efficiency of the incident solar radiation.

V. Project I - Solar Power Generation

The objective of this project is to design a power generation plant capable of producing 10 MW of electricity. The power plant is located in a tropical region along a coast with plenty of sunshine. The ocean water is 27°C at the surface but drops to 12°C at a depth of about 500 meters. On the average, 1,200 W/m^2 of solar insolation is available. Three options are being considered:

a. Photovoltaic solar cells using poly-crystalline silicon cells with an efficiency of around 15%.
b. A closed OTEC system using ammonia as the working fluid. Ammonia has a maximum temperature of 25 degrees and a minimum temperature of 15 degrees centigrade. 30% of the power generated is wasted through various frictional losses.
c. A solar heliostat composed of plane and parabolic mirrors. The heliostat can heat water to steam at 500°C.

Calculate:

1. For option a, the total area of photovoltaic cells required.
2. For option b, the theoretical maximum and actual efficiency of the OTEC plant.
3. For option c, calculate:
    a. The ideal Carnot efficiency (Eq. 5-1),
    b. The total amount of heat that must be disposed into the atmosphere.

Hint: Recall that efficiency is the ratio of work delivered per heat input, and that conservation of energy requires that the part of energy not converted to work must be disposed into the atmospheric sink.

Project II - Solar Photovoltaics for Homes

You are asked to evaluate the economics and the environmental merits of building a solar electric system for your home instead of buying electricity from your local utility. Obtain a copy of your electric bill (desirably for a peak month) and note the following:

a. The monthly electric bill ($)
b. The total electric consumption for the month (kWh)

Calculate:

1. Average cost of unit electricity ($/kWh)
2. Average power (kW)
3. Average solar insolation for your area. Insolation maps and tables are generated from historical data and are readily available for many cities. US data can be found from the National Renewable Energy Laboratory web site at (*http://rredc.nrel.gov/solar/pubs/redbook*).
4. Pick a commercial manufacturer and determine cost, peak and average power delivered, and efficiency. How much solar cell do you need? What is the cost?
5. Calculate peak power available from your solar system.
6. Estimate the total cost by adding the cost of the cell, the inverter, and other control devices you may need. Add 50% for the cost of installation.
7. What is the payback period?
8. Assuming your utility company uses natural gas to produce electricity, how much less carbon dioxide are you producing annually (tons/year) when you choose the solar option?

Nuclear Energy

CHAPTER 11

I ask you to look both ways. For the road to a knowledge of the stars leads through the atom; and important knowledge of the atom has been reached through the stars. ~ Sir Arthur Eddington (1882 - 1944)

On December 2, 1942, the world's first nuclear reactor went critical on the floor of an abandoned squash court at the University of Chicago. At 3:26 p.m. that afternoon, scientists achieved the first self-sustaining nuclear chain reaction inside what was known as Chicago Pile-1, initiating the atomic age. This opened the possibility of providing power from the energy locked safely inside the atom (Figure 11-1). To promote nuclear energy, electric companies assured the public of a source that provided power so cheap that there would be no need to even meter it. This optimism and excitement was soon tarnished as the hazards, environmental costs, and the dangers associated with the accidental release of radiation became apparent.

Figure 11-1
This 12-foot bronze sculpture by Henry Moore marks the site at the University of Chicago where the first sustained nuclear reaction was achieved

Today, the same optimism exists toward the future development of controlled fusion - that all our energy needs will be satisfied and this source of clean, renewable energy will substitute for all other forms of energy.

Atoms

The entire physical world is made of atoms. The word atom is derived from the Greek word *atomos* which means "indivisible." It was coined by ancient Greek philosophers, who thought of the atom as the smallest possible constituent of a substance. We have since learned that atoms are not indivisible, but are rather made of even smaller particles. Different elements have atoms that contain different numbers of these smaller particles.

Science has promised us truth: an understanding of such relationships as our minds can grasp. It has never promised us either peace or happiness.
~ Gustav Lebon (1841-1931)

Today, the idea of the atom has been modified to specify the smallest particle of a chemical element that still exhibits all the chemical properties unique to that element. The atoms of these elements react with one another and combine in different ways, forming a virtually unlimited number of chemical compounds. When two or more atoms combine, they form a molecule. For example, two atoms of hydrogen combine with one atom of oxygen to form a molecule of water. In other instances, atoms of one element can change their entity and form new elements. In these reactions, called nuclear reactions, the elementary particles within the nucleus of the atom are rearranged. Nuclear reactors and our sun rely on such reactions to produce energy.

Atoms: The Facts

Did You Know That ...?

- A row of 100 million atoms of hydrogen would only be about one centimeter long.
- Eight of the elements known to us make up 98.5 % of the earth's crust by weight: oxygen makes up 46.6%, silicon 27.7%, and aluminum 8%, with Fe, Ca, Na, K, and Mg composing the rest.
- Hydrogen is the most abundant element in the universe.
- Of 92 naturally occurring elements, 76 are solids, 11 are gases, and 5 are liquids at room temperature.
- Nearly all of the mass of an atom is contained in the nucleus, which has a density of around 100 million tons per cubic centimeter.

Inside the atom

You can picture the nucleus of an atom as a minuscule cluster of particles surrounded by a whirling cloud of much tinier particles, the electrons. The nucleus itself is composed of neutrons and protons, collectively known as nucleons. Protons are positively charged, whereas neutrons, as the name implies, carry no charge. It is the number of protons that gives the element its chemical identity; in other words, no two different elements have the same number of protons (Figure 11-2). Although the nucleus contains most of the mass of the atom (about 99%), the volume it occupies is very small, leaving a relatively large empty space between the nucleus and the electrons.

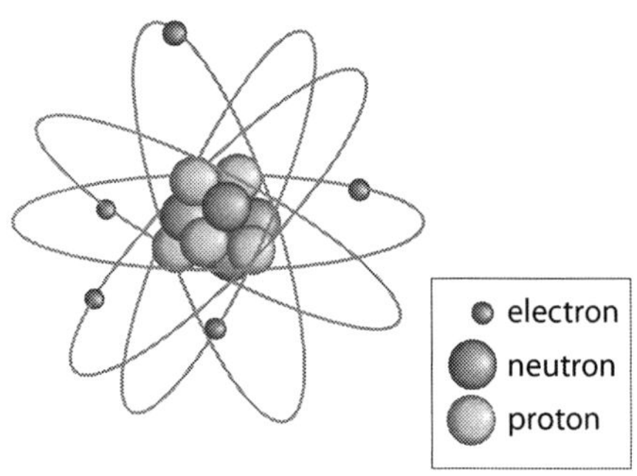

Figure 11-2
Every atom consists of a nucleus surrounded by electrons.

Hydrogen, the lightest of all elements, has only a single proton. Helium has two. Carbon, the basis of all life and the main constituent of fossil fuels, contains 6 protons. Oxygen has 8. The Uranium atom, with 92 protons, is the heaviest of all natural elements that make up our planet. Besides these natural elements, other elements have been discovered, but only in a laboratory environment.

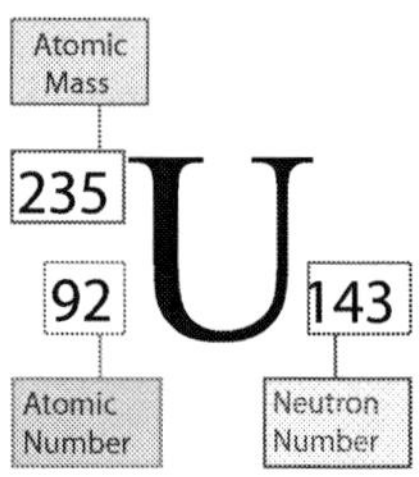

Figure 11-3
Uranium-235

Every atom is designated by its atomic number (P), which represents its number of protons and by its atomic mass (A), which is equal to the total number of neutrons (N) and protons ($^{A}_{P}X$). Almost all uranium naturally present in the universe have 92 protons (and also 92 electrons)and 146 neutrons (P = 92, N = 146, and A = 238) and is represented as($^{238}_{92}U$)or simply ^{238}U. The nucleus of a small fraction, however, has three less neutrons (but still 92 protons and 92 electrons); as it will be seen, this form of uranium called uranium-235(^{235}U) is very unstable and plays an important role in construction of both nuclear reactors and nuclear bombs(Figure 11-3).

Different atoms make up everything in the universe. A list of these elements along with their atomic masses and numbers are summarized in the Periodic Table shown on the next page.

Binding Energy

The binding energy of a nucleus is a measure of how tightly its protons and neutrons are held together by nuclear forces. The binding energy per nucleon, the energy required to remove a neutron or a proton from a

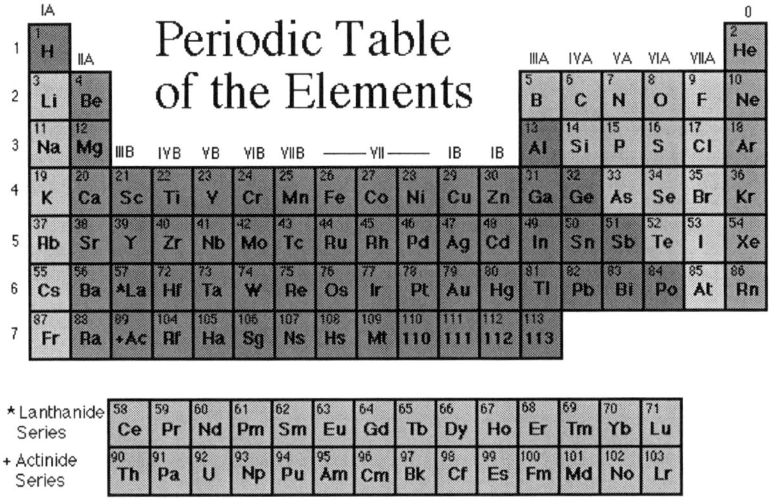

nucleus, varies with mass number and is shown in Figure 11-4. The figure implies that when a heavy nucleus (such as that of uranium) splits or two light nuclei (such as the nuclei of deuterium and tritium) coalesce, more stable nuclei form and energy is released. The former example describes fission, while the latter represents the basic mechanism that occurs in fusion reactions.

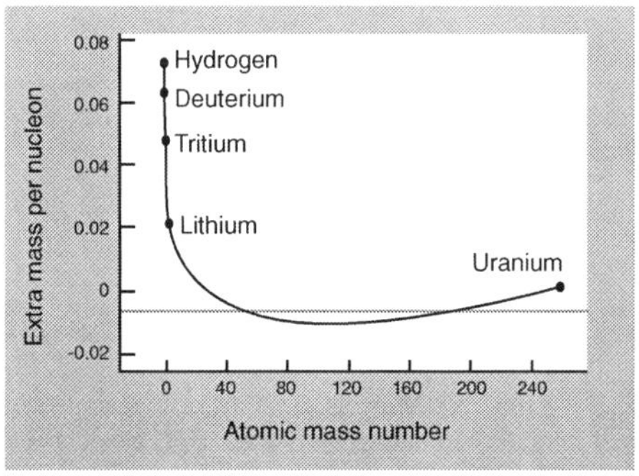

Figure 11-4
Energy can be liberated when the nuclei of light elements combine (fuse) to form heavier elements, or the nuclei of heavy elements are split to form lighter elements (fission). The energy release per unit mass of fuel from fusion is much more than that of fission.

Nuclear versus Chemical Reactions

There is a fundamental difference between chemical reactions (such as the burning of coal or natural gas) and nuclear reactions. Consider the chemical reaction:

$$CH_4 + 2O_2 \longrightarrow CO_2 + 2H_2O + \Delta E$$

Methane + Oxygen ⟶ Carbon dioxide + Water vapor + Heat

Note that after the reaction is completed, we still have one atom of carbon, four atoms of oxygen, and four atoms of hydrogen. The number of molecules, however, is not the same in a chemical reaction since atoms are rearranged to form new molecules. In this reaction, one molecule (or one mole) of methane reacts with two molecules (or two moles) of oxygen, and results in one molecule (or one mole) of carbon dioxide and two molecules (or two moles) of water. The total mass of reactants (methane and oxygen in our example) and products (in this case carbon dioxide and water vapor) also remain the same. Chemical energy appears because of the reconfiguration of atoms and the lower energies in newly created bonds, as compared to that in bonds which have been broken.

In chemical reactions, atoms are conserved - the total number of atoms of each element remains constant.

Question: Why do elements making up a chemical compound

combine in definite proportions?

Answer: John Dalton, an English scientist, pointed out that all elements are made of atoms that look like billiard balls and that these atoms must bind with each other in full or not bind at all. Although the description of atoms as rigid balls is not quite right, the explanation was a great step forward in understanding the nature of atoms.

In a nuclear reaction, atoms lose their identity. For example, when uranium-235 is hit with a neutron, two lighter elements such as krypton and barium are formed and two or three neutrons are released. In this case, the total number of elementary particles (electrons, protons, and neutrons) is conserved. The mass of the product is slightly smaller than the mass of the reactants. Energy is released because the configuration of elementary particles forming the nucleus of the fragments has a lower binding energy (mass) than that of the nucleus of the atom undergoing fission.

In nuclear reactions, the total number of nucleons is conserved - the total number of neutrons and protons remains constant.

Another notable distinction between chemical and nuclear reactions is the huge difference in the amount of energy each releases. This is because the binding energy that holds a nucleus together is far greater than the energy that holds electrons to a nucleus. For example, the fission of one kilogram of uranium provides as much energy as 2.4 million kilograms of the best quality coal. Even though nuclear fuel contains only 3-5% fissionable material, the tremendous energy released from a nuclear reaction is apparent.

Question: Throughout history, alchemists have tried to transform copper, mercury, and other metals into gold, only to experience disappointment. To what do you attribute the cause of their failures?

Answer: Alchemists mistakenly tried to perform this task by means of chemical reactions. As we now know, the process requires the deconstruction of the nucleus and rearrangement of the elementary particles (protons and neutrons) into new elements. This is only possible using nuclear reactions that were unknown to them at the time!

Isotopes

To make atoms electrically neutral, all that is required is that the number of protons and electrons be equal. The number of neutrons, however, can change, so it is possible for two atoms to have a different number of neutrons for the same number of protons (different A, but same P). These different atoms of the same element are known as isotopes. Uranium and hydrogen, the heaviest and lightest of all the naturally-occurring elements and the fuel of choice for nuclear fission and fusion reactions, are both found naturally in three isotopes. Figure 11-5 shows the three isotopes of hydrogen.

${}^{234}_{92}U$ (0.006%)	${}^{1}_{1}H$ (99.80%)
${}^{235}_{92}U$ (0.714%)	${}^{2}_{1}H$ (0.15%)
${}^{238}_{92}U$ (99.28%)	${}^{3}_{1}H$ (0.05%)

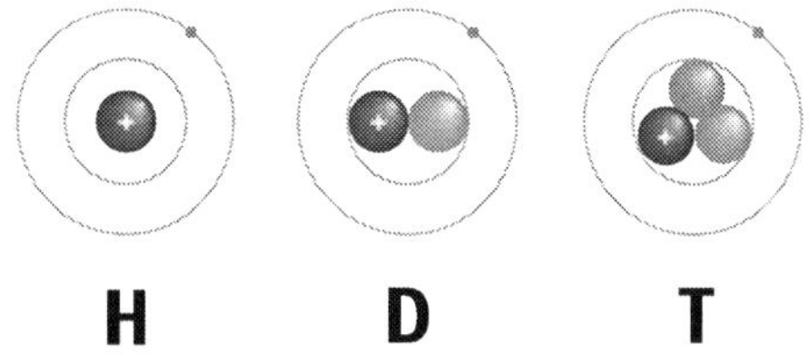

Figure 11-5
There are three isotopes of hydrogen that differ only in the number of neutrons they possess. The hydrogen isotope has only one proton, the deuterium isotope has one proton and one neutron, and the tritium isotope has one proton and two neutrons.

The numbers in parenthesis give the relative abundance of the natural occurrence of each isotope. Since different isotopes are only different in the number of neutral charges they contain, they are chemically identical. As we will see, these isotopes may have significantly different nuclear characteristics. **Radioactive materials** are those isotopes which have unstable atomic nuclei and may undergo spontaneous decay, forming lighter isotopes (called *daughters*) and releasing highly energetic ionizing radiation in the process. For example, the isotope of uranium-238 decays into several isotopes before it becomes stable lead (Figure 11-6).

All elements with atomic masses heavier than bismuth are naturally radioactive. It is, however, possible to artificially produce isotopes of lighter elements. Half-life is an important quantity in measuring the effect of harmful radiation from a radioactive source.

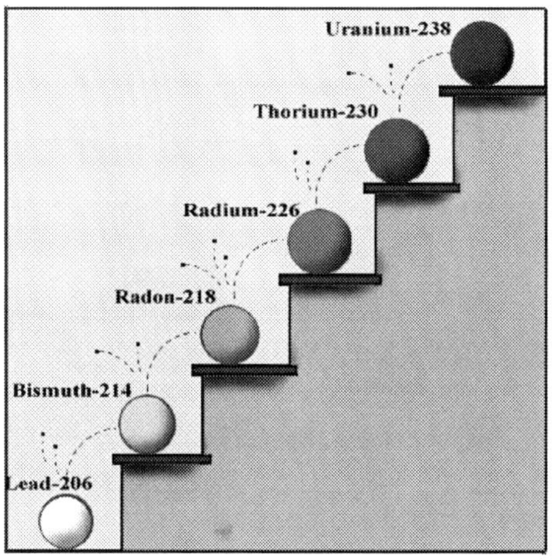

Figure 11-6
Decay chain of isotope of uranium-238

Half-life

Isotopes have an interesting property; they decay to form other more stable isotopes. The rate of decay is expressed by the isotopes' half-lives. A half-life is the time in which the concentration of an isotope decays to one half of its original concentration. The rate of emission from a radiation source reduces to 1/2 in one half-life, to ¼ in two half-lives, to 1/8 in three half-lives, and to $1/2^n$ in n half-lives. Half-lives of some of the isotopes of interest to us are uranium-235 (about 0.7 billion years), plutonium-239 (24,360 years), cesium-137 (30 years), and strontium-90 (28.8 years). Iodine-131 has a half-life of only 8.04 days.

> **Example 11-1:** A radiation leak released cesium-137 radionuclide. How long would it take until the radioactivity of cesium drops to 1/1000 of its initial value? The half-life of ${}^{137}Cs$ is 30 years. (*Hint:* $2^{10} = 1024$)
> **Solution:** About ten half-lives or 300 years.

In Chapter 1, we derived the equation for the doubling-time of a quantity that grows exponentially with time. Examples given were population growth, interest accumulating on a savings account, and energy consumption. The same equation can be shown to apply for the radioactive decay of isotopes, except decay rate (negative growth rate) and half-life are used instead of growth rate and doubling time. Following the approach used in equation 1-1, we can write:

An Even Smaller World!

Digging Deeper ...

Recent experiments with large particle accelerators have indicated that both protons and neutrons are composed of even smaller particles called quarks that have fractional charges and held (glued) together by gluons. Some believe that quarks and electrons are themselves made up of tiniest particles called strings. If the string theory proved to be valid then all matters can be thought of a collection of these strings.

$$T_{1/2} = \frac{70}{Percentage\ Rate\ of\ Decay} \quad (11\text{-}1)$$

Example 11-2: Using data from the previous example, calculate the rate of decay for cesium-137?
Solution: Using equation 11-1, rate of decay is 70/30 = 2.33% per year.

Fission

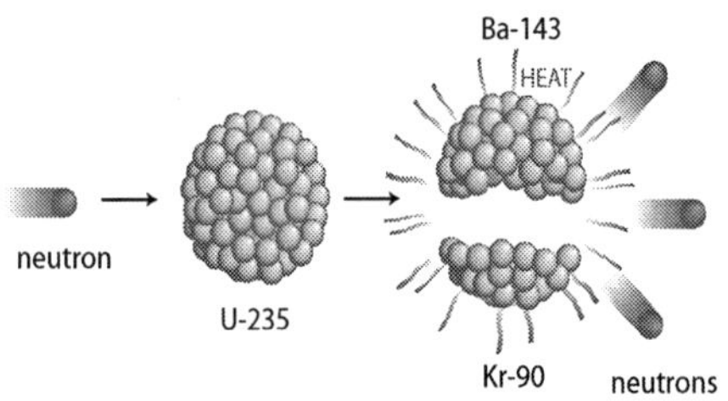

Figure 11-7
The fission process. When U-235 is bombarded by a neutron, it breaks into smaller fragments and 2 or 3 neutrons releasing a huge amount of energy.

Fission is the splitting of an atomic nucleus. Very heavy nuclei, like isotopes of uranium and plutonium, are easiest to split. The process involves bombardment of these atoms by small sub-atomic particles like neutrons, which splits them into two fission products; two to three neutrons and excess energy in the form of gamma rays are also produced. These neutrons collide with more uranium atoms, thus initiating a self-sustaining chain reaction that results in more fissions and the continuous release of enormous amount of energy (Figure 11-7). Any substance capable of sustaining a fission chain reaction is known as a fissile material. The materials most suitable for fission are ^{235}U and ^{239}Pu. These radioactive isotopes are fissionable by what are termed as *thermal neutrons.* The isotopes ^{238}U (and ^{232}Th) are fissile if bombarded by *fast neutrons.*

The physics of nuclear energy is simple. When a neutron collides with an atom of the isotope uranium-235, a highly excited atom of uranium-236 is formed, which immediately fissions into two lighter nuclei. Fission reactions can produce any combination of lighter nuclei, as long as the number of protons and neutrons in the products add up to the fissioning nuclei. Among possible reactions are:

$$^{235}U + n \longrightarrow {}^{236}U \longrightarrow {}^{143}Ba + {}^{90}Kr + 3\,n + energy\ (\gamma\ ray)$$

$$^{235}U + n \longrightarrow {}^{236}U \longrightarrow {}^{140}Cs + {}^{93}Ru + 3\,n + energy\ (\gamma\ ray)$$

$$^{235}U + n \longrightarrow {}^{236}U \longrightarrow {}^{134}Xe + {}^{100}Sr + 2\,n + energy\ (\gamma\ ray)$$

The profound feature of fission is that the mass of the products (fission fragments and neutrons) is less than the mass of a ^{236}U atom. The mass difference (or the mass defect) appears as the kinetic energy of the fission fragments in an amount determined according to Einstein's famous $E = mc^2$ formula, where m is the mass defect and c is the speed of light in a vacuum (300,000 kilometers per second).[1]

[1] Speed of light is usually denoted by c, from latin *"celeritas"* meaning rapidity of motion or action.

Carbon Dating

Digging Deeper ...

Archeologists use a simple approach to determine the age of fossilized objects buried underground a long time ago. A technique known as carbon dating uses the ratio of an isotope of carbon, carbon-14, to that of common and stable carbon, carbon-12, to estimate the age of a fossil that was buried many decades or centuries before. All living organisms naturally contain an isotope of carbon, ^{14}C, which is produced when the earth's upper atmosphere is bombarded by cosmic radiation that breaks nitrogen down into unstable carbon-14. Because C-14 is chemically identical to stable C-12, they attach to complex organic molecules through photosynthesis in plants, and enter the food chain. Since all living matters undergo a continual carbon exchange process through inspiration or respiration of carbon dioxide, they contain the same ratio of $^{14}C/^{12}C$ (1.28×10^{-12}). When a tree is cut down or an animal dies, the carbon exchange process stops, and C-14 starts to decay, whereas stable C-12 remains largely unchanged. The decrease in the ratio of $^{14}C/^{12}C$ can therefore be used as a measure of the age of a fossil or an artifact.

Example: A group of British scientists have studied the charcoal remains from Stonehenge and found it to contain the carbon isotope ratio of 8×10^{-13}. Estimate the possible age of the Stonehenge site.
Solution: Assuming the charcoal is leftover from social activities during the Stonehenge era, we can write:

$$\frac{^{14}C}{^{12}C} = 1.28 \times 10^{-12} \times 2^{-t/5700}$$

Substituting for the carbon ratio ($^{14}C/^{12}C = 8 \times 10^{-13}$), we get t = 3,800 years.

It must be noted that the accuracy of the carbon dating technique decreases with less sensitive instruments and with the age of the sample. With a half-life of 5,700 years, an initial C-14 population will decay to 1/16 of its original value in 22,800 years – about the limit of detection possible with current instruments. For estimating the age of older fossils and rocks, scientists use other radioactive isotopes (See table below).

Parent Isotope	Daughter Isotope	Half-Life
Carbon-14	Nitrogen-14	5,760 years
Aluminum-26	Magnesium-26	700,000 years
Uranium-235	Lead-207	700 million years
Potassium-40	Argon-40	1.25 billion years

Question: In the example of nuclear fission of uranium-235 given above, the mass appears to be the same on both sides of the reaction (235+1 = 143+90+3 = 134+100+2 = 236). Thus, it seems that no mass is converted into energy. In this case, where does the energy come from?
Answer: The statement is not entirely correct. Actually, the mass of a nucleus is more than the sum of the individual masses of its protons and neutrons, and contains the extra mass equal to the binding energy that holds the protons and neutrons of the nucleus together.

Question: Carbon has two stable isotopes, ^{12}C and ^{13}C. The isotope ^{14}C is unstable and decays to ^{14}N, with a neutron changing into a proton. Describe the reaction and how it can be used in dating fossils and other archeological artifacts.[2]
Answer: Both ^{14}C and ^{14}N have the same atomic mass. The difference

[2] Radioactive dating is not limited to biological systems. A similar technique called pottasium-argon dating can be used to determine the age of volcanic rocks that contain potassium. When volcanic liquids called lava-- was solidified into a rock, it contained an isotope of potassium K-40 which decays into argon gas. The technique works by melting the rock and measuring the concentration of the trapped argon gas to estimate the rock's age.

is in the number of protons and neutrons in their nuclei. ^{14}C with eight neutrons and six protons is less stable than tightly bound ^{14}N which has seven protons and seven neutrons. The half life of the decay is 5730 years which is used in determining the age of fossils and other biological organisms (See box "Carbon Dating").

Example 11-3: What is the energy equivalent of one gram of graphite, that is, the energy available if all its mass were annihilated?
Solution: The energy release is calculated as $E = mc^2 = (0.001\ kg)(3x10^8\ m/s)^2 = 9x10^{13}\ J$, or the energy-equivalent of 10,000 tons of TNT!

Nuclear Fuel

Most nuclear reactors use uranium as fuel. Uranium is a common element on earth, formed during the earlier stages of its formation. At the time our planet was formed 4.5 billion years ago, about 75% of all uranium was in the form of U-238 and the remaining 25% was in the form of isotope U-235. As uranium decayed ($T_{1/2}$ = 700 million years), the ratio of U-235 to U-238 dropped considerably. Today only 0.7 % of uranium found in nature is U-235. The non-fissile isotope U-238 makes up the rest.

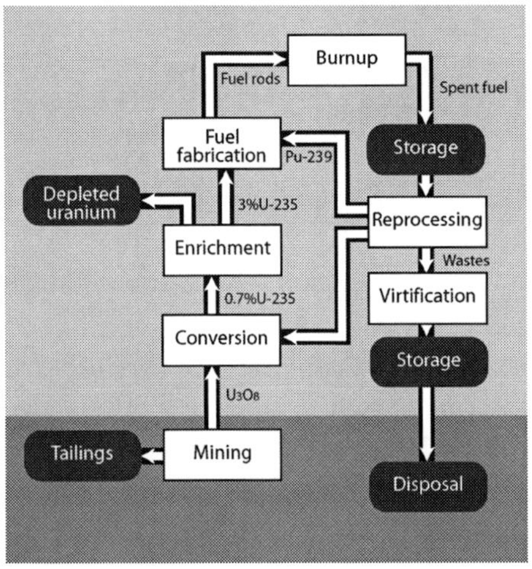

Figure 11-8
Nuclear fuel cycle from extraction to disposal.

Uranium is scattered throughout the earth's crust in most rocks and soils, as well as in many rivers and in seawater. Uranium is not an unlimited resource and, like fossil fuel, has only a limited lifetime. With 24% of the world's total uranium resources, Australia is the leading producer. Other countries with vast uranium reserves are Kazakhstan (16%), Canada (10%), and South Africa (7%). The United States, with 340,000 tons, has about 5% of the earth's uranium reserves. Eighty-five percent (85%) of all US uranium reserves are in New Mexico and Wyoming.

a Processing

From extraction to disposal, uranium fuel undergoes many changes. Various stages of the nuclear fuel cycle are shown in Figures 11-8 and 11-9 and consist of:

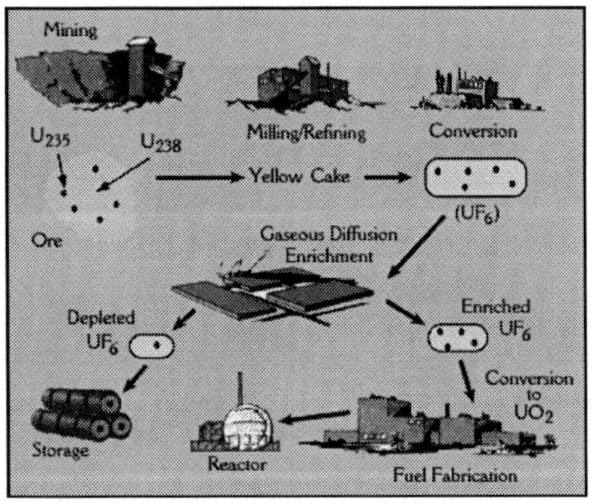

Figure 11-9
Uranium from extraction to storage.

1. *Mining and milling* - Like coal and other minerals, uranium must be mined and hauled away to a mill where uranium ore is crushed and ground to a fine slurry that is poured into an acid, which dissolves the uranium, but not the rest of the crushed rock. The acid solution is dried into a yellow powder called *yellow cake.* The leftover rock is known as *tailing.*
2. *Conversion* - The yellow cake is then shipped to a conversion plant where it is purified and chemically converted to uranium hexafluoride (UF_6) gas.
3. *Enrichment* - The flow stream has very little uranium-235, the fuel for nuclear fission. During the enrichment process, about 85% of uranium-238 is removed from the flue gas to raise the concentration

of U-235 isotopes from the 0.7% which is naturally present in the ore to the 3-5% required for use as nuclear fuel.[3] The enrichment is carried by gaseous diffusion or gaseous centrifuge. In gaseous diffusion, the gas is pumped through filters with holes large enough to allow uranium-235 atoms to pass through but not the slightly bigger atoms of uranium-238. In a gaseous centrifuge plant, gas is sent through hundreds of centrifuges that spin uranium hexafluoride gas at very high speeds, separating the lighter uranium-235 hexafluoride molecules from the heavier uranium-238 ones.

4. *Fabrication* – Enriched UF_6 gas is sent to a fuel fabrication plant where it is converted to uranium dioxide (UO_2) powder and pressed into small ceramic pellets that are then stacked inside long zirconium or stainless steel tubes to form fuel rods. Finally, fuel rods are bundled together to make the fuel assemblies used in nuclear reactors. Some U-235 remains in the tail and is called depleted uranium. Because of its high density, depleted uranium has been used in yacht keels, antitank missiles, and artillery shells.
5. *Burn-up* – A reactor core contains several hundred fuel assemblies where the fission process takes place and fuel is "burned," producing heat in a process called a chain reaction. Fission products are plutonium and various fission-fragments.
6. *Spent Fuel Storage* – Over time, the uranium fuel becomes depleted and the concentration of fission fragments begins to build up in the reactor. As more and more neutrons are absorbed, a point is reached where old fuel is no longer efficient and must be replaced with new fuel. The spent fuel and other fission products remain highly radioactive and continue to release heat and radiation long after they are removed from the reactor core. Typically, spent fuel is unloaded into a storage water pond immediately adjacent to the reactor. This allows the fuel to cool and the short-lived isotopes to decay and radiation levels to decrease by as much as 90%(Figure 11-10). Spent fuel is held in such pools from several months to many years before it is sent out to be reprocessed or dried (vitrified) and deposited in storage facilities. Each year, over 10,000 tons of spent fuel is generated by the world's current 441 operating nuclear plants, of which less than one-third is reprocessed for recycling as mixed-oxide (MOX) fuel.[4] The remainder is placed into interim storage facilities.
7. *Reprocessing* - The fissionable part of the fuel rod assembly is only a few percentages. About 95-97% of the uranium is still intact after fuel is burned. Plutonium makes up 1% of its mass and the rest are other radioactive elements. During reprocessing, uranium and plutonium

Figure 11-10
Storage pond for spent fuel at a reprocessing plant

[3] Enrichment is a physical process which relies on the small mass difference between atoms of two isotopes. No chemical methods can separate uranium-isotopes, as all isotopes are chemically identical. The two main enrichment processes are diffusion and centrifuge. In the diffusion technique, the mass differences between different isotopes result in different rates of diffusion through a membrane. In centrifuge, a similar separation occurs as the uranium substrate is spun at a very high speed. Today, most uranium enrichment plants use centrifuges to separate lighter uranium-235 hexafluoride from heavier uranium-238 hexafluoride. Successive operations allow production of more and more highly concentrated uranium-235.

[4] In the last few years many European countries have been using mixed uranium with plutonium extracted from their surplus stockpiles of nuclear weapons as a nuclear fuel (called MOX for mixed oxide fuel). The United States has not been using MOX to fuel its nuclear reactors yet, but is planning to build a MOX facility at a DoE site at Savannah River, South Carolina. The project is a part of the disarmament process and is aimed at reducing the nuclear threat by minimizing the number of nuclear weapons and disposing of nuclear-grade plutonium which has accumulated over many years of weapons manufacturing and reactor operation. Critics of the plan warn that opening the large stockpiles of plutonium to commercial sites make power plants more attractive to thieves and terrorists, because the plutonium can be separated from the MOX fuel rather easily and used for constructing an atomic bomb. By 2010 it is expected that MOX fuels will power some 15-20% of the world's power reactors.

Units Commonly Used by Nuclear Scientists

FYI ...

The mass of nucleons is usually represented in atomic mass units.
1 amu = 1/12 of the mass of atom ^{12}C = $1.66x10^{-27}$ kg

- mass of proton = 1.00728 amu
- mass of neutron = 1.00866 amu
- mass of electron = $5.486x10^{-4}$ amu

The energy of nucleons is usually expressed in electron volts (eV or Mev)
1 eV = energy that an electron would gain if it were accelerated through an electrical potential difference of 1 volt ($1 eV = 1.6x10^{-19}$ J and 1 MeV $=10^6$ eV)

- Energy equivalent of 1 amu = 931.5 MeV

are separated out from the rest, returned to the conversion plant, and blended with additional enriched uranium to build new fuel rods. During WWII, nuclear reactors were primarily designed for manufacturing large quantities of plutonium. The existing stockpile of plutonium is sufficient to fuel many breeder reactors. All US reprocessing plants were shut down at the end of the twentieth century and no new reprocessing plants are believed to be necessary in the foreseeable future.

8. *Vitrification* – High-level liquid waste is dried and stored in special containment vessels called casks. Casks are currently made of stainless steel alloy, but ceramic might be a more favorable material. Ceramics do not rust and have good resistance to radiation and heat.
9. *Disposal* – The canisters are stored in an underground permanent repository.

Example 11-4: How much U-238 must be removed from the uranium oxide (UO_2) ore to enrich the U-235 concentration to 3.5%?
Solution: Initially, only 0.7% of the uranium oxide is U-235, while the remaining 99.3% is U-238. If X is the fraction of U-238 to be removed from the mixture, then its concentration has reduced to 99.3(1-X) percent and we have:

$$\frac{[U\text{-}235]}{[U\text{-}238]} = \frac{0.7}{99.3(1\text{-}X)} = 3.5\% \text{ or } X = 0.8$$

That means 80% of the U-238 must be removed and discarded. This is a major source of radioactive waste.

Example 11-5: How much energy is released from the fission of 1 kg of natural uranium enriched to 3.5% U-235? Assume that each atom of uranium releases 200 MeV when undergoing the fission process.
Solution: Assuming natural uranium is mainly U-238, then the number of nuclei in one kg of uranium is

$N = (1000\ g)(6.02x10^{23}\ atoms/mole)/(238\ g/mole) = 2.5x10^{24}$

Assuming that only 3% of these nuclei participate in fission reaction, the total energy release is

$E = (0.03x2.5x10^{24})(200x10^6)(1.6x10^{-19}) = 2.4x10^{12}\ J = 667\ MWh$

Plutonium

Plutonium is only slightly heavier than uranium, but unlike uranium that is abundant in mines, there is no plutonium left in nature. This is because, compared to uranium, plutonium has a very short life and has therefore decayed to lighter, more stable elements. Plutonium, however, can be manufactured in nuclear reactors. When uranium-238 – which comprises the bulk of nuclear fuel – is collided with neutrons it turns to highly radioactive uranium-239 with a half-life of only 23 minutes. The resulting isotope is called neptunium-239, which is also radioactive and decays further to the isotope plutonium-239. Plutonium is not only the fuel for breeder reactors but also the ideal fuel for making nuclear bombs. Plutonium is also highly toxic and causes death within a few hours or days if ingested. In bulk quantities, plutonium is not very dangerous, but if vaporized or aerosolized it can cause great damage and must therefore be carefully safeguarded from falling into the hands of terrorist organizations.

Table 11-1. Nuclear Generation of Electricity by Country*

Country	*Reactors in Operation*	*Total GW_e*	*% Electricity*
U. S.	103	98.2	20
France	59	63.0	78
Germany	17	20.3	31
Russia	31	21.7	16
Japan	55	47.6	29
China	9	6.6	2
Lithuania	1	1.2	70
World	441	368	16

**Source: IAEA General Conference, Nuclear Technology Review, July 2006.*

Nuclear Reactors

As of January 1, 2006, there are 441 nuclear power reactors in the world producing 368,000 MW_e, or one-sixth of the world's electricity (Table 11-1). The United States has the greatest number of nuclear reactors, but the fraction of electricity they generate is small compared to that in some European countries. For example, while nuclear energy accounts for 20% of US and 50% of Europe electricity. In total, there are 31 countries that use nuclear power to meet some of their electricity demands.[5] Most nuclear reactors were constructed during the 1960s and early 1970s. Following the accidents at Three Mile Island (TMI) in 1978 and Chernobyl in1986 and the mounting volume of nuclear waste, public confidence in the use of nuclear reactors has declined. In the United States, the last nuclear plant went online in 1996, and in many other countries the number of new facilities being constructed has declined considerably. In contrast, in a rush to join the club of nuclear nations, many countries are building new nuclear reactors (Figure 11-11). In light of continued concerns regarding the storage and disposal of nuclear waste, even with the addition of these reactors, it is unlikely that overall nuclear power generation will increase in the near future, and some even predict that it may decline (Figure 11-12). However, the increasing dependence of the industrial world on imported oil, the volatility in prices of crude oil and natural gas, and the concern over environmental issues may tilt the balance in favor of nuclear energy sometime in the future.

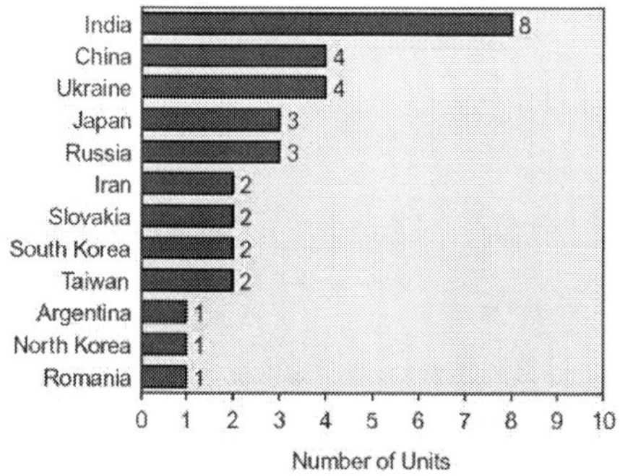

Figure 11-11
Nuclear power reactors under construction as of January 2003.
Source: IAEA, Data Series 2, "Power Reactor Information System", January 2003.

A nuclear power plant produces electricity in almost exactly the same way that a conventional (fossil fuel) power plant does. We discussed the basic components of a power generating station in Chapter 5. In a conventional thermal power plant, fuel (coal or oil) is burned to create heat, which boils water in a boiler (steam generator) to create steam. The steam is expanded in a turbine; the turbine runs a generator, which in turn

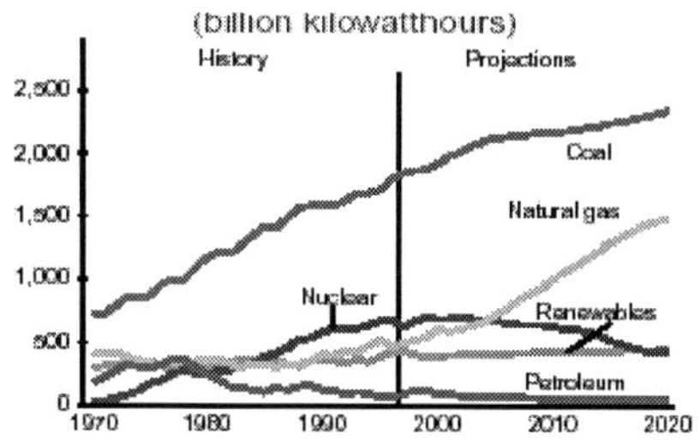

Figure 11-12
The history and predicted consumption of various forms of energy. It appears only demand for nuclear energy would be in decline.

[5] International Atomic Energy Agency (*http://www.iaea.org/programmes/a2*).

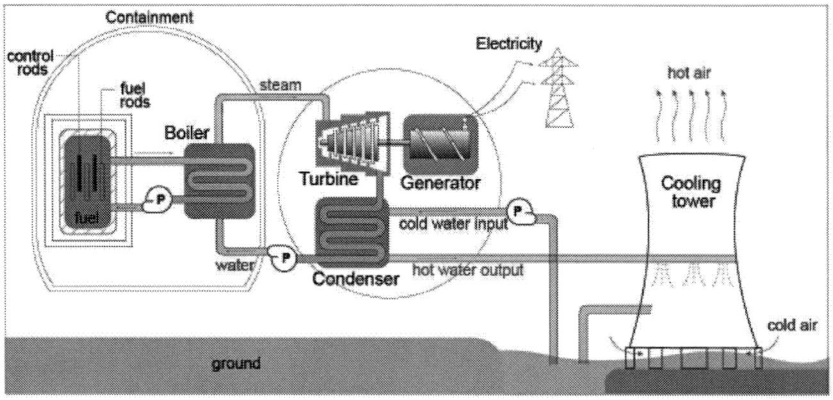

Figure 11-13
A typical nuclear reactor has several primary components. Inside the reactor "core" are the fuel rods, the control rods, the moderator, and the coolant. Outside the core are the turbines, the heat exchanger, and the main cooling system.

generates electricity. A nuclear power plant uses the same basic principles, except that the boiler is replaced by a reactor core where nuclear fuel undergoes fission processes to produce heat (Figure 11-13).

Reactor Core

The reactor core is the heart of any nuclear power generation facility. Inside the reactor core, fission takes place and provides energy to produce the steam required by the power plant. A reactor vessel made of several inches of carbon steel surrounds and protects the reactor core. The pressure vessel itself is housed in a containment structure made of several feet of thick concrete reinforced with thick steel bars. The reactor core consists of four main components.

Figure 11-14
Fuel Assembly- Fuel rods are made of zirconium alloys grouped into fuel assemblies which are then placed in the reactor core. Inside each fuel rod are hundreds of pellets of uranium fuel stacked end to end. The rods are each about 3.5 meters long and about a centimeter in diameter. Zirconium is a hard, corrosion-resistant metal and is permeable to neutrons.

a. **Fuel Assembly** Nuclear fuel consists of pellets (usually about 1 cm in diameter and 1.5 cm long, about the size of your fingertip) of ceramic uranium oxide arranged in long tubes in the fuel rods, which are grouped into bundles (fuel assemblies) in the reactor core (Figure 11-14). The bundles are submerged in a coolant inside a pressure vessel. Depending on the type, nuclear reactors have a number of fuel assemblies that must be replaced occasionally as they deplete and fission products buildup.
b. **Moderator** A fast neutron is capable of causing fission in either U-235 or U-238. In lieu of the much larger concentration of U-238, it is unlikely that a fast neutron will hit a U-235 atom and cause fission. For a neutron to be captured by U-235, it must be sufficiently slowed down. Under these conditions neutrons have more times to be pulled by the nuclear forces imparted by the uranium nuclei and there is a higher probability to be captured. The moderator is the material that slows down the fast neutrons to thermal neutrons (neutrons with an average kinetic energy similar to molecules of gas of the same temperature). Without a moderator, the fission chain reaction cannot be sustained. With ordinary water as a moderator, uranium-235 must be enriched to about 3%. It happens that natural uranium (at 0.7% concentration) will also work if instead of ordinary water either

graphite or heavy water (D_2O) is used as a moderator. The Canadian deuterium reactor called (CANDU) is of this type.

c. **Control Rods** It was previously explained that a fission reaction splits a uranium atom into two smaller fragments and two or three neutrons. To sustain a steady and controlled chain reaction, one of these neutrons is needed to bombard the next uranium atom, causing the next fission and continuing the chain reaction. This means that in order to prevent a possible run away reaction, the concentration of neutrons in the reactor must be controlled. Control rods are responsible for removing the extra neutrons and control the power output from the reactor. To control the rate of reaction, the control rods, made of neutron-absorbing material alloys such as cadmium, hafnium, or boron, slide up and down between the fuel rods; these materials absorb neutrons, but do not cause fission. To produce more heat (power up the reactor), the rods are raised and more of the uranium bundles are exposed; to create less heat, the rods are lowered. To shut the reactor down in the case of an accident or to change the fuel, the rods are lowered all the way.

d. **Coolant** To carry away the heat of the reaction, coolants are needed. Water is usually the material of choice. Some gas-cooled reactors use carbon dioxide or helium gas, while breeder reactors use sodium, potassium, bismuth, or other liquid metals.

a Classification of Nuclear Reactors

In addition to commercial reactors used for the generation of electricity, nuclear reactors are used in research, in ships, and in submarines. Depending on the type of coolants used, commercial fission power reactors can be divided into water, gas, or sodium cooled reactors. Depending on whether ordinary or heavy water is used, water reactors can be classified as light water or heavy water reactors. Table 11-2 gives summary of all nuclear reactors currently in operation throughout the world.

Table 11-2. Nuclear Reactors in Operation (2008)

Type	Number	%
PWR	268	61
BWR	106	24
HWR	40	9
HTGR	23	5
Breeder	4	1
PBMR	0	0
Total	**441**	**100**

Source: World Nuclear Association (http://www.world-nuclear.org)

Besides their intended use for generating electricity, nuclear plants have been used for the propulsion of large surface naval vessels and submarines. The United States, the United Kingdom, Russia, and France all have nuclear-powered submarines. The Soviet government built and operated the first nuclear-powered icebreaker in the Arctic. Other countries have constructed nuclear cargo ships, but high operating costs and restrictive port regulations have prevented their commercialization. In addition to the stated applications, a variety of small nuclear reactors at power levels around one megawatt have been built for education, training, and research.

Light Water Reactors (LWR)

Light water reactors (LWR) typically use enriched uranium as nuclear fuel, ordinary water as the moderator, and may be of either pressurized or boiling water types. A *pressurized-water reactor* (PWR) operates at a pressure high enough so that water never boils. *Boiling-water reactors* (BWR), operate at a somewhat lower pressure, allowing the water to boil within the core. The steam produced in the reactor is piped directly to the turbine before it is condensed and pumped back to the reactor. Since the steam becomes radioactive, it is necessary to shield the entire steam loop, including the turbine, from the outside by placing it inside the containment building. All US commercial nuclear plants (including the power reactors at Three Mile Island) are light-water reactors, using water as both the coolant and the moderator. About two-thirds of these are pressurized water reactors; the remaining third are boiling water reactors.

Heavy Water Reactors (HWR)

Heavy water reactors use heavy water (D_2O) instead of ordinary (light) water (H_2O) as the moderator. The major advantage of heavy water reactors is that they can use natural uranium instead of enriched uranium to operate. Because deuterium is a more stable isotope of hydrogen, it has less affinity to absorb neutrons[6], and there is a higher probability of collision with uranium-236 and a higher rate of conversion to plutonium. Since no enrichment facilities are needed, these reactors are considered dangerous from a proliferation perspective. Worldwide, there are 40 reactors of this type, mainly in Europe and Russia. There are no commercial heavy water reactors in the United States.

High Temperature Gas-Cooled Reactors (HTGR)

These reactors operate with natural or enriched uranium, use graphite as the moderator, and employ helium as the coolant. Helium is much less corrosive than steam, and is thus less likely to cause a leak in the pipes. The hot helium gas can directly drive a gas turbine or be used through a heat exchanger to heat water to steam that drives a steam turbine. Because of the high operating temperature (750°C) common with this type of reactor, efficiency is higher - around 40% compared to the 33% achieved by the water-cooled reactors.

Fast Breeder Reactors (FBR)

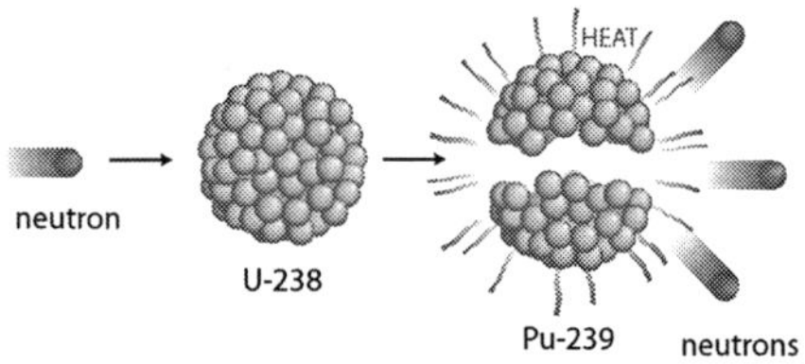

Figure 11-15
In fast breeder reactors the fuel is Pu-239 which is made by bombarding non-fissionable U-238 with fast neutrons

Like fossil fuel, uranium reserves are limited.[7, 8] It is possible to extend the supply of nuclear fuel if non-fissionable component of nuclear fuel is converted into fissile materials. By doing so, more fissile plutonium nuclei is produced than consume, hence the term "*breeding*". The breeder system uses uranium-238 as its fuel. It can however, transform into fissile plutonium-239 through collision with fast neutrons (Figure 11-15). The sequence of reactions is:

$$^{238}U + n \longrightarrow {}^{239}U \longrightarrow {}^{239}N \longrightarrow {}^{239}Pu$$

[6] In other words, the neutron cross-section of heavy water is significantly smaller than in light water.
[7] World Nuclear Association website (*http://www.world-nuclear.org*).
[8] The estimate is of course based on low cost availability of uranium fuel and may be extended if the market can bear higher electricity prices.

Uranium-238 + Neutron ⟶ Neptunium-239 ⟶ Plutonium-239

Because neutrons are not slowed down, no moderator must be used, which rules out water as a coolant. A suitable coolant is a liquid metal like sodium which has excellent heat transfer properties and a very high boiling point, which provide significantly better cooling and allow operation at essentially atmospheric pressure eliminating need for a heavy pressure vessel.[9] The main concern with these reactors is that, in the presence of water or even moisture in the air, liquid metals become highly explosive. Therefore, breeder reactors require special care to assure the safety of reactor operation.

The first large-scale plant of this type, called *Super Phoenix*, went into operation in France in 1984. From the beginning, the plant was the focus of anti-nuclear and environmental groups, who eventually succeeded in closing down the plant in 1998. Currently, the only reactor of this type, BN-600, is operating in Russia producing 600 MW_e. India has started construction of a 500 MW_e prototype reactor, scheduled to be operational by 2010. Many factors including the availability of sufficient uranium fuel for the near future, safety concerns, the fear of terrorism, and nuclear proliferations have greatly dampened the enthusiasm for these reactors and many countries have stopped work on fast breeder technology all together.

Pebble Bed Modular Reactors (PBMR)

The reactor core is made of thousands of billiard-ball sized pebbles, each consisting of several thousand enriched uranium (up to 10%) microspheres (kernels) the size of small sand grains (0.9-mm). Each microsphere is coated with several layers of graphite that act as moderator and a silicon carbide outer shell. Unlike today's nuclear reactors that boil water to drive steam turbines, pebble bed reactors use an inert gas like helium to drive gas turbines directly. Helium coolant enters the reactor and heated to about 900°C before it enters the turbines to generate electricity. After the fission is completed, the coating acts as a casket isolating and retaining the radiation long enough for the uranium to decay to reasonably safe limits (Figure 11-16). Shutdown will be done similar to other reactors, by inserting the control rods. These reactor are small, modular, flexible in design, and can be competitive with fossil fuels.

Figure 11-16
Pebble Bed Modular Reactor

Proponents of this type of reactor claim that this design is intrinsically safe and that there is no possibility of catastrophic nuclear accidents or fire; even if all helium coolants were lost, the temperature would never rise above 1,500°C (well below the 3,000°C needed to melt the uranium). Furthermore, the silicon carbide and graphite coating holds up better than the zirconium-alloy jackets of conventional fuel rods. So even in the worst case scenario where all cooling water has stopped, all contaminations stay within the pebbles walls and reactor automatically

[9] Matzke, H., "Development status of metallic, dispersion and non-oxide advanced and alternative fuels for power and research reactors," IAEA-TECDOC-1374, September 2003.

shuts off. The opponents agree that, although the reactor is inherently safer than the current water reactors, it may still pose a danger similar to the Chernobyl nuclear accident (to be discussed later in this chapter). A crack in the reactor could allow air into the reactor core and cause the graphite to burn.

No commercial reactors of this type are operating anywhere yet. China operates an experimental PBMR with a larger 200 MW_e "demonstration" reactor planned. South Africa is planning to construct several commercial plants to eventually produce 4,000 - 5,000 MW_e of electric power, but because of substantial financial risks, its fate is uncertain.[10] No PBMR is operating in the United States.

Retiring of Nuclear Reactor

Nuclear reactors are usually designed to last about 40 years, after which embrittlement and corrosion of the reactor vessels could pose a danger. By the year 2033, all 103 nuclear reactors currently operating in the United States will have reached the end of their original 40-year license period and must be shut down or must be upgraded for new licenses.[11] Nuclear reactors cannot be simply shut down and demolished as most non-nuclear facilities are. The building, equipment, and the leftover fuel are radioactive and will continue to generate heat and emit radiation long after the reactor has stopped operation. Most often following the reactor close down, the plant enters a storage mode during which it is basically left alone but guarded for many years or decades to allow the intense short-lived radioactive material to decay and become less active. Another approach is to entomb the reactor under heavy protective layers of lead and concrete and leave it there until they are permanently disposed. The final phase of retirement is the dismantling or decommissioning of the plant by tearing it up using special robots and moving and storing its critical parts in a permanent storage site. Nuclear waste disposal will be discussed in Chapter 12 in more detail.

Reactor Safety

One of the most frightening aspects of nuclear power is the possibility of something going wrong with disastrous consequences. In one scenario, a malfunction of the control rods or a loss of coolant results in a "runaway" reaction and an uncontrolled release of energy that melts the fuel rods and causes them to fuse. The fused rods would then burn through the reactor vessel and the containment floor. Such a meltdown would cause leakage of dangerous radioactive materials into the environment. Two of the most famous (or infamous) nuclear accidents occurred at the Three Mile Island (TMI) reactor no. 2 in the United States and the Chernobyl reactor no. 4 in Ukraine.

[10] Pebble Bed Modularized Reactor, Centurion, Republic of South Africa (*http://www.pbmr.co.za/*).
[11] Farber, D., and Weeks, J., "A Graceful Exit? Decommissioning Nuclear Power Reactors," *Environment*, 43, No. 6, July/August 2001.

Three Mile Island

Although various safety features make serious nuclear accidents unlikely, an accident did occur on March 28, 1979 at the Three Mile Island Nuclear Power Station near Harrisburg, Pennsylvania (Figure 11-17). The incident was triggered by a sudden pressure drop in the core because a relief valve failed to close. Failing to notice the valve malfunction, operators turned off the emergency cooling pumps that went into operation automatically when the water started draining. This and a number of additional operator errors eventually led to the loss of some coolant, which ultimately resulted in severe core damage and the release of large volume of fission products into the containment structure. The reactor suffered a partial meltdown of the uranium fuel rods. To prevent overpressure, some of the radioactive gases was vented out. Fortunately, majority of the radioactive release stayed within the containment, and very little radiation escaped outside the reactor. Although human exposure to radiation was relatively minor, the economic cost and the psychological stress on the public was significant.

Figure 11-17
Three Mile Island Nuclear Power Plant
Photo Courtesy: Nuclear Regulatory Commission.

The TMI accident is considered the worst nuclear disaster in US history. Under mounting pressure from the public, Congress enacted legislations for more stringent regulations for design, construction, and operation of nuclear reactors. In addition, the nuclear industry formed its own watchdog, the Institute of Nuclear Power Operation (INPO) that oversees safety standards and assures that power plants follow these standards and maintain active and continuous training of nuclear operators. Following the Chernobyl accident, the World Association of Nuclear Operators (WANO) was formed to fulfill a similar role to that of INPO on a global basis.

Chernobyl

On April 26, 1986, about 7 years after the TMI incident, the controversy was set ablaze again, this time at the Chernobyl nuclear power plant located about 130 km north of Kiev, the capital of Ukraine. One of the four nuclear reactors exploded and released more than 50 tons of radioactive material - 10 times that of bomb dropped on Hiroshima - into the environment. The scientific consensus is that the accident was the product of a flawed reactor design coupled with inadequate training and serious mistakes made by the operators of the plant.[12] Unlike the TMI, the Chernobyl accident was not a result of loss of coolant, but followed a sequence of events that led to a fire to the carbon moderator.

The effects of the disaster at Chernobyl were widespread. The immediate

[12] One fundamental difference between the Chernobyl reactor and the water-cooled reactors operated in the United States and elsewhere is the role that steam plays during an accidental loss of coolant. In water-cooled reactors, steam may accumulate to form pockets known as voids. With excess steam, more voids are created and water becomes less effective as a moderator; as a result, the chain reaction is not sustained and less power is produced which tends to shut down the reactor. When moderator and coolant are kept separate (as was the case in Chernobyl), any loss of coolant reduces the cooling capacity without affecting the moderator and the rate that neutrons are released. As the reactor heated, the rate of the fission chain reaction increased, until the cooling water turned to steam and exploded. This in turn, accelerated the chain reactions and increased power output. More power means additional steam, less cooling, less neutron absorption, and still more power. With a lack of proper safety precautions the process may continue to dangerous levels. Water-cooled nuclear reactors are therefore inherently safer because they do not face this risk. In addition, the lack of containment structures in the Chernobyl plant, similar to those present in the American, European, and Japanese nuclear plants, resulted in substantial releases of radionuclide and added to the severity of the accident.

Figure 11-18
Chernobyl nuclear power plant after the nuclear explosion.

casualties were among the clean-up crew, firefighters, and pilots who died from acute radiation exposure. Environmental damages included the destruction of vast areas of agricultural and farmland and poisoning of major surface and underground water reservoirs. Large areas of Ukraine, Byelorussia, and Russia were contaminated, resulting in the relocation of roughly 200,000 people from within 30 kilometers of the plant. Inhabitants of surrounding areas could not drink water or eat food produced locally. In Europe, many plants and animals were also contaminated and had to be destroyed. The economic damage from the accident is estimated at more than 13 billion dollars (Figure 11-18).

Initially, prevailing winds carried the radioactivity northwest from the plant across Byelorussia and into Poland, Hungary, and Sweden. Radiation levels in many parts of Europe rose well above normal before wind carried and scattered the radionuclide a long way away from the reactor site. Long-term effects on public health have been difficult to determine and are subject to considerable controversy, but many believe that the effects will be felt for many years in terms of widespread birth defects and various forms of cancer. Some believe that in Russia alone, 10,000 people will die as a direct result of radiation exposure from the Chernobyl accident over a 70 year period, the average lifespan of humans at the present. Worldwide the number is estimated at 25,000. Most, however, argue that the health problems caused directly by radiation are not easy to measure; many are caused by poor nutrition, lower health care conditions, and the anxiety and stress produced by fear of radiation exposure. What is clear, however, is that the rate of cancer and other deadly diseases among the more than half a million workers who participated in the Chernobyl cleanup has been significantly higher. Environmental and health consequences of radiation exposure are discussed in Chapter 12.

> **Question:** A common fear among many ordinary people is that a reactor meltdown may result in an explosion similar to those of nuclear bombs. Could nuclear reactors detonate in this fashion?
> **Answer:** No! Explosions of this type are mostly a myth. The fuels used in the construction of nuclear weapons are plutonium and highly enriched uranium (90% or higher). Commercial nuclear reactors use low-grade fuels containing only 3-5% fissionable uranium-235. The remainder is uranium-238, which acts to absorb the neutrons and prevent chain reactions of the type necessary for nuclear bombs.

Figure 11-19
An advertising poster featuring the movie "The China Syndrome."

The China Syndrome
In 1979, the release of the movie "The China Syndrome" had, and continues to have, a stalwart effect on the public view of nuclear power plants (Figure 11-19). The themes discussed in the film remain relevant, as the issues it raised over the safety of nuclear power are still a source of debate. The title refers to a worst-case scenario of a US reactor core meltdown where the contents of the reactor would have enough heat to

melt a hole through the earth, all the way to China.[13] Although scientists agree that this scenario is highly unlikely, the idea of a nuclear accident does give nuclear power an ominous image that is difficult to refute.[14]

Perhaps the most effective aspect in the film's power of persuasion was its unintentional release date. The film debuted in March 1979, less than two weeks before the Three-Mile Island accident. The real-life headlines were so similar to the movie's plot, says China Syndrome executive producer Bruce Gilbert, that he thought "someone had seen the picture and sabotaged the plant."

Together, the film and the parallel crisis sparked a move to pull the plug on the nuclear-power industry. In the following months, several power plants were shut down as safety precautions while plans to open others were scrapped. The latest global public opinion poll conducted by IAEA shows that only 28% of the population consider nuclear power safe. 25% consider it to be dangerous and all existing plants should be shut down.[15]

Nuclear Power and the Environment

The major environmental concern with nuclear plants is the disposal of nuclear waste, which will be discussed in the next chapter. Putting this problem aside, after solar and wind, nuclear plants are among the cleanest of all available electricity-generating alternatives. Because no carbon is present in the fuel, it does not generate carbon dioxide or other green house gases. In fact, compared to coal power plants that produce around 1000 grams of CO_2 per kilowatt-hour of electricity generated and natural gas that produce around 500 grams of CO_2 per kilowatt-hour of electricity generated, nuclear plants release only 5 grams per kilowatt-hour electricity. Furthermore, nuclear generating stations occupy much less land space than solar or wind power plants of similar capacity. In smaller countries such as France and Belgium, there is a great incentive to prefer nuclear power over fossils and even renewable resources such as solar and wind.

Economics of Nuclear Power

Relative to other power generating plants, nuclear reactors require large initial capital cost. The operating cost is usually low, and over the life of the plant, nuclear can compete with alternative sources of energy for base load power. If the savings in the cost of pollution abatement and health-related issues is included, nuclear power can turn out to be even cheaper than fossil power.

Nuclear Power and Public Opinion

The debate over nuclear energy dates back over 50 years. 1950s marked the golden era of nuclear energy where it was supposed to produce unlimited

[13] A hole initiated in the United States extended all the way through the earth would not end up in China, but somewhere in the Atlantic Ocean.
[14] Sauter, M., *Entertainment Weekly*, March 20 1998 Issue.
[15] "Global Public Opinion on Nuclear Issues and the IAEA," Nuclear Technology Review 2006, pp.7, IAEA GC(50)/INF/3.

amount of energy, too cheap to be metered. Although the dream of "almost free" electricity never materialized, the oil embargo of 1970s provided the new ammunition to nuclear enthusiasts and nuclear energy became once more the cornerstone of US energy policy. The nuclear accidents at TMI in 1979, and Chernobyl in 1986 along with release of the movie "China Syndrome" dampened the public enthusiasm and nuclear power became a "dirty" name, seemingly nobody wanted to be a part of it. The public reaction forced the government to ban construction of new nuclear reactors that is still in effect. The new surge in the price of petroleum, desire for a greater independency from foreign oil, and the rising awareness on threat of CO_2 emission and global warming has, however, renewed the debate once again. The new polls suggest that over 50% of the population want to keep the nuclear option open and part of the overall US energy policy. Similar trends can be seen in Europe and Asian countries.

Fusion

Right after WWII, nuclear technology was perceived to offer a future of cheap, plentiful energy that would replace increasingly scarce fossil fuels. In hope for the transition of nuclear power from wartime to peaceful uses, the general public and even environmentalists embraced this technology, as they foresaw a major reduction in air pollution and reduced strip mining. Nuclear plants do not generate harmful emissions often associated with burning fossil fuels (See Chapter 8), and there appear to be no significant adverse effects to water, land, habitat, species, and air resources.

Figure 11-20
Fusion is responsible for energy released by suns and stars.

With the passing decades, however, reservations about nuclear energy began to grow as greater attention was focused on the issues of nuclear safety, weapons proliferation, high construction costs, and problems associated with waste disposal took away much of the glory of nuclear plants. At the start of the twenty-first century, some European countries began to reduce their dependence on nuclear energy. Currently, the US is generating only 20% of its electricity needs by nuclear power; this share is expected to decline even further, as there have been no new orders for nuclear plants since 1978. Except for in Asia, where new nuclear power plants are still being constructed, the growth for nuclear power is expected to be limited.

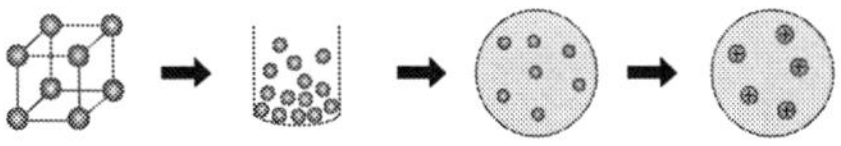

Figure 11-21
Four states of matter

Fortunately, as demand for nuclear fission reduces, a new technology called "fusion" is rekindling an interest in nuclear energy. Fusion is the same process that powers our sun (Figure 11-20). The enormous amount of energy released by the sun in the form of radiation is a result of the conversion of some of its mass to energy, as predicted by Einstein's equation $E = mc^2$. Unlike in fission reactions, atoms are not split apart, but instead are fused together, hence the name **nuclear fusion**. In the case of our sun (and all other stars), massive gravitational forces compress and heat deuterium and tritium gases to form plasma. Plasma is sometimes referred to as the fourth state of matter (Figure 11-21) and is formed when gases are heated to a temperature that is high enough to cause ionization. This process knocks electrons out of the atom, causing it to be left with a positive charge. As the plasma is heated further, eventually a point will be reached where the positively charged atomic nuclei overcome their mutual electrostatic (Coulomb) repulsion and fuse together. In the process, helium, neutrons, and a tremendous amount of energy are produced. Researchers inspired by such reactions are working feverishly to replicate these processes on earth. If successful, fusion reactors have the potential of providing unlimited energy with fewer disadvantages than fission reactors.

a Physics of Fusion

The thermonuclear fusion in the sun involves a series of reactions starting with the fusion of two atoms of hydrogen nuclei (protons) to produce deuterium and a positron (positive electron). The deuterium subsequently interacts with a proton to produce a helium-3 isotope. Finally, two He-3 atoms are fused together to produce a He-4 nucleus (alpha particle). The reactions involved in the solar thermonuclear process are far too slow to be useful for producing energy here on earth. A better approach would be to use deuterium or a mixture of deuterium and tritium. Two specific reactions of particular interest are:

$$D + D \xrightarrow[\text{100 million degrees}]{} {}^{3}He + {}^{1}n + 3.3\ MeV$$

$$D + T \xrightarrow[\text{45 million degrees}]{} {}^{4}He + {}^{1}n + 17.6\ MeV$$

Unfortunately, these isotopes will not fuse to each other at ordinary temperatures and pressures because they are all positively charged. The attractive forces exist only over very short distances; the repulsive forces between the positive nuclei and negative electrons are too strong to allow these nuclei to bring them close enough unless their average speeds (temperature) are raised. The D-D reaction has an ignition temperature of 100 million degrees, much higher than the ignition temperature of 45 million degrees needed for a D-T reaction. If normal hydrogen is used as the fuel for fusion, the required temperature would be even greater.

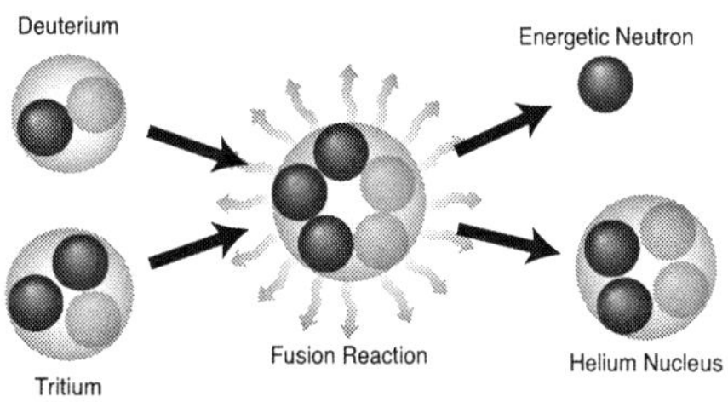

Figure 11-22
In a thermonuclear reaction, an atom of deuterium and an atom of tritium fuse to produce an atom of helium. A neutron and a great amount of energy in the form of gamma rays are produced. A temperature of 45 million degrees is required to initiate the ignition.

The D-T reaction is currently the reaction of choice (Figure 11-22). Deuterium is naturally present in seawater, but at a very low concentration–one atom for every 7,000 atoms of hydrogen. Tritium is also very rare and must be manufactured. One possible source of tritium is lithium, which is plentiful both in the earth's crust and in seawater. Indeed, one gallon of sea water contains enough hydrogen isotopes for fusion to equal the energy which would be released by burning 300 gallons of gasoline. Large amounts of tritium can be obtained if an isotope of lithium (Li-6) is bombarded by neutrons in a reactor.

$${}^{1}n + {}^{6}Li \longrightarrow {}^{4}He + {}^{3}T$$

neutron + lithium ⟶ helium + tritium a

Fusion Dilemma

The basic challenges facing the commercial development of fusion reactors are (a) to achieve ignition with a net gain in the energy, beyond the break-even point; and (b) to find a method that supports fusion chain reaction in a sustained manner. The break-even point occurs when the amount of energy released by the nuclear reaction exceeds the amount of energy required to raise the temperature (and pressure) of the reactants to the level needed to initiate fusion (namely, Q of 1). These requirements can be reduced to three tasks of acquiring:

1. *Very high temperatures* (In excess of 100 million degrees). This is needed to overcome the repulsive forces of the positively charged nuclei.
2. *Very high compressions.* This increases the collision frequency and the rate of reaction.
3. *Long residence times.* This is needed to keep the reactants in proximity long enough to sustain the reaction; i.e. to increase energy production at a rate greater than that needed to initiate the reaction.

In 1991, the Princeton Tokomak was the first fusion reactor to reach this critical break-even point for about two seconds, opening the door for further research toward development of commercial nuclear fusion reactors. The current world record for a sustained high-temperature, high-pressure plasma is held by JT-60, at 24 seconds.

To make the reaction sustainable, we need to find a way to allow this reaction to proceed at a much larger scale and in a continuous way. Obviously, no vessel exists that can tolerate the extreme temperatures and pressures required for the fusion reaction to take place.

Three approaches have been investigated: a) gravitational confinement, b) inertial confinement, and c) magnetic confinement. Gravitational confinement is what happens in the sun, where gases are compressed and heated by gravitational forces. Gravitational confinement cannot be duplicated on earth. Inertial confinement mimics a short-lived micro-miniature star; a spherical capsule containing a deuterium-tritium (D-T) mixture is heated and compressed by an array of powerful lasers directed toward the pellet for a few nanoseconds. In the third approach, magnetic confinement, hot ionized gases are confined in a toroidal path that is held away from the reactor walls by strong magnetic fields.

Fusion: The Facts

FYI...

- Each day, our sun loses 5×10^{16} kg of its mass and produces 3.6×10^{31} joules of energy.
- The deuterium contained in one liter of sea water can generate as much power as 2.5 barrels of petrol.
- Relative to a fission reaction, fusion produces little radioactive material.

Inertial confinement was first used in the design of the Shiva[16] reactor at Lawrence Livermore National Laboratory (LLNL) in California. The original lasers were later replaced by NOVA lasers which were ten times more powerful (Figure 11-23). Very strong laser pulses strike and uniformly heat the deuterium-tritium pellet (made of substances such as lithium deuteride, the active ingredient of a hydrogen bomb) placed inside a vacuum sphere. The laser beams blast and incinerate the pellet and compress the gas to very high temperatures and pressures that are required to initiate fusion.[17] The next generation high powered lasers are being constructed to deliver at least 60 times more energy than any previous laser system. The National Ignition Facility (NIF) -- scheduled to be completed in 2010 – plans to deliver two million joules (2 MJ) of ultraviolet laser energy by focusing its 192 giant lasers to about 4 billionths of a second (producing 500 trillion watts of power) on a tiny 0.5-millimeter-diameter capsule in the center of its target chamber, creating conditions similar to those that exist only in the cores of stars and giant planets and inside nuclear weapons.[18]

Figure 11-23
Fusion inertial research facilities in Livermore, California (NOVA).

The preliminary results seem to indicate that this approach will probably not be as successful as the magnetic confinement approach used in a number of other facilities throughout the world (JT-60 in Japan, Tokomak Fusion Test Reactor (TFTR) in Princeton (Figure 11-24), and EU's Joint European Torus (JET) located in England).

Figure 11-24
The Tokomak Fusion Test Reactor (TFTR) in Princeton, New Jersey. The design was originally proposed in the early 1950's by Igor Tamm and Andrei Sakharov of Moscow University. Electric current produces a magnetic field spiraling around a torus or donut where the plasma forms a continuous circuit.

Magnetic confinement is a process where hot hydrogen plasma is circulated inside a huge coil of wire (magnetic bottle) that is cooled until it becomes a superconductors. If a current flows through this coil, it creates a magnetic field that confines the plasma within the torus shaped cavity. The plasma is continuously heated by the magnetic field as it circulates through the coil. Once it reaches the fusion temperature, deuterium and tritium fuse to form helium and neutron. The excess energy is carried by neutrons and deposited in a lithium jacket and produce electricity.

The 500 MW *International Thermonuclear Experimental Reactor* (ITER, "the way" in Latin) being based on Tokomak design is a joint, €10 B ($16 B) cooperative effort between the European Union, Japan, the US, Russia, China, India, and South Korea. ITER will be built in Cadarache, France, for operation in or near 2016 (Figure 11-25).

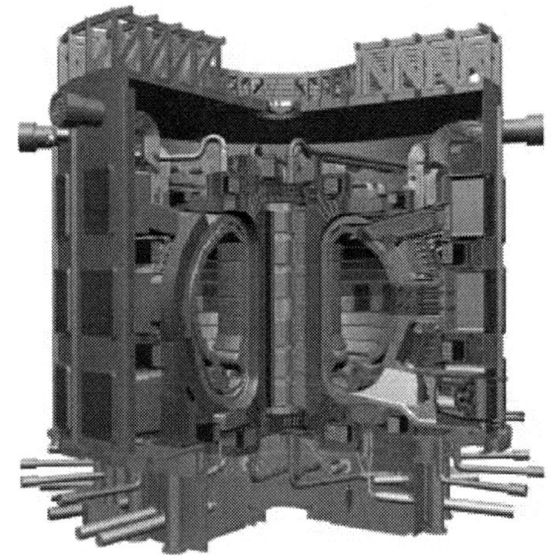

Figure 11-25
The ITER device.
Source: ITER, http://www.iter.org.

The initial goal is to sustain a deuterium-tritium chain reaction and generate ten times as much energy as was put into it, to a Q of 10. Furthermore, research will be conducted to test critical issues of fusion safety (such as finding materials that can withstand the high-energy neutrons created by the ITER, coolants, etc.) that must be resolved before the full-scale

[16] Shiva often called the Destroyer or The Lord of the Dance is the Hindu God of many hands and is attributed with great power and physical prowess.
[17] Maniscalco, J. A., *"Inertial Confinement Fusion,"* Ann. Rev. Energy, 5; 33-60, 1980.
[17] Yamanaka, C., "Inertial confinement fusion: The quest for ignition and energy gain using indirect drive," Nucl. Fusion 39 825 (1999).

commercial reactors can be built.

Cold Fusion

When people talk about fusion, they are mostly referring to "hot fusion". There has been much talk about the possibility of "cold fusion" in the last couple of decades. In 1989, Stanley Pons of the University of Utah, and Martin Fleischmann from the University of Southampton in the United Kingdom, claimed that they had achieved nuclear fusion in a simple photochemical cell.[19] Cold fusion takes advantage of an elementary particle called a *muon*. Muons have negative charges, and when they collide the nucleus they cancel positive charges of protons, thus eliminating the need for extremely hot temperatures to overcome the repulsive forces of hydrogen nuclei.[20]

These experiments generated enormous publicity around the world, but were later denounced by leading experts who could not verify and reproduce the results. Although there are still some researchers who believe that cold fusion is possible, no credible experimental proof has yet to be presented!

Future of Fusion

Because of a wide range of technical difficulties that must be overcome, it is difficult to predict when fusion will become practical. As with many other research efforts, economics will have a major impact on the degree of progress that can be achieved. Because of its complexity and cost, fusion research requires not only cooperation between private industries, utility companies, and the government, but also a high degree of international collaboration. The optimistic view is that commercial fusion power plants can be constructed just before the earth's fossil fuels are depleted, somewhere around the years 2045-2060.

Summary

Nuclear power has been shown to be an economical, clean source of energy with no emission of carbon dioxide, acid rain, and other chemical pollution. Public concern about radiation associated with nuclear fuel production, waste disposal, nuclear proliferation, and now terrorism has, however, made nuclear fission controversial. Furthermore, the nuclear accidents in Three Mile Island and Chernobyl and the resulting fallout of radio nuclides heightened the fear that what happened by accident could also happen by design. According to the latest poll carried out on behalf of the European Commission[21], only 7% of the public is "totally in favor" of nuclear power.

[19] Editorial (April, 2004c): "*DOE Warms to Cold Fusion*". Physics today.
[20] Rafelski, H. E., et al., "*Cold fusion: muon-catalysed fusion,*" J. Phys. 9: At. Mol. Opt. Phys. **24** (1991) 1469-1516.
[21] Special Eurobarometerm Radioactive Waste, commissioned by DG TREN, European Commission, September 2006.

Many believe the problem is not of a technical nature, but of a social nature instead. The public has lost confidence in the government and political and economic influence by various lobbyist groups. Unless concrete measures devoid of political influence are taken to assure the public of nuclear safety, for many, the nuclear option may remain one of the last choices to meet our energy needs.

Unlike fission, the physics of fusion make it inherently safer. A fusion reactor cannot go through a meltdown, and the waste generated by fusion is less radioactive and has a shorter half-life, therefore it is easier to dispose. The fusion fuel, deuterium, is found in ordinary seawater and is cheap and abundant. However, existing technical problems make the reality of fusion technology a distant dream.

Cold fusion, if proven possible, eliminates much of the technological difficulties associated with hot fusion and does so at a very low cost. This would level the playing field and provide the less developed and poorer countries with an inexpensive source of energy. Because of its high costs, hot fusion will potentially be available only to rich and developed countries, giving these countries the power to monopolize the energy industry.

Additional Information

Books

1. Bodansky, Nuclear Energy Principles, Practices, and Prospects, Second Ed., Springer, 2004.
2. Seaborg, G., T., Peaceful Uses of Nuclear Energy, University Press of the Pacific, 2005.

Periodicals

1. *International Journal of Nuclear Engineering and Design,* Direct Science Elsevier Publishing Company, devoted to the Thermal, Mechanical, Material and Structural Aspects of Nuclear Fission.
2. *Journal of Fusion Energy,* Springer Netherlands. It features articles pertinent to development of thermonuclear fusion.

Government Agencies and Websites

1. Federation of American Scientists *(http://www.fas.org/nuke/intro/nuke/index.html).*
2. International Atomic Energy Agency *(http://www.iaea.org).*
3. DoE Office of Nuclear Energy, Science & Technology *(http://www.ne.doe.gov).*

Non-Government Organizations and Websites

1. American Nuclear Society, *(http://www.ans.org).*
2. World Association of Nuclear Operator (WANO) *(http://www.wano.org.uk).*

Exercises

I. Consult the periodic table of elements and answer the following questions:

1. Consult the periodic table to determine the atomic numbers of the following elements:
   a. Helium
   b. Iodine
   c. Chlorine
   d. Lead
   e. Barium

2. The atomic structures of gold and mercury are represented as ($^{197}_{79}Au$) and ($^{201}_{80}Hg$). For each of these atoms, determine:
   a. The atomic number
   b. The atomic mass
   c. The number of protons
   d. The number of neutrons

II. Select the term that best fits the definition given:

________1. Nuclear reaction involving splitting atoms.

________2. Nuclear reaction in which two light atoms are joined together.

________3. The elementary particle that differs in number between different isotopes of an atom.

________4. The elementary particle in number equal to the number of protons in a nucleus.

________5. Sequence of reactions that release neutrons that cause additional atoms to fission.

________6. Material used to slow down the neutrons and improve their chance of capture by nuclear fuel.

________7. Component of the reactor core responsible for absorbing the excess neutrons and prevent the runaway reactions.

________8. The process of concentrating uranium-235 to make it suitable for nuclear fuel.

________9. The type of reactor that uses uranium-238 as its fuel.

________10. The type of reactor that uses graphite as the moderator and helium as the coolant.

III. Essay Questions

1. What are the differences between chemical and nuclear reactions? In each case, which of the following quantities are conserved?
   a. Number of nucleons
   b. Number of electrons
   c. Number of atoms
   d. Number of molecules

2. An atom of the element iron can be represented as $^{56}_{26}Fe$. Determine the number of neutrons, the number of protons, the atomic mass and the atomic number for iron.

3. What is a half-life? How many half-lives does it take for the activity of a radioactive isotope to drop by a million times?

4. What does NIMBY stand for?

5. Explain the fuel cycle, from mining to storage.

6. What were the causes of the TMI and Chernobyl accidents? What are the differences in design of the two reactors?

7. What is a chain reaction? How can a chain reaction be stopped? What will happen if chain reaction continues indefinitely?

8. What is the magnetic confinement? Inertial confinement?

9. A radioactive substance with a half-life of 5 years contains 6,000 radioactive atoms. How many atoms in the sample are expected to remain radioactive after 20 years?

10. A quantity of radioactive substance has lost 15/16 (94%) of its radioactivity after one minute. What is its half-life?

11. If there is a nuclear power plant near you,
    - What is it called?
    - When did it start operating?
    - What type is it?
    - What is its electric generating capacity?

12. Name three advantages and three disadvantages of nuclear power plant.

13. What is fusion? How does fusion release energy? Where does fusion occur in nature? What are the basic fusion reactions?

14. What are the main obstacles to development of fusion reactors?

15. Name three approaches proposed as possible alternatives to storing nuclear wastes. What are advantages and disadvantages of each?

IV. Multiple Choice Questions

1. The mass of a nucleus is___________ the total mass of its separate constituent particles.
   a. Equal to
   b. More than
   c. Less than

2. The nucleus of __________is fissile, while that of ___________is fertile.
   a. ^{235}U, ^{238}U
   b. ^{235}U, ^{239}Pu
   c. ^{238}U, ^{235}U
   d. ^{239}U, ^{235}U
   e. ^{238}U, ^{239}Pu

3. Two isotopes of the same element can have the
   a. Same number of electrons but a different number of protons
   b. Same number of neutrons but a different number of protons
   c. Same number of protons but a different number of neutrons
   d. Same atomic mass but a different atomic number
   e. Same atomic mass and same atomic number

4. The energy released in a nuclear reaction (fission or fusion) is associated with
   a. A decrease in mass
   b. An increase in total mass
   c. An increase in total binding energy
   d. No change in total mass
   e. None of the above

5. The reason that alchemists could not produce gold is that
   a. They did not use the right chemicals
   b. They did not know how the chemical reacted
   c. They did not know about the structure of atom
   d. They did not have a nuclear reactor
   e. Both c and d

6. Energy released from a nuclear reaction is the result of
   a. Reconfiguration in the molecular bonds
   b. Change in the number of electrons
   c. Change in the number of atoms
   d. Change in the binding energy of the nucleus
   e. All of the above

7. The breeder reactor may eventually substitute light and heavy water reactors because
   a. It is safer than the BWR and PWR
   b. We need to use ^{235}U instead of ^{238}U
   c. We need to use ^{238}U instead of just ^{235}U
   d. It has a higher efficiency
   e. It is easier to construct

8. The known supplies of uranium (U-238) in the United State are estimated to be
   a. 400,000 tons
   b. 1 million tons
   c. 4 million tons
   d. 1 billion tons
   e. Practically unlimited

9. The largest reserves of uranium lies in
   a. The United States
   b. Kazakhstan
   c. Russia
   d. South Africa
   e. Australia

10. Nuclear fission can be best described as
    a. The direct conversion of nuclear energy to electrical energy
    b. A chemical reaction involving uranium atoms
    c. A process where uranium atoms ionize to produce energy
    d. The splitting atoms accompanied by release of

a large amount of energy
e. The splitting atoms by electron bombardment

11. In the equation $E = mc^2$, E represents
    a. The potential energy of an object of mass m
    b. The kinetic energy of an object of mass m
    c. The total energy of an object of mass m
    d. The internal energy of an object of mass m
    e. None of the above

12. Nuclear chain reactions can be controlled by
    a. Controlling the number of electrons released by the fission process
    b. Controlling the number of protons released by the fission process
    c. Controlling the number of neutrons released by the fission process
    d. Controlling the number of photons released by the fission process
    e. Controlling the electromagnetic radiation released by the fission process

13. How long would it take for the strength of a 10,000-curie radioactive source to drop to only 10 curies?
    a. 33 years
    b. 75 years
    c. 150 years
    d. 300 years
    e. 10,000 years

14. Which of the following statements applies to nuclear reactors?
    a. A neutron smashes the nucleus of a uranium atom with no neutrons produced.
    b. A neutron smashes the nucleus of a uranium atom with 2-3 additional neutrons released on average. Each neutron goes on to smash additional nuclei.
    c. A neutron smashes the nucleus of a uranium atom with 2-3 additional neutrons released on average. All except one neutron is captured.
    d. Is essentially the same as a nuclear bomb.
    e. Differs from a nuclear bomb only in the amount of energy that it releases.

15. Which of the following statements is correct?
    a. Both moderator and control rods absorb excess neutrons.
    b. Both moderator and control rods slow the neutrons down.
    c. The control rods slow the neutrons down, whereas the moderator captures the excess neutrons.
    d. Depending on the reactor type either moderator or control rods are used.
    e. The moderator slows the neutrons down so they can be captured by the fuel rods.

16. Which one of the following is a principal disadvantage of nuclear power?
    a. Dangerous waste can be produced.
    b. Greenhouse gases are not produced.
    c. Greenhouse gases are produced.
    d. It is not safe to operate.
    e. The cost is much higher than a fossil fuel reactor.

17. A suitable fuel for a light water reactor is
    a. Natural uranium containing 0.7% ^{235}U
    b. Uranium enriched to contain at least 3% ^{235}U
    c. Uranium enriched to contain at least 90% ^{235}U
    d. Uranium enriched to contain 99% ^{235}U
    e. Pure ^{235}U

18. A suitable fuel for a heavy water reactor is
    a. Natural uranium containing 0.7% ^{235}U
    b. Uranium enriched to contain at least 3% ^{235}U
    c. Uranium enriched to contain at least 90% ^{235}U
    d. Uranium enriched to contain 99% ^{235}U
    e. Pure ^{235}U

19. A suitable fuel for a fast breeder reactor is
    a. U-235
    b. U-238
    c. Pu-239 reprocessed from U-238
    d. Deuterium
    e. Hydrogen

20. A suitable coolant for most fission reactors is
    a. Ordinary water
    b. Heavy water
    c. Graphite
    d. Sodium
    e. Helium

21. A suitable moderator for most fission reactors is
    a. Water or graphite
    b. Liquid sodium or argon
    c. Water or liquid sodium
    d. Carbon dioxide or helium
    e. Boron or cadmium

22. "Chain reaction" is a common occurence in
    a. Nuclear reactors and weapons
    b. Rumors and legends
    c. Spread of diseases like cancer and smallpox
    d. Population growth
    e. All of the above

23. Most European nuclear reactors use ___________ as a moderator.
    a. Graphite
    b. Ordinary water
    c. Heavy water
    d. Sodium
    e. Helium

24. Which one of these statements is true about the moderator and the control rods in a nuclear power station?
    a. The moderator and control rods slow the neutrons down.
    b. The moderator and control rods absorb the neutrons.
    c. The moderator slows the neutrons down, while the control rods absorb the neutrons.
    d. The control rods slow the neutrons down, while the moderator absorbs the neutrons.
    e. None of the above.

25. The reactors at the Three Mile Island plant are
    a. Boiling light water reactors
    b. High temperature gas reactors
    c. Pressurized heavy water reactors
    d. Pressurized light water reactors
    e. Fast breeder reactors

26. The reactors at the Chernobyl plant are
    a. Water-cooled with water moderators
    b. Water-cooled with graphite moderators
    c. Gas-cooled with water moderators
    d. Gas-cooled with graphite moderators
    e. Graphite cooled with water moderators

27. The principal cause of the explosion in the nuclear accident at Chernobyl was
    a. An uncontrolled chain reaction in the fissile material
    b. A chemical reaction between elements released by the overheated reactor
    c. Excess pressure of steam in the reactor
    d. A massive electrical short-circuit
    e. All of the above

28. Which organization is in charge of inspecting and ensuring safeguard of nuclear power plants around the world?
    a. US Atomic Energy Commission
    b. International Atomic Energy Commission
    c. Defense Nuclear Agency
    d. Department of Energy
    e. Department of Defense.

29. Fusion reaction requires that plasma be heated to very high temperatures because
    a. Enough energy is needed to split the atoms
    b. Combining particles are positively charged
    c. Combining particles are negatively charged
    d. Combining particles have opposite charges
    e. The higher the temperature of the plasma, the higher is the fusion efficiency

30. Most nuclear fission reactors use fuel rods enriched to
    a. About 0.7% uranium–235
    b. 3% or greater in uranium–235
    c. 90% or greater uranium-235
    d. 3% or greater in uranium–238
    e. 3% or greater in plutonium–239

31. The main advantage of plutonium breeder reactors over conventional hot water reactors is
    a. Reduction of the operation cost of the reactor
    b. Reduction of thermal pollution
    c. Extension of the lifetime of uranium reserves
    d. Reduction of the amount of plutonium produced
    e. Reduction of the threat of nuclear weapons proliferation

32. The energy released in a nuclear fission reaction is proportional to
    a. The mass which disappears
    b. The mass of the neutron
    c. The mass of the proton
    d. The mass of ^{235}U
    e. The number of neutrons emitted

33. What is the fuel used in most present-day nuclear power plants?
    a. Coal
    b. Uranium-238
    c. Uranium-235
    d. Platinum-196
    e. Plutonium-239

34. Most nuclear reactors in operation today are of this type.
    a. Coal
    b. Fission
    c. Fast breeder
    d. Hot fusion
    e. Cold fusion

35. The coolant most suitable in a fast breeder reactor is
    a. Water
    b. Graphite
    c. Liquid sodium
    d. Carbon dioxide
    e. Helium

V. True or False?

1. All the atoms in a material have the same number of neutrons in their nuclei.

2. The only difference between various isotopes of an atom is in the number of neutrons in their nuclei.

3. The nuclei of a normal hydrogen atom is called a proton.

4. All the atoms in a material have the same number of electrons.

5. Most instances of cancer in the United States are caused by man-made radiation.

6. All nuclear materials remain highly toxic for thousands of years.

7. For an equal radiation dose, man-made radiation is more toxic to humans than naturally occurring radiation.

8. About half of all electricity in the world is generated by nuclear reactors.

9. Heavy water is ordinary water that has been enriched by the addition of deuterium.

10. The function of a control rod is to slow down fast neutrons to thermal neutrons.

11. The easiest way to shut down a reactor is to remove the fuel rods.

12. In case of a rapid temperature rise in the core and the need for rapid reactor shutdown, all the control rods are lowered immediately.

13. Nuclear power plants can explode like a nuclear bomb.

14. On average, people receive more radiation from nuclear power plants than from building materials and radon gas in homes.

15. Most high-temperature, gas-cooled reactors use helium as a coolant.

16. The only commercial pebble bed reactors operating today is in the United States.

17. Breeder reactors use uranium-238 instead of uranium-235 for fuel.

18. The first cold fusion experiment was successfully concluded in 1989.

19. It is easier to achieve fusion by fusing atoms of D-T than D-D.

20. The mass of a nucleus is less than the sum of the

masses of its elementary constituents.

VI. Fill-in the Blanks

1. Isotopes have the same number of _______ but a different number of ________.

2. The mass of an atom is _______ than the sum of the individual masses of its protons and neutrons. The extra mass is a result of the __________ that holds the protons and neutrons of the nucleus together.

3. As a result of isotopic decay, unstable carbon-14 changes to stable ___________________.

4. It is possible to extend the remaining supply of uranium by converting U-238 to Pu-239 and using it as the fuel for _________ reactors.

5. Isotopes of an element that can be split through fission are called __________.

6. The four main components of conventional nuclear reactors are the fuel rod, __________, moderator, and ___________.

7. __________ has the greatest number of nuclear reactors; ________________ and Lithuania are fulfilling the greatest percentage of their electricity need using nuclear energy.

8. The process in which high-temperature liquid waste is dried is called ______________.

9. Uranium ore is commonly referred to as _________________.

10. When the temperature in the core rises, ______ rods are lowered and the energy output decreases because fewer _________ are available to sustain chain reactions.

11. The principal causes of major nuclear accidents are the production of an excessive number of _________________ and the loss of coolants.

12. All US commercial reactors are of the _______ ________ water reactor type.

13. A breeder reactor produces more _______ than it uses.

14. The most serious nuclear accident that has ever happened was in Chernobyl in the Republic of __________.

15. PWRs and BWRs are both __________ water reactors.

16. The agency responsible for granting operating licenses and oversee a safe operation of nuclear reactors in the United States is ______________.

17. Fusion reactors are not expected to be commercially available before ______________.

18. Compared to natural gas power plants, nuclear plants produce _________ amount of greenhouse gas emissions.

19. Uranium separated from rocks and concentrated by a series of chemical processes is called the ________________.

20. The fuel made of recycled uranium and plutonium oxides from spent fuel is called _____________.

VII. Project - Half-Life of Pennies

In this experiment you will use a statistical approach to gain a better understanding of the concept of half-life. For this experiment you need 100 pennies (or other coins), a large tray, a pencil, and a piece of paper.

Procedure:

1. Lay all pennies heads up in the tray.
2. Shake the tray for a few seconds; remove all the coins which have turned tails up. Count the number of the remaining pennies in the tray (those which stayed heads up). Enter the data in a table similar to Table 1.
3. Shake the tray some more; remove all the coins that have turned tails up. Enter the number of heads in the table.
4. Repeat this procedure until no pennies are left on the tray.

5. Now repeat the whole experiment two more times.

Data reduction:

a. On graph paper, plot the number of pennies remaining in the tray versus the trial #. Make the y-axis represent the number of heads and x-axis the trial #.
b. On the same graph, repeat the procedure for the second and third attempts. Use different symbols or colors to distinguish between different attempts.
c. Take the average number of heads for each trial and enter it in Table 4. Plot this graph superimposed on top of the earlier graphs.

Now answer these questions:

1. How reproducible were your data?
2. How does this experiment illustrate the behavior of radioactive isotopes? Which set of results represents the best model of half-life?
3. How many trials does it take for the population of pennies to reduce to 50? To 25?
4. What is the rate of decay of pennies in the tray?
5. Compare your results with those of three other classmates. What do you conclude?

Table 1. Half-Life of Pennies

	1	*2*	*3*
Trial #	No of Heads	No of Heads	No of Heads
0			
1			
2			
3			
.....			

Nuclear Radiation

CHAPTER 12

Any dose of radiation is an overdose. ~ George Wald (1906-1997)

Only two things are infinite, the universe and human stupidity, and I'm not sure about the former. ~ Albert Einstein (1879-1955)

Unlike fossil fuel combustion that produces toxic air pollutants, carbon dioxide, and other greenhouse gases, nuclear energy is clean and produces no such contaminants. Nuclear energy, however, generates other products that pose far greater threats to the environment if not handled properly. The radiation released during the Three Mile Island nuclear accident in 1979, and the much greater catastrophic Chernobyl accident in 1986, catalyzed a major debate about the safety of nuclear reactors. The buildup of nuclear arsenals (primarily by the United States and Russia and, to a lesser extent, by China, France, England, Pakistan, India, Israel, and now North Korea), the threat of terrorism on nuclear stockpiles, and the issues related to storage of nuclear wastes have all raised concerns about the environmental aspects of nuclear energy and radiation. In this chapter, we will examine various sources of radiation, their effect on health, and the consequences of nuclear energy in times of war and peace.

Ionizing Radiation

All forms of radiation can be categorized into two types: ionizing and non-ionizing. **Ionizing radiation** has enough energy to eject electrons from atoms and produce positively charged ions. Materials capable of emitting ionizing radiation are radioactive. Cosmic rays, gamma rays, x-rays, and high-energy UV lights are examples of ionizing radiation. Visible, infrared (heat), microwave, and radio waves do not have sufficient energy to cause ionization and thus are called **non-ionizing radiation**. Non-ionizing radiation was covered earlier and will not be discussed any further.

Our body is largely transparent to both low-energy radar and high-energy x-rays and gamma radiations, but readily absorbs visible and infrared. Although visible and infrared absorptions are quite harmless, even small amounts of x-ray and gamma ray absorption can harm our bodies.

Sources of Ionizing Radiation

Ionizing radiation can come from both natural and man-made sources. **Natural (background)** sources of radiation are cosmic radiation and terrestrial radiation. Cosmic radiation is due to interactions between earth's magnetic field and charged particles coming from the sun and other distant stars. Cosmic radiation is stronger at higher elevations. Sources of terrestrial radiation are uranium and products of its decay such as thorium, radium, and radon, which are found in soil. Our body also contains radioactive materials such as potassium-40, carbon-14, and lead-210 when we are born.

Man-made (anthropogenic) sources of radiation come from nuclear reactors, consumer products, irradiated food, medical checkups, and radiation therapy. Precluded from this list is possible exposure due to the accidental release of nuclear radiation from nuclear reactors, testing of nuclear bombs, or nuclear wars. It is therefore reasonable to assume that the geographical location we live in, our lifestyle, and the state of our health all affect the amount of radiation to which we are exposed. Because the earth itself contains radioactive materials, the food and water we consume may also be affected. Even the materials from which our houses are made contribute. For example, stone and bricks are more radioactive than wood and aluminum. Consumer products such as televisions and computer screens, luminous watches and dials, smoke detectors, and airport x-ray machines (See box "X-ray Vision") also contribute to the radiation doses we regularly receive.

In addition to the sources mentioned above, there are other sources such as alpha particles, beta particles, and neutrons which may occur naturally in the outer atmosphere or may be produced in nuclear fission processes.

a Measuring Radiation

Depending on whether we refer to radiation emitted by a source (**activity**), absorbed or deposited in an object (**exposure**), or the biological damage it causes (**dose**), different units of measurement are most commonly used (Figure 12-1).

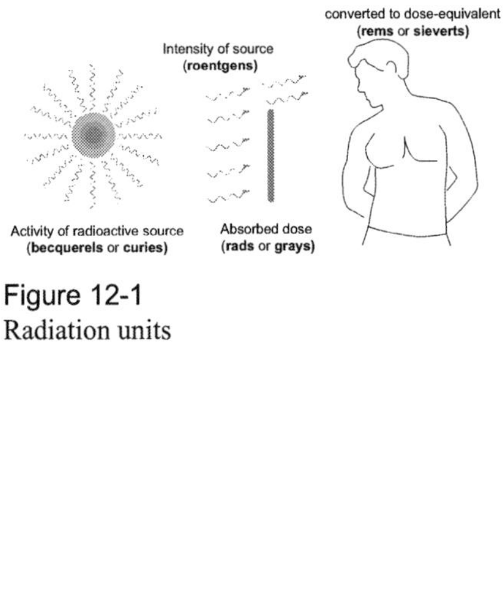

Figure 12-1
Radiation units

Becquerel is the unit most commonly used for ionizing radiation; it is simply the number of nuclei disintegrations per second. A larger unit is the *curie*, which represents the amount of radiation emitted by one gram of radium, or 3.7×10^7 nuclei per second. Both curie (Ci) and becquerel (Bq) are measured by an appropriate radiation detection device such as Geiger counters and are used when the amount of radiation emitted from a source is of interest. They tell nothing about the radiation dose that is actually absorbed by objects. *Roentgen* (R) is the measure of the radiation intensity of x-rays and gamma rays and refers to the number of ionizations produced in air.

Rad (radiation absorbed dose) gives the amount of radiation absorbed and deposited in a matter (exposure) — be it bone, fat, muscle, or concrete, irrespective of the effects this deposit has on the material. For water and soft tissues, rad and roentgen are approximately equal. Another commonly used unit of absorbed dose is a *gray*, which is equal to 100 rads. Equal doses of different types of radiation cause different degrees of biological damage.

X-ray Vision

FYI ...

Soon after the discovery of X-rays in 1895, scientists constructed devices that looked into broken bones and located bullets in patients' bodies.[i] Critics, however, envisioned this device as a threat to privacy, fearing some might design instruments that could see through walls of houses and peek through women's clothing. The New Jersey legislature even tried to pass a law banning x-rays from use in opera glasses so as to protect the stars' modesty. Although the quality of x-ray images was rather poor at the time, new scanners are now being developed that reveal much more than meets the eye.

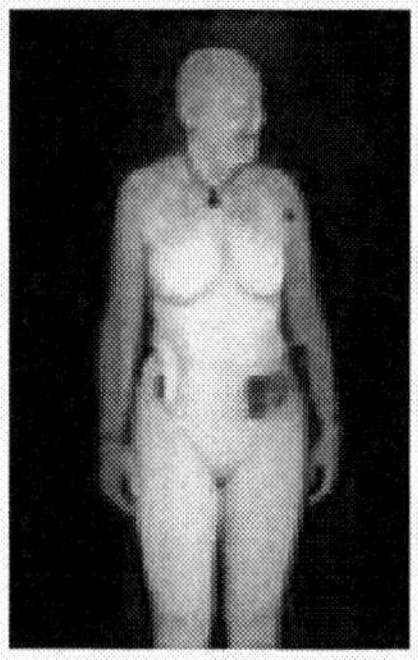

Backscatter x-ray machines are being considered at airport security checkpoints. This woman appears to be carrying a gun and a grenade.

[i] The discovery of the x-ray by Roentgen brought excitement and public interest along with many speculations on its possible applications. For example, one proposal suggested making an invisible ink that can only be read by x-ray. Another claimed to have taken photographs of his thoughts; one company marketed anti x-ray underpants, while others sent the hollow eyepieces of their opera glasses to be fitted with x-ray. One customer was even seeking to buy one kilogram of x-ray.

Rem (roentgen equivalent man) is a measure of the damage caused by one rad of radiation in the human body (dose). It takes into account the effectiveness of both the source and the living tissue. Since rem values are normalized to different tissue types, one rem of any type of radiation does the same biological damage. For x-rays and gamma rays, the rem and rad are taken to be equivalent. Alpha particles have 20 times the risk of x-ray and gamma ray radiation, so one rad of alpha particle radiation is equivalent to 20 rem (both singular and plural). Since one rem of radiation is a large quantity, the unit most often used is the millirem (mrem), which is one thousandth of a rem. Another commonly used unit is the *sievert*, which is a measure of the biological effect of one gray of gamma ray. One sievert is equal to 100 rem.

Radiation Dosimetry

Individuals receive radiation from natural sources and from various activities over their lifetimes. The cumulative effect is not equivalent to the effect of an equal amount of radiation received at once or over a short time. *Chronic exposure* refers to relatively low doses of radiation for periods of months and years. This is typically experienced in normal daily activities such as handling nuclear materials and living next to a nuclear waste dump, or is accumulated as a result of years of radiation therapy and routine medical checkups. *Acute exposure* refers to a large dose of radiation received over a short time, such as that resulting from a nuclear accident or a nuclear war.

Although people cannot sense radiation directly, sensitive instruments called dosimeters can measure the amount of radiation they receive. Dosimeters are small sensors that a person can carry to measure the total dose of radiation received over a period of time. Geiger counters are normally used to detect the presence of radiation. In the United States, each healthy person receives an average of 300 mrem of radiation per year. The amount can be significantly higher for those who receive radiation therapy or who work in nuclear power plants, radiology units, and some research laboratories. The US Environmental Protection Agency (EPA) is responsible for setting limits on the amount of radiation humans can receive. According to these guidelines, the maximum annual permissible dosage is 500 mrem for the public. Furthermore, it has set a limit of 5,000 mrem per year, not to exceed 1,250 mrem in any given three month period, as the safety limit for adults working with radioactive materials from man-made radiation sources. The EPA guideline for the maximum one-time radiation dose for emergency workers volunteering for lifesaving work is 75 rem. You can estimate the amount of radiation you receive every year by filling in the dosimeter chart given below.

Effects of Radiation on Health

There are over 300 different products derived from fission of uranium and plutonium isotopes. Some have very short half-lives and practically disappear after a few seconds. Others, such as strontium-90 and cesium-137 with half-lives of about 30 years, enter the body by inhalation and the ingestion of contaminated foods and slowly decay by the continuous emission of beta and gamma radiation. Iodine-131 has half-life of only 8 days, so it accumulates in victim's thyroid gland very fast. This is why doctors prescribe the normal (nonradioactive) iodine pills to saturate the body before radioactive isotope has a chance to be absorbed. This of course has no protective effect on other types of radioactive materials.

Radiation Dosimetry

FYI ...

You can get a good approximation of the amount of radiation you receive every year from various sources. Here is how to estimate:

1	Where do you live? ________ Elevation (meters) ______ Cosmic radiation at Sea level Elevation x 0.015	28 mrem ______
2	If you live within 10 km of a nuclear or coal power plant, add 0.3	______
3	What is your house made of? Brick or concrete (add 70), wood (add 30)	______
4	Ground radiation (US average) Radon (consult EPA sources) Water, food, and air (US average) Watching TV (1 mrem/hr) Number of hours a day _____ x 1	26 mrem ______ 28 mrem ______
5	How many miles do you fly every year? Number of miles _____ x.001 or Numbers of hours of flying _____ x 0.5	______
6	How is your health? Number of chest x-rays _____ x 10 Number of dental x-rays _____ x 5 Number of mammograms _____ x 30 Others (chemotherapy, varies)	______ ______ ______ ______
	Total	

When molecules of living organisms are exposed to ionizing radiation, they become unstable; chromosomes and strands of DNA (deoxyribonucleic acid) break apart to form new molecules. DNA is a double helix of thousands of atoms strung along carbon-linked struts that contains all the information needed to produce and control living tissues. When mutations occur in reproductive cells (eggs or sperm), the changes can be passed on to subsequent generations.[1]

Effects of ionizing radiation on human health are hard to assess and are highly dependent on the dose, the duration of exposure, and the type of cells involved.[2] For example, if a body receives a massive amount of radiation in a short time, such as what might occur following a nuclear

Ionizing Radiation

Digging Deeper ...

Alpha particles are charged particles made up of two protons and two neutrons (the same as a helium nucleus). An example of alpha particle emission is the radioactive decay of uranium and radium. An alpha particle can travel only a few millimeters in the air and is easily stopped by a sheet of paper. Although alpha particles cannot penetrate human skin, they can do serious damage if swallowed. As alpha particles slow down, they encounter and attach themselves to electrons and form helium.

Beta particles are electrons ejected from the nucleus of a decaying atom. Beta particles cannot be stopped by a sheet of paper or human skin, but need a thicker shield (like wood) to stop them.

Gamma and x-rays are high-energy electromagnetic radiation produced by radioactive decay of unstable nuclei. Gamma rays have no charge or mass and can travel thousands of meters in air, easily penetrating the human body or even lead. X-ray is similar to gamma ray except they are 10-100 times less energetic. They can penetrate the soft tissues but are stopped by bones.

Neutrons are heavy, uncharged particles resulting from nuclear fission.

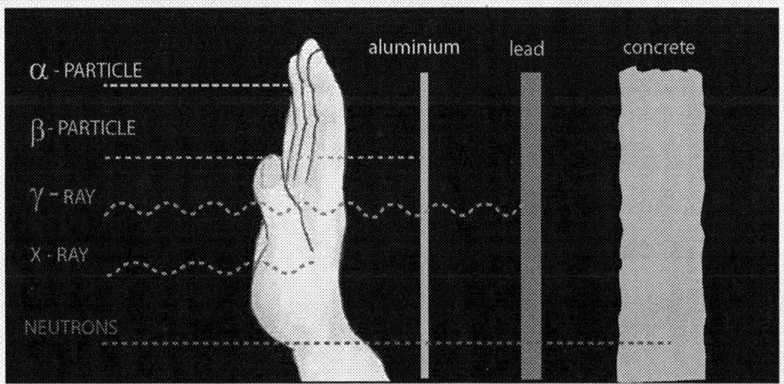

Type	Symbol	Charge	Composition	Penetrating Power	Shielding Guideline
Alpha	α	+2	Helium nucleus	Low (1)	Stopped by paper.
Beta	β	-1	Electron	Medium (100)	Stopped by aluminum, wood, or Plexiglas.
Gamma	γ	0	Photon	High (1000)	Stopped by lead or concrete.
Neutron	ν	0	Neutron	High (1000)	Slowed by water.

[1] The Los Alamos national laboratory maintains a website that provides information on radiation health physics and a radiation exposure calculator (*http://newnet.lanl.gov/dosecalc.htm*).
[2] US Environmental Protection Agency (*http://www.epa.gov*).

accident or an atomic blast, the effects are immediate – usually resulting in radiation sickness and death within hours or days. On the other hand, chronic exposure to low-level radiation may not be observed for many years or until future generations. Rapidly-growing cells like bone marrow and soft tissues such as ovaries, testes, and lenses of the eyes are most susceptible to destruction by radiation. Furthermore, depending on the general state of health of the individual, the same amount of radiation may result in different symptoms.

It is difficult to quote a number as the threshold at which death by radiation exposure is a certainty. Most nuclear scientists, however, agree that a single radiation dose of 500 rads or higher will almost certainly result in death within hours or days. If exposed to radiation doses of 100-250 rads, symptoms will include nausea, vomiting, and diarrhea, and many may eventually die as a result. No death is expected with a radiation dose of 100 rads or less, although in future years the risk of development of cancer, leukemia, cataracts, or sterility is significantly higher.

> **Question:** Does a single 10 rem dose cause the same amount of damage as 10 doses of 1 rem spread over many years?
> **Answer:** Many believe that genetic damage is probably independent of the dose rate since all doses cause non-reversible mutations. The somatic effects of several smaller doses are less, however, because time allows for some repair between administration of small doses.

Most data on high-level radiation doses are available from the two nuclear bombs dropped on Japan and the nuclear accident in Chernobyl. Over 200,000 people perished in a few days following the Hiroshima and Nagasaki bombings, and numerous others suffered from cancer, miscarriages, birth defects, and stillbirths in the years after. Even third-generation offspring of bombing survivors show higher rates of leukemia, thyroid, and colon cancers. Data from the Chernobyl accident point to similar catastrophes. More than 28,000 square kilometers of prime farmland were contaminated and another 200,000 people were forced to leave their homes as a result. According to the World Health Organization (WHO)[3], rates of thyroid cancer have climbed five to 30-fold depending on the indivisual's distance from the power plant. There has also been a considerably higher rate of spontaneous miscarriages.

The effect of low-level radiation is much harder to assess, as factors that can affect one's health increase with time. However, it has been established that low-level radiation is responsible for increased rates of several types of cancer. Iodine-131, a byproduct of nuclear reactions, concentrates in the thyroid gland and ovaries. Strontium-90 attacks bone marrow, increasing the risk of leukemia and other blood diseases. Plutonium-239 is one of the most toxic substances known. If inhaled, one thousandth of

[3] World Health Organization (*http://www.who.int/en*).

a gram can cause massive fibrosis of the lungs. Another radioactive waste, cesium-137, is mostly absorbed by the liver, kidneys, and sexual organs.

Useful Radiation

Radiation is not always associated with ills and disaster. Radiation has many uses in diagnosis and treatment of diseases, as well as industrial applications. Radiation can be used in numerous research and diagnostic tools, such as x-rays, MRI, and CAT scanners. Radioactive solutions can be ingested by a patient as a tracer to allow mapping of the blood stream for detection of tumors and restricted blood vessels, as well as photography of a particular internal organ. Radiation therapy can also be used in the treatment of disease, primarily to kill cancerous cells. Other applications include crop improvement and protection, manufacturing of consumer products such as watches, and ionization smoke detectors, and in the sterilization of such products as cosmetics and medical supplies. Foods are also irradiated to kill germs and other microorganisms, to aid in preservation, and to increase their shelf life. Radiation is a convenient tool for dating art and antiques, even dating such things as the age of the earth or a meteorite. Industrial uses of radiation include process monitoring, desalination, welding, detection of cracks and seams, and numerous others.

Question: Radiation therapy involves irradiating and killing the cancerous cells without damage to the healthy cells. How does radiation distinguish normal cells from cancerous cells?
Answer: Patients will be subjected to crossed beams of radiation that pass through the cancerous region. Alternatively, the patient can be rotated as he is radiated by a single beam of radiation. This way, the normal, non-cancerous, cells are exposed to much less radiation than the damaged cells that receive continuous exposure.

Nuclear Wastes

Uranium undergoes several changes as it is mined, processed, burned, reprocessed, and eventually discarded and stored in a repository. Along the way, uranium fuel leaves a footprint by producing and leaving behind various levels of radioactive waste. Depending on the amount and type, radioactive waste can be classified as low level, intermediate level, or high level. Low- level nuclear wastes are those with relatively short half-lives. They come from hospitals, medical and research laboratories, x-ray machines, and contaminated clothing and equipment. Many commercial products such as watches, ionization smoke detectors, eyeglasses, dental porcelain, luminescent products, instrument dials, signs, and markers contain some radioactive materials. Low-level wastes do not need any casks and are suitable for shallow depth burial or incineration.

The sources of intermediate-level nuclear waste are nuclear fuel processing,

Healthy Radiation[i]

FYI ...

A small but growing number of studies suggest that very low doses of ionizing radiation may actually have some health benefits due to stimulation of certain enzymes. Radon treatments have been used in treating severe pain in joints and rheumatic arthritis. This phenomenon is termed radiation hormesis. It has even been suggested that a significant fraction of cancer deaths are preventable by increasing our low dose radiation exposure.

[i] *Luckey, T.D., "Hormesis with Ionizing Radiation," CRC Press, Baca Raton, 1980.*

enrichment plants, nuclear weapons facilities, and contaminated parts of nuclear power plants before decommissioning. Intermediate-level wastes are usually solidified in concrete and, depending on their type, buried in shallow or deep underground repositories.

High-level nuclear wastes are spent fuel and other products from the reactor core and, if it has been decommissioned, the reactor itself. A typical power plant generates about 20-30 tons of high-level radioactive wastes each year, mainly cesium-137, strontium-190, and technetium-99. Plutonium-239 is the most serious byproduct of breeder reactors.

Disposal of Nuclear Wastes

Waste management is one of the most controversial and pressing issues tarnishing the image of the nuclear industry. Nuclear waste has not only presented challenges for the present generation, but has also endangered the welfare of future generations for many tens of thousands of years.

> **Example 12-1:** The concentration of strontium-90 in high-level liquid wastes is about 1 curie (Ci) per gallon. If the concentration is to be dropped to 10^{-9} curie per gallon for it to become safe, how long do we need to wait?
> **Solution:** Stronium-90 has a half-life of 30 years. In 10 half-lives (300 years), the concentration drops by 1000; in 20 half-lives (600 years), the concentration drops by a million, and in 30 half-lives (900 years), the concentration drops by a billion to an acceptable level. Examples like this help us understand the concern environmentalists have for the storage of nuclear waste.

As of yet, there is no completely reliable method of permanently disposing of nuclear waste. All current methods are only interim measures. In addition, there is no reliable information on the total amount of nuclear waste we have produced so far. No matter how small or large a role nuclear energy plays in meeting future energy needs, developing technologies that can safely dispose of the intensely radioactive byproducts remains a top priority. Here are some promising ideas.

Permanent subterranean storage – The most commonly favored method for disposal is the placement of waste into deep geological repositories. Factors affecting this determination are soil stability, proximity to large water reservoirs and run-offs, seismic activity, and the local population. An ideal site must be completely dry, with no possibility of moisture percolating through the cracks and corroding the alloys. Salt has been shown to be an effective barrier to radiation; therefore lands with large salt deposits located in geographically stable landmasses are believed to be the best choice.

The US government has considered many sites and has decided on a repository near Yucca Mountain. This Nevada desert site, with its thick section of porous volcanic rock, will serve as the US's first permanent underground repository for more than 40,000 metric tons of nuclear waste that has already been accumulated. The plan includes processing the spent fuel in steel canisters and inserting them into holes drilled in the rock floor of caverns hundreds of meters below the surface.

Figure 12-2
Yucca Mountain, the site of the permanent US nuclear waste disposal facility.
Source: Thinkquest.com

Residents of nearby communities have vehemently opposed this construction (Figure 12-2). Some jokingly refer to Yucca Mountain as "Yucky Mountain" or "Yuck-a-Mountain." Other opponents charge that the government has overlooked seismic and volcanic activity and other potential dangers in the area in the rush to find a location. Still others claim that since the Department of Energy has set its mind on Yucca as the permanent repository and has already spent billions of dollars on the Yucca project, they are unwilling to consider other sites or other storage alternatives. As a result, incentives for seeking more innovative solutions and better options have been gradually diminishing.

In the meantime, nuclear wastes are piling up across the nation. Waste levels are getting so high that even if the Nevada facility were constructed, it would not be large enough to house all of the waste. Originally planned to open by 1998, the date has now been pushed back to 2010, although many experts doubt that even this date is realistic.

Entombment under the seabed – This plan is similar to the plan outlined above, but in this case, canisters are dropped from ships and allowed to come to rest on the deep ocean floor far below. The advantage of this plan is the use of seafloors, which are far away from the shore and from

Alchemists and Transmutation

FYI ...

Transmutating metals to gold has been the aspiration of alchemists for over 4,000 years and independently by ancient Egypt and China. The origin of the word alchemy is "*khem*" which means fertility. The Egyptians' original desire was to find a way to preserve bodies in anticipation of the eventual return of their dead to life. Khem was the technique used for mummification. The prefix -*al* was added to the word when Arabs occupied Egypt in the 7th century. The Arabs believed that all metals were made up of mercury and sulfur in different proportions, and gold was the "purest" of all metals. So it was ideal to turn (transmute) metals from their lowest form to the most perfect form: gold. The interest in alchemy eventually led to the discovery of new elements and the start of the science of chemistry. Alchemists tried to achieve their goal through various chemical reactions. What they did not have was an understanding of the structure of atoms and a nuclear reactor!

people. However, the potential leakage could be catastrophic, and there are international treaties that bar the disposal of radioactive wastes at sea.

Nuclear transmutation – This method converts (transmutes) the radioactive waste materials with very long half-lives to short-lived or non-radioactive products by bombarding them with elementary particles such as neutrons. For example, by absorbing a neutron, technetium-99 ($T_{1/2}$= 210,000 years) becomes technetium-100, which then rapidly decays to rubidium-100. The Ru-100 isotope is non-radioactive, with a half-life of only 17 seconds. Another example is the transmutation of iodine-129 ($T_{1/2}$ = 17 million years) into the stable element xenon-130. An intense beam of fast-moving protons produced by a linear accelerator hits a lead or tungsten target and knocks out a large number of neutrons which then attacks the radioactive waste. In the process, additional energy is released, a portion of which is used to run the accelerator; the remainder would be fed into the regional power grid.

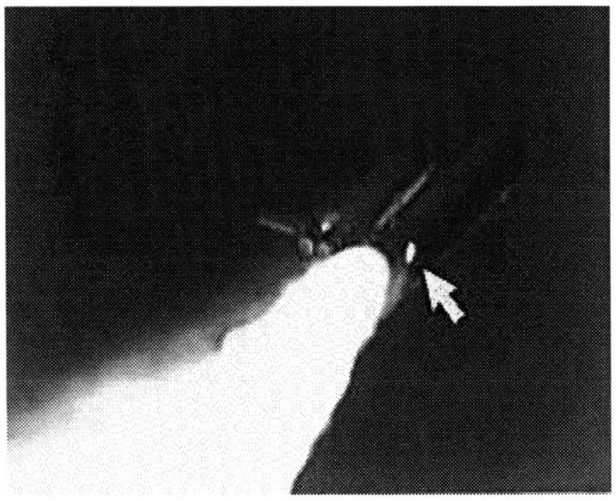

Figure 12-3
On January 28, 1986, America was shocked by the destruction of the space shuttle Challenger and the death of its seven crew members. If nuclear material were being carried on-board, the scope of the catastrophe would have been far greater.

In addition to these, there have been other proposals that are considered impractical. Among the less serious proposals are:

Taking nuclear waste into space or to the moon – The major concern in regards to this method is an explosion during the launch of the rocket containing nuclear waste. In light of the explosions of the space shuttles Challenger in 1986 (Figure 12-3) and Columbia in 2003, this fear seems ever more relevant. As space launch technology becomes more and more reliable, it may become possible to find a nearly fool-proof method to send the waste products into orbit around the sun or bury them on the moon and other planets.

Storage under polar icecaps – The concern surrounding this proposal is that a large amount of heat generation would result in melting of major icecaps with irreversible environmental consequences.

"We knew the world would not be the same. A few people laughed, a few people cried, most people were silent. I remembered the line from the Hindu scripture, the Bhagavad-Gita. Vishnu is trying to persuade the Prince that he should do his duty and to impress him takes on his multi-armed form and says, 'Now, I am become death, the destroyer of worlds.' I suppose we all felt that one way or another." ~ Robert J. Openheimer, Physicist and Director of the Manhattan Project (1904-1967).

Nuclear Weapons

Alarmed by the military advances of Hitler's army, and fearing that Germany was developing its own nuclear weapon, Einstein wrote a letter to President Roosevelt in 1939 informing him of possibility to "set up a nuclear chain reaction in a large mass of uranium, by which vast amounts of power and large quantities of new radium-like elements would be generated." [4] In response to Einstein's letter, Roosevelt ordered a team of physicists to initiate research on uranium fission and whether a bomb could be built in a short time. The task proved to be difficult as researchers soon found that uranium-238 could not sustain a chain reaction; uranium-235 was a possibility, but only if it was enriched to very high concentrations; separating the two isotopes was extremely difficult as they were chemically identical, and their masses differed by a mere 1%.

[4] Later in his life Einstein admitted that signing the letter to Mr. Roosevelt was the most tragic mistake in his life, and that he had never believed that the President would ever use the atomic weapon.

As the work continued, a second possible path to building the bomb was suggested by scientists at Lawrence Radiation Laboratory at Berkeley who produced a new man-made element, first identified as element-94 and later named plutonium, that fissioned more easily and could be produced easily in large quantities; it is formed when uranium-238 captures a neutron, becoming neptunium-239, which is unstable and rapidly decays to plutonium-239.

On December 2, 1942, researchers headed by Italian-émigré Enrico Fermi of the University of Chicago achieved the first self-sustaining chain reaction in a graphite and uranium pile. Soon thereafter, Robert Oppenheimer, a professor of physics at the University of California at Berkeley led a team of nuclear physicists in what became to be known as the "Manhattan Project" to develop the atomic bomb. The effort eventually succeeded in the first nuclear explosion test named ***Trinity*** in Alamogordo Bombing range near Los Alamos, New Mexico. The test conducted on July 16, 1945 used plutonium and had a yield of 21 kilotons of TNT (Figure 12-4).

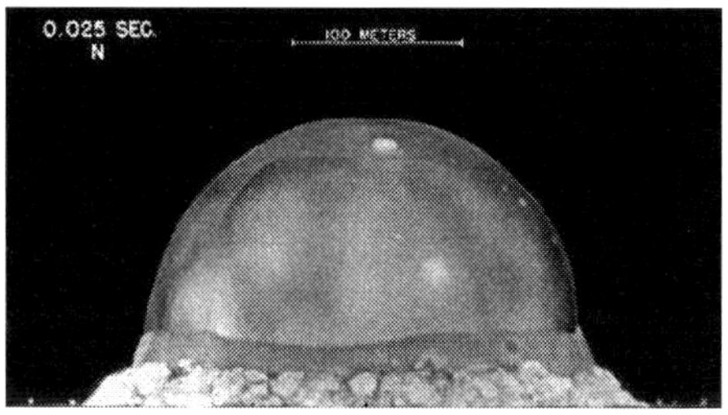

Table 12-4
The "Trinity"

Less than a month later, at 8:15 on the morning of August 6, 1945, the United States detonated the first atomic bomb on Hiroshima, Japan, which leveled the city and killed 140,000 people instantly. Germany was not a target, as it surrendered earlier in May. In addition, many tens of thousands of people died from cancer in the years that followed.[5] That bomb, nicknamed "***The Little Boy***," (allegedly named after US President Roosevelt) used uranium-235 and had the destructive power of 13,000 tons of TNT (Figure 12-5).[6] Three days later a bigger bomb, nicknamed "***The Fat Man***," (allegedly named after British Prime Minister, Winston Churchill) was dropped over Nagasaki, killing another 75,000. It was fueled by plutonium-239 and had the destructive power of 21,000 tons of TNT.[7] Shortly thereafter, Japan surrendered and WWII was officially ended.

Figure 12-5
The "Little Boy," Hiroshima Japan, August 6, 1945. The bomb was fueled by U-235 and had a yield of only 13 kilotons.

Following Trinity test, and bombing of Hiroshima and Nagasaki, the US conducted numerous other tests over Marshall Islands in the Pacific and underwater. In 1946, then US President Truman suggested establishment of an international agency that oversaw all nuclear activity. Not surprisingly, the Soviet Union, a non-nuclear power objected, arguing that all atomic weapons must be abolished before such agency would have the legitimacy. Three years later, in August 1949, the Soviets tested their first fission bomb. The Cold War had begun.

As the Cold War intensified, so did the demand for bigger and more deadlier nuclear bombs. The advent of the Soviet bomb had reduced

[5] "*The Atomic Bombings of Hiroshima and Nagasaki: The Manhattan Engineer District,* The Manhattan Engineer District," *1946,* The World Wide School, 1997. Copies can be downloaded from *http://worldwideschool.org/library/books/hst/northamerican/TheAtomicBombingsofHiroshimaandNagasaki/legalese.html.*

[6] Tri-nitro-toluene (TNT) is a chemical compound used commonly in road construction and for demolition. It is non-radioactive and its explosion does not result in the release of any nuclear products. Nuclear destructive power is often expressed in terms of kilograms or tons of TNT. One ton of TNT=4.18×10^9 J.

[7] In comparison, the destructive power of the planes that flew into and destroyed New York's Twin Towers in 2001 was equivalent to only 1 kiloton of TNT.

the US absolute superiority to only a numerical advantage. The Hungarian émigré physicist Edward Teller promoted the development of thermonuclear (hydrogen) bomb. In 1950, Truman ordered accelerated development of the bomb. On November 1, 1952, US detonated "Mike," the first H-bomb at Eniwetok in the Pacific. Three years later, the Soviet Union followed suit and detonated its first hydrogen bomb with a force equivalent to 1.6 megaton of TNT.

Uncontrolled Nuclear Fission: The A-Bomb

As we discussed before, the difference between a nuclear explosion and a nuclear reactor is in the ability to control the chain reaction in the fission process. Unlike a nuclear reactor, where the rate of neutron production must be carefully controlled, there is no such effort in a nuclear bomb.

Critical Mass

A small mass of pure fissile material, such as uranium-235 or plutonium-239, would not sustain a chain reaction. Too many neutrons leak out through the large empty volume surrounding nuclei. A large enough mass of these materials is needed to ensure that enough neutrons are generated to compensate for the loss through the void. This mass is dependent upon the size, shape, and purity of the isotopes being used.

The minimum amount of fissile material (of a given shape) required for maintaining a chain reaction is known as the **critical mass**. In an atomic bomb, a mass of fissile material greater than the critical mass must be assembled and held together for about a millionth of a second to permit the chain reaction to propagate and the bomb to explode. Plutonium has a very small critical mass, whereas uranium must be enriched substantially before it can serve as a weapons-grade material. Naturally occurring uranium has only 0.7% of the fissile uranium-235 isotope and must be enriched to a minimum 90% before it qualifies as weapons-grade. For this reason, plutonium is favored over highly enriched uranium (HEU). Building weapons with HEU is, however, easier, as no reprocessing facility to separate plutonium from spent reactor fuel is needed. The critical mass necessary to construct a bomb using Pu-239 is only the size of a baseball. Construction of a plutonium bomb is more difficult, however, as a byproduct of plutonium processing operation, Plutonium-240 is highly unstable and cause predetonation which reduces the effectiveness of the bomb.

It is far easier for countries to develop secret weapons using enriched uranium fuel; the necessary facilities can be disguised as ordinary chemical plants, and they do not produce signatures that can be readily identified. Plutonium, on the other hand, requires waste-reprocessing facilities that, unlike the uranium enrichment plants, cannot be easily hidden.

Weapon Design

To make detonation possible, the mass of the fissile material must become critical. This is accomplished by bringing two or more subcritical masses together. The simplest mechanism for assembling a supercritical mass is to shoot one piece of the material against another in a gun tube. A heavy material, called a tamper, surrounds the fissile mass. The tamper (also called neutron reflector) reduces the number of neutrons that can escape and prevents the bomb's premature explosion (Figure 12-6a). An alternative method (called implosion method) is to detonate a conventional explosive that surrounds the fissionable material. The rapid increase in pressure compresses the fissile mass, preventing neutrons from escaping through the void and reducing the critical mass necessary to initiate the fission reaction (Figure 12-6b).The implosion device is very difficult to build, as the implosion has to be highly symmetrical. The Little Boy was a gun-type; the Fat Man was of the implosive design.

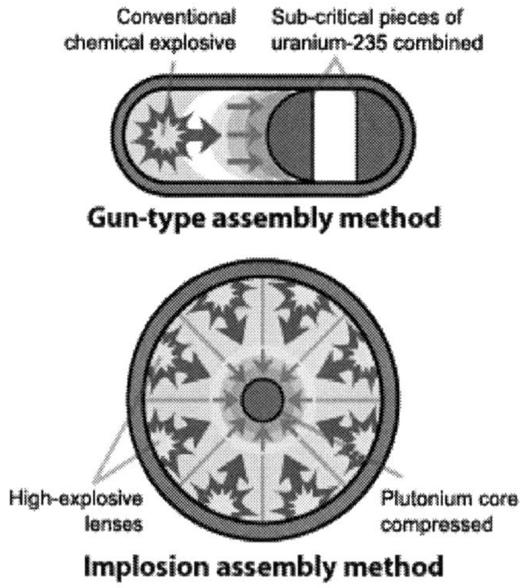

Figure 12-6
Construction of A-Bombs
a) In the gun method (top), a conventional propellant is used to shoot a subcritical mass of fissile material (uranium-235) into another subcritical mass, making the combined mass supercritical.
b) In the implosion method (bottom), a conventional explosive compresses a fissile material (plutonium-239) releasing more neutrons and starting the fission.

In addition to the A-bomb, which is a result of fission reactions, radioactive materials can be used in conventional bombs and warheads. Two examples are dirty bombs and depleted uranium warheads.

Dirty Bombs

The term *"dirty bomb"* usually refers to any device that generates a significant amount of radioactive waste without actually undergoing a fission reaction. These may include conventional weapons which, upon explosion, spread radioactive, biological, or toxic materials and may be delivered as an aerosol or simply by wrapping the materials around dynamite. These weapons do not require weapons-grade materials and even the relatively common materials used in radiological medical equipment would be enough to cause catastrophic results and an extensive loss of life.

Depleted Uranium

Depleted uranium (DU) is the byproduct of the reprocessing of spent nuclear fuel that has been enriched for use in nuclear reactors or weapons--essentially pure uranium-238. Much like natural uranium, depleted uranium is both toxic and radioactive and, because of its long half-life (4.5 billion years), it is practically indestructible. DU is 1.7 times denser than lead and, when turned into metal and used to make shells, it penetrates heavy steel and concrete with relative ease. The search for more effective weapons has led to the development of armor-piercing shells made of depleted uranium. These weapons were used for the first time in the 1991 Gulf War after the Iraqi invasion of Kuwait. Since then, they have been used extensively in other regional conflicts in areas including the Balkans (1994, 1999), Afghanistan (2001), and again in Iraq (2003). The use of this material in ammunitions has remained highly controversial (Figure 12-7).[8] When a DU round hits its target, as much as 70 percent of the

[8] Catalinotto ,J. Metal of Dishonor-Depleted Uranium : How the Pentagon Radiates Soldiers & Civilians with DU Weapons, 2nd. Ed., Independent Publishers Group, Chicago, 1995.

Figure 12-7
A reporter using a radiation detector examines an Iraqi tank destroyed by depleted uranium bullets.

projectile burns on impact and forms an aerosol of radioactive particles. The uranium dust can then be spread by the wind and may become part of the food chain. Studies by the United Nations have concluded that DU is at least partially responsible for the increase in the rate of cancer among Iraqi children and in the number of spontaneous abortions among pregnant Iraqi women.[9] DU may even be linked to the Gulf War Syndrome that is affecting US veterans of the 1991 Gulf War.[10,11]

Uncontrolled Nuclear Fusion: The H-Bomb

Hydrogen or *thermonuclear* bombs are probably the most deadly of all nuclear weapons. The large energy release per unit mass of fuel makes H-bombs very efficient. This allows for the building of bombs with much greater yields than fission bombs. The Soviet scientist Andrei Sakharov first proposed the idea in 1948, but it was the Americans who detonated the first H-bomb, code named Ivy Mike, on November 1, 1952. The bomb used liquid deuterium as fuel and had a yield of 10.4 MT, about 700 times more powerful than the A-bomb dropped on Hiroshima. A second 15-MT bomb, code named Bravo, was tested on the island of Bikini near the Marshall Islands in 1954. Since then, several hydrogen bomb tests have been conducted. The Soviet Union conducted tests in 1953 and 1961; the British detonated an H-bomb in 1957. France conducted a test in 1960, as did China in 1964. Fortunately, hydrogen bombs have never been used in a military conflict.

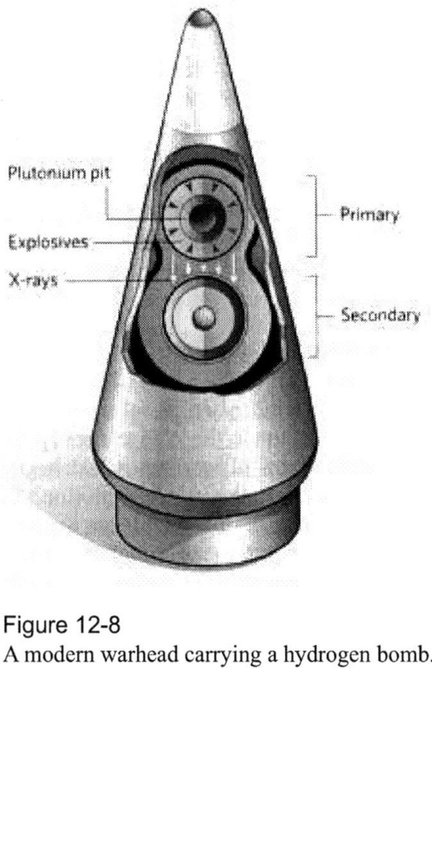

Figure 12-8
A modern warhead carrying a hydrogen bomb.

An H-bomb is essentially a three-stage weapon (Figure 12-8). In the first stage or trigger, a small amount of plutonium is detonated *(explosion)*, as in a conventional A-bomb. In the second stage, the immense energy in the form of x-ray-radiation compresses and heats up a solid mixture of lithium deuteride (a source of tritium) and deuterium[12] located in the center of the bomb to millions of degrees *(implosion)*, triggering the thermonuclear reaction. The fusion reaction produces an ample supply of neutrons so powerful they can initiate fission *(explosion)* of otherwise unfissionable casing of depleted uranium. This is the third stage and the one that produces most of the radioactive fallout.

> **Question:** Why doesn't the sun blow up like a hydrogen bomb?
> **Answer:** The main difference between the sun and the hydrogen bomb is that the sun contains very few heavy hydrogen isotopes. The proton-proton reaction that results from normal hydrogen atoms proceeds about a billion times more slowly than a strong deuterium-tritium reaction at the same temperature and pressure.

Neutron Bombs

Neutron bombs are small (less than one kiloton) thermonuclear bombs

[9] Fifty-sixth Genral Assembly, Uniteed Nation Press Release GA/SHC/3639, 2001.
[10] Peterson, S., "Remains of toxic bullets litter Iraq," *Christian Science Monitor*, May 15, 2003.
[11] Hastings, D., "Is an Armament Sickening U.S. Soldiers?," *Associated Press*, August 12, 2006.
[12] The first device used liquid helium at cryogenic temperatures.

designed to produce a high-intensity burst of radiation over a relatively small area. Instead of absorbing the neutrons inside the weapon, the neutrons are allowed to escape into the environment. Because neutrons are absorbed by the air, neutron bombs have limited yields, and so are effective over relatively small areas. Since heat and blast effects are minimal, they work primarily on biological systems and are dubbed as anti-personnel bomb. Because of this many consider these weapons as highly immoral that could be used prematurely in regional conflicts which can lead into full-scale nuclear war.

I do not know with what weapons World War III will be fought, but World War IV will be fought with sticks and stones.
~ Albert Einstein

Effects of Nuclear War

Nuclear bombs can result in a wide range of destruction in any of the following ways:[13,14]

1. **Blast:** Blast damage from a nuclear weapon comes from the overpressure in the air and from winds which result from the pressure. Right after the bomb detonates, a shock initiates at the center and moves outward at a rate several times faster than the speed of sound, and a large overpressure develops across the shock wave.[15] Depending on the yield and distance from the blast, the blast wind can reach several thousand kilometers per hour and overpressure can exceed several atmospheres. Most buildings suffer moderate to severe damage when subjected to only 35 kilopascals (0.35 atmospheres). As for the human body, the shock causes a pressure wave to travel through the body, with the most damage occurring in junctions between materials of different densities, such as the interface between bones and muscles or between tissues and air. Most eardrums rupture at around 100 kPa (one atmosphere), and lungs are damaged at around 70 kPa (0.7 atmospheres) overpressure. Roughly, 50% of the energy of the low-altitude atmospheric detonation is released in the initial shockwave.

2. **Heat:** Depending on yield, fireballs as large as one mile in diameter, with temperatures of several thousand degrees, are possible immediately after the explosion. Such fireballs practically vaporize everything near ground zero and causes extensive, rapidly spreading fires. As the weapon explodes, it produces a brilliant flash of light and a heat wave beyond the tolerance of the human retina, causing severe eye injuries, including blindness. 35% of the energy of the nuclear blast is in the form of heat, and visible radiation.

3. **Radiation:** Both short and long term effects of ionizing radiation are important. Shortly after the blast, the heat of the fireball vaporizes weapon residue into small particles that are quickly drawn up into

[13] Glasstone, S. "The Effect of Nuclear Weapons," US Atomic Energy Commission, 1962.
[14] A good sorce of information on nuclear weapon effects is the Federation of American Scientists. See http://www.fas.org/nuke/intro/nuke/effects.htm
[15] For a 10-megaton fission bomb, overpressure can reach one thousand atmospheres close to ground zero (GZ), but drops to about 15 atmospheres when it becomes a fireball about one mile in radius. The wind speed can reach 1700 mph (2700 km/hr). The overpressure becomes negligible at around 20 miles from ground zero, where wind has slowed down to about 55 mph (90 km/hr).

the stratosphere. Larger particles settle and contaminate the earth, while smaller particles are carried by the stratospheric wind across the globe, eventually showing up as radioactive fallout. If detonation occurs over water, sea salts will act as condensation nuclei to seed the clouds, causing highly radioactive rainout. In addition, the large flux of neutrons released during the first few seconds after detonation travels outward and is absorbed by the soil, air, water, and other materials, making them radioactive. In turn, these radioactive materials release gamma and beta radiation over an extended period of time.

4. **Electromagnetic pulse:** In addition to various radioactive isotopes, nuclear blasts emit a significant amount of x-rays into the atmosphere. X-rays can ionize the air at high altitudes and produce large numbers of fast-moving electrons. The moving electric charge produces a pulse so powerful that long metal objects would act as antennae and electrify all electronic devices on its path. The resulting voltage and the associated high current could destroy unshielded electronics and interfere with electric signals, resulting in failures of critical medical and transportation equipment. The ionized air also disrupts radio traffic that would normally bounce from the ionosphere. Ionizing radiation and EMP account for the remaining 15% of the detonation energy.

Nuclear Arms Treaties and International Conflicts

Following WWII and the explosion of the two powerful nuclear bombs on Hiroshima and Nagasaki, the Allies, mainly the United States and Soviet Union, engaged in a rapid buildup of nuclear weapons. So many weapons were amassed that if one country decided to use its nuclear option, the entire planet could have been destroyed many times over in a matter of hours. At the peak of the nuclear weapons stockpile in 1967, the US and USSR (Union of Soviet Socialist Republics or Soviet Union) maintained over 40,000 active warheads.[16] The rapid proliferation of the nuclear arsenal, and with it the escalation of Cold War rhetoric, required a new way of thinking and bold initiatives to save the world by preventing an accidental nuclear war.[17]

Figure 12-9
Boeing B-52 Stratofortress is a long-range heavy bomber capable of carrying multiple nuclear or precision guided conventional ordinances.

In 1970, the five great nuclear powers (the US, USSR, UK, France, and China) signed a non-proliferation treaty which mandated that those countries would "pursue negotiations in good faith on effective measures relating to the cessation of the nuclear arms race at an early date and to nuclear disarmament." In exchange, nations without nuclear weapons pledged never to acquire them and, as signatories, open their nuclear facilities to the Vienna-based International Atomic Energy Agency (IAEA), to monitor nuclear program activities and assure no nuclear weapons are being developed.

[16] Nuclear Notebook, Bulletin of Atomic Scientists, March/April 2000. [17] An excellent website on nuclear arm treaties can be found at *http://www.nautilis.org/library/security/natotreaty.html.*

[17] A good reference on the history of nuclear arms treaties can be found at *http://www.nautilis.org/library/security/natotreaty.html.*

In the years that followed, the US and the Soviet Union negotiated the **Strategic Arms Limitation Treaty (SALT-I, 1971 and SALT-II, 1979).** These treaties aimed at limiting the number of strategic offensive weapons on each side.[18] Furthermore, the **Anti -Ballistic Missile Treaty (ABM)** constrained the numbers of launchers and interceptors available to each country. The underlying assumption prevalent during this time was that if each country had enough nuclear power to survive an enemy's first strike and was still able to retaliate with overwhelming nuclear response, then neither country would undertake such madness. The policy commonly referred to as **Mutually Assured Destruction (MAD)** left no doubt that both countries would be destroyed in the event of any nuclear attack.

In 1983, US President Ronald Reagan announced his intention to initiate a program for research and development of a space-based system to defend the nation from attack by strategic ballistic missiles. The **Strategic Defense Initiative (SDI)**, popularly referred to as "Star Wars," would supposedly eliminate the threat of a nuclear arms confrontation by installing lasers and other space-based defensive systems that detected and destroyed all incoming missiles. The critics charged that this would be in violation of the ABM treaty and would encourage the militarization of space and destabilize the nuclear balance of power. Furthermore, SDI was based on untested technologies and was unable to defend against cruise missiles, airplanes, and several other possible delivery systems. Following the dissolution of the Soviet Union in 1991 and the end of the cold war, the selection of a new administration temporarily put the SDI program on hold.

The reality of the new world order and the demise of communism provided a unique opportunity not only to limit the nuclear arsenal, but also to actually reduce it. **The STrategic Arms Reduction Treaties (START-I, II, and III)** are the first comprehensive arms-control agreements that require a reduction of offensive nuclear weapons. Under these treaties, the United States, the former Soviet Union, and its successor states (Russia, Ukraine, Belarus, and Kazakhstan) had to cut down their nuclear warheads and delivery systems that included intercontinental bombers and land-based and submarine-launched ballistic missiles. Based on these treaties, the US and Russia agreed to reduce the total number of strategic nuclear warheads to about 2,000-2,500 warheads each by 2007. Although some progress has been made in reducing the size of nuclear arsenals, talks have reached impasse as NATO announced plans to build a missile defense system in Poland and Czech Republic. There are new proposals by the Obama administration to restart talks to reduce the size of nuclear stockpiles.

[18]Nuclear arsenals can be divided into either strategic or tactical categories. Strategic nuclear weapons are weapons designed to target cities and other large areas. Tactical nuclear weapons are smaller weapons used to destroy specific military targets. Depending on delivery method, weapons are classified as bombs, ballistic missiles, cruise missiles, artillery shells, or hand-held. Ballistic missiles are long-range missiles that use ballistic orbital trajectories, whereas cruise missiles use low altitude trajectories and are suitable for short range and smaller payloads. Artillery shells and hand-held devices are purely for tactical use.

Terrorism

The end of the Cold War and the resulting reduction in tensions between the Former Soviet Union and the United States brought about a new spirit of cooperation. This resulted in a number of treaties which have reduced the number of nuclear warheads each country possesses, and with them, the threat of nuclear war. The end of the Cold War also raised a new threat: terrorism. As nuclear warheads are dismantled, an abundant amount of weapons-grade nuclear fuel and accessories which must be disposed of accumulate. In addition, thousands of tons of uranium have been refined and converted to plutonium or enriched to weapons-grade material. Less than 8 kg of plutonium is all that is needed to make a Nagasaki-sized bomb. The possibility exists that terrorist groups with no access to reprocessing plants could steal or buy enriched uranium and plutonium fuels on the black market and disperse them throughout cities and other population centers or use them in making crude and dirty bombs. Equally important is the danger of attack on or accidental spillage of nuclear waste en route to a proposed permanent repository site. Nuclear reactors themselves could also be the subject of sabotage and therefore must be designed with many layers of safety in the event of a loss of coolants, damage to the containment structure, or the possibly of the reactor control rooms being taken over by terrorists and their internal collaborators. The key defense against large radioactive leaks to the outside is to keep the containment intact. The containment dome is shaped such that it can withstand internal pressure resulting from the buildup of gas pressures. Unfortunately, existing containments may not withstand external overpressures that may result from major rocket attacks or car bombs.[19] In light of the September 11 terrorist attacks on New York and Washington D.C.,[20] nuclear reactors may be particularly vulnerable.

Summary

Nuclear radiation has found many uses and, depending on its application, can be either the deadliest enemy known to mankind or a source of hope in diagnosing and treating many formerly incurable diseases. If operated safely, nuclear reactors generate electricity with practically none of the adverse environmental consequences of fossil power plants. The waste, however, could have tremendous environmental impact for many thousands of years. The obvious solution is not to keep radioactive waste for such long times, but to find solutions to rid us of these wastes within the next few decades. In the meantime, natural radioactive decay has cooled the waste, making it easier to handle.

[19] For an excellent overview of nuclear terrorism risks see: Leventhal, P. L., and M. M. Hoenig, "Nuclear Terrorism: Reactor Sabotage and Weapons Proliferation Risks," Contemporary Policy Issues, July 1990, V. 8, No. 3, pp.106-121.

[20] Early in the morning of September 11, 2001, Al Qaeda terrorists hijacked four commercial jets and crashed three of them into the World Trade Center in New York and the Pentagon in Washington D.C. The fourth aircraft was heading toward the White House, but crashed in Pennsylvania. Over 3,000 innocent lives were lost, 265 aboard the airplanes and the rest in the buildings that collapsed as a result of the tragic incident.

Additional Information

Books

1. Petty, G. W., A First Course in Atmospheric Radiation, Sundog Publishing, 2006.
2. Wagner, H., and Ketchum, L., Living with radiation: the risk, the promise, The Johns Hopkins University Press, 1989.
3. Glasstone, S. The Effect of Nuclear Weapons, US Atomic Energy Commission, 1962.

Periodicals

1. *Journal of Radiological Protection*; the official journal of the Society for Radiological Protection and publishes articles on all aspects of radiological protection, including non-ionizing and ionizing radiation.
2. *Nuclear Instruments and Methods;* Elsevier Science Direct. papers on design, manufacturing and performance of scientific instruments with an emphasis on large scale facilities.

Government Agencies and Websites

1. US Environmental Protection Agency (*http://www.epa.gov*) is responsible for issuing standards and guidance to limit human exposure to radiation.
2. US Nuclear Regulatory Commission (*http://www.nrc.gov*) is responsible for implementing the EPA's exposure standards through regulation of nuclear power reactors. It also oversees licensing of facilities that posses or dispose nuclear materials.
3. US Department of Energy (*http://www.doe.gov*) is responsible for disposal of nuclear waste from civilian power plants.
4. US Department of Defense (*http://www.dod.gov*) is responsible for the safe handling and storage of nuclear weapons.
5. US Department of Transportation (*http://www.dot.gov*) is responsible for the packaging and transport of radioactive materials.
6. US Food and Drug Administration (*http://www.fda.gov*) regulates medical procedures that use radiation.

Non-Government Organizations and Websites

1. International Atomic Energy Commission (*http://www.iaea.org*) is the United Nation's agency responsible for promoting peaceful applications of nuclear science and technology for critical needs in developing countries.
2. Nuclear Energy Institute (*http://www.ne.doe.gov*).
3. World Nuclear Association (*http://www.world-nuclear.org*).

Exercises

I. Essay Questions

1. What is the difference between isotopes of U-235 and U-238? Which one is more stable? Which one has a longer half-life?

2. Which types of electromagnetic radiation are ionizing? Which kinds are non-ionizing?

3. Name three sources of natural background radiation.

4. Name three sources of man-made radiation.

5. What are the risks involved in the decommissioning of a nuclear plant?

6. What is depleted uranium, what is it used for, and why is it dangerous?

7. What are the differences between an A-bomb and an H-bomb?

8. What are some of the hazards radiation poses to human health? How can radiation help in improving human health?

9. How does a nuclear bomb work? Draw a schematic of a nuclear bomb.

10. What is a dirty bomb?

11. What are similarities between fission and fusion bombs? How are they different?

12. What is transmutation? What are its applications?

II. Multiple Choice Questions:

1. Which of the following is considered to be ionizing radiation?
   a. Microwaves
   b. RF
   c. Visible
   d. Infrared
   e. High frequency UV

2. What percentage of human radiation exposure comes from natural sources?
   a. More than 99%
   b. About 80%
   c. About half
   d. Less than 10%
   e. Less than 1%

3. The amount of radiation damage to a living tissue is expressed in
   a. Roentgen
   b. Rad
   c. Curie
   d. Rem
   e. Becquerel

4. Which of the following units is usually used to express the strength of an x-ray source?
   a. Roentgen
   b. Rad
   c. Rem
   d. Curie
   e. Becquerel

5. Radiation emitted by a microwave oven is
   a. Ionizing
   b. Extremely dangerous
   c. Visible
   d. Radio waves of very high frequency
   e. Absorbed by nonmetallic substances like glass, ceramics, and paper products

6. What is (are) the mechanism(s) responsible for major destruction and loss of life following a nuclear attack?
   a. Shock
   b. Heat of the fireball
   c. Short term immediate radiation
   d. Long term, low-level radiation
   e. All of the above

7. The maximum permissible occupational dose for nuclear industry workers in the US is ________ per year.
   a. 5000 mrem
   b. 1250 mrem
   c. 500 mrem

d. 500 rads
e. 500 curies

8. The maximum permissible occupational dose for the general public in the US is _________ per year.
   a. 500 mrem
   b. 125 mrem
   c. 0.5 rads
   d. 500 rads
   e. 500 curies

9. Alpha particles are
   a. The same as nuclei of helium
   b. The same as electrons
   c. Electromagnetic radiation with no electric charge and no mass
   d. The same as protons
   e. None of the above

10. Gamma radiation is
    a. High energy electrons
    b. High energy neutrons
    c. High energy protons
    d. High energy helium nuclei
    e. High energy electromagnetic wave

11. The beta particle is
    a. The same as proton
    b. The same as electron
    c. The same as positron
    d. The same as neutron
    e. High-intensity electromagnetic rays

12. Lab coats and gloves provide shielding from
    a. Alpha radiation
    b. Alpha and beta radiation
    c. Alpha, beta, and gamma radiation
    d. Neutrons
    e. None of the above

13. Which of the following have the highest penetrating power?
    a. Electrons
    b. Neutrons
    c. Alpha particles
    d. Beta particles
    e. Gamma rays

14. The agency responsible for setting the standards for safe radioactive exposure in the United States is
    a. The Nuclear Regulatory Commission
    b. The Department of Energy
    c. The Environmental Protection Agency
    d. The International Atomic Energy Agency
    e. The White House

15. The agency responsible for enforcing the safe operation of US nuclear reactors is
    a. The Nuclear Regulatory Commission
    b. The Department of Energy
    c. The Environmental Protection Agency
    d. The International Atomic Energy Agency
    e. The White House

16. The reason that alchemists could not transform mercury and other metals into gold was that
    a. They had no knowledge of the structure of atoms
    b. They did not use the right chemicals
    c. They did not have the right catalysts for making the necessary reactions possible
    d. The cost was extremely high
    e. All of the above

17. Consider a substance with a half-life of 30 years; what fraction remains after 90 years?
    a. 1/16
    b. 1/8
    c. 1/4
    d. 1/3
    e. 3/4

18. A nuclear physicist measures the nuclei disintegration of a substance. His detector counts $8x10^6$ integrations in the first hour, but drops to $1x10^6$ in the second hour. What is the half-life of the radioactive source?
    a. 8 hours
    b. 30 minutes
    c. 20 minutes
    d. 7.5 minutes
    e. Cannot determine

19. The best solution for disposing of nuclear waste is
    a. Sell it to other countries
    b. Dispose it in the landfills
    c. Reuse them as nuclear fuel
    d. Store and monitor the waste
    e. All of the above

20. What is currently being done with US nuclear waste?
    a. It is reused as nuclear fuel.
    b. It is enriched to make more bombs.
    c. It is disposed in nuclear landfills.
    d. It is sold.
    e. It is held in temporary storage facilities.

21. What was the name of the project to develop nuclear weapons?
    a. The Little Boy
    b. The Fat Man
    c. Manhattan
    d. Trinity
    e. Mike

22. What was the name of the bomb dropped on Nagasaki?
    a. The Little Boy
    b. The Fat Man
    c. Manhattan
    d. The Peacemaker
    e. The Last Resort

23. The fuel used in the nuclear bomb dropped on Hiroshima was
    a. TNT
    b. Hydrogen
    c. Uranium
    d. Plutonium
    e. None of the above

24. Who was the US president who authorized the dropping of the nuclear bomb on Japan?
    a. Harry Truman
    b. Franklin Roosevelt
    c. Dwight Eisenhower
    d. John F. Kennedy
    e. Winston Churchill

25. Who was the scientist in charge of design and fabrication of the atomic bomb dropped on Japan?
    a. Albert Einstein
    b. Enrico Fermi
    c. Robert J. Oppenheimer
    d. Glean Seaborg
    e. Edward Teller

26. In what form is most energy from a nuclear blast released into the atmosphere?
    a. Shock
    b. Heat
    c. Prompt radiation
    d. Radioactive fallout
    e. All of the above

27. Plutonium is of concern in proliferation of nuclear weapons because
    a. It is cheaper to construct a bomb using plutonium than uranium
    b. Plutonium bombs are much more destructive than uranium bombs
    c. Plutonium has a much smaller critical mass than uranium
    d. Plutonium is much more abundant than uranium
    e. Enriching plutonium is much easier than enriching uranium

28. An electromagnetic pulse generated by nuclear explosions is most effective in
    a. Killing people
    b. Destroying other nuclear weapons
    c. Damaging building
    d. Harming the environment
    e. Damaging electrical equipments and communication devices

29. START nonproliferation treaty stands for
    a. START arms reduction now.
    b. STrategic Arm Reduction Treaty
    c. STop Arm Restructuring Treaty
    d. STop ARming the Terrorists
    e. None of the above

30. Background radiation levels
    a. Are constant world-wide

b. Decrease with elevation
c. Are nearly zero except when in proximity of nuclear reactors
d. Depend on altitude and type of rock or soil in the area
e. Are highest in snow and ice

III. True or False

1. Infrared and visible radiations are the primary sources of ionizing radiation.

2. Ionization radiation refers to those electromagnetic waves with enough energy to eject electrons from atoms.

3. The range of electromagnetic radiation with wavelengths slightly shorter than those of visible light is called infrared.

4. Rad is an abbreviation for radiation.

5. Alpha particles are much potent source of ionizing radiation than x-rays and gamma-rays.

6. The human body is naturally radioactive.

7. The three types of radiation in order of decreasing penetrating power are alpha, beta, and gamma.

8. An alpha particle has the atomic mass of four.

9. Relative to the surrounding areas, coal mines have elevated level of radiation.

10. All nuclear materials remain highly toxic for thousands of years

11. The main factors affecting the choice for a proper, permanent nuclear waste repository are soil stability, seismic activity, and proximity to large water reservoirs.

12. Up to now, most nuclear waste has been disposed of by nuclear transmutation.

13. Because of their enormous destructive powers, atomic weapons are commonly referred to as dirty bombs.

14. Depleted uranium is uranium depleted of all its harmful radiation.

15. The best way to get rid of nuclear waste is to use it as nuclear fuel.

16. Strategic nuclear weapons refer to weapons designed to target cities and other large installations.

17. There are no major differences between a reactor meltdown and a nuclear bomb explosion.

18. The main consequence of electromagnetic pulse emission from nuclear blasts is the rapid increase in the number of brain cancers.

19. X-rays are produced when a target is bombarded with an electron beam.

20. Unlike tactical nuclear weapons, strategic weapons are designed to inflict maximum casualties.

IV. Fill-in the Blanks

1. Alpha, beta, and gamma radiation can physically dislodge a/an ____________ from atoms in their path.

2. ______________ are the least penetrating of all ionizing radiations.

3. The most penetrating type of radiation is _____________.

4. The radiation emitted from a source is measured in ___________ or ____________.

5. Any radiation dose less than _______ millirem is considered an acceptable level for general public.

6. A radiation exposure of over __________ rads will certainly cause death.

7. A __________________ is used to detect radiation.

8. Electromagnetic radiation in the frequency range just higher than x-rays is called ______________.

9. Uncontrolled nuclear fusion is the basis for manufacturing the ________________bomb.

10. Neutron bombs are small ____________ bombs designed to produce a locally intense bust of radiation.

IV. Project I - Radiation Dose

To estimate the annual radiation dose your body receives, fill out the radiation dosimetry questionnaire in page 282 and answer the following questions:

1. What is the annual radiation dose you receive?
2. Do you receive considerably more radiation during any particular season? If so, why?
3. Do you meet the radiation exposure limit set by the regulatory agencies in your country of residence? Explain.
4. Is there any reasonable way to reduce the level of your radiation exposure?

Project II - Energy Alternatives (Group project)

There is a ballot on whether a nuclear, a fossil, or a solar power plant is to be built near the city you are living in. You are asked to assume one of these roles and act either as a proponent (pro) or opponent (con) in the following list and address the issues from that point of view.

Pro as the	Con as the
parent	child
oil company person	conservationist
electric utility	stockholder
scientist	economist
economist	scientist
business person	consumer
consumer	business person
regulatory agency	electric utility
electric utility	consumer

Electricity

CHAPTER 13

If it weren't for electricity we'd all be watching television by candlelight

~ George Gobal

Electricity is another elusive word meaning different things to different people. Power companies use electricity to mean electrical energy; many trade books talk of electricity as the flow of electrons, while others talk of electricity as being synonymous with electric charge. Whatever electricity is, it is closely associated with the imbalance between positive and negative charges within matter.

Unlike other forms of energy, electricity is not a source of energy, but rather a convenient and cost-effective way to carry energy between two points. For example, we can use the chemical energy of coal to operate a power plant. Electrical transmission wires carry the electricity that is produced by the plant to our homes, where it is used to provide thermal energy through an electric heater, give off light through an electric light bulb, produce power to run an elevator or operate a TV, or emit sound energy through a speaker. Along the way, a small portion of the energy is lost to the atmosphere as waste heat. The versatility of electricity, exhibited by its ability to carry out so many tasks, is one reason it is considered to be a very high quality form of energy.

Worldwide, 36% of all the energy consumed is used to produce 14 trillion kilowatt-hours of electricity. In the United States, half of all energy consumed is in the form of electrical energy, for a total installed capacity of 900 gigawatts or about one quarter of the world's total.[1]

In this chapter, we will become familiar with the laws governing the flow, production, storage, and transmission of electricity, as well as issues of electrical safety.

a Electricity and Magnetism

Electricity

The word "electricity" is derived from the Greek word *electron*, meaning amber. The Greeks knew that a piece of amber rubbed against a cat's fur would collect charges. Furthermore, they observed that identical objects touching the amber repelled each other. The same thing happened if the amber and fur were replaced with another pair of materials, glass and silk. However, if glass was brought next to amber, or if fur was brought next to silk cloth, the pairs would attract.

[1] Energy Information Agency, EIA, US Department of Energy (2004 data).

Benjamin Franklin (1706-1790) suggested that an imbalance results when objects having an excess of an invisible fluid (amber, glass) were brought next to objects deficient in this fluid (fur, silk); in an attempt to reach equilibrium, this invisible fluid would move from one object into the other and the two would attract. We had to wait another century until it was discovered that this fluid was, in fact, a stream of electrons that flows from one object to another when they rub. However, Franklin erroneously presumed a lack of the invisible fluid rather than a surplus of electrons; opposite to his assertion, electrons transferred from fur and silk to amber and glass.[2]

Today, we understand the phenomenon by noting the existence of a charge imbalance between different atoms. As we discussed in previous chapters, atoms consist of negatively charged electrons orbiting a nucleus composed of positively charged protons and neutrons with no charge. For an atom to be neutral, the total number of protons and electrons is equal, resulting in a net zero charge. We also know that, just like with energy, electric charges are conserved. Electric charge can transfer from one material to another, but no new charge is created or destroyed.

Electric charges may remain stationary or flow through an electrical conductor. The study of charges at rest is called *electrostatics*. Electrostatic charges may be caused by friction (shuffling one's foot across a carpeted floor), by contact (touching a metal door-knob), or by induction (lightning). Electrostatic charges are not of concern in this study and will not be discussed further. When motion of charges is involved, electric currents are produced that flows through circuits, powering devices ranging from simple battery-operated toys and home appliances to state of the art electronics.

a Electrical Circuits

To have a current between two points, we need a difference in electrical potential. This electrical potential can be supplied either by a battery (between the two electrodes), or by an electric utility company (between the two slots of a house's electrical outlet). As long as there is no path between the two points (open circuit), or there exists a path that is non-conductive (such as connecting two point charges with a glass rod), no charge is transferred and no current flows. When the path is established by flipping a switch or by inserting a load, such as an electric motor or a light bulb, electricity flows. Unless a supply of charges from an external source (a battery or electric outlet) maintains the electric potential, the current eventually stops. In a battery, electric potential is maintained as it discharges through a chemical reaction. Household electrical outlets maintain electric potential by electrical generators driven by wind, water,

[2] When Benjamin Franklin made his conjecture regarding the invisible fluid, he assumed that fluid flowed from glass to silk. In reality, silk has an excess of electrons and glass is deficient in electrons. If we had designated electrons as having positive charges, the flow of current would have been from higher to lower concentration of electrons, in line with the assertion that excess fluid moved from one object to the other.

Electrical and Gravitational Forces

Digging Deeper ...

There is a striking resemblance between the way electrical and gravitational forces function. Gravitational force follows Newton's law of universal gravitation and is proportional to the product of the two masses and inversely proportional to the distance between them. Electrical force follows Coulomb's law and increases with the product of two charges and is inversely proportional to the distance between the two. The same way gravity is responsible for a satellite orbiting the earth, electrical force is responsible for electrons orbiting nuclei. Unlike gravitational forces that are always attractive, electrical forces can be either attractive or repulsive.

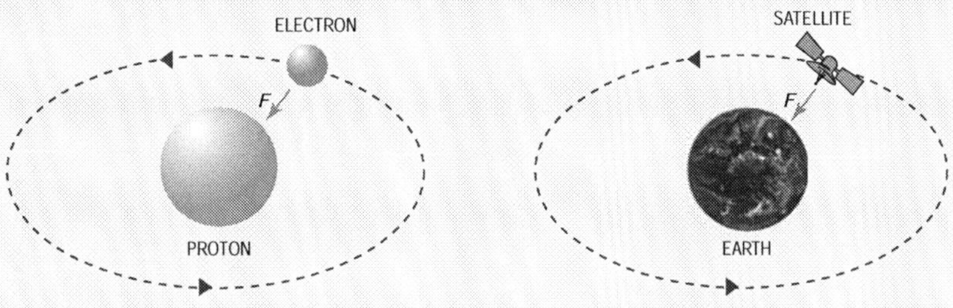

The electric force holds electrons in orbit.

The gravitational force holds satellites in orbit.

gas, or steam turbines.

The amount of current that can flow through a conductor (wire) depends, not only on the electric potential across the wire, but also on its resistance along the path. This is given by **Ohm's Law**, which is expressed as:

$$V = I \cdot R \qquad (13\text{-}1)$$

where V is the voltage, I is the current, and R is the resistance.

The unit of electric current is amperes (A), defined as the quantity of charge in coulombs (C) flowing in one second. One coulomb is equal to the amount of charge possessed by 6.2×10^{18} electrons. In other words, one ampere of current represents the flow of 6.2×10^{18} electrons per second.

$$1\ \text{A} = 1\ \text{C/s}$$

Similar to gravitational potential energy, which is a measure of the work needed to raise a weight above the ground, the volt (V) is potential electric energy and is equal to the work needed to raise the level of one coulomb of charge from its ground level.

$$1\ \text{V} = 1\ \text{J/C}$$

The resistance is measured in ohms (Ω) and is the attribute of a conductor that controls the amount of current that can flow.

$$1\ \Omega = 1\ \text{V/A}$$

More conductive wire material, bigger cross sectional area, and shorter length mean less resistance to electric flow and greater capacity to carry a current.

A current can either be uniform (direct current or DC) or varying in time (alternating current or AC). **Direct current** is the direct flow of electrons through a conductor. **Alternating current** results from the back and forth movement of electrons in a wire. The electrons themselves move very little, but their periodic motions result in the transfer of energy to adjacent electrons. This is similar to a longitudinal wave along a spring coil. Although atoms of the spring move only a little, a wave is propagated downstream, carrying the energy from one end of the spring to the other end. The electrical potential of an alternating current is created by moving a coil of wire in a magnetic field.

Usually the voltage provided by an electric utility company cannot be easily modified. Domestic electric service delivered by utility companies to US homes is 240 volt/60 hertz.[3] This voltage is required for powering electric cooking ranges, clothes dryers, air conditioners, and other large appliances. The 240 V service is split at the power panel into the familiar 120 V circuits used for most household appliances which are plugged into wall outlets. Europe and Asia use 220 volt/50 hertz and appliances that are designed accordingly.

Electrical Power

The electrical energy supplied by a generator or battery is either converted to work (as in an electric motor) or is dissipated to heat (as in an electric heater or an iron). The rate at which the energy is converted is called power, and is given by multiplying the rate at which electron flows (current) by the energy that is carried by each electron (voltage):

$$P = V.I \qquad (13\text{-}2)$$

Substituting for voltage from equation 13-1 we get:

$$P = R\,I^2 = \frac{V^2}{R} \qquad (13\text{-}3)$$

These equations show that energy dissipated as heat by an electrical conductor increases as the square of the current.[4] In a resistance heater, the goal is to maximize the amount of heat that can be dissipated, so we would prefer to draw high current. Since voltage is constant, equation 13-3 suggests we need to minimize resistance. The easiest way to control current in a particular circuit is to insert special components, called resistors, in

[3] Hertz is the unit of frequency (cycles/s).

[4] Equations 13-1 to 13-3 above are for direct currents. The same equations work with alternating currents, except current and voltage values must be substituted with "root-mean-squared" values given by:

$$V_{rms} = \frac{V_0}{\sqrt{2}} \text{ and } I_{rms} = \frac{I_0}{\sqrt{2}}$$

In this equation, V_0 is the peak voltage, and I_0 is the peak current. Ohm's law and average power dissipated through AC circuits are rewritten as:

$$V_{rms} = I_{rms} \cdot R$$

$$P = V_{rms} \cdot I_{rms} = \frac{V^2_{rms}}{R}$$

various series or parallel configurations. Resistors are made of materials with low electrical resistivity but which can withstand high temperatures and are commercially available for a large range of resistances – from a fraction of an ohm to many megaohms.[5]

If electric potential and current are given in volts and amperes, power will be in watts. One watt equals the energy used, in joules, in one second.

$$1\text{ W} = 1\text{ J/s}$$

One kilowatt is 1,000 W, and 1 megawatt is 1,000 kW or 10^6 W.

When discussing electricity consumption and production, it is customary to express power in kilowatts and time in hours. Since energy is the product of power and time, electrical energy is often expressed in kilowatt-hours.[6] Table 13-1 shows power ratings for several household appliances.

Table 13-1. Power Consumption for Several Household Appliances

Appliance	Power, W	Appliance	Power, W
Clock	2	Air Conditioner	800 - 1000
Radio	70 - 400	Coffee Maker	900 - 1200
Television (color)	65 - 200	Iron	1200 - 2400
Freezer/Refrigerator	600 - 1000	Dishwasher	850 - 1400
Hair Dryer	1200 - 1800	Toaster	800 - 1400
Washer (clothes)	350 - 500	Microwave Oven	750 - 1100
Vacuum Cleaner	1000 - 1400	Dryers (clothes)	1800 - 5000
Home Computer	100 - 400	Oven range	10000 - 12000

Source: US DoE Office of Energy Efficiency and Renewable Energy, http://www.eere.energy.gov.

Example 13-1: Calculate the power dissipated by the lamp in the circuits shown on Ex 13-1.
Solution: Case (a): $I = V/R = 18/3 = 6$ A; $P = V \times I = 18 \times 6 = 108$ W
Case (b): $I = V/R = 36/3 = 12$ A; $P = V.I = 36 \times 12 = 432$ W
Note that increasing the battery voltage by a factor of 2, from 18 V to 36 V, increased the current by a factor of 2 and power by a factor of $2 \times 2 = 4$.

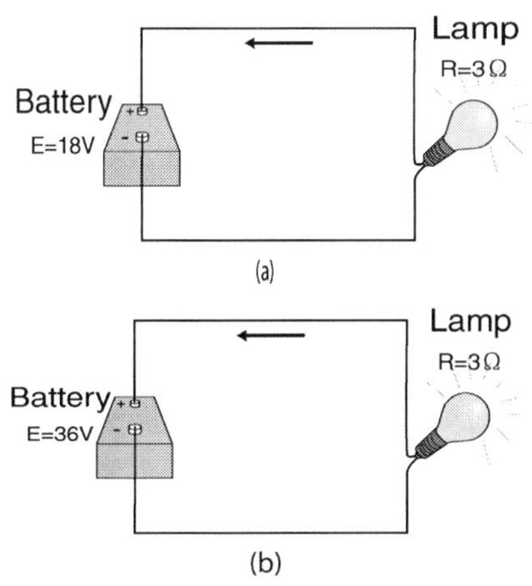

EX 13-1

Example 13-2: A 100-W light bulb operates in an American household for 4 hours every night.

a. How much current is drawn by the light bulb?
b. What is the resistance in light bulb filament?
c. How much power is consumed by the light-bulb?
d. What is the monthly cost of operating this light bulb if the electricity is charged at a rate of \$0.10/kWh?

Solution: American electrical power is delivered at 120 volts.

[5] Note that resistivity is not the same as resistance. Resistivity is a property of material, whereas resistance represents the impedance to the flow of electricity and depends not only on the resistivity of the material, but also on size and geometry.
[6] 1 kWh = 1,000 W x 3,600 s = 3.6×10^6 J.

a. The current drawn is: $I = P/V = 100\ W/120\ V = 0.833\ A$
b. The resistance is calculated as: $R = V/I = 120\ V/0.833\ A = 144\Omega$
c. The power is calculated as: $P = VI = 120\ Vx0.833\ A = 100\ W$
d. Monthly energy consumption by the light bulb is:
$E = (0.1\ kW)x(4\ hr/day)x(30\ days) = 12\ kWh$. The monthly electrical cost is 12x0.10 = $1.20

Question: In the example above, which draws more current: the wire leading to the light bulb, or the bulb filament? Which gets hotter?
Answer: The flow of electrons is exactly the same in all parts of the circuit. As the electrons try to overcome the greater resistance in the thin filament, they heat it. The wires connecting to the outlet are significantly thicker than that of the filament and therefore barely warm as they allow the same amount of current to pass through.

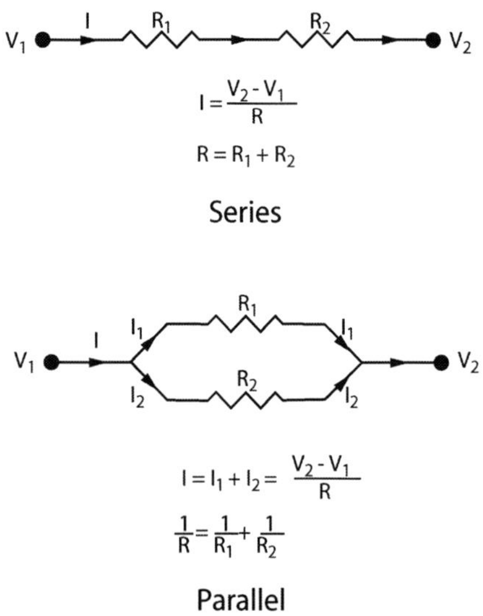

Figure 13-1
Series (top) and parallel (bottom) resistances.

Series and parallel Circuits

Different electrical devices can be assembled in a circuit in series, in parallel, or in a combination of the two. In a *series configuration*, the output current of one device will be the input current of a second device, and so on; the same current passes through all devices (Figure 13-1a). Since the total load increases for the same voltage potential, according to Ohm's law, the current must decrease. In a *series circuit*, resistances add up:

$$R_{total} = R_1 + R_2 + \ldots R_n \qquad (13\text{-}6a)$$

Since the current passes through both resistances, the voltage across the two is the sum of the voltages across the individual resistances.

$$V = V_1 + V_2 + \ldots V_n \qquad (13\text{-}6b)$$

The main disadvantage of a series configuration is that if one of the devices burns out we have an "open circuit," in which current stops and none of the devices in the series circuit will work.

a Open and Short Circuits

Digging Deeper ...

Neither voltage nor current can provide power by itself. No current flows in an open circuit, although there is a finite voltage across the source terminals. Consequently, power is zero. Likewise, a circuit shorted by connecting terminals with a superconducting wire (zero resistance) draws a very large current even though there is no voltage difference, and again, no power is produced.

Question: In a flashlight, a battery provides power to turn on a light bulb. What happens if the light suddenly burns out? What happens if two ends of the battery are accidentally connected with a piece of conducting material?
Answer: In the first instance, the circuit will be broken (open) and the flow of electrons stops. Short-circuiting the circuit by directly connecting the poles of the battery with a piece of wire allows a sudden release of energy with no substantial resistance to slow down the flow. This is actually very dangerous, as the high current can result in an electric shock. The rapid discharge generates heat, which can cause the battery to explode.

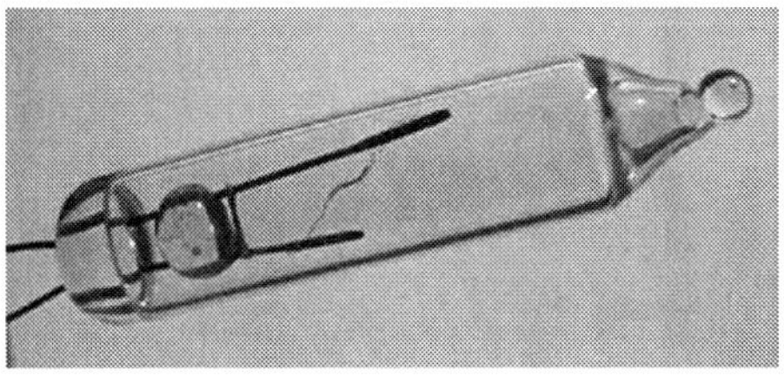

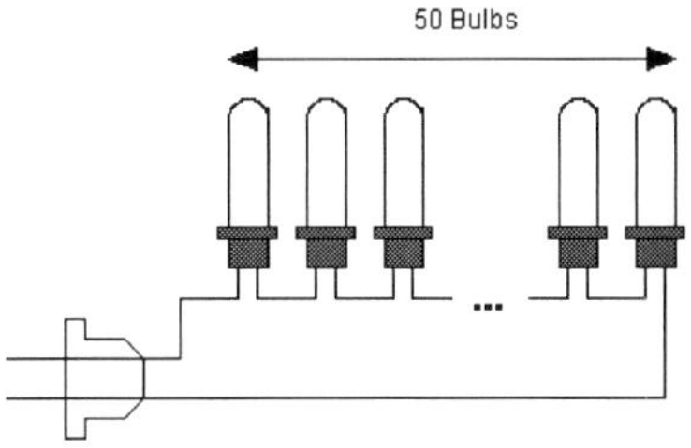

EX 13-3

Example 13-3: Typically, Christmas tree decoration lights are arranged in a series configuration. Assume a strand of 50 mini-incandescent bulbs is connected to a 120 V outlet; calculate the current drawn by the strand and the total power consumed. A typical bulb has a resistance of 8 ohms.
Solution: The voltage across each bulb is 1/50th the voltage drop across the strand, or 120/50 = 2.4 volts. The current drawn is I = V/R = 2.4/8 = 0.3 A. The power consumed is P = 50.VI = 50x2.4x0.3 = 36 W. As explained above, when the bulbs are put in series, if one bulb burns out the entire strand will go out. Today's Christmas lights, however, will stay on even if one or more lights are burned out. The trick is that new bulbs have an internal shunt resistance connecting two poles. The shunt is coated with a high-resistance material that prevents shorting the filament. When a light burns out, the heat causes the coating to melt thus reducing the shunt resistance and allows the current to pass through the rest of the strand so the remaining lights will stay on.

In a *parallel configuration*, the same voltage potential is applied to all devices. The current necessarily divides. Since the same voltage is supplied to all electrical devices, each device works independently of others and a burned out device does not affect the operation of other devices. In a *parallel circuit*, currents add up:

$$I_{total} = I_1 + I_2 + \dots I_n \qquad (13\text{-}7a)$$

In other words, parallel resistances act to divide the current. It is left as an exercise to show that several resistances placed in parallel can be replaced by an equivalent resistance equal to:

$$\frac{1}{R_{total}} = \frac{1}{R_1} + \frac{1}{R_2} + \dots \frac{1}{R_n} \qquad (13\text{-}7b)$$

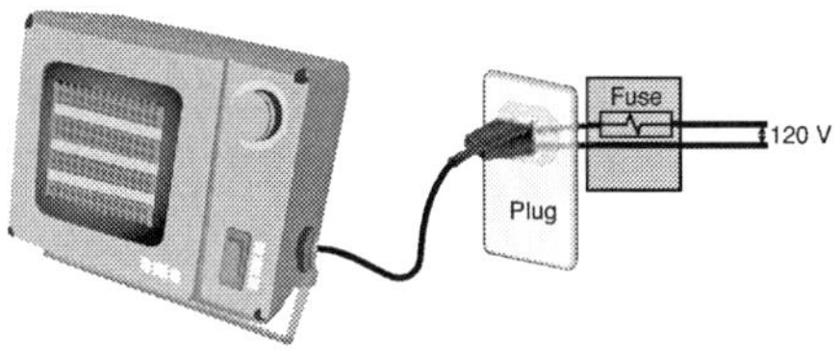

EX13-4

Example 13-4: A 1,200-W electrical heater, ten 100-W electric light bulbs, a 300-W refrigerator, and a 1,500-W microwave oven operate simultaneously. Assuming that the household circuit is 120 V and all appliances are placed in parallel, what is the total current drawn?
Solution: Applying Ohm's law (I = P/V) for each device and using power ratings given in Table 13-1, we can write:

Electric Heater:	1200/120 = 10.0 A
Ten 100-W light bulbs:	10x100/120 = 8.33 A
Refrigerator:	300/120 = 2.50 A
Microwave Oven:	1500/120 = 12.50 A
Total current drawn:	33.33 A

This is a large current and may be a fire hazard. Houses are usually equipped with 10-20 A circuit protection. If current exceeds these values, a fuse will burn out or a circuit breaker will trip, disconnecting the circuit. In house wiring, several circuits are used, each equipped with a separate circuit breaker. In the example above, at least two circuits each capable of carrying 20 A maximum are needed if all appliances are to be used simultaneously. Indeed, US electric codes require a microwave, garbage disposal, and refrigerator to have a 20 A circuit separate from lighting circuits and wall outlets for other appliances.

Example 13-5: A three-way light bulb can be constructed by putting two resistances in parallel. Whether the first, second, or both resistances are placed in circuit, the light can be dim, bright, or very bright. What are the values for the two resistances if the light bulb can be operated to dissipate 50/100/150 W of electricity?

Hydraulic and Electric Analogy

FYI ...

The flow of the electrons in an electronic circuit is analogous to the flow of water in a hydraulic circuit. Consider a hydraulic circuit in which water is mechanically pumped from a reservoir at a lower elevation to another at a higher elevation. The water is then used to drive a waterwheel, where it releases its energy in the form of shaft work. This is analogous to an electrical circuit, in which electrons are pumped up by a battery to a higher electric potential, flow through the wire with negligible resistance, and pass through a load to a lower potential when the circuit is closed. The drop in electric potential across the load is used to run an electric motor or to heat the filament of a light bulb to emit light.

The greater the difference in elevation (gravitational potential) between the two ends of a pipe and the less resistance to the flow of water through the pipe, the more water can flow in a given time. Similarly, in a conductor (wire), the greater the electric potential between the two ends of a wire and the less the resistance to the flow of electricity through the wire, the larger the amount of charge that can be carried in a given time. In a pipe, we can reduce resistance by making the tubes shorter, smoother, and of larger diameter. In a wire, we can reduce resistance by making the wire shorter, making it from material with a lower electrical resistivity, and by using a thicker wire.

Open electrical circuits can be viewed as analogous to closing the valve leading to a waterwheel; whereas shorting the battery would be analogous to removing the waterwheel and allowing all the potential energy in the water reservoir to be suddenly converted to the kinetic energy of a waterfall.

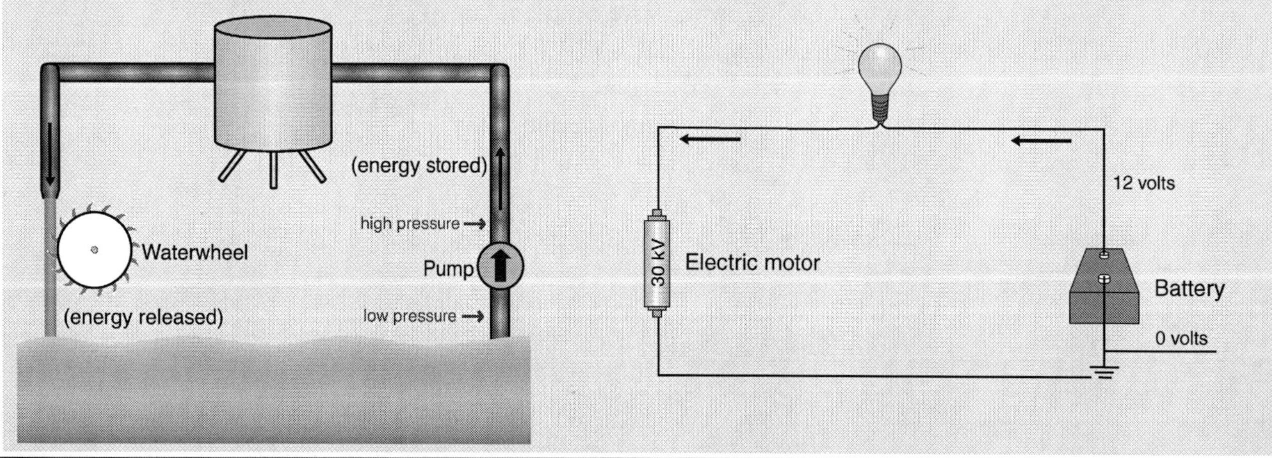

Solution: The value of resistances and currents are calculated as:

$$\frac{1}{R} = \frac{1}{R_1} + \frac{1}{R_2} = 0.0124$$

$$R = 80.67\ W$$

$$P = \frac{V^2}{R} = \frac{110^2}{80.67} = 150\ W$$

The total current is: $I = I_1 + I_2 = 0.45 + 0.91 = 1.36\ A$

At the brightest position, we have

$$R_1 = \frac{V^2}{P_1} = \frac{110^2}{50} = 242\ W;\ I_1 = \frac{V}{R_1} = \frac{110}{242} = 0.45\ A$$

$$R_2 = \frac{V^2}{P_1} = \frac{110^2}{100} = 121\ W;\ I_2 = \frac{V}{R_2} = \frac{110}{121} = 0.91\ A$$

as expected.

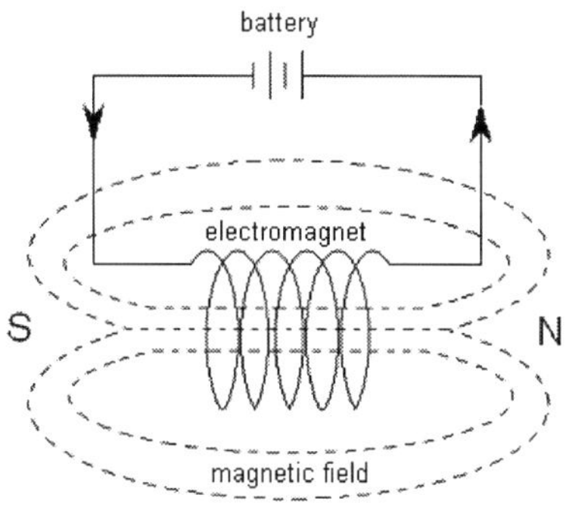

Figure 13-2
Electromagnet

Magnetism

Electricity and magnetism are two phenomena which occur hand in hand; we cannot have one without the other. This was demonstrated by Hans Christian Oersted (1777-1851), who showed that a compass needle brought near a wire carrying electrical current deflects. Similarly, if an electric current is passed through a looped wire (called an electromagnet), a magnetic field is created around the wire (Figure 13-2). The greater the current, the stronger the magnetic field will be. Electromagnets are used in a number of devices, including simple on/off switches and practically all electromechanical devices, from small toys and hair dryers to large electric motors in elevators and cable car trolleys.

Motors and Generators

Electric motors are devices that convert electrical energy into useful mechanical energy. The principle of operation of an electric motor is simple. If the coil of an electromagnet is placed between the poles of a magnet, the positive side of the wire and the north pole of the magnet are of the same charge and repel each other. The same is true of the negative side of the wire and the south pole of the magnet. The result is that the magnetic field exerts torque on the wire, which is deflected by half a turn and stops (Figure 13-3a). The electromagnet can now be rotated only if the direction of the current in the loop reverses, i.e. the polarity of the battery is changed. This is commonly accomplished by placing a commutator between the battery and the loop (Figure 13-3b). A commutator is a ring split in half – each side is in contact with a brush. As the wire loop continues to spin the polarity of the brushes reverse to assure that the loop and magnet remain of the same polarity. As long as a current is applied, the loop spins continuously. The rotors of practical motors are made of not one, but thousands of loops wound around a soft iron armature.

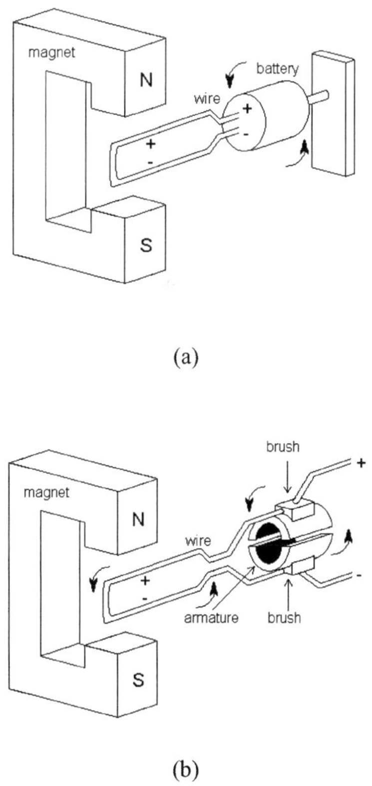

Figure 13-3
Electrical motors.
(a) Without a commutator, the loop is rotated by only half a turn and stops; (b) The loop can be made to spin continuously with a commutator.

Electric motors can be driven by either direct current (DC) or alternating current (AC). The difference between DC and AC motors is in the way the magnetic fields are created. *In DC motors,* this is done by means of an electromagnet or a permanent magnet. *In AC motors,* the magnetic field is created by passing an alternating current through a stator such that the polarity changes just when the armature is lined up with the poles of similar polarity. Because AC motors do not use brushes, they are simpler to construct and require little maintenance.

One of the main features of electric motors is their ability to produce torque as soon as they start. This is in contrast to internal combustion engines, where no torque is delivered until the engine attains a certain speed. In fact, this is why all cars running on petroleum require starter motors to operate. Furthermore, with only one moving part, electric motors are much simpler and have a much longer lifetime than internal combustion engines. Because of their ability to deliver peak torque at or near stall, electric motors are widespread in trolley cars, elevators, cranes, forklifts, and electric railroad locomotives. We will discuss the application of electric motors in electric and hybrid vehicles in the next chapter.

Electric generators are devices that convert the rotational energy of turbines or spinning shafts into electrical energy. Electric generators work opposite to motors - a magnet is turned by some external means to induce current through a wire. In a typical power plant, a turbine shaft is directly connected to a magnet that spins inside wire coils.

Depending on their application, there are different types of generators on the market. In *synchronous generators,* the rotor turns at exactly the same frequency as the electric grid; for a 60 Hz grid, it makes 60 revolutions per second, or 3,600 rpm (50-Hz or 3,000 rpm in Europe). These generators are often used in coal, oil, or nuclear power plants where fuels can be burned at a controlled rate, making the turbine rotate at a fixed velocity. In *asynchronous generators,* the magnetic field of the rotor and the electric field of the stator do not synchronize, but the rotor falls behind (slips). Wind turbines take advantage of variable slip generators by adjusting the slip (adjusting the resistance in the rotor winding) to allow the turbine to run faster as wind speed increases. *Pole changing generators* allow operators to choose the number of stator poles in order to change rotational speed and power of the turbine.

Generation, Transmission, and Distribution of Electricity

The delivery of electricity from production to consumption requires efficient generating facilities, access to reliable networks, powerful transformers and relaying stations, accurate metering, and other procurement services such as scheduling and dispatching (Figure 13-4).

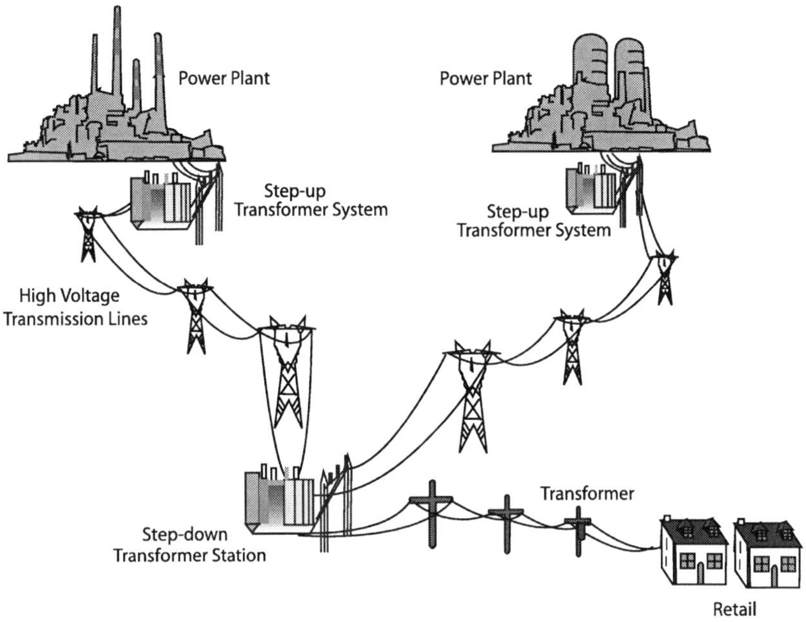

Figure 13-4
Electrical power distribution from generating plants to consumers.

At all points, the quality of electricity must be assured to maintain its frequency and voltage stability. This is particularly important in large networks in which electricity is continuously added and consumed at various nodes within the grid. Without this, a potential failure can propagate quickly, damaging a large part of the network and causing a blackout over a large geographical area.[7]

Generation

The primary method of electricity generation is by power plants. No matter which energy source is used, the energy produced is transformed into the rotational energy of a turbine, which drives a generator that produces electricity. Wind farms utilize the kinetic energy of air streams to power wind turbines. Hydroelectric and tidal plants use the potential energy of falling water to run water turbines. The potential energy of ocean waves can be used either to rotate a water wheel directly or to compress a column of air and drive a gas turbine.

The generation of electricity by thermal power plants is accomplished in three steps: the conversion of chemical or nuclear energy of the fuel into thermal energy (heat) inside a combustion chamber, the conversion of thermal energy into mechanical (rotational) energy, and the conversion of mechanical energy into electrical energy by an electric generator. In coal, oil, and natural gas power plants, the source of energy is the energy trapped in the chemical bonds of the fossil fuel. In nuclear power plants, the binding energy of the nucleus provides the energy. In solar-thermal and geothermal power plants, thermal energy is directly available.

[7] The largest power failure in US history occurred in August, 2003, following a problem resulting from a fallen tree in Ohio on an unusually hot summer day when demand was especially high. This created an overload that triggered a series of power failures across the grid. To prevent equipment damage, more than a dozen nuclear power plants and over 80 fossil fuel generating stations in the United States and Canada were automatically shut down within nine seconds. Over 50 million people lost power.

Besides mechanical and thermal power plants, electricity can be generated directly. Fuel cells produce electricity from electrochemical reactions between hydrogen and oxygen. Photovoltaic cells convert the energy in sunlight directly to a flow of electrons through an external circuit. Thermophotovoltaics work in a similar fashion except that, instead of using the visible light from the sun, they convert infrared radiation or heat from furnaces into electricity.

Magneto hydrodynamic (MHD) power generators omit the intermediate step by converting the thermal energy of fossil fuels directly into electrical energy, without first converting it to mechanical energy. Hot gas from coal combustors are seeded with some seeding agent (usually potassium) to turn into a hot plasma gas. A strong magnet separates out the charges, which drift toward and are collected by electrodes on opposite channel walls. An external circuit can be setup to take advantage of the charge gradient and produce electricity. Because of the low level of heat rejection, efficiency is higher than that of conventional thermal power plants and the demand for cooling water is lower. The hot exhaust gas is used to boil water to steam and produce additional electricity by conventional methods.

Example 13-6: According to the data published by the Energy Information Agency (EIA)[8], the world's total energy production in 2001 was estimated at 403 quads. The same reference gives total electricity generation at 14 trillion kilowatt-hours. What fraction of the world's total energy is converted to electricity? (1 quad = 1.055 EJ = 2.93×10^{11} kWh.)

Solution: Assuming a thermal power plant efficiency of 33%, the total energy needed to produce 14 trillion kWh of electricity is:

$$\left(\frac{14x10^{12}}{0.33}\,kWh\right)\left(\frac{1\ quad}{2.93x10^{11}\,kWh}\right) = 145\ quads$$

This is 36% of the world's total energy production.

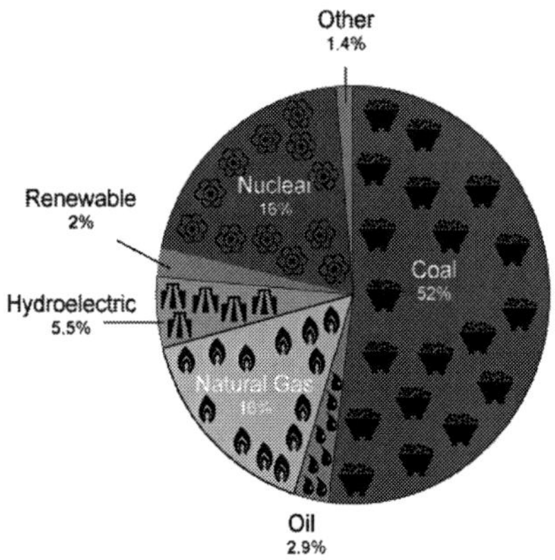

Figure 13-5
2004 U.S. Electricity Generation by Fuel Source.
Source: EIA, Monthly Energy Review, March 2005.

In the United States, 52% of power plants use coal, 16% use gas, 3% use oil, 20% are nuclear, and the remaining 10% are hydroelectric, solar, wind, or geothermal (Figure 13-5). Worldwide, about 64% of all electrical power generation is from fossil fuels, 18% is hydroelectric, 16% is nuclear, and only 2% is from geothermal and renewable energy sources. As of yet, no commercial MHD plants are built and the volume of electricity generated by fuel cells is insignificant.

Load and Capacity

The demand for electricity, called load, is not always constant and changes with geological location, season, even time of the day. In the United States, for example, daily electricity demand usually follows a pattern called

[8] EIA, Monthly Energy Review, March 2005.

"the load curve," which is low at nights and early mornings, gradually increasing during the day. It peaks in the early afternoon when industrial activities are the highest and drops back in the late afternoon or early evening (Figure 13-6 top). On an annual basis, overall demand is higher in summer, when a large number of air conditioners are operating, than in winter. The opposite is true for Canada, where extreme cold winters demand more electricity for heating than cooling. Depending on climate, different regions may experience their peak loads during either summer or winter months. Many regions of the world show curves with two maximums, one in summer and another in winter (Figure 13-6 bottom). The annual load curve can be constructed by summing up the average daily consumption over each month and plotting it over the period of one year.

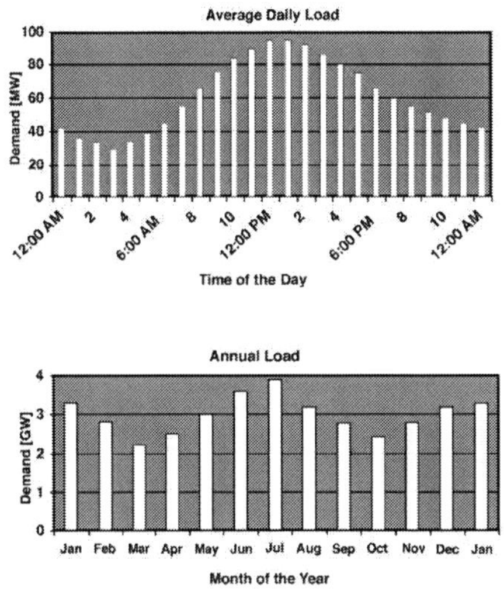

Figure 13-6
Average daily and annual electrical loads.

Just as demand for electricity varies, so does overall capacity of electricity generation facilities. It depends not only on the number of plants that are operational, but also on whether or not they are operating at their full capacity. When electricity is produced from renewable energy sources such as solar or wind, environmental parameters such as temperature and humidity and factors such as time of day, cloud cover, wind speed, and wind direction also become important.

One way to meet the maximum load is to design a large power plant capable of satisfying the peak demand. This obviously is not acceptable, because the cost is prohibitively high and much of the power plant's capacity would remain idle most of the time. The other extreme is the baseload capacity; power would seldom drop below it and the plant's capacity wouldn't go unused. This is the minimum capacity a power plant must meet at all times and sets the low limit for power plant size. The optimum capacity would obviously be somewhere between the two limits.

Power plants' total electrical output is measured in megawatts and expressed as installed, peak, baseload, or reserve capacity. **Installed capacity** is the maximum electricity that can be generated if all power plants are operated simultaneously and at their full capacities. The installed capacity in the US was 900 gigawatts in 2002. **Peak capacity** is the maximum amount of electricity that is needed in a given period. In any single day the peak demand occurs in the early afternoon when industrial consumption is highest. The peak demand during a year usually occurs during the hottest summer or coldest winter months. **Baseload capacity** is the minimum amount of electricity delivered continuously at any time during the year. **Reserve capacity** is the additional capacity that is needed during periods of unusually high demand or during a period of maintenance where some equipment is not operating.

Normally nuclear, coal, or hydropower (from run-of-river) plants supply the baseload demands. These plants are not suitable for rapid shut downs and start-ups so, except for regularly scheduled maintenance, they remain operational continuously throughout the year. During periods of peak consumption, baseload capacity is not sufficient and additional power generating capacity is needed. Because peak demands last only a few hours at a time and often occur with little or no warning, the additional power should be brought on line or taken off line quickly. In these instances, gas turbines and diesels are much smaller in size and capacity, and any variation in demand can be easily satisfied by adding or removing one or more of these equipments. However, the cost of electricity production using these devices is higher than that of conventional power plants.

Every year there may come times when demand exceeds the installed capacity (that of the baseload power plants and all available gas turbines and diesels combined). In such instances, the power company buys power from other large utility companies or from individual farmers, homeowners, and other independent generation producers. If total supply is still not enough and the shortage persists, the power company may be forced to stop or limit power delivery to some customers until demand stabilizes and full capacity is restored.

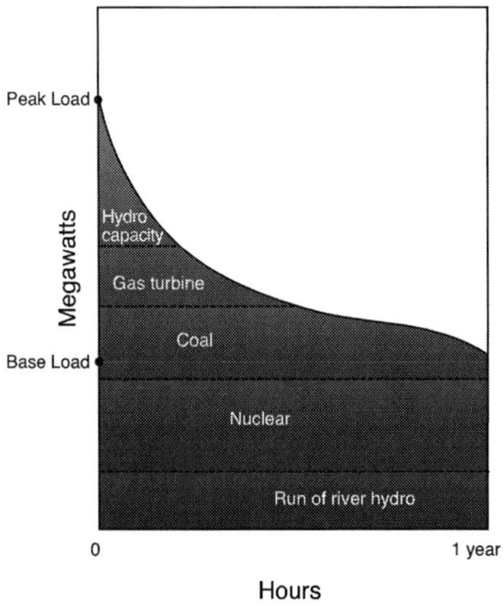

Figure 13-7
Load Duration Curve.

To stabilize demand, power utilities adjust prices according to a predetermined schedule. Elaborate algorithms can be envisioned where prices are changed according to the time of day and the quality of power (e.g. the current in the line). The "real-time" prices could then be transmitted to appliances. The end user could program the gadget to schedule its operation.

A common practice used to calculate power generation capacity requirements is to calculate the number of hours that power plants must meet a particular demand and plot it in descending order. A sample of such a graph, called the annual load duration curve (LDC), is shown in Figure 13-7.[9] The abscissa is the time in hours (T = 8,760 hours or 1 year), and the ordinate is the total electrical load measured in MW.

Example 13-7: A utility company must meet the electricity demand for a community with daily and annual loads represented in Figure 13-7. Calculate:

a. The total daily energy production in kWh.
b. The average daily power in MW.
c. The baseload capacity in MW.

Solution:

a. Total daily electrical output can be calculated by adding hourly productions. Mathematically, this is the area under the average daily load curve. Referring to Figure 13-7 and adding the hourly loads, daily consumption is determined to be 1,520 megawatt-hours.

[9] Thomas, B. G., and Hall, D., "Probabilistic production costing under integrated operation agreement and joint power agency financing," *Energy Economics*, pp. 200-208, July 1992.

b. The average daily power is calculated by dividing the total daily consumption by 24; i.e., 1,520 MWh/24 h = 63.3 MW
c. The baseload capacity is the power generation capacity that satisfies minimum demand, in this case 30 MW.

Capacity Factor

Power plants do not utilize 100% of their capacity all the time. In addition to occasional down-time for repair and maintenance, there are times that demand is not sufficient for a plant to run at full capacity. Other factors such as a lack of necessary resources, moderate weather conditions, or economic sluggishness can also affect the demand.

The capacity factor represents the degree of utilization of a particular power plant over a certain period. It is defined as the ratio of the actual amount of electricity produced within a specified time and what would have been produced if the machinery had operated at full capacity during the same time period.[10]

$$\text{Capacity Factor} = \frac{\text{Actual power produced in a given period}}{\text{Maximum power that could have been produced during the same period}}$$

The actual amount of production is calculated as the area under the load curve. The maximum electricity that the plant could produce is calculated by multiplying the peak capacity by the number of hours in the period under consideration. In the United States, the average capacity factors for electricity generation by nuclear, coal, solar, and wind plants are 76%, 67%, 25%, and 40%, respectively.[11]

Example 13-8: A 500-MW power plant operates at full power 45% of the time, 80% capacity for 50% time, and is shut down for the remaining 5% of the time. What is the capacity factor for this power plant?
Solution: The total power produced is:
(500MW)[(1.00)(0.45)+ (0.80)(0.50)+(0.0)(0.05)] = 425 MW
The capacity factor for this power plant is 425/500 = 0.85 (or 85%)

Example 13-9: To meet the peak capacity demand, a main power plant is equipped with an additional 10-MW from gas turbines. Assume that on average each gas turbine operates at full power 10 hours a day for two months in summer and 8 hours a day for one month in winter. What is the capacity factor for each additional turbine?
Solution: Capacity factor can be calculated by dividing the actual energy use by the energy use if turbines operated 100% of the time.

$$\text{Capacity Factor} = \frac{(10\,MW)(10\,hours)(60\,days) + (10\,MW)(8\,hours)(30\,days)}{(10\,MW)(24\,hours)(365\,days)} = 9.6\%$$

[10] Instead of Capacity Factors, power companies often give the "Utilization Factor," which is the ratio of the actual energy produced and the energy produced during the time the machines operated at partial load capacity, not counting the down-times.
[11] Hall, Darwin, Private Communications.

Optimal Size of the Power Plants

Although at first glance it may seem that a power plant must be sized to meet the peak demand, from an economic standpoint, this is probably not the best option. Oversizing the plant will result in only short intervals when it utilizes its full capacity, thus leaving much of its capacity idle the rest of the time.

It is therefore reasonable to size a plant to meet, at minimum, the baseload demands, with additional power generation capabilities to meet peak demands. To reduce stresses on the system it is common to reduce peak loads and redistribute them to off-peak hours. This approach, called **load leveling**, is done most effectively by restricting access while providing customers with incentives to cut demand. For example, a utility company may offer its customers a cheaper rate if they shift usage from peak hours to hours where demand is lower. Another option, energy storage, is attractive when the marginal cost of the energy storage system is lower than the marginal cost of constructing additional plants. This is usually done by storing the excess capacity during periods of off-peak production and using it during times of peak demand. The last and possibly best option is to reduce consumption through conservation. Energy conservation measures alleviate stress on the grid, reduce load and emissions, and indirectly reduce the chance of power blackouts. Federal and state agencies can encourage conservation by providing tax incentives, rebates, and subsidies to companies that invest in energy-saving practices or allocate funds for energy related research or to individuals who purchase cleaner, more energy efficient appliances.

Transmission

After electricity is generated, it must be sent through transmission wires. The transport of electricity across an electrical network is fundamentally different from transport of a good along a transportation network. In transportation networks, cargo moves from one location to another along specific routes, usually the most direct route. Any interruption along a particular cargo route will affect only that route and possibly a few routes in its vicinity. Unlike cargo, electrical signals travel along the paths of least resistance, not necessarily along the shortest line connecting the two geographical locations; sometimes many thousands of miles are covered before a signal reaches its destination. An interruption along an electrical transmission line can affect power delivery at a point that is a long distance away from the original failure. In addition, as different generating stations go on and off the grid, either by choice or due to an equipment malfunction, loads continuously change, possibly affecting the stability of the system and causing fluctuations in the frequencies and voltages at various points on the grid. Because wires link everything, all generators must spin at exactly the same rate and in complete sync with each other.

To prevent the overload of power lines, a reliable control strategy and power conditioning system must keep current in each line within safe limits. Also, it must allow for additional capacity to absorb extra flow in case of a sudden failure somewhere else on the network. A generator located in the wrong place will not be able to meet an increase in demand because to do so would push the flow on some lines over their safe limits.[12]

For the efficient transmission of power, electricity must be in the form of an alternating current at very high voltages. Early power stations generated and sold DC electricity to their customers. The major problem with DC power is its high transmission loss, which limits its application to short distances. As time and technology progressed, power stations switched to AC power, which could be transmitted long distances with little loss of power. Today, almost all electric power generated in the United States is in AC form.[13]

To minimize energy losses through the network, current must be reduced as much as possible. Resistance losses depend only on the type of material and the current passing through it, but not on the voltage. As we saw in Equation 13-5, for the same power, voltage and current are inversely related; therefore, to lower the electric current we must raise the voltage. This is commonly done using a step up transformer which raises voltages to 300 - 500 kV before power is transmitted through the grid. At the city of destination, voltages are dropped back to convenient levels (around 12-30 kV) using a reduction transformer and further reduced to 120V (220V in Europe) before being distributed among various end users.

Superconductors may also play a role in reducing transmission losses. Superconductors are materials that lose their electrical resistances at temperatures below a certain limit. Most materials exhibit this behavior at temperatures very close to absolute zero, although some alloys of yttrium, barium, and copper become super conductors at much higher temperatures. If room temperature superconductor materials are found, much of the losses due to long-distance transmission would be eliminated.

Distribution

The distribution network refers to the last mile delivery of electricity to homes and offices. Many household appliances use AC directly, although there are instances when DC power is needed. Rectifiers are devices that convert AC to DC power. They contain diodes that work by allowing current to flow in one direction, but not the opposite. The digital world is a prominent example, as computers, telephones, digital satellites, and

[12] Blumstein, C., Friedman, L. S., and Green, R. J., "The History of Electricity Restructuring in California," Center for the Study of Energy Market (CSEM), UCEI, Berkeley, CA. 2002.
[13] The conversion from DC to AC was not a simple process. Until the middle 1880s, Thomas Edison's power company supplying DC power dominated the utility market over his arch competitor George Westinghouse. Westinghouse used the newer and superior AC technology invented by Nikola Tesla which could be transmitted across long distances with very little losses. Rather than challenging him on technical merits, Edison started a smear campaign against Westinghouse using the argument that AC is dangerous and that its use should be avoided at all costs. He even set up stations using Westinghouse's AC generator, burned and killed animals in public, and later lobbied the New York legislature to use AC power in designing the electric chair. Edison's plan to bring about the demise of Westinghouse was eventually unfolded and AC technology was adopted in power transmission.

other communication devices all work by direct current. Automobile accessories and all other battery-operated devices also use direct current. Power generated by photovoltaics and fuel cells is in the form of DC and is therefore more convenient to use when operating these devices. On the other hand, when electricity is generated for transmission over commercial networks using photovoltaics and fuel cells, it must undergo DC to AC conversion using inverters before it can be connected to the grid.

The electrical operator matches electricity consumption with production, mostly by bringing new generation capability online when needed and taking it offline when demand falls. The operators can also decide to get electricity from certain power companies to lower costs and better serve their customers. To attract customers, power and transmission companies find incentives to invest in technologies that improve their performance and make them more competitive.

Storage

Not all the energy produced by electric generators can be used immediately after it is generated. Electrical utilities take advantage of lower demand during off-peak periods by storing excess electricity and using it during peak periods, when load exceeds the generator's capacity. This reduces the load on the system during peak hours, allowing the main generators to work at constant power, near full capacity, and under optimal conditions, thus reducing the average capital and operating costs by cutting the number of power generating stations. The important parameter in evaluating energy storage devices is their energy storage densities per unit mass (kJ/kg) and per unit volume (kJ/m^3).

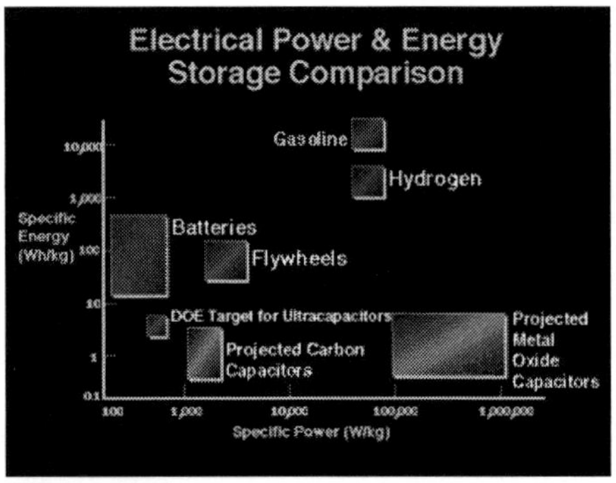

Figure 13-8
Specific power and energy for several storage systems.

The most common methods of storing energy are pumped storage plants, storage batteries, capacitors, and flywheels. In addition, electricity can be stored by producing hydrogen that can be recovered when it is burned or used to operate fuel cells. Figure 13-8 shows a comparison between energy and power densities for various energy storage devices.

Pumped Storage Plants

In pumped storage plants, the water downstream of a dam or hydropower generating station is pumped back into a storage tank at a higher elevation to be used to generate electricity when it is needed. Pumped storage plants were discussed in detail in Chapter 4.

Compressed Air

Instead of pumping water as is used in pumped storage plants, it is possible to compress air to a high pressure and store it in sealed underground caverns. During peak hours, air is mixed with natural gas in a conventional gas turbine combustion process to generate electricity. The McIntosh plant in Alabama has the largest compressed air storage facility in the United States (Figure 13-9).

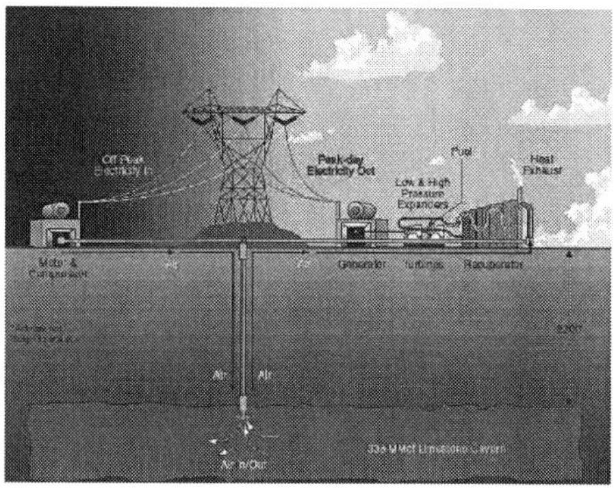

Figure 13-9
Alabama Electric Cooperative in McIntosh, uses air compressed and stored in an underground cavern to supplement electrical generation capacity during peak periods.

Batteries

Batteries are devices that store electric energy as chemical energy (during charging) and release it as electric energy during discharge. A battery consists of a number of voltaic cells. Each cell produces a certain voltage that depends on the type of cell and materials used. When several cells are arranged in a series configuration in a plastic casing, they are collectively called a battery.

Each cell consists of two unlike metal electrodes immersed in an electrolyte which reacts with them. The positive electrode (cathode) produces negative ions (molecules with extra electrons in their outer shells), while the negative electrode (anode) produces positive ions (molecules which lack electrons in their outer shells). If the two ends of the electrodes are connected through a wire and external load, an electric current path is established that allows electrons to flow from the negative to the positive electrode (Figure 13-10). The circuit is closed by a flow of ions through the electrolytes. In small batteries, electrodes are small rods inserted in a pool of electrolyte. In larger batteries, electrodes are in the shape of thin metal plates. The most widely used battery is the lead-acid battery, where the electrolyte is a solution of sulfuric acid and water and electrodes are pure lead and lead-dioxide.

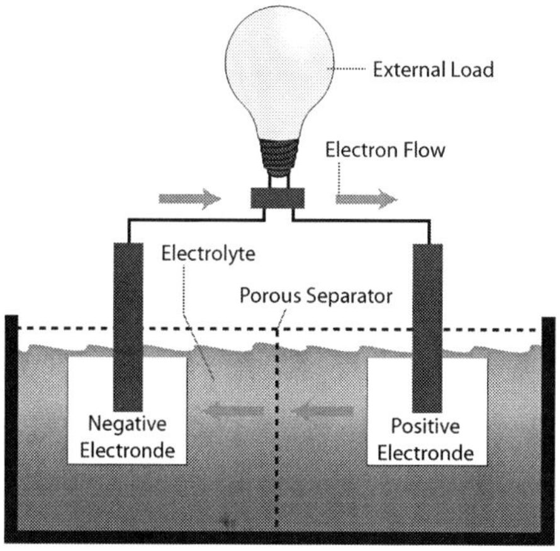

Figure 13-10
Lead-Acid Battery.

Depending on whether cells can be recharged or not, batteries are classified as primary and secondary. In **primary batteries** (such as those used in most flashlights) the chemical reactions that supply current are irreversible. Once these batteries are discharged, they must be replaced. In **secondary batteries** (such as car batteries) chemical reactions are reversible; by supplying current in the opposite direction, the chemical reactions are reversed and depleted electrode materials are restored.

Depending on their application, batteries can be classified as deep-cycle or shallow-cycle. **Deep-cycle batteries**, such as those used for golf-carts, electric vehicles, backup power, or renewable energy systems, are allowed to deplete up to 80% of their charge many hundreds of times. Automotive batteries (also called SLI, for starting, lighting, and ignition) are **shallow-cycle batteries** and cannot be allowed to discharge more than 2-5% of their capacity without being damaged. In addition to deep- and shallow-cycle batteries there are **marine (hybrid) batteries** used on boats and other marine applications which can be discharged up to 50%.

Fuel Cells

Another electrochemical device capable of storing energy is a fuel cell. The principle of operation of fuel cells is given in Chapter 14 and will not be repeated here.

Ultracapacitors

Unlike batteries which store energy by chemical reactions, ultracapacitors (also called super capacitors) store energy as static energy (Figure 13-11). A capacitor is made of two conducting plates filled with an insulator called the dielectric. The major differences between a battery and an ultracapacitor are their energy and power densities. Typically, batteries are best suitable for storing a lot of energy and little power, whereas capacitors can provide large amounts of power but store little energy.

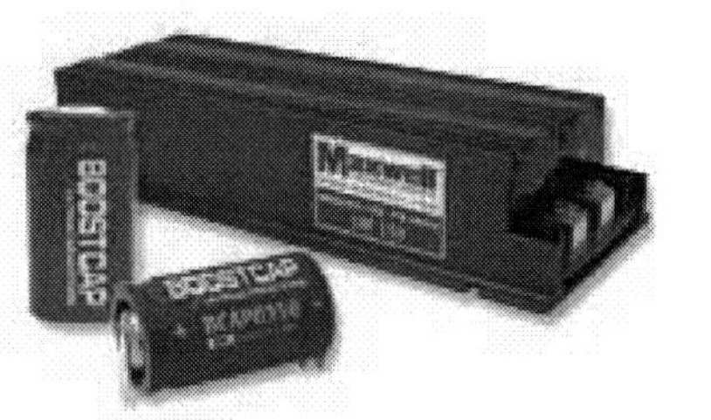

Figure 13-11
Ultracapacitor.
Image courtesy of Maxwell Technologies, Inc.

Because of their ease of operation, low maintenance requirements, lack of negative environmental impacts, and virtually indefinite number of cycles (charging and discharging), ultra capacitors have been of intense practical interest in hybrid and electric vehicles for capturing the energy produced during regenerative braking, which improves fuel efficiency by 10-15%.

Flywheels

Flywheels are large wheels that store energy by rotating at very high speeds. Traditionally, flywheels have been used for energy balance in internal combustion engines. Recent advances in material technology have made it possible to design flywheels from lighter composite materials that can run at much higher speeds - up to 100,000 rpm - making them ideal for storing surplus energy from power plants to be delivered later during peak electrical demand. The storage capacity of flywheels can reach those of batteries with the advantage that they can be energized and de-energized faster; the cost is higher however .

Flywheels have also found applications in fuel cells, gas turbines, and other prime movers. Flywheels are advantageous over conventional lead-acid batteries because they have high efficiencies, much longer life, operate at any ambient conditions, and can deliver energy over a much shorter time, i.e., they have a very high specific power. If desired, flywheels can be used to charge batteries, making them suitable for providing steady power to electric and hybrid vehicles or to produce bursts of energy during acceleration and on steep grades.[14]

Hydrogen

Strictly speaking, hydrogen is not a fuel source, but a carrier and a means of energy storage. To produce hydrogen, one requires at least as much energy as is produced when it is burned in a combustion process, or the electricity it generates when operating a fuel cell. In other words, we store the energy contained in other energy sources in the form of hydrogen, and then use the hydrogen to produce power. Issues related to hydrogen production and storage will be covered in detail in the next chapter.

[14] Rosen, H., and Castleman, D. R., "Flywheels in Hybrid Vehicles," Scientific American, p. 75-77, October 1997.

a Structure of the Electrical Power Industry

How much control the government should have over the pricing and regulating of power companies depends on how one sees energy and other natural resources. Many countries consider energy to be a basic public good, vital to society and essential in everybody's life; it cannot be replaced by other means and therefore cannot be traded like other commodities. It must be offered in a predictable manner and be made available to everybody based on need rather than profit. Energy regulations must therefore fall under government jurisdiction for setting prices, assuring its reliability and quality of service, and enforcing established environmental laws. In these countries energy sources are nationalized and all aspects of production from recovery, processing, production, distribution and sales are strictly controlled by their governments.

The proponents of the free market economy favor complete deregulation of the energy industry and argue that, to increase competition, energy must be treated as a commodity that follows the same laws of supply and demand that other commodities are subjected to. Any attempt to regulate the market will stifle competition, killing incentives for power companies to invest in new and more energy-efficient generation facilities and to upgrade existing transmission infrastructure. Deregulation will allow power companies to buy and sell electricity in an open market, but also empowers smaller utilities and individual energy producers to compete with the larger utility firms. This provides a more effective load leveling and reduces stress on generators. For example, a rural farmer or a single family homeowner can install photovoltaic cells on his roof or operate a wind turbine in his backyard (See Chapter 3). In this scenario, the homeowner is both a buyer and a seller of electricity; he can buy electricity at night or during a period of low wind velocity and sell his excess power during the daytime or at high wind velocities. In one plausible scenario, even an electric or hybrid car can be a source of power generation when it is parked in the garage during peak time. As soon as power demand drops, the batteries can be recharged again. Power companies can also benefit from these small entrepreneurs because they do not need to install new expensive equipments during times when demand peaks. The savings can be transferred to customers, lowering the price of electricity.

The opponents of the deregulation argue that when electricity is treated as a commodity, maximizing profits – not efficiency – becomes the ultimate motive. When profit is not a motive, electricity is allowed to move freely in the network along the path of the least resistance. With profit as the primary motive, depending on transmission rights and costs, distributors determine over which lines power is transmitted, even though this may not be the best route as far as demand over the entire network dictates; electricity might travel thousands of miles around the country before it reaches its destination.

Some have proposed a middle-ground where the government allows competition but sets a maximum price that can be charged to consumers. In other instances, some subsidies are provided to assure reasonable profits to the utilities. England, Wales, Norway, and Argentina have followed this approach and have rearranged the electricity infrastructure by breaking up the utility monopolies into separate generation, transmission, and distribution entities.[15] Many Eastern European countries undertook similar restructuring as part of the privatization of state-owned assets.

Electricity Infrastructure in the United States

Until a few years ago, private firms and investor-owned utility companies had total control over the electricity market and over setting prices in specific geographical areas.[16] These companies were normally vertically integrated; they produced electricity, carried it through transmission lines, and distributed it among customers. Except for the national transmission network that is shared by all utilities, these companies owned or controlled a major portion of generating stations, transmission lines to and from the national grid, and other ancillary services. The role of the government was to oversee their operation to assure the industry's compliance with environmental and safety regulations and to limit their monopolizing power, thus preventing them from illegal practices such as hoarding and market manipulation.

The task of overseeing US energy industries including interstate commerce, wholesale power transactions, transmission, and infrastructure regulation is assigned to the **Federal Energy Regulatory Commission (FERC)**.[17] The role of local and state authorities has been limited to levying taxes and regulating rates utilities can charge customers while giving the utilities a fair rate-of-return on their investments. Below we discuss the electricity infrastructure in California. Other states have infrastructures that function in similar manners.

In California, electricity infrastructure consists of four principal components: electricity generating facilities, the interstate electricity transmission grid, transmission lines connecting the generating facilities to the grid, and the distribution lines that connect the electricity grid to the end electricity users. Californian receive their electricity service from one of three types of providers (See the text box for definitions of commonly used terms below):

- Investor-owned utilities (IOUs), which provide about 2/3 of retail electricity service. The state's three largest electricity IOUs are *Pacific Gas and Electric*, *Southern California Edison*, and *San Diego Gas and Electric*. Each IOU has a unique, defined geographic service area. The

[15] Jokow, P., L., "Restructuring, Completion and Regulatory Reform in the U.S. Electricity Sector," Journal of Economic Perspectives, 22, Number 3, Summer 1997, pp. 119-138.
[16] Ibid.
[17] Federal Energy Regulatory Commission, FERC (*http://www.ferc.gov*).

Commonly Used Terms

FYI ...

FERC (Federal Energy Regulatory Commission) - In charge of overseeing US energy industries including commerce, transmission and infrastructure, as well as issuing permits to all hydroelectric generating facilities such as dams.

EC (Energy Commission) - State agency in charge of enforcing, research, and implementing energy conservation progress. EC issues permits to large thermal electricity generation facilities, primarily natural gas-fired power plants.

ESP (Electric Service Provider) - A company that provides electricity service directly to customers who have chosen not to receive service from the utility that serves their geographic area.

IOU (Investor-Owned Utility) - A privately owned electric utility that provide electricity to a specific geographic service area for profit. The Public Utilities Commission regulates the IOU's rates and terms of service.

POU (Publicly-Owned Utility) - A local government agency, governed by a board—either elected by the public or appointed by a local elected body—that provides electricity service in its local area.

PUC (Public Utilities Commission). The state agency that regulates various types of utilities, including IOUs and ESPs. These conditions on electricity rates and service are known as "terms of service."

RPS (Renewables Portfolio Standard) - Requirement that electricity providers increase their share of electricity from renewable resources (such as wind or solar power) according to a specified time line.

rates that IOUs can charge their customers are determined by the California Public Utilities Commission (PUC).

- Publicly-owned utilities (POUs), which provide 1/4 of retail electricity service. POUs are governed by a board elected by public or appointed by a local elected body. POUs are not regulated by PUC. Rather, they set their own terms of service. California's major publicly owned electric utilities include the *Los Angeles Department of Water and Power* and the *Sacramento Municipal Utility District.*
- Electric service providers (ESPs), which provide the remaining retail electricity service. These ESPs generally serve large industrial and commercial customers. The ESPs also provide electricity to some state and local government agencies, such as several University of California campuses and some local school districts. Like POUs, ESPs set their own terms of service.

Energy From Renewable Sources

Current law requires IOUs and ESPs to increase the amount of electricity they acquire (from their own sources or purchased from others) that is generated from renewable resources, such as solar and wind power. This requirement is known as the renewable portfolio standard (RPS). Each electricity provider subject to the RPS must increase its share of electricity generated from eligible renewable resources by at least 1 percent each year so that, by the end of 2010, 20 percent of its electricity comes from renewable sources. Current law does not require POUs to meet the same RSP that other electricity providers are required to meet. Rather, each POU set their own renewable energy standards and schedule to meet those standards.

a Electrical Safety

Although electricity appears to be the cleanest form for storing energy, there are a number of health and safety issues that must be considered. Electrical power lines produce a magnetic and electric field that may pose a health problem or interfere with surrounding electronic devices. Inappropriate use of electricity could also be dangerous and, at times, even life-threatening. It is therefore necessary to understand the harmful effects of electricity and the proper procedures for safe handling of electricity in a manner that can minimize long-term health problems and risk of electrical shock.

Health Effects of Electromagnetic Radiation

The health effects associated with overhead power lines and household electrical appliances have been one of the most controversial topics debated among scientists.[18] Strong magnetic forces are associated with high currents, such as those near thick wires and equipments which draw high currents. A current passing through a wire creates a magnetic field that encircles the wire (Ampere's law). The magnetic field's strength is highest closer to the wire and drops off as the square of the radial distance from the wire. It is very difficult to shield the magnetic field.

Strong electric fields are associated with the presence of strong electric charges, such as around high-voltage equipment. A person standing underneath a power line experiences an electric field perpendicular to the ground. The closer the power line is to the ground and the higher its voltage, the higher the electric field's strength. For example, a single 115 kV cable running 15m above the ground causes an electric field of E = 500 V/m near the ground. Depending on the frequency of the AC line, the direction of the field lines reverses 50-60 times per second, causing rapid changes in the direction of the electric field through a body and exposing it to ionization radiation. Unlike magnetic fields, electric fields can be shielded by conductors such as metals. We discussed the effect of ionizing radiation and its ability to cause symptoms from headaches to cancer in detail in Chapter 12. Although the connection between power line fields and cancer is an area of continuing research, so far there are no scientific studies that point to a consistent, significant link between cancer and power line fields. A summary of health effects associated with electrical shocks is given in Table 13-2.

Table 13-2. Health Effects of Electrical Shocks*

Current through the body trunk	Effect on average human
< 1 mA	No sensation
3-10 mA	Tingling. Person can let go.
10-30 mA	Muscle contraction, person cannot let go
30-50 mA	Painful. Severe muscular contractions. Breathing difficult
50-100 mA	Ventricular fibrillation, probable death
> 100 mA	Fatal

** Center for Disease Control & Prevention website http://www.cdc.gov/niosh/*

Physiological Effects of Electricity

When current passes through living tissue, it experiences resistance and dissipates its energy as heat. Depending on the magnitude of current and tissues involved, electricity can manifest itself in several forms, from low heat and slight tingling to severe burns, paralysis, and even death.

[18] For more information on effects of EMF exposure on health see the Word Health Organization's International EMP Project (*http://www.who.ch/programmes/peh/emf/emf_home.htm*).

The most significant hazard associated with electric shock is its damage to a person's nervous system. The nervous system consists of a series of nerve cells called neurons that are responsible for coordinating all of the body's movements – from the beating of the heart to the blinking of eyes – by passing information from various organs to the brain. This is done by releasing chemicals called neurotransmitters which generate tiny electrical signals that can travel the nervous system. Currents from external sources can be strong enough to override the neurotransmitters, preventing them from carrying out their normal functions. If this happens, volitional signals cannot be transmitted and affected muscles will involuntarily contract. This effect is particularly dangerous when a person touches a bare electrical wire. Fingers have the least resistance to current flow and can easily bend involuntarily, clenching into a fist that grabs the wire. The victim becomes immobilized and unable to let go of the wire, making the shock even more dangerous. The involuntary contraction of muscles is called tetanus in medical terminology. The condition persists as long as the current flows.

Shock Severity

The length of time, the type of tissues involved, and the magnitude of the current determine the severity of hazard associated with electrical shock and the extent of damage it causes.

The best way to reduce the electrical shock from a live circuit is to add resistance to the path of the current. Rubber gloves and boots increase the resistance, thus reducing the current for the same voltage difference. Grounding makes an excellent means of protection by providing a path of least resistance through the ground (and not the body). The current passing through the victim's body is determined by the body's resistance, which varies greatly from one organ to another and whether the skin is dry or wet. For the same voltage difference, the current passing through wet skin or a sweaty hand will be higher. When the path of current is from hand to hand or from hand to foot, vital organs such as the heart, lungs, or spinal cord are affected, and the shock effect is the most severe.

Question: A common phrase often heard in regard to electrical safety is, "It's the current, not the voltage that kills." Why then, are we often

Safety Tip: The Third Prong

FYI ...

Faulty electrical devices can deliver dangerous and sometimes lethal shocks. In the most common scenario, frayed insulation causes the high-voltage wire of the device to short (become connected) to the case. If a person standing on the ground touches the case, he completes a path for the current to reach the ground. Since a person's resistance is typically much less than the resistance within the device, the person presents the path of least resistance and will therefore draw a large current. This kind of hazard can be avoided by the use of a three-wire system. A wire to the case of the appliance connects to the third prong of a plug, while the third hole in the outlet is connected to the ground. In the event of faulty wiring in the appliance, the current is routed to the ground through the prong and not through the person because the case and the ground have essentially no resistance between them.

warned of the danger of high voltage?

Answer: Strictly speaking the sentence is correct. It is the current that is dangerous. The voltage only pushes the current through bodily resistance. The question is where do all those currents come from? High voltages mean the potential for creating high currents for a given resistance path through a body - higher voltages can be directly translated into higher currents.

Question: When is the effect of electrical shock most severe? With a 120 V or a 220 V? With direct current (DC) or alternating current (AC)?

Answer: If everything else is the same, higher voltages result in higher currents; therefore voltages in the US outlets are safer than those in Europe. It is difficult to quantify whether DC or AC is more dangerous. Direct current is generally considered to pose less of a shock hazard, but it produces more severe burns. A person shocked with an alternating current is more likely to go into heart fibrillation.

Example 13-10: In the diagrams to the left, determine which instance provides the safest situation for the bird or person involved.

Solution: For current to flow through a circuit, an electrical potential is required. A bird sitting on the wire experiences practically no voltage drops between its feet and therefore is immune to potential danger, no matter how high the current is. On the other hand, if the bird were to touch both the high- and low-voltage cables at the same time, it would then draw a lethal current through its body. To assure birds' safety, the separation distance between the power cables is chosen to exceed the wingspans of most birds.

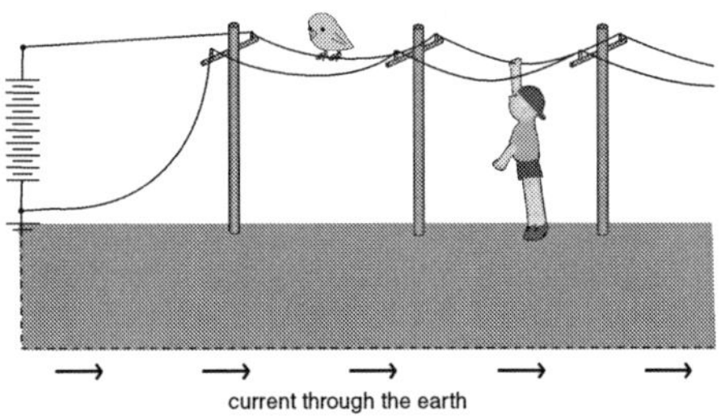

(a)

(b)

EX 13-10

This is not true for the boy, however. In diagram (a) the boy holds the bare wire with one hand. Since he is standing on the ground (which by convention has zero voltage), there is a voltage difference between his hand and foot. As a result, electricity flows through his hands and body and eventually reaches his feet, closing the circuit and shocking the boy. The downed power line in diagram (b) causes a large electric potential between the points where wire touches the ground and the nearest pole where the transmission line is grounded. Thus there is a voltage differential between the feet of the boy standing somewhere in between, and he would be shocked. Probably the best way to avoid shock is to keep your feet close together, stand on one leg, or run away from the power line. Running has the same effect of having one foot on the ground, preventing a large voltage drop between the victim's feet.

Safety Tip: Circuit Breakers

FYI ...

Household circuit breakers are normally designed to trip at currents of around 15-20 amperes or higher, much more than the lethal limit of 100 mA. They are intended to prevent electrical fires, not protect you from shock.

Example 13-11: In the example above, would it make a difference if the boy were bare-footed? What about if he touched the wire with both hands?
Solution: If the boy wears shoes with thick, insulated rubber soles, then he is protected from electric shocks. The problem arises if any moisture, dirt, or other conducting substances (such as a metallic strip) provide a path of least resistance and allow the electricity to bypass the sole directly to the body. Leather soles provide much less resistance and are not nearly as effective. Some ground surfaces are better insulators than others. Asphalt contains some oil, which makes it a better insulator than most dirt, concrete, and rocks.
If the boy holds the wire with both hands, the contact area doubles and two parallel pathways are available for the current to flow. The overall resistance is only one-half of the resistance from one hand, and twice as much current would flow though the body. It is a good practice to keep one hand in a pocket when working around electrical devices!

Example 13-12: A bird is sitting on a piece of bare copper wire carrying 100 amperes. It is estimated that the copper cable has a resistance of 20 ohms/kilometer. Assuming the bird's feet are 10-cm apart, what is the voltage potential established through the bird's body?
Solution: The resistance between the bird's feet is calculated as 0.10 m x 20 ohms/1000 m = 0.002 ohms, and the potential difference is $\Delta V = I \times R = 100 \times 0.002 = 0.20$ volts; this is not enough to do any harm! As mentioned earlier, power cables are often setup so the separation between the high and low-voltage cables exceeds the bird's wingspan.

Summary

Electrical energy is the driving force behind much of our technological innovation. Not only does it provide lighting and heating to our rooms, but it is also essential in running our household appliances as well as the heavy industrial machinery which powers modern societies.

Most of the electricity generated today uses coal, nuclear, or hydro power. As demand for electricity continues to increase, new resources must be exploited. Designing more efficient appliances and better utilization of existing resources can also help extend the lifetime of our valuable fossil reserves and maintain the quality of the air and environment.

Additional Information

Books

1. Bureau of Naval Personnel, Basic Electricity, Dover Publishing Company.
2. The Environmental Effects of Electricity Generation, IEEE, 1995.

Periodicals

1. *The Electricity Journal*, Direct Science Elsevier Publishing Company, This journal addresses issues related to generating power from natural gas-fired cogeneration and renewable energy plants (wind power, biomass, hydro and solar).
2. *International Journal of Electrical Power and Energy Systems*, Direct Science Elsevier Publishing Company.
3. H*ome Power Magazine (http://www.homepower.com).*

Government Agencies and Websites

1. Federal Energy Regulatory Commission (*http://www.ferc.gov*).
2. Energy Information Agency, Department of Energy (*http://www.eia.doe.gov/fuelelectric.htm*).
3. California Energy Commission (*http://www.energy.ca.gov/electricity*).

Non-Government Organizations and Websites

1. National Council on Electricity Policy (*http://www.ncouncil.org*).
2. Southern California Edison *(http://www.sce.com).*
3. Pacific Gas and Electric (*http://www.pge.com*).

Exercises

I. Essay Questions

1. What is Ohm's Law? How does power transmitted through a circuit change with current? With voltage across the battery?

2. What is the difference between a series and a parallel circuit? What are the advantages of one over the other? Which type of wiring is more convenient in a house?

3. Describe the principal of operation of an electric motor and an electric generator. When do you use an AC or a DC motor. A synchronous and an asynchronous generator?

4. Define installed, peak, baseload, and reserve capacities. What kind of power plant is suitable to meet the needs for each of these?

5. Identify and describe the major components of a coal-burning electric power plant. How does it differ from a nuclear or a geothermal plant?

6. Describe the advantages and disadvantages of generating electricity using renewable sources of energy such as wind and solar.

7. What causes an electric shock? What are the basic safety precautions for avoiding electric shocks?

II. Multiple Choice Questions:

1. Which of the following statements is not correct?
   a. Atoms are made of negatively charged electrons and positively charged nuclei.
   b. In neutral atoms, there are as many protons as there are electrons.
   c. The total number of charges in the universe is constant.
   d. Charge and mass are two properties of matter.
   e. Unlike charges which are always attractive, two masses separated at a distance can attract or repel each other.

2. A neutral atom with 10 protons must have
   a. 10 neutrons
   b. 10 electrons
   c. 10 neutrons and 10 electrons
   d. 10 neutrons and any number of electrons
   e. 5 neutrons and 5 electrons

3. A live wire (a wire in which current flows) is placed directly over a magnetic compass. As a result, the needle of the compass will
   a. Not move
   b. Point in the direction perpendicular to the wire
   c. Point in the direction parallel to the wire
   d. Fluctuate about an axis perpendicular to the wire
   e. Fluctuate about an axis parallel to the wire

4. According to Ohm's Law, the current in a circuit is
   a. Independent of the voltage
   b. Linearly proportional to the voltage
   c. Proportional to the square of the voltage
   d. Inversely proportional to the voltage
   e. Inversely proportional to the square of the voltage

5. According to Ohm's Law, the current in a circuit is
   a. Independent of the resistance
   b. Linearly proportional to the resistance
   c. Proportional to the square of the resistance
   d. Inversely proportional to the resistance
   e. Inversely proportional to the square of the resistance

6. Two light bulbs, a 100-W and a 200-W, are connected in parallel to the two terminals of a battery.
   a. Both light bulbs receive the same amount of current.
   b. The 100-W light bulb receives more current.
   c. The 200-W light bulb receives more current.
   d. The entire current passes through the 100-W bulb.
   e. The entire current passes through the 200-W bulb.

7. Two light bulbs, a 100-W and a 200-W, are connected in series to the two ends of a battery.
   a. Both light bulbs receive the same amount of current.

b. The 100-W light bulb receives twice the current.
c. The 200-W light bulb receives twice the current.
d. The 100-W light bulb receives four times the current.
e. The 200-W light bulb receives four times the current.

8. Consider two lamps that are identical but for one exception; one has a thicker filament than the other. Which has the higher resistance?
   a. Thicker filament
   b. Thinner filament
   c. Both about the same
   d. It depends on the current
   e. It depends on the voltage across them

9. How many 100-W light bulbs can be simultaneously turned on in a household before the circuit breaker fuse burns out? The maximum current through the circuit breaker is 15 amperes and the supply line is at 120 volts.
   a. 1
   b. 5
   c. 6
   d. 8
   e. 18

10. Ten identical bulbs are placed in a series circuit. One light burns out. What will happen?
    a. The rest of the bulbs will be turned off.
    b. The rest of the bulbs will burn out.
    c. The rest of the bulbs get dimmer.
    d. The current in the circuit reduces by 10 percent.
    e. The rest of the bulbs keep glowing as if nothing has happened.

11. The kilowatt is a unit of
    a. Energy
    b. Power
    c. Voltage
    d. Torque
    e. Force

12. The kilowatt-hour is a unit of
    a. Energy
    b. Power
    c. Either power or energy
    d. Torque
    e. Electric potential

13. Which of the following represent(s) energy
    a. Joule
    b. BTU
    c. kWh
    d. eV
    e. All of the above

14. Which of the following represent(s) power
    a. Joule/min
    b. Barrels of oil per day
    c. Kilowatt
    d. Horsepower
    e. All of the above

15. What is the unit used by utility companies when they sell the electricity used in our homes?
    a. Kilowatts
    b. BTU
    c. Volts
    d. Kilowatt-hours
    e. Therms

16. Electrical energy can be produced by
    a. Mechanical energy
    b. Chemical energy
    c. Radiant energy
    d. Magnetic energy
    e. All of the above

17. Electrical power can be calculated by multiplying
    a. Current by voltage
    b. Current by resistance
    c. Current by resistance-squared
    d. Resistance by voltage
    e. Resistance by voltage-squared

18. To maximize power dissipation through a given resistance heater, we must
    a. Increase current
    b. Increase resistance
    c. Decrease current

d. Decrease resistance
e. Increase both current and resistance

19. The relationship between voltage and current in an electric line is like
    a. Flow rate to volume of the fluid in a hydraulic line
    b. Pressure to flow rate in a hydraulic line
    c. Pressure to volume of the fluid in a hydraulic line
    d. Flow rate to pipe diameter in a hydraulic line
    e. Flow rate to pipe length in a hydraulic line

20. The relationship between electrical resistance and flow resistance is like
    a. Wire length to pipe length
    b. Wire diameter to pipe diameter
    c. Wire material to pipe material
    d. Current in the wire to flow rate in the pipe
    e. Energy carried by electricity in a wire to energy carried by water in a pipe

21. Most electrical power transmission is in the form of an alternating current because
    a. Direct currents cannot be transmitted in electrical wires
    b. It is cheaper to produce
    c. It is safer to use
    d. It can be transmitted with less transmission loss
    e. It can travel faster

22. In the United States, most electricity is generated by
    a. Burning fossil fuels
    b. Nuclear power
    c. Solar energy
    d. Hydroelectric
    e. Wind and biomass

23. The failure of California to deregulate energy was mainly due to
    a. Insufficient refineries and generating stations
    b. Rapid increase in number of trucks and sport utility vehicles
    c. Out-dated power plants
    d. Malicious conspiracy by energy suppliers, politicians, and operators
    e. All of the above

24. The agency responsible for regulating interstate electric commerce is
    a. Federal Energy Regulatory Commission
    b. Power Exchange
    c. Department of Energy
    d. Department of Commerce
    e. Independent System Operators

25. The baseload of daily energy consumption represents
    a. The fraction used in early morning
    b. Average energy consumption during the day
    c. The load supplied by nuclear or fossil fuel
    d. The load required to provide minimal needs of consumers
    e. None of the above

26. Baseload power plants are used
    a. Primarily during the nighttime
    b. Primarily during the daytime
    c. Day and night
    d. During peak times
    e. During emergencies

27. Gas turbines are best suited for power generation
    a. At all times
    b. Only during daytimes
    c. Only during nighttimes
    d. During peak hours and emergencies
    e. For baseload power production

28. In the US, when is the peak electricity demand in summer?
    a. 6:00 am - 9:00 am
    b. 9:00 am - noon
    c. Noon - 3:00 pm
    d. 3:00 pm – 6:00 pm
    e. 6:00 pm – 9:00 pm

29. The severity of an electric shock depends on
    a. The health and physical condition of the victim
    b. The amount of current and length of time it passes through the body

c. The path the current takes through the body
d. The type of electricity
e. All of the above

30. If you think of a battery as a pump charging electric current, then you can think of the battery voltage as
a. The flow rate of fluid
b. The flow velocity
c. The pressure
d. The water
e. None of the above

31. Consider two lights, A and B, each screwed into 120-volt sockets. The two bulbs are identical except that the filament in B is thicker than the filament in A. Which statement is true?
a. Both A and B have the same brightness.
b. A will be brighter because it has the highest resistance.
c. A will be brighter because it has the least resistance.
d. B will be brighter because it has the highest resistance.
e. B will be brighter because it has the least resistance.

32. The amount of current that passes through the body as a result of an electrical shock depends on
a. The voltage in the line
b. The resistance of the body and clothing
c. The source of electricity
d. The contact area and contact pressure with the live wire
e. All of the above

33. Ultracapacitors are best suited when we need
a. A lot of power over a very short time
b. A lot of power over a long time
c. A lot of energy over a very short time
d. A lot of energy over a long time
e. A lot of energy and power over a short time

34. A magnet that is moved inside a coil of wires produces
a. A magnetic field
b. Electricity
c. Gravity
d. A spark
e. Nothing

35. How much of the energy of coal burned in a power plant is discarded as heat?
a. An insignificant amount
b. 10-20 percent
c. 20-40 percent
d. 40-60 percent
e. More than 60 percent

II. True or False?

1. Lightning is the result of static electricity.

2. Even currents as low as 50-100 mA can cause death, especially if the path of current is through the heart.

3. An alternating current results from the back and forth movement of electrons in a wire.

4. Power companies usually charge electricity in kilowatt-hours.

5. For the most economical way of transmitting power, it is best to use extremely high voltages.

6. Base-load energy is cheapest to produce; peak-load electricity is the most expensive.

7. Fuel cells are efficient devices for the direct conversion of solar energy to electricity.

8. Photovoltaic and thermophotovoltaics are two names for the same device.

9. Ultracapacitors are best when we want a lot of energy, but little power.

10. Both high currents and high voltages present danger to shocks.

Safety Quiz:

11. You should not touch a person who was electrocuted unless you have unplugged the appliance.

12. It is only current that can kill. Voltage does not matter.

13. During a storm, a tree standing alone in the middle of an open field is more vulnerable to lightning than a tree in the middle of a city.

14. You can get a shock by touching just one live conductor, or even a charged object such as a fence.

15. You need not worry about electric shock if you work around low voltage equipment.

16. The best approach to electrical safety is to add resistance to the path by wearing rubber gloves, boots, and other safety gear.

17. It is best to run extension cords under a carpet so no one trips over them.

18. You should not unscrew a light bulb with a wet hand, no matter the wattage.

III. Fill-in the Blanks

1. The study of charges at rest is called ___________ ___________.

2. The product of electric potential across a resistance and the current passing through the resistance gives the _____________ dissipated by that resistance.

3. When resistances are placed in parallel, ___________ are added.

4. To transmit power to long distances, it is best to use ______________ current.

5. Shock severity depends on the length of time, the type of tissues involved and the magnitude of the _______________ passing through the body.

IV. Match the List

1. Match the list of variables from the column to the right with the units given in the column to the left.

a. Force	watt
b. Charge	ohm
c. Power	joule
d. Weight	volt
e. Mass	coulomb
f. Electric Potential	ampere
g. Electrical Resistance	pascal
h. Current	kilogram
i. Heat	newton
j. Pressure	newton

2. Match letters on the diagram with components in the figure:

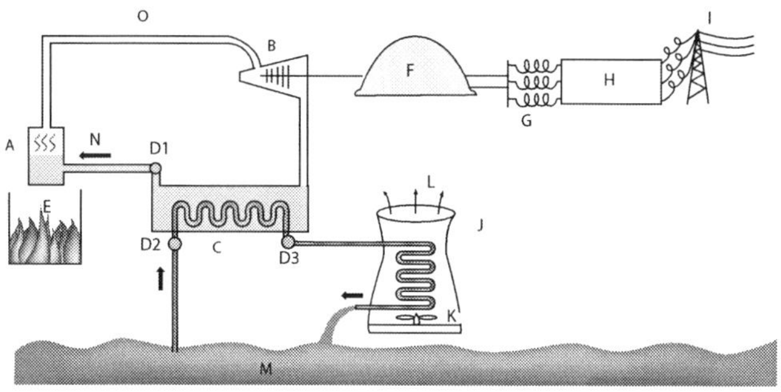

a. Ambient air	i. Fan
b. Water	j. Generator
c. Steam	k. Transformer
d. Boiler	l. Heat source
e. Condenser	m. Heat sink
f. Pump	n. Low-voltage line
g. Turbine	o. High voltage lines
h. Cooling tower	

IV. Project I – Electrical Power Generating Station

The migration of new industries and increased population requires the construction of new power generation facilities with a capacity of $10\ MW_e$ near your city. A panel consisting of consumer groups, oil and gas company representatives, coal miners, economists, scientists, business leaders, regulatory agencies, electric utilities, and environmentalists are meeting to debate the merit of various proposals. Several options are being considered:

a. Purchase necessary additional power from a neighboring state or country.
b. Use a combination of renewable energy alternatives such as wind, wave, and solar.
c. Expand the existing power plant facilities using

coal, oil, or natural gas.

d. Build a new fossil plant in the city suburbs, away from major population centers.
e. Construct a 15 MW combined heat and power (CHP) geothermal power plant.
f. Construct a nuclear fission power plant.

You are assigned the task of assuming the role of a reporter who is to report on various arguments being made in favor of or against each option. Please detail the findings by writing a report summarizing the following:

1. What are the geographical considerations that preclude using one or more of these options?
2. Which remaining options make the most economic sense? Consider the initial cost of construction, cost of producing and distributing power, maintenance costs, and the potential increase in employment.
3. Which option makes the most environmental sense?
4. What are the political implications of this decision, if any?

Project II - Health Hazards associated with Power lines and mobile phones.

There has been a wide array of studies on possible risks associated with the long-term health effects from electrical systems which was previously assumed to be safe. Some studies have claimed an increase rate of cancer among those who live in proximity to the high-voltage power lines, or using mobile phones frequently. None of these studies are conclusive, in part because of the difficulty in carrying laboratory experiments, or isolating the effect from many other factors which may contribute to the health risks. In case of mobile phones, data are limited, as these devices found their way in our daily life only about a decade ago. In light of the long latency period for developing tumors, the linkage of these devices to cancer can only be marginally documented.

In this project you are asked to search the internet for scientific studies that link or unlink the electromagnetic radiation to safety and health. Write a one-page report outlining the recent findings and answer the following questions:

1. How does electricity and magnetism interact with matter, specifically to various human tissues?
2. Would you expect a greater effect at higher currents? At higher voltages?
3. Are there needs for concern? If so, are there steps that can be taken to minimize the health risks, what are they?

Transportation

CHAPTER 14

"The Stone Age did not end because people ran out of stones- it ended because they found something better." ~ Zaki Yamani, Saudi Arabia's Oil Minister

The world's transportation systems are expanding every day, and with expansion comes the rapid depletion of natural resources, ever increasing environmental degradation, more traffic congestion, and a higher frequency of accidents.

The United States, with four million miles of highways, railroads, and waterways, 130 million passenger cars, 88 million buses and trucks, 230,000 aircrafts, 1.4 million miles of oil and gas pipelines, 40,000 ships, numerous railroad cars and boats, and 20,000 airports, has the largest transportation system in the world.[1] Every day about 13 million barrels of petroleum, more than 2/3 of total US petroleum consumption and 27% of its entire energy use, are required to operate this huge system (Table 14-1).[2] Because most of these fuels are imported, the US must seek ways to reduce its reliance on foreign oil either by increasing fuel efficiency and switching to alternative fuels or by reducing total miles traveled by investing in mass transit systems and carpooling, bike riding, and walking.

Table 14-1. Transportation Energy Use by Mode, 2005

Modes	*Trillion BTU*	*Percent of total*
Highway	**22,043**	**81.8**
Automobiles	9,140	35.0
Light Trucks[a]	8,108	26.5
Motorcycles	27	0.1
Buses	191	0.7
Heavy Trucks	5,342	19.5
Non-Highway	**5,721**	**18.2**
Air	2,348	8.4
Water	1,300	3.9
Rail	657	2.4
Pipeline	842	3.5
Total	**27,065**	**100**

[a] Including SUVs
Reference: Davis, S. C., Diegel, S. W., Table 2.6. "Transportation Energy Data Book," ORNL-6978, Edition 26, 2006..

Although most transportation issues are global in nature, because of the sheer size of the US transportation system, its role as a major technological powerhouse, and due to the fact that it is the biggest contributor to environmental pollution, this chapter focuses mainly on US data.

Power Requirements

We discussed the concept of energy and power in some detail in Chapter 2. Briefly, energy is what is needed to carry out a given task without concern for the time it takes to accomplish it. Power is the rate at which the energy is used and therefore determines the speed at which the task is performed. Without sufficient energy, we cannot go very far without having to refuel. If everything else remains the same, a larger fuel tank (more energy) makes it possible to drive longer distances, whereas a larger engine (more power) allows a vehicle to reach its destination faster, carry a greater load, or climb steeper roads. A major concern in any design of land, air, or marine vehicle is the power required for operation under various conditions.

[1] Transportation Statistics, Annual Report 2003, US Department of Transportation, Bureau of Transportation Statistics, Washington, DC 20590, 2003.
[2] Davis, S. C., and Diegel, S. W., Transportation Energy Data Book, Oak Ridge National Laboratory, Edition 24, 2004.

US Transportation: The Facts

Did You Know That ...?

- There are over 220 million vehicles in the United States; this equals roughly 760 cars for every 1000 people.
- There are enough roads in the United States to circle the globe 157 times.
- Europeans and Japanese pay about three times as much for a gallon of gasoline as Americans.
- US carbon emissions alone are more than those of China, India, Germany, and Japan combined.
- According to the 2000 US Census, 88% of all people used personal vehicles to commute to work, of which 76.3% drove alone. Only 5.2% of people used public transportation, nearly all of whom lived in low-income neighborhoods.
- Road accidents cost about $90 billion and kill 40,000 Americans annually. The social costs associated with the loss of productivity, health, and road repairs approach one trillion dollars a year.
- As a result of congestion in US metropolitan areas each American wastes an average of 50 hours every year in traffic.
- Transportation accounts for 2/3 of all US petroleum consumption.
- In 2001, Americans used 100 million gallons of fuel to travel a distance of 2.8 trillion vehicle-miles, an average of 9,800 miles per capita.
- Annually, American cars burn enough gasoline to equal their own weight.

Land Vehicles

The most popular method used to power a vehicle is to burn fuel (mainly fossil fuels) in an onboard engine. Alternatively, energy can be stored in some other form (batteries, fuel cells, flywheels, or compressed air) and used as is needed. No matter which approach is used, to propel land vehicles, power is needed for four purposes.[3]

1. *Overcoming the rolling friction of the tires* - A heavier vehicle, a rougher road, and faster movement all result in a greater rolling resistance. Other parameters that affect rolling friction are the size and inflation pressure of the tires and the type, material, and age of the treads. At low speeds a vehicle's rolling resistance is large relative to other resistances, and most power is dissipated at the wheels. For tracked vehicles such as tanks, heavy construction equipment, and rail systems, rolling friction is much greater and dominates all other types of resistance.[4] Newer trains using magnetic levitation have no contact with tracks and rolling friction is virtually eliminated.[5]
2. *Overcoming aerodynamic (wind) resistance* – Aerodynamic resistance is the result of the interaction between a moving vehicle and the fluid through which it travels. The magnitude of the force increases with the density, the square of the vehicle's velocity relative to oncoming wind, the projected area of the vehicle in the direction of travel (called the frontal area), and the shape of the vehicle. The effect of body shape on drag is usually expressed in terms of the coefficient of the drag. The drag coefficients for some common vehicles and body shapes are given in Table 14-2. Streamlining reduces the coefficient of drag and is particularly effective at high speeds where wind resistance

Table 14-2. Drag Coefficients

Car	C_D	*Shape*	C_D
Trucks	0.70	Flat Plate	1.17
Sedans	0.55	Cylinder	1.10
Porsche 924	0.34	Sphere	0.41
GM Sunraycer	0.13	Teardrop	0.04

[3] Here we concern ourselves mainly with passenger cars. Similar analysis can be carried out for other types of land vehicles such as buses, trucks, tanks, or railcars.
[4] Wong, J. Y., Theory of Ground Vehicles, Second Edition, John Wiley and Sons, Inc., 1993.
[5] With a top speed of 430 km/h (267 mph), China's maglev train is considered to be the world's fastest commercial train. An experimental Japanese mag-lev train set the current speed record of 581 km/h (361 mph) on a test track near Tokyo in 2003.

is dominant. For most passenger cars, drag coefficients vary from 0.3 to 0.7 and aerodynamic resistance becomes significant at speeds exceeding 30-50 km/hr (20-30 mph).

3. *Climbing (overcoming gravitational resistance)* – As a vehicle climbs a slope, it must overcome gravity. The resistance increases with the grade of the road and the weight of the vehicle. When the vehicle is going downhill, this force is negative; therefore the total power required to propel the vehicle downhill is actually less than what is needed on a flat surface.
4. *Accelerating* – The acceleration force is equal to the vehicle's mass times the acceleration. When the vehicle is decelerating (braking) this force is negative, which helps to reduce the overall power (energy) requirements. We will see later that electric and hybrid vehicles can regain a large fraction of the energy that would otherwise be lost during deceleration by regenerative braking.

In addition, engines must provide enough power to operate a number of accessories such as heater, air conditioner, lights, wipers, horn, power steering, and a variety of microprocessors.

Of the four types of resistance, only the first two contribute significantly to power requirements during cruising. Climbing and acceleration are mainly important during city driving, in stop-and-go traffic, and when passing other vehicles on highways. During idling, engine provides no useful work, and all energy goes to overcome engine friction. Whether a vehicle is cruising at a constant speed on a flat plain or accelerating over an incline, to reduce power it is desirable to reduce various resistances. This can be done:

a. *By making vehicles lighter.* Heavier cars waste energy by flexing and heating up the tires. Both rolling and acceleration forces increase with vehicle mass and are therefore reduced proportionally to weight in lighter vehicles. There are new fiber composite materials (carbon, glass, and Kevlar foam) that are many times stronger than steel and weigh only one third to one half as much. Composites also make it possible to build frameless "monocoques,"[6] making manufacturing easier with considerable savings in both material and energy. Reducing the frame weight by a certain amount makes the vehicle lighter by more than that amount; the synergistic effect resulting from lighter frames makes it possible to have a lighter suspension to carry the weight, a smaller engine, and less fuel to move it.
b. *By making vehicles more aerodynamic.* Sleeker, sportier shaped bodies and smoother underbodies reduce air friction, allowing cars to move faster and consume less fuel. Convertible cars, cars with rolled down windows, and cars with large frontal areas are considerably less aerodynamic.[7]

[6] A type of vehicle construction in which the body is integral with the chassis.
[7] Katz, J. Race Car Aerodynamics, Robert Bentley, Inc., 1995.

c. *By proper maintenance.* Fuel economy can be considerably improved by keeping tires inflated, air filters clean, and the engine tuned. Driving at the cruising speeds, using accessories less frequently, and avoiding fast braking and acceleration can also help.
d. *By using hybrid technology.* Hybrid vehicles have two modes of propulsion, usually an internal combustion engine and an electric motor. The vehicle operates as an electric car during city driving and in stop-and-go traffic, but is essentially a conventional gasoline or diesel vehicle during cruising and highway driving. Since the power required during cruising is low, a much smaller engine is needed. The electric motor delivers additional torque when accelerating or climbing steep grades. Thus, the fuel economy of these vehicles is significantly improved.
d. *By using alternative fuels.* Contemporary automobiles are highly inefficient; almost 80% of the fuel energy is lost in the exhaust and dissipated to the environment as waste heat. Certain fuels have been shown to have marginally better efficiencies than gasoline and diesel oil, but are of interest mainly due to their reduced emissions.

Marine Vehicles

As ships and other marine vehicles move through water, they experience both water and air resistances. The resistance through water is the greater of the two and depends on many factors including ship speed, hull form, and water temperature. The total hull resistance consists of several components:

a. Skin friction drag due to friction between the hull and water
b. Form drag due to pressure forces acting on the hull
c. Wave drag due to energy lost in creating and maintaining the ship's characteristic bow and stern waves
e. Wind resistances

Frictional losses are the function of the hull's wetted surface area, surface roughness, and viscosity. At low speeds, viscous resistance is dominant and can account for up to 50-80% of total resistance. As speed increases, the wave-making resistance increases rapidly and eventually dominates all other resistances. Air drag can also play a role and may contribute between 4 to 10% to total ship resistance, depending on speed, the shape of the ship above the waterline, and the area of the ship exposed to the air. Wind and current resistances can be significant in rough waters and when a ship runs into strong headwinds.

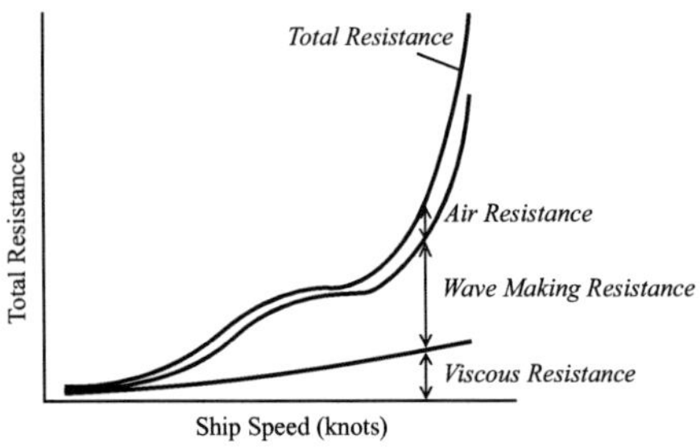

Figure 14-1
Components of hull resistance

The resistance is not proportional to velocity, but increases as the square of velocity at low speeds and more rapidly as velocity to the fifth power at higher speeds (Figure 14-1). The power required to propel a ship through water is the product of total hull resistance and ship speed (P = F.V). Therefore, the power required can be proportional up to ship speed raised to the 6th power! The fuel consumption rate also increases accordingly; it takes much more fuel to travel a given distance at a faster speed than traveling the same distance at a slower speed.

Aircrafts

The resistance forces acting on aircrafts are similar to those discussed for marine vehicles and consist mainly of pressure and viscous forces. Pressure forces act normal to the surface and arise from differences in air pressure along the body. Viscous or shear forces are surface resistances close to the airplane body due to the fact that it moves in a viscous fluid. The magnitudes of these resistances vary widely depending on the airplane's shape, size, speed, and altitude.[8] Normally, the resultant aerodynamic forces are resolved into two components: that perpendicular to the flight path *(lift)* and that parallel to the flight path *(drag)*. During cruising, the engine must be able to overcome drag forces. Maximum power is needed during takeoff and increases with the lift and the rate of the climb.

Example 14-1: To ferry a space shuttle from Edwards Air Force Base in California to its launching site at NASA's Kennedy Space Center in Florida, it is bolted on the back of a modified Boeing 747-400 jumbo jet. The volume of fuel used for this journey is huge, as it will take one gallon of fuel to travel only one length of the plane (231 feet). Because of many refueling stops along the way, the trip takes two days to complete. Calculate the amount of fuel it takes to complete the 2,600 mile journey.

Solution: The fuel efficiency of the Boeing 747-400 is around 0.2 mpg. Because of the heavy payload and loss of much of the aerodynamic advantages, with the space shuttle piggy-backed, the fuel efficiency drops to only 231 feet per gallon (0.04 mpg). For the 2,600 mile trip, 2,600 mi/(0.04 mi/gal) = 65,000 gallons of fuel is needed. The Boeing 747-400 can carry 32,750 gallons (215,000 liters) of fuel, which is about half the required fuel. For safety reasons, planes carry at least twice the amount of fuel required to complete a trip. This means the plane has to make at least 3 or 4 stops for refueling.

Ex 14-1

Internal Combustion Engines

Internal combustion engines have found major applications in many industries; particularly in transportation, home appliances, and stationary power generation systems. Depending on their size, they can deliver power in the range of a few to several thousand kilowatts. The principle of operation of internal combustion engines is simple -- a mixture of fuel and air is burned inside a combustion chamber and the product of the combustion is expanded, turning the energy of fuel into useful (shaft) work. Examples of internal combustion engines are gasoline engines, diesels, and gas turbines.

[8] Schaufele, R. D., The Elements of Aircraft Preliminary Design, Aries Publications, Santa Ana, Ca, 2000.

There is no single person who can be named as the inventor of the internal combustion engine; the process was an evolutionary one that started with Christian Huygens in 1680 who designed (but never built) an internal combustion engine working with gun powder. Over the next two centuries many inventors perfected combustion engines, among them Nicholas Otto (1876), who proposed a four-step "Otto" cycle, Gottlieb Daimler (1885), who invented the prototype gas turbine, and Karl Benz (1889), who built the first practical four-stroke, two cylinder engine and the first four-wheeled automobile. Wilhelm Maybach (1890) then used Benz's design to build the first four-cylinder, four-stroke engine.[9]

Figure 14-2
1908 Ford Model T

The first automobile manufacturing company in the United States was established by Henry Ford in 1903. The company named its automobiles according to the letters of the alphabet. In 1908 it introduced the Model-T (Figure 14-2), the first mass-produced vehicle, and offered it at prices affordable to a great number of people. With increased reliability and cheap, widely available petroleum, the number of privately owned cars increased rapidly.

To meet their transportation needs, Europeans and Americans followed different paths. Europeans invested mainly in trains, trolleys, and subways. American policy makers, under pressure from the booming automotive industry, promoted private car ownership. Some American car companies went as far as buying and dismantling streetcars and replacing them with diesel buses, which in many instances were poorly designed, effectively forcing the public to purchase private cars for transportation.

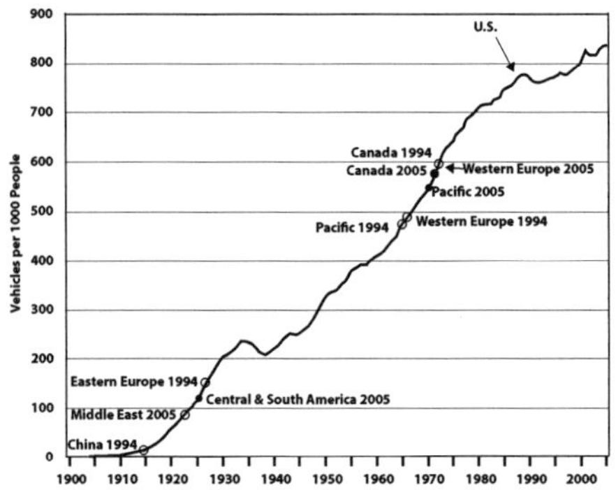

Figure 14-3
Vehicles per thousand people for various regions in the world.
Source: US Department of Transportation, Federal Highway Administration, Highway Statistics 2007, Washington D.C. (http://www.fhwa.dot.gov)

Following WWII, US interstate highways expanded rapidly, literally paving the way for automobiles as the principal source of transportation.[10] Of the 700 million cars in the world, over one-third belong to Americans. Today, Europeans have the same ratio of cars to people as the US had a quarter century ago. Similarly, present per capita car ownership in the former Soviet Union is comparable to that of the US in 1923 (See Figure 14-3). In 2003, Americans traveled more than 4.1 trillion miles, over 87% of which was by automobiles, SUVs, and other light trucks. As the data presented in Table 14-3 indicate, the miles traveled by commuters using rail systems totaled only about 0.6% of that traveled by passenger cars. Bus and train occupancy rates were very low; as a result, fuel efficiency measured as BTU/passenger-miles were roughly the same for passenger cars and other modes of transportation. Public transportation in Europe is, however, significantly more efficient as trains and buses run with much heavier occupancy rate; four to five- fold increases in efficiency (730 BTU/passenger-miles for trains and 1050 BTU/passenger-miles for buses) are common in most European countries.

[9] Brittanica Online Encyclopedia *(http://www.britannica.com).*
[10] See for example Jane Holtz Kay, Asphalt Nation: How the Automobile Took over America and How We Can Take It Back, *New York: Crown Publishers, 1997,* p. 171.

Whether the European or American model of transportation is superior is a matter of taste and public policy. In the United States, cars are a necessity and are well-entrenched in the American lifestyle. Distances between homes and offices and between different cities are much longer, and public commuter systems are vastly inadequate. As a result, the use of public transportation is largely limited to the low-income and the disadvantaged. In Europe, on the other hand, distances are shorter, the infrastructure is more developed, and a vast network of railways and other public transportation systems for commuting and long distance travel already exist. Affordable and easily accessible public transportation has removed the need for personal vehicles for most people, making cars a luxury - nice to have for special occasions, leisurely evening activities, and weekend excursions.

Spark Ignition (Gasoline) Engines

Most passenger cars operating today are spark ignition; their engines work by allowing a mixture of fuel and air prepared in a carburetor to burn and expand inside the cylinders. The downward motion of a piston is converted to rotational motion of the crankshaft by means of connecting rods and eventually, by way of a transmission, to the wheels.

The *rotary (Wankel) engine* is a special type of spark ignition engine in which the piston-cylinder assembly is substituted by a three-edged rotor mounted off-center of a specially designed housing. Unlike conventional engines where the four strokes are accomplished by linear up and down motions of the pistons, rotary engines work by varying the volume as the rotor rotates in the housing. The main advantages of rotary engines are their lower mass and volume as compared to reciprocating engines of comparable power rating and their simpler construction. Furthermore, because rotary engines deliver power directly in rotational form, their overall efficiency is higher. The main disadvantage is that rotors are more difficult to seal, which makes hydrocarbon emission higher. Basic operation of reciprocating and rotary engines is described in Figure 14-4.

Table 14-3. US Passenger Travel and Energy Use, 2004

Mode	*Number of Vehicles*	*Passenger-miles (billions)*	*% of total travel*	*Load Factor (persons/ vehicles)*	*BTU/ passenger-miles*
Automobiles	136 million	2,669	56	1.6	3,496
Light trucks	81 million	1,479	31	1.7	4,329
Motorcycles	5.4 million	11	0.2	1.1	2,272
Buses[a]	695,000	21	0.3	8.7[c]	4,318[c]
Planes[b]	219,000	549	12	90	3,959
Rails	18,600	31	0.6	24	2,978

[a] Includes transit, intercity, and school; [b] General aviation; [c] Data available for transit buses only

Ref: Davis, S. C., Diegel, S. W., Table 2.12. "Transportation Energy Data Book," ORNL-6978, Edition 26, 2006.

Charged Ignition (Diesel) Engines

Instead of compressing the air/fuel mixture to high temperatures and pressures as is done in gasoline engines, diesel engines operate by compressing air alone. Because there is no fuel present, air can be heated to pressures and temperatures well over the ignition temperature of fuel without concern for engine knock. Unlike gasoline engines, diesels require no carburetors or spark plugs. Instead, tiny droplets of diesel fuel are injected directly into the cylinder, where they mix and react with already heated air and burn (Figure 14-4). Since injection times are relatively long, the piston travels an appreciable distance before fuel is cut off, keeping the pressure in the cylinder nearly constant.

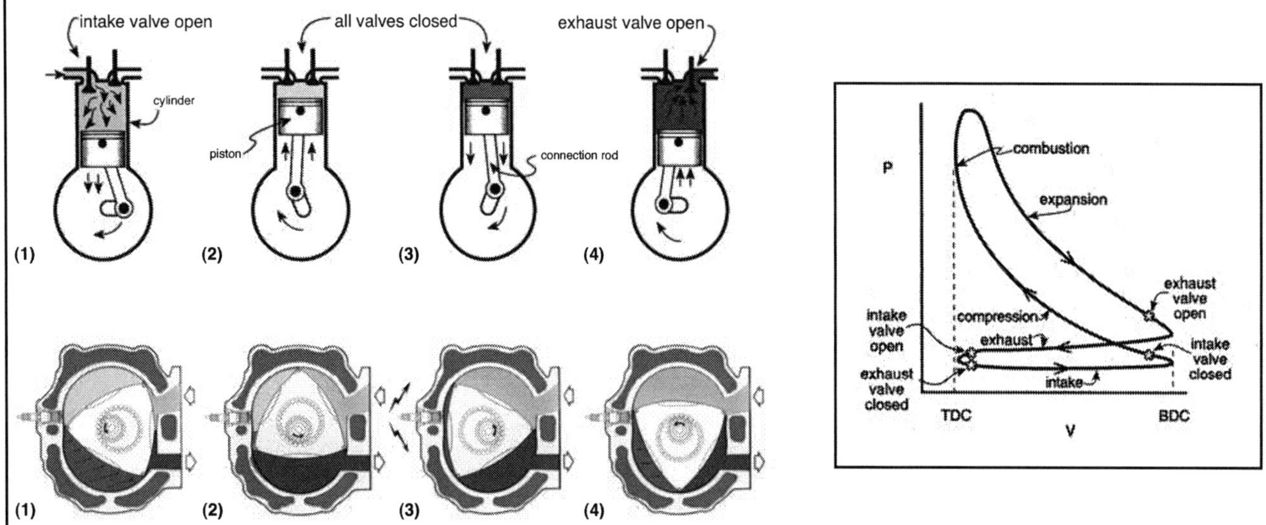

Four-stroke Otto/Rotary Engines

1. *Intake stroke:* the intake valve (port) opens and the air-fuel mixture is sucked into and fills the cylinder.
2. *Compression stroke:* The intake valve closes, and the crankshaft helps the piston to move up (or the rotor to rotate), compressing the mixture to a high temperature and pressure.
3. *Power stroke:* A spark plug ignites the mixture. The pressure suddenly rises, forcing the piston to move up (or the rotor to rotate), expanding gases and providing shaft work. At the end of the power stroke, the exhaust valve opens and the pressure suddenly drops to near atmospheric conditions (blow-down).
4. *Exhaust stroke:* The piston moves upwards (or the rotor continues its rotation and clears the exhaust port), exhausting the burned gases.

Four-stroke Diesel Engine

1. *Intake stroke:* As the piston slides down, the intake valve opens and pure air is sucked into and fills the cylinder.
2. *Compression stroke:* The intake valve closes, and the crankshaft helps the piston to move up, compressing the air to a high temperature and pressure. The temperature is now above what is necessary for self-ignition of the fuel.
3. *Power stroke:* At TDC, an injection pump gradually injects fine fuel droplets into the combustion chamber. As fuel droplets meet particles of hot air they burn, increasing the pressure and forcing the piston to move down, expanding gases and proving shaft work. At the end of the power stroke, the exhaust valve opens and pressure suddenly drops to near atmospheric conditions (blow-down).
4. *Exhaust stroke:* As the piston makes the second round of upward motion, burned gases are exhausted. The exhaust valve then closes in preparation for the cycle to begin again.

Figure 14-4

Four-stroke internal combustion engines for reciprocating (top) and rotating (bottom) engines.

Generally speaking, diesels are less efficient than gasoline engines with similar compression ratios. However, because diesels are designed to operate at higher compression ratios than spark-ignition engines (20-22 compared to 8-9 for spark ignition engines), they have higher efficiencies (~40% compared to ~30% for the petrol engines). Their main drawbacks are that they are bulkier, emit more, have a higher capital cost, and accelerate more slowly. These characteristics make diesels particularly attractive in stationary applications and for buses and large trucks. Because of the stricter air pollution standards, diesels are less common in the United States than Europe. New technological inventions are underway that makes diesels cleaner, and thus sales of diesels are likely to increase in the US in the near future.

Gas Turbines

The basic operation of a gas turbine is relatively simple. Air is drawn through a nozzle and compressed in a compressor to a very high pressure and temperature before entering a combustion chamber. Fuel is injected through a fuel injector into the combustion chamber where it mixes with the air and burns. Depending on the application, kerosene, natural gas, or jet fuel can be used. The product of combustion is expanded and runs a turbine that provides power. The shaft work can be used to propel a helicopter, provide thrust to a jet engine, drive a tank, or turn a generator and produce electricity (Figure 14-5).

The main advantages of gas turbines over gasoline and diesel engines are their relatively low emissions, very high power density (kW/m^3) and specific power (kW/kg), multi-fuel capability, and smooth, vibration-free power delivery. New ceramic materials allow gas turbine operation at higher temperatures with efficiencies exceeding those of conventional engines by as much as 50%.

Transportation Fuels

The most common fuels used for automobiles are still gasoline and diesel oil. They are liquid, are relatively inexpensive, and have good combustion characteristics. Furthermore, the necessary infrastructure for production,

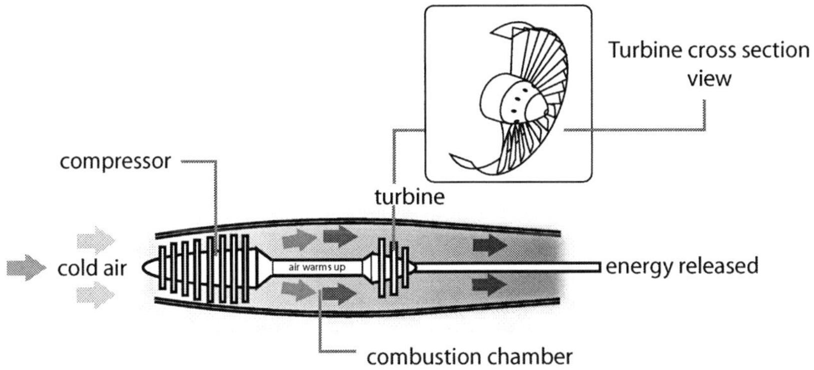

Figure 14-5
Gas turbine operation follows the Brayton cycle, consisting of four processes:

storage, and distribution of these fuels is fully in place. These fuels, however, are becoming increasingly more scarce and have been associated with most of our air pollution problems.

In the search for better fuel economy and reduced emissions, much research has been done on the use of alternative fuels. However, these fuels have not found wide acceptance and their use has been limited to fleet and government vehicles. Suitable alternative fuels can only find wide-scale acceptance if they meet many of the same requirements as gasoline and diesel fuels. To get a better understanding of the issues in using alternative fuels, various transportation fuels as well as their advantages and drawbacks are summarized below.

Gasoline (Petrol) is the most common fuel used in vehicles. It is a colorless and volatile liquid made up of many hydrocarbons, but is conveniently represented as a single compound with a molecular structure approximated as C_8H_{17}. Like most other liquid fuels, gasoline is derived primarily from crude oil. To aid combustion and reduce hydrocarbon emissions, it is advantageous to add some oxygen to gasoline. The product, called *reformulated gasoline*, has been found to reduce smog in cities with a large concentration of cars. Additionally, adding some detergents to gasoline will help to prevent the build-up of engine deposits, keeping engines working smoothly and burning fuel cleanly.

Diesel oil, like gasoline, is a mixture of light distillate hydrocarbons. Its boiling point is somewhat higher than that of gasoline, allowing for engine operation at an increased compression ratio with less concern over the engine knock, common in gasoline engines.

Kerosene is a mixture of heavier hydrocarbons; it is mostly used as heating oil. Kerosene is lighter than gasoline and has a greater heating value. It is used as jet fuel in pure distillate (JP5) or is blended with gasoline (JP4). Kerosene is often represented as a single hydrocarbon with the formula $C_{12}H_{26}$.

Biodiesel is derived from vegetable oils, seeds, animal fats, and algae. Biodiesels are, however, more viscous than fossil fuels; this makes cold-temperature starts more difficult. Because biodiesel is a biomass, it is biodegradable, less toxic, has fewer emissions, and does not contribute to global warming. Biodiesel is marketed as pure biodiesel (B-100), or as a blend of 20% biodiesel and 80% petroleum diesel (B-20).

Natural gas is about 90% methane and 10% heavier hydrocarbons. Its high octane number (ON = 130) allows engines to operate at compression ratios as high as 12:1, compared to 8:1 for gasoline engines. The main problems with methane are its toxicity and lack of odor, which makes leaks undetectable. Furthermore, methane-powered cars have inferior power and energy densities, resulting in roughly 20% less fuel efficiency than gasoline and diesel engines and a relatively low cruising range.

Because of its low energy density, natural gas must be compressed or liquefied before it can be used for vehicles. To produce *compressed natural gas* (CNG), methane must be compressed to 15-25 MPa (2,400-3,600 psi) and stored in special containers. At ordinary room temperatures, it is impossible to liquefy methane by compression alone and must be cooled to cryogenic temperatures. CNG is generally considered a relatively clean fuel, as the emission of carbon monoxide, hydrocarbons, and particulate matter is substantially less than that of gasoline or diesel fuel. Reduced emissions are offset, however, by an increase in the emission of unburned methane, a potent greenhouse gas which traps heat about twenty times more effectively than carbon dioxide. At this time, methane emission accounts for 13% of all greenhouse gases, but it would become a major concern if natural gas were to become a dominant fuel for transportation.

Liquefied Petroleum Gas (LPG) is natural gas leftover after methane is removed. It is mostly propane, but some butane and higher hydrocarbons are also present. Unlike methane, propane can be liquefied at room temperature and relatively low pressure; this makes it suitable for storage in light fuel tanks and gives cars a driving range comparable to that of gasoline.

Methanol (also called methyl or wood alcohol) is the simplest alcohol. Many consider methanol to be the fuel of choice. It is liquid, like gasoline, but has a higher octane rating (ON = 100), and can be produced by utilizing a variety of methods from gasoline, natural gas, and other fossil fuels in large quantities and at a low cost; it may also be produced from the distillation of wood chips, garbage, and animal manure. Since methanol has a high H/C ratio, it is a relatively clean fuel and, except for aldehydes, methanol emissions are considerably lower than those of gasoline.

Unfortunately, methanol has several drawbacks which limit its use as a viable transportation fuel. Methanol has an invisible flame. To enhance flame luminosity, methanol is usually mixed with a small percentage of gasoline and is often used as M85 (85% methanol, 15% gasoline) instead of pure methanol (M100). Methanol is so toxic that even ingesting a small amount can cause blindness or death. Another problem is that methanol is corrosive to aluminum and other materials commonly used in seals and pipe fittings for transporting gasoline. Therefore, methanol cannot be transported through existing pipelines and must be trucked. Methanol has only half the energy density of gasoline, which means for a given driving range about twice the amount of fuel will be needed. Because of these limitations, the number of methanol vehicles has declined steadily in recent years. Because methanol can be converted to hydrogen at relatively low temperatures, it remains an attractive fuel for producing the hydrogen needed to operate fuel cell vehicles.

Ethanol (also called ethyl or grain alcohol) is made through the biochemical conversion (fermentation) of sugar and perennial grasses (such as *miscanthus*), or by the hydration of ethylene from petroleum. Ethanol can be used directly in fuel cells, used in the pure form or mixed with gasoline to power internal combustion engines.

Since ethanol is a biomass, there is no net carbon dioxide emission and it does not contribute to global warming. Like gasoline, ethanol is a liquid, making it possible to use in existing fuel tanks, fuel pumps, and fueling stations. Engine performance is also comparable to that of engines using gasoline. NOx emissions are lower and formaldehyde higher, but other emissions are about the same as gasoline and slightly greater than methanol.

Ethanol has a lower heating value than gasoline and burns at a lower A/F (9:1 as compared with 15:1 for gasoline). So, for a given engine more fuel can be introduced per cycle, compensating for its lower heating value, and an overall higher power output is possible. Ethanol has a very high octane rating (ON = 113) and is therefore most suitable for use in high-compression engines. Ethanol, like methanol, burns with an invisible flame. To increase visibility, ethanol is often blended with gasoline to produce gasohol (a mixture of 10% ethanol and 90% gasoline), E85 (85% ethanol, 15% gasoline), and E95 (95% ethanol, 5% gasoline). E85 is usually used for light-duty applications, whereas E95 is best for heavy-duty vehicles. Gasohol is not common in the United States.

Hydrogen is considered by some to be the fuel of the future; it is the simplest and most abundant element in the universe. Pound for pound, hydrogen has the highest energy content of all fuels, and when burned in oxygen the only by-product is water. Hydrogen is also an ideal fuel to produce electricity – through fuel cells – directly eliminating carbon dioxide and other pollutants that are associated with burning of fossil fuels.

There are, however, a number of factors that prevent hydrogen from being utilized as the most promising transportation fuel, and some even have dismissed hydrogen as hype. Unlike petroleum and methane, hydrogen cannot be mined but must be produced in a number of ways, ranging from hydrocarbon reformation, to coal gasification and pyrolysis, to the electrolysis of water. Although hydrogen is a clean fuel, converting conventional fuels to hydrogen does not necessarily improve air quality. Cleanliness depends on how the hydrogen is produced. If hydrogen is

[11] If hydrogen is burned in an internal combustion engine then, besides water vapor, some NOx will also be produced. The high temperatures of hydrogen flames cause air to be ionized into atomic nitrogen and oxygen, which subsequently react to produce nitric monoxide. When hydrogen is used in fuel cells, the reaction is at low temperatures, and the only product is water vapor.

produced by reforming ethanol, then the large amount of energy required for the distillation of ethanol will take away most of the hydrogen advantage; if hydrogen is produced by reforming fossil fuels, then most of its environmental benefits disappear as well. Renewable sources such as solar, wind, and hydropower are ideal because they can electrolyze water in large quantities without polluting the atmosphere. Nuclear sources are another option.

Currently, most commercial hydrogen is produced by steam reformation of natural gas. Natural gas is a particularly good candidate for producing hydrogen at the wellhead. Instead of piping crude or natural gas, large reformers can be installed to strip off the hydrogen and ship it through the pipelines. The byproduct of the reformation process, mainly carbon dioxide, can then be sequestered by injecting it back into the gas field, raising the pressure and improving extraction efficiency (See Chapter 7, Enhanced Recovery).

Hydrogen safety and storage are two other concerns most often raised in regard to hydrogen. Hydrogen is highly explosive and burns easily with an invisible flame. When compared to most other gaseous fuels, hydrogen's flammability range is unusually broad, from 4 to 75 percent. Natural gas, in contrast, burns between 5 and 15 percent and requires much less energy to ignite. Table 14-4 compares the ignition characteristics of hydrogen with those of methane and gasoline.

Table 14-4. Ignition Characteristics of various gaseous and liquid fuels

	Hydrogen	*Methane*	*Gasoline*
Ignition Temperature	585 °C	540 °C	230-480 °C
Flammability Limits	4-75% by vol.	5-15% by vol.	1.4-7.6% by vol.
Explosion Limits	20-65% by vol.	6-14% by vol.	1-7.6% by vol.
Energy Density	120 MJ/kg	50 MJ/kg	44 MJ/kg
	8.5 MJ/L	40 MJ/L	21 MJ/L
Ignition Energy	0.02 mJ	0.2 mJ	0.2 mJ

Source: Flynn, T., Cryogenic Engineering, Second Ed., CRC Press, 2005.

Hydrogen storage is also a problem, as hydrogen must be compressed to very high pressures, cooled to very low temperatures, or stored as chemical or metal hydrides, adding considerable weight to the storage container. Hydrogen has a low energy density; one kilogram of hydrogen contains about the same amount of energy as a gallon of gasoline and about one third that of natural gas. Another concern is hydrogen embrittlement. Due to the small size of its molecules, hydrogen is far more prone to leak out, causing pipes to deteriorate faster. Existing distribution channels used for transport and delivery of gasoline are not suitable for hydrogen and the cost of developing new systems is very high.

Aside from these considerations, some concerns have recently been raised in regards to the effect of potential gas leaks to the environment. Normally, hydrogen is stored at high pressures and about 10% of all hydrogen manufactured is leaked through the atmosphere. Being so light, hydrogen rapidly rises through the atmosphere, eventually reacting with oxygen to form water vapor. The additional water makes the stratosphere wetter, cooling the lower atmosphere, particularly in the polar regions where most hydrogen is converted to water vapor. This would disrupt the ozone layer, causing 7-8% more depletion over the poles.[12]

Emissions from Internal Combustion Engines

When automobiles were marketed in the early twentieth century, they were seen by many as a clean form of transportation, doing away with the nuisance of horse feces in large urban cities. In the decades that followed, as the numbers of vehicles increased, the impact of this technology on air quality and health became more pronounced and cars became the main source of environmental pollution. Today, motor vehicles are responsible for roughly four-fifths of all carbon monoxide emissions and nearly half of all hydrocarbons and nitrogen oxides released into the atmosphere globally (Table 14-5). In addition, motor vehicles produce carbon dioxide, a potent greenhouse gas and a major contributor to the production of ozone in the lower atmosphere. Compared to reciprocating engines, gas turbines produce lower emissions. This is because gas turbines operate under ultra-lean combustion. Furthermore, gas turbines do not respond well to transient operations and must operate in a continuous mode.

Table 14-5. US Total emission of criteria pollutants in 2002*

Source	*Total Emission*	*Mobile*	*Stationary*	*Industrial*	*Others***
CO	108 million tons	82%	3.8%	2.3%	12.6%
NOx	20 million tons	55%	38%	3.9%	2.2%
HC (VOC)	16 million tons	42%	6.6%	41.5%	10.2%
PM_{10}	21 million tons	2.2%	6.2%	3.1%	88.5%
$PM_{2.5}$	7 million tons	6.1%	17.9%	7.1%	69.0%
SO_2	14 million tons	4.4%	86.3%	8.9%	0.3%
NH_3	7 million tons	6.3%	1.4%	3.4%	88.8%

* To the closest percent
** Waste disposal, recycling, etc.
Source: US EPA, National Emission Inventory Air Pollutant Emission Trends website http://www.epa.gov/oar/oapqs

The main sources of engine emissions are unburned hydrocarbon from the crankcase, carburetor, and fuel tank, as well as tailpipe emissions. Most emissions come from the exhaust pipes and include carbon monoxide, nitrogen oxides, soot, and various volatile organic compounds. Exhaust pipe emissions are largely a result of incomplete combustion processes and are affected by many factors. Three parameters have profound effects on how combustion proceeds and how much pollution is formed. These

[12] NaTrompme, et al., "Potential Environmental Impact of a Hydrogen Economy on the Stratosphere," *Science*, 300, 1740-1742, 2003.

parameters are referred to as the three Ts of combustion - temperature, time, and turbulence. *Temperature* determines the rate of reaction and heat release. If everything else remains the same, higher temperatures increase production of nitric oxides and limit the production of hydrocarbons, carbon monoxide, and soot. Temperature is highest near the stoichiometric conditions and for fuels with high hydrogen to carbon (H/C) ratios. Stoichiometric conditions occur when an air/fuel mixture burns to completion and when, under equilibrium conditions, the only products of combustion are carbon dioxide and water. The *time* that reactants have in a flame reaction zone determines the extent to which reactions go to completion. Greater residence times generally favor more stable molecules (carbon dioxide, nitrogen, water vapor) and lower overall emissions. *Turbulence* affects the rate of mixing of fuel and air molecules. Better mixing prevents localized pockets of very rich or very lean mixtures, thus allowing the reaction to complete and produce fewer contaminants.

A brief discussion of major pollutants from automobiles and other internal combustion engines and factors affecting their formation is given below.

Nitric oxide results from a reaction between oxygen and nitrogen heated to combustion temperatures. As we may expect, maximum NOx occurs near stoichiometric conditions where temperature is at its peak and sufficient oxygen is available. Too much air dilutes the mixture and reduces the flame temperature; with too much fuel, little oxygen is left over for reacting with nitrogen. The best way to reduce NOx emission is by lowering the combustion temperature and limiting the availability of air. Recirculating a portion of exhaust gases back to the intake manifold is shown to be effective in reducing NOx emission by lowering the air to fuel ratio and cooling the flame. Unfortunately, exhaust gas recirculation (EGR) valves must operate with a rich mixture, which promotes production of hydrocarbons and carbon monoxides. Another method to reduce NOx emissions is to add some hydrogen to the gasoline/air mixture. The drawback is backfiring, a result of hydrogen's low ignition temperature and lower power rating.

Carbon monoxide results from incomplete reactions. It is formed when there is insufficient oxygen, temperature, or time to oxidize all carbon to carbon dioxide. The obvious way to reduce carbon monoxide emissions is by increasing the compression ratio (increasing temperatures) or by maintaining a lean fuel mixture. Advancing the spark raises peak temperatures and increases the time available for reaction, both of which inhibit carbon monoxide formation.

Hydrocarbon emission can be divided into two categories, unburned and partially burned. Roughly half of all hydrocarbon emissions come from unburned vapor during fill ups, evaporation from the carburetor, and from other hot surfaces when the engine is shut off. Other sources of hydrocarbon emission are crankcase blow-by and emission from leaky valves, piston rings, and gaskets. The hydrocarbon release during fueling is reduced by installing vapor recovery nozzles on gas pumps. The crankcase emission is practically eliminated by closing off the vent to the atmosphere and installing a positive ventilation valve (PCV) to recycle blow-by back into the engine's intake.

Particulates– Besides gaseous pollutants, most internal combustion engines produce some solid carbon particles called soot. Soot particles are large clusters of carbon atoms generated in the combustion chambers and aggregated to sizes ranging from a few nanometers to hundreds of microns. Soot can also be formed from the lubrication of oil next to cold surfaces and from the rapid expansion cooling during the power stroke. Although some soot is generated in gasoline engines under heavy loads, diesels are by far the main source of soot. This is due to the relatively poor mixing of the fuel and air in the combustion chamber. Diesel particulate problem has largely been solved in Europe with superior fuel injection systems and specially manufactured catalytic filters. In the US, improved fuel injection systems have helped remove odor and visible smoke. However, because of the high sulfur content of the US diesel fuel, catalytic filters have yet to be deployed.

Lead was a major source of particulate emission prior to the 1980s, during which time it was added to gasoline as an anti-knocking agent. Knock occurs as a result of premature ignition in engines with high compression ratios. A small amount of lead additive had the effect of increasing the octane number (and ignition temperature), which allowed building engines with higher compression ratios and better efficiencies. Lead, however, is a poisonous metal with detrimental health effects. It also contaminates catalytic converters that are designed to reduce exhaust emissions. Because of this, use of lead additives was gradually phased out and, except for very old cars, all cars operate with lead-free gasoline now.

In addition to carbon particles, many combustion systems such as industrial boilers and kilns produce sulfate. Because sulfur in gasoline and diesel fuels is removed before they are sold, few sulfate particles are formed in automobiles.

Catalytic Converters

Vehicular exhaust emissions not only contain carbon dioxide and water vapor, but also a significant amount of carbon monoxide, nitrogen oxides, and hydrocarbons. Generally speaking, nitric oxides are produced during cruising and acceleration (driving mode) where the mixture burns at its maximum temperature near stoichiometric conditions. On the contrary, during braking and when the engine idles (stopping mode), some exhaust enters the intake manifold, causing carbon monoxide, hydrocarbon, and particulate emissions to be high.

For the best performance - maximum efficiency and minimal environmental pollution - it is ideally desirable to convert all carbon into carbon dioxide, all hydrocarbons into water vapor, and all nitrogen into molecular nitrogen. To convert carbon monoxide and hydrocarbon we need a high temperature and an oxidizing atmosphere. The problem is that these same conditions provide a favorable environment for the oxidation of nitrogen to nitric oxides. Three-way catalytic converters can achieve successful modifications of CO and HC while at the same time reducing NOx to molecular nitrogen. This is achieved in two steps:

- Convert all nitric oxides into nitrogen in a reducing atmosphere (fuel rich).

$$NO \xrightarrow{Rh} N_2 + O_2 \tag{14-1a}$$

- Convert all hydrocarbons and carbon monoxide into carbon dioxide and water in an oxidizing atmosphere (air rich).

$$CO + HC \xrightarrow{Pt,\,Pd} CO_2 + H_2O \tag{14-1b}$$

Generally, catalytic converters are made of a stainless steel canister mounted along the exhaust pipe. Inside the container is either a porous ceramic honeycomb or is filled with ceramic pellets embedded with small particles of alumina with catalytic materials deposited on their surface. Rhodium is the catalyst of choice for reducing, and platinum and palladium work best for oxidizing reactors. Three-way converters employ redox (for reduction/oxidation) catalysts, which are essentially made of two reactors, a reducing reactor followed by an oxidizing one.

$$CO + HC + NO \longrightarrow \boxed{\textit{reducing reactor}} \longrightarrow CO + HC + N_2 \longrightarrow$$

$$\longrightarrow \boxed{\textit{oxidizing reactor}} \longrightarrow CO_2 + H_2O \tag{14-1c}$$

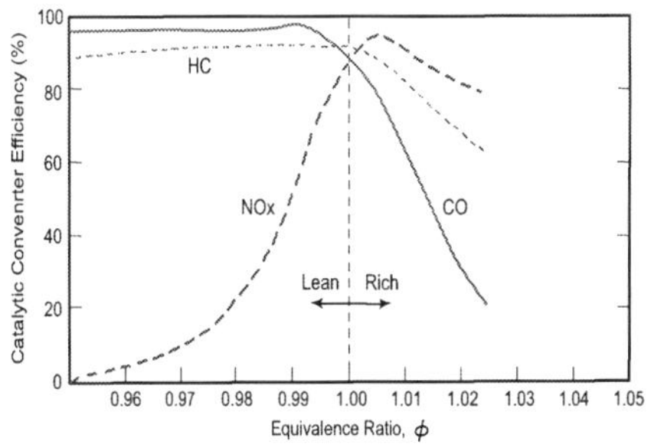

Figure 14-6
Conversion efficiency of the catalytic reactors as function of the fuel/air ratio. NOx removal efficiency is very low under lean conditions where efficiency is highest for CO and HC. In a narrow region near the stoichiometric ratio the conversion efficiency is high for all three contaminants.

Figure 14-6 shows the conversion efficiencies of the catalytic converter for CO, HC, and NOx at different fuel/air ratios. As can be seen, for redox reactors to work, they must operate within a narrow range of air fuel ratios (around stoichiometric). When in good working conditions, these reactors reduce in excess of 98% of CO and 95% of HC and NOx emissions.

Emission Standards

Prior to 1970, there were no standards regulating automobile emissions. In 1970, the Environmental Protection Agency (EPA) was established and, as its first act, it regulated the emissions of carbon monoxides to less than half their previous level. Complying with this standard was easy; all that was needed was to adjust the carburetors in order to provide a slightly leaner mixture. The result, however, was that the concentration of nitric oxides in the atmosphere climbed to unreasonable levels, forcing the EPA to adopt new emission standards for the control of both nitric oxide and carbon monoxide. As the number of cars grew and air quality deteriorated, the EPA became more stringent. The Clean Air Act Amendments (CAAA) of 1990 defined two sets of standards, Tier-1 and Tier-2, for light-duty vehicles (vehicles under 6,000 pounds such as passenger cars, sport utility vehicles, minivans, and pickup trucks).[13] Tier-1 regulations have already been implemented. Tier-2 standards began being phased-in in 2004 and are to be completed by 2007. Because emission standards are expressed in grams of pollutants per mile, light trucks, SUVs, and larger engines will have to utilize more advanced emission control technologies in order to meet the standards. Table 14-6 shows the minimum EPA standards for vehicles in the United States. California has devised its own standards which have always been more stringent than the Federal standards. Europeans and Japanese impose their own standards, which are generally comparable to the US standards. There is currently no particulate emission standard for passenger cars and light trucks. For big diesel trucks and buses, starting with model year 1994, the EPA has required a reduction in particulate emissions by 90%.

Table 14-6. US Federal emission standards for gasoline and diesel passenger cars

	HC g/mile	*CO g/mile*	*NOx g/mile*
Uncontrolled	11.0	80	4.0
1970	2.2	23	--
1972	2.9	28	--
1975	1.5	15	3.1
1981	0.41	3.4	1.0
2004	0.41	3.4	0.4

Source: 40 CFR 86.000.8 Office of Air and Radiation, US EPA (http://www.epa.gov).

Combustion Efficiency

We learned in Chapter 5 that for better thermal efficiencies, combustion must be carried out at a high temperature. This is accomplished by using better fuels, and by further compressing the air/fuel mixture. Diesel engines can operate at significantly higher compression ratios than gasoline engines. In gasoline engines, however, there is a limit on the amount that a mixture can be compressed before temperature exceeds the ignition point. If this happens, the entire charge is exploded at once in what we commonly call engine knock. In addition to the uncomfortable noise it creates, knocking reduces the life of the engine. Most modern gasoline cars have compression ratios of 8-10 and thermal efficiencies of around 20-25 percent; that means that roughly three quarters of the fuel energy is lost through the exhaust, releasing a substantial amount of air pollutants into the atmosphere. One way to reduce knock in gasoline engine is to use fuels with higher octane numbers. This has been done by switching to alternative fuels (like methanol), or by mixing gasoline with fuels of high octane numbers (such as gasohol). A team of investigators at MIT have recently reported a factor of two improvements of fuel efficiencies by reducing engine size but operating with a turbocharger. **Turbochargers** are devices used to pre-compress air, allowing more air to be ingested into the cylinders. The fuel burned will be proportionally higher and more power is obtained.

The knocking problem was solved by injecting a precisely controlled

[13] US Environmental Protection Agency website (*http://www.epa.gov*).

amount of ethanol into the combustion chamber at exact instance when mixture reaches its maximum pressure. The injected methanol cools the mixture, effectively raising the fuel octane number to 130 – comparable to that of natural gas vehicles and high performance racing cars.[14]

Following the oil shock of 1973, the US government has introduced the **Corporate Average Fuel Economy (CAFÉ)** standard, which has gradually doubled the 14 miles per gallon (mpg) average of 1975 to 27.5 mpg in 1990. CAFÉ standards have been frozen at this level ever since.

Even though cars are becoming consistently more efficient, the savings in fuel has, unfortunately, not translated into reduced consumption. Instead, better fuel efficiency along with the cheap price of petroleum has triggered an upsurge in the use of sport utility vehicles, pickups, and minivans (cumulatively called light trucks). For example, the US Department of Transportation reported that between 1990 and 2000, vehicle miles traveled by passenger cars have increased by around 15%; during the same period, the total vehicle miles traveled by light trucks have increased by 60% (Figure 14-7).[15] Even those who own the more compact economy cars have not helped reduce overall energy use, as the continuous drop in gasoline cost has encouraged more and more use of private vehicles.[16] It is not clear that further increases in efficiency will ultimately lead to a lower rate of gasoline consumption.

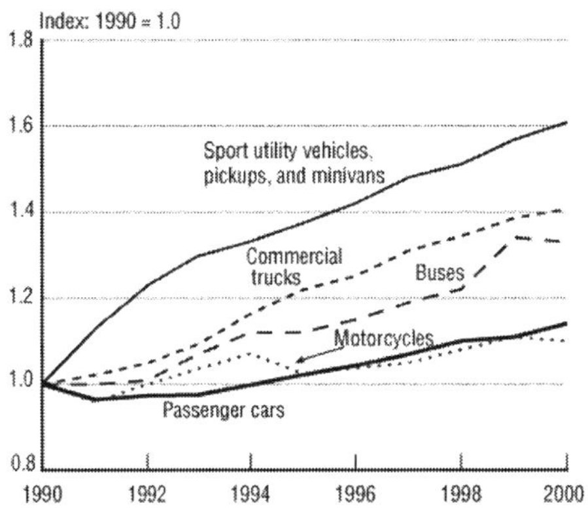

Figure 14-7
Sport utilities had the highest rate of increased use in the last decade.
Source: Transportation Energy Data Book, Ed. 25,2005.

Making cars more fuel efficient does not necessarily save us energy. What is needed most are better designed neighborhoods that reduce our need for commuting. Unfortunately, Americans have opted for decentralization - moving away from population centers to suburbs. In the last three decades, 86% of the nation's growth was suburban. This resulted in heavier reliance on personal transportation and increased the number of miles traveled. European cities, on the other hand, have become more centralized. The larger population density was accommodated by better architectural planning, reducing the need for cars and transit. In Europe, 40-50% of trips are taken walking and biking and about 10% are by transit. In contrast, in the United States, 87% of trips are by private cars and only 3% are by transit.[17]

Example 14-2: Which one is a more efficient (economical) mode of transportation, a car or an airplane?
Solution: Jet aircrafts use huge amounts of fuel but also carry a lot of passengers at a very fast speed. Cars, on the other hand, are relatively slow and carry only a few passengers. However, they are not big gas-guzzlers like jet aircraft. For example, a Boeing-747 needs 12,000 liters of kerosene per hour to lift its 300-ton body. Assuming

[14] Bullis, K., "Better Than Hybrids," Technology Review, April 2006.
[15] Davis, S. C., and Diegel S. W., "Transportation Energy Databook," Ed., 24, US DoE, ORNL-6983, 2004.
[16] Paul McCready points out that in 1986 dollars, fuel cost for driving 25 miles has dropped from $4 in 1929 to $3 in 1949, $2 in 1969, and $1 in 1989.
[17] Gibbs, W. W., "Transportation's Perennial Problems," *Scientific American* 277(4):54-57, October 1997.

a density of 0.8 kilogram per liter of kerosene, the Boeing-747 uses about 3% of its weight in fuel every hour it flies. The aircraft has the capacity to carry 400 passengers at a speed of 560 mph (900 km/hr). This is about 12,000/(400x900) = 0.033 liters per passenger-km. A typical passenger car, on the other hand, gets about 30 mpg (9 km/l). Even if there are four passengers riding together in the car (which is very rare in the United States), we consume 1/(4x9) = 0.03 liters per passenger-km. This is roughly the same as the airplane, though it travels at a much slower speed. Therefore, flying is a more efficient way to travel. Haven't birds known that for a very long time?

Alternative Fueled Vehicles

To increase efficiency, improve air quality, and reduce dependence on foreign oil, the US, Japan, and many European governments have passed laws and enacted taxes and other incentives to promote the use of alternative fuel vehicles. Two US laws, the 1990 Clean Air Act Amendments and 1992 Energy Policy Act, require certain fleets (government agencies, buses, taxis, etc.) to acquire (purchase, lease, convert) vehicles that operate on fuels other than petroleum as a portion of their fleet. The US Department of Energy has established programs to work with automotive manufacturers and various state agencies to achieve these objectives. The task of administering these programs is delegated to the EPA.

Table 14-7. Number of alternative fuel vehicles in use during 2004

Fuel Type	Number	Annual % change since 1993
LPG	194,000	1.3
CNG	144,000	12
LNG	3,000	20
M85	4,600	-14
E85	146,000	66
Electricity	56,000	13
Total	**548,000**	**9.3**

Reference: Davis, S. C., Diegel, S. W., US Oak ridge National Laboratory, "Transportation Energy Data Book", Ed. 25, Table 6-1, ORNL-5198, 2006.

As of 2003, over half a million alternative fuel vehicles were in operation in the United States (Table 14-7). This is in addition to about 20 million vehicles that have been using various blends of biodiesel fuels. The most common fuel by far was propane (LPG), followed by natural gas (CNG), and ethanol (E85). In addition, about 46,000 electric vehicles were on the road, mainly in California and Arizona.[18] The number of vehicles that use alternative fuels, as classified by the Department of Energy, is expected to increase and is shown in Figure 14-8.

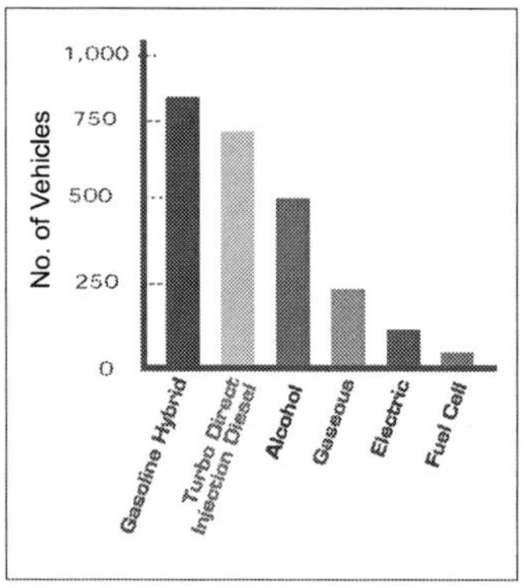

Figure 14-8
Projected sales of advanced technology light-duty vehicles in 2020 by fuel type expressed in thousand vehicles sold.
Source: National Renewable Energy Laboratory Website (www.nrel.gov/).

The best possible fuel is, of course, pure hydrogen. Since no carbon is present, with the exception of trace amounts of hydrocarbon and carbon monoxide (the result of lubricating oil being swept into the combustion chamber), no such emissions are produced. The only products are water and a small amount of nitric oxide. Hydrogen, however, costs considerably more than petroleum fuel – about five times more for the same amount of energy. In addition to cost and inherent problems with production and handling, no commercial hydrogen-powered passenger cars are yet available. Daimler-Chrysler has two prototypes; NECAR-4 is running on liquefied hydrogen, and NECAR-4a is using compressed hydrogen. GM has developed the HydroGen, which uses liquefied hydrogen. BMW's H2R hydrogen-powered concept car uses a modified 6-liter, 12-cylinder internal combustion engine for its propulsion. In addition, BMW has introduced a modified V-12 engine that can be fueled either by gasoline

[18] Alternatives to Traditional Transportation Fuels, Fuel Data Center, US DoE, Energy Information Administration, in Transportation Energy Data Book, Ed 23-2003.

or by liquid hydrogen stored in a pressurized tank placed in the trunk of the automobile (Figure 14-9). The car can deliver 232 horsepower (210 kW) and has a range of 300 kilometers using hydrogen alone. The engine will automatically switch to gasoline to extend the range for an additional 650 km. In addition to the engine, an onboard 5-kW fuel cell is used to operate all the electrical functions as well as auxiliary systems like the air conditioner.

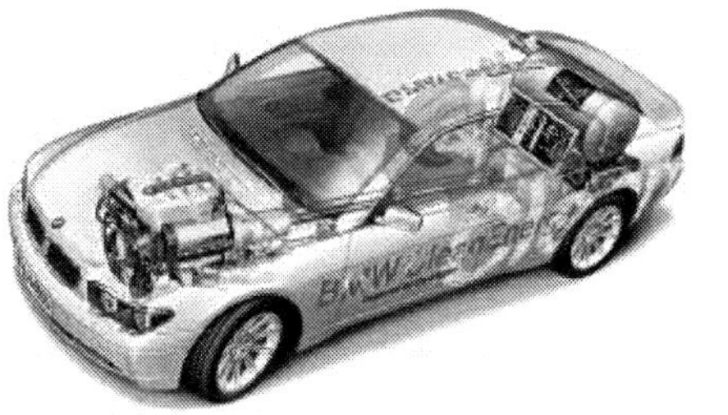

Figure 14-9
BMW 745hl, The first hydrogen-powered car using liquid hydrogen.

In addition to these vehicles, the US government has partnered with the three big automakers (Ford, General Motors, and Daimler-Chrysler) to initiate the **Freedom Cooperative Automotive Research (FreedomCAR)** program.[19] Its aim is the advancement of high-efficiency vehicle technology, focused on fuel cells and hydrogen produced from renewable sources. The long-term goal is to develop a hydrogen-based economy to clean the environment and reduce the US dependence on foreign oil.

Electric Propulsion

At the second half of the nineteenth and early twentieth centuries were the advents of innovations, as horse-drawn carriages were replaced with steam locomotives, petrol engines, and electric trolleys. In 1900, of the 4,200 automotives sold in the US, 40% were steam, 38% electric, and 22% were powered by gasoline.[20] By 1912, electric cars were dominating the automobile market, surpassing all other vehicles equipped with internal combustion engines.

The demise of electric cars came with the invention of the starter motor by Charles F. Kittering in 1911. The Kittering starter was a very small electric motor that produced enough power to crank an engine for a very short time, eliminating the need for much more tedious hand cranks. The popularity of gasoline engines continued to increase as they become more reliable, eventually making the steam engine obsolete and electric vehicles (EV) inferior. Mass production of Ford's Model-T, with affordable prices and a range twice that of the best electric cars available, helped make internal combustion engines even more popular.

As electric vehicles became less and less attractive and the number of internal combustion engines grew, so did the concern for automotive pollution. The sudden increase in the price of gasoline in early 1970, following the Arab oil embargo, revived the interest in electric vehicles. The main problems that have precluded a wide acceptance of electric vehicles are their higher cost, limited range, and the inconvenience these vehicles present in comparison to internal combustion engines. Most electric vehicles use lead-acid batteries, which are heavy and of relatively

[19] For additional information about the FreedomCAR project, visit the websites *http://www.eere.energy.gov/hydrogenfuel*, *http://www.eere.energy.gov/hydrogenfuelcells*, and *http://www.eere.energy.gov/vehiclesandfuels*.
[20] Shacket, S., The Complete Book of Electric Vehicles, Domus Books, 1981.

low power density. This means that, for the same weight, EVs have a much shorter range than gasoline and diesel engines. Furthermore, charging the batteries takes a long time and, unlike gasoline filling stations that are widely available throughout the world, battery charging stations are few and far between. The obvious challenge to electric vehicles' popularity is to develop compact batteries with similar energy density to gasoline at a reasonable price.

Battery-operated Electric Vehicles (BEV)

Lead-acid batteries are the primary power source for many of today's plug-in electric vehicles. More advanced nickel batteries (such as nickel-iron, nickel-cadmium, and nickel-metal hydride) offer better ranges, shorter recharge times, and longer lifetimes, but certain safety concerns and higher costs have precluded their widespread use. Cadmium is very toxic, while nickel-iron batteries tend to produce a buildup of hydrogen during charging. The battery with the greatest potential is the nickel-metal hydride (Ni-MH), which is nontoxic, has about four times the range of lead-acid batteries, and can be recharged in a few minutes. These batteries are currently used in the Honda EV Plus and Toyota RAV4-EV. A close competitor to the Ni-MH battery is the lithium-metal hydrite (Li-MH) battery. These batteries have a very high specific energy but must be operated at temperature of about 70° C. Li-MH batteries are currently being tested in the Nissan Altra.

Currently, the relatively low price of gasoline probably remains the most important impediment to wide acceptance of electric cars in the United States. As the price of gasoline rises, it is expected that electric cars will claim a larger fraction of the total vehicle market. Out of 230 million cars operating in the United States, about 40% are used as a second car, and 87% of automobile trips are less than 50 kilometers. Electric vehicles can satisfy some of these needs as around-town cars. Furthermore, EVs may have utility as fleet vehicles in inner city short route buses, delivery vehicles, airports, and mail delivery services.

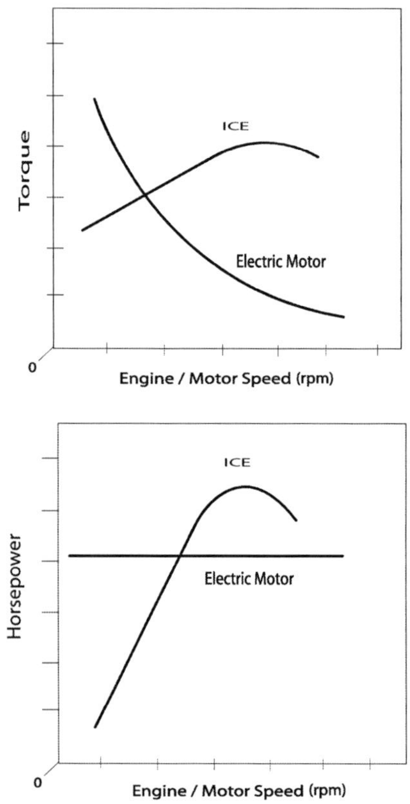

Figure 14-10.
Performance characteristics of electric motors versus internal combustion engines

Advantages and Disadvantages of Electric Vehicles

The major advantages of electric over conventional vehicles are:

1. *Better torque characteristics.* Electric motors are simpler and have torque characteristics that match those demanded by vehicles more closely (high torques at low speeds and acceleration, relatively low torques during cruising). Internal combustion engines, on the other hand, deliver maximum torque at an optimum cruising speed (Figure 14-10). Because little or no torque is delivered during start ups and at lower speeds, internal combustion engines need starter motors. Electric motors produce the highest torque at zero speed when it is most needed. An electric motor can therefore be attached directly to

the drive wheels and can accelerate the vehicle from rest to the desired speed without the need for a transmission or torque converter. EVs' power trains are simpler and don't usually need more than one or two gear ratios. Reverse gears are also absent because their function can be achieved simply by reversing the polarity of the electrical input.

2. *Regenerative Braking.* Electric motors can run as a generator by running in reverse. Electric vehicles take advantage of this feature by employing regenerative braking, where up to 50% of the kinetic energy of the vehicle can be reclaimed during urban stop-and-go traffic to recharge the battery. The experimental data shows that, depending on design and the driving cycle, regenerative braking extends driving range between 5 to 15 percent.
3. *Lower noise and emission.* Electric vehicles are much quieter during operation and do not consume any power or produce any emission when stopped. This is not true with gasoline cars, which continue to consume a substantial amount of fuel and produce pollution even when they idle. Some emissions, such as hydrocarbons, are actually higher during idling than when cruising at optimal speeds.
4. *Superior efficiencies.* Electric vehicles have efficiencies in the range of 40-45% compared to efficiencies of 18-25% common for most conventional vehicles.[21] However, when losses associated with the generation of electricity and transport are considered, depending on how the electricity is generated, EV efficiency advantages are only 10-30%.

The major disadvantages of electric vehicles over conventional vehicles are:

1. *Batteries have a relatively small capacity to store energy.* Compared to gasoline, which has an energy density of 12,000 watt-hours per kilogram, lead-acid batteries have only 40 watt-hours per kilogram. Lower battery storage capability results in limitations on the distance electric vehicles can travel before they must be recharged. More batteries add to vehicle weight which indirectly limits performance. With current technology, EVs that are equipped with lead-acid batteries offer a range of 60-80 kilometers and take six to eight hours to charge. Ni-MH batteries offer a range of 400 kilometers and can be recharged faster. The EVs available in the market are, on average, 300 to 1000 kilograms heavier than similar conventional vehicles.
2. *Infrastructure does not exist.* There is no network of recharging stations. Charging must be accomplished in much faster times that are currently possible and in all types of weather conditions.
3. *Electric vehicles cost more.* Cost of the ownership (initial purchase price plus costs associated with maintenance and repair) are higher for electric vehicles.

[21] It should be noted that the efficiencies quoted here are first law efficiencies, the fraction of the input energy (electricity for EV and petrol for ICE) that is used to drive the vehicle.

Fuel Cell Vehicles (FCV)

Figure 14-11
In NECar-5 by Daimler-Benz methanol is reformed to produce hydrogen to run a PEM fuel cell.

Fuel cells produce electricity through a chemical reaction between hydrogen and oxygen. The fuel cell was invented in 1839 by the British physicist Sir William Grove but, because of inherent problems associated with its operation, was soon forgotten. It was only in the 1960s that the first fuel cell was first employed aboard Gemini and Apollo capsules to replace heavy batteries used in previous space missions. Ever since, fuel cells have been used in a number of applications such as in airplanes, stationary power generators, and now automobiles.

In 2002, the Honda FCX and Toyota FCHV-4 became the first fuel cell vehicles to enter the market. Daimler-Chrysler has also introduced a limited number of their newest fuel cell model in 2006 (Figure 14-11).

a Principle of Operation

Fuel cell operation can be accurately described as reverse electrolysis. In electrolysis, an electric current is passed across two electrodes immersed in an electrolyte solution, splitting water into its constituents – hydrogen and oxygen. The fuel cell works in the opposite direction; hydrogen and oxygen (usually from air) recombine in an electrochemical reaction to form water and an electric current is produced. The reaction is different from combustion, as it usually occurs at lower temperatures and electricity (instead of heat) is the output. Since fuel cells are not heat engines, their efficiencies can be quite high and are not constrained by Carnot principles. Efficiencies of around 50% are typical.

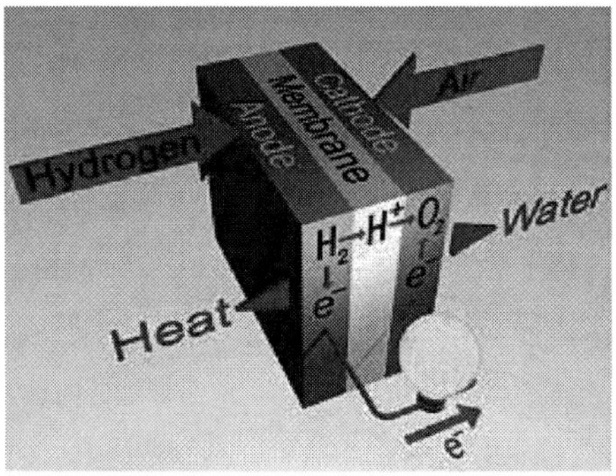

Figure 14-12
Principle of operation of a PEM Fuel Cell. *Source: NASA Website (http://www.nasa.gov/).*

The most common fuel cell for passenger cars and light-duty trucks is *Proton Exchange Membrane* (PEM). It consists of a positive and a negative electrode separated by a membrane or electrolyte (Figure 14-12). Hydrogen gas is introduced at the negative electrode (anode) where a catalyst strips electrons from the hydrogen atom. Hydrogen ions migrate through the membrane toward the positive polymer electrode (cathode) and react with oxygen to form water. Without the membrane, hydrogen and oxygen mix and react directly in a chemical reaction, generating heat instead of electricity. The membrane works by allowing the hydrogen ions (protons) to pass through while blocking electrons. Electrons are then diverted through an external wire before arriving on the other side where they are recombined with protons and oxygen to complete the reaction. As long as the flow of hydrogen gas is not interrupted, the electrical current is maintained, which can turn a light on or be used to power an electric motor. The PEM operates best at temperatures of around 80° C.

To summarize, the reaction at each electrode and the overall reaction can be written as:

$$\begin{aligned} &\text{Anode:} && 2\,H_2 \longrightarrow 4\,H^+ + 4\,e^- \\ &\text{Cathode:} && O_2 + 4\,e^- + 4\,H^+ \longrightarrow 2\,H_2O \\ &\text{Overall Reaction:} && 2\,H_2 + O_2 \longrightarrow 2\,H_2O \end{aligned} \qquad (14\text{-}2)$$

Besides PEM fuel cells, three other technologies show promise for various transportation applications. The *Phosphoric Acid Fuel Cell* (PAFC) is most appropriate for locomotives, heavy-duty trucks, and urban transit buses. These fuel cells are already used for stationary power generation. Two other technologies, *Solid Oxide* and *Direct Methanol* fuel cells, are mainly of interest for on-board fuel processing and reformer technologies. The major problem with all these fuel cells is that they operate at much higher temperatures and at very high costs.

"Water will one day be employed as fuel, that hydrogen and oxygen which constitute it, used singly or together, will furnish an inexhaustible source of heat and light..." ~ Jules Verne (1828-1905)

Reformers

Fuel cells suitable for a car must be small, have a high power density, refuel in a short time, store sufficient hydrogen for a range of 500 kilometers or better, and be available at a cost comparable to conventional internal combustion engines. When fuel cells use pure hydrogen as fuel, there are no emissions except for the pollutants produced in generating the hydrogen. Producing hydrogen in large quantities and storing it in convenient containers is not always practical, however. Hydrogen has a low volumetric energy density, and storing it requires very high pressures or cryogenic temperatures. Hydrogen can also be stored as metal or chemical hydrides. Compressed gas occupies the largest volume, while hydrides are the heaviest of all options. Storing hydrogen in liquid form is impractical and not an option at this time. To travel 500 kilometers, a fuel-cell car requires three kilograms of hydrogen. If stored at 20 MPa (3000 psia) pressure tanks, it occupies a volume of 190 liters. Furthermore, the current lack of convenient refueling stations is a big impediment in marketing fuel cell vehicles using pure hydrogen.

Rather than relying on storing hydrogen and using it directly, a number of car companies are concentrating on developing on-board reformers. Reformers use a variety of fuels to produce hydrogen by reformation, where it is subsequently used in the fuel cell. Gasoline, methanol, or any other fuel can be used. Reforming natural gas does not seem to be the best option as natural gas is most suitable to be used in combined cycle turbines to generate electricity relatively cleanly and at a high efficiency. Methanol is considered to be a favorite because it is liquid at room temperature and can be produced from biomass. Although methanol has lower energy content than gasoline, the higher efficiency of methanol reformation can compensate for lower energy density and vehicle range remains relatively the same.

Advantages and Disadvantages of Fuel Cell Cars

The advantages of fuel cells are many. They are efficient, operate virtually vibration-free and silently, have a high energy density, have no moving parts, and produce little or no emissions. Furthermore, fuel cells have no exhaust and thus have a low thermal signature. They can be customized to any shape and for any power demand, which makes these devices ideal for numerous applications. They will eventually replace a major portion of internal combustion engines and power future cell phones, computers, camcorders, and other cordless devices. The major barriers to commercial development of fuel cell cars are the public's concern over safety and the cost. If hydrogen is used directly as the fuel instead of being produced by an on-board reformer, then other issues such as hydrogen distribution, delivery, and storage would also become a problem.

The issues of hydrogen storage were briefly discussed above, namely, the low density of hydrogen even at very high pressures or at liquid temperatures. At present, hydrogen can be stored at 1-2% by weight, at best. To make fuel cell cars practical, the Department of Energy has set a goal of 62 kg hydrogen per cubic meter for storage space. Metal hydride and carbon nanofibers are two promising technologies.

Figure 14-13.
Fuel leak simulation. Hydrogen-powered vehicle is on the left, and gasoline-powered vehicle is on the right. Notice that after 1 minute, hydrogen flame is subsiding, whereas gasoline flame is engulfing the entire vehicle. *(Ref. 22)*

Concern about hydrogen safety is partially justified, but mostly over-exaggerated. Hydrogen gas is highly explosive, and its flame is invisible. Images of the catastrophic explosion of the hydrogen-propelled "Hindenburg Zeppelin" in 1937 are still alive in many people's minds,[22] even though some recent studies have shown that the cause of explosion was not the hydrogen, but the electrostatic charge accumulated by an oily substance used to treat the cotton skin of the airship. The lack of flame visibility can be easily resolved with certain additives. New advances in composites allow the design of hydrogen tanks that can withstand major crashes and falls from buildings several stories tall. In addition, hydrogen sensors are available that can detect and shut down hydrogen flow as the smallest leaks are detected. Finally, since hydrogen is the lightest of all gases, hydrogen flames are highly buoyant and spread in the form of jets diffusing upward and away from the body of the car and its occupants, as demonstrated by several crash and fire simulations (See Figure 14-13).[23]

Lack of a hydrogen distribution and delivery system is another important factor delaying the introduction of fuel cell equipped passenger cars. To make fuel cells viable, a sizable investment in hydrogen generation and distribution infrastructure is required. This is an uphill battle for hydrogen enthusiasts, who have to compete with already established gasoline infrastructure. The United States now has more than a dozen hydrogen stations, mostly in California, used only for demonstration and research

[22] Hindenburg Zeppelin was a giant hydrogen-filled airship (at 245 meters, it was the largest ever to fly) that operated regularly in 1930s taking transatlantic passengers between Germany and both North and South America before it was destroyed by a fire in 1937.
[23] Swain, M. R., "Fuel Leak Simulation", Proceedings of the 2001 DoE Hydrogen Program Review, NREL/CP-570-30535.

purposes.

Fuel cells produce power at considerably higher costs than internal combustion engines-- at around $4 a watt compared to $0.05 a watt. To make fuel cells a viable alternative to batteries and internal combustion engines, their cost must be brought down, they must work virtually maintenance free, and their safety must be improved. This cannot happen unless fuel cell-operated devices can be produced at a fraction of what it costs today. The prices cannot fall unless fuel cells can be mass-produced, which happens only when prices are already low and sufficient infrastructure (such as hydrogen filling stations) is in place - a classic chicken and egg problem![24]

Emissions from Electric and Fuel Cell Cars

Electric and fuel cell vehicles don't have tailpipe emissions. This does not, however, mean that these cars are truly zero emission vehicles (ZEV); there are still some emissions from the tailpipe to the smoke stacks where electricity or hydrogen is generated. For fuel cell vehicles, if hydrogen is produced by on-line reformers, additional emissions are introduced. Nevertheless, electric and fuel cell vehicles are far cleaner and pose fewer health risks than internal combustion engines. Among the reasons are:

1. There are options other than fossil fuels for generating electricity or hydrogen. In fact, if nuclear energy or renewable sources such as wind, solar, or hydropower are used, no pollution is produced and these vehicles would truly be zero emission vehicles.
2. Most power plants are located in rural areas and areas away from city centers and other population centers. The health impact is expected to be lessened as pollutants are diluted and scattered over larger distances. Furthermore, it is easier to control emissions from relatively few power plants and stationary sources than from millions of automobile sources on the road.
3. The majority of trips are during city driving when vehicles are not fully warmed up and a significant portion of travel is in stop and go traffic. Conventional vehicles produce a large amount of HC and CO during cold starts, acceleration, and braking.

All in all, substituting EVs and FCVs for conventional vehicles would significantly reduce urban emissions. According to one study,[25] if all conventional vehicles were replaced with electric vehicles, overall hydrocarbon and carbon monoxide emissions would drop by 95-99% and nitric oxides by up to 90%. Ozone and the volatile organic compounds that contribute to urban smog will be eliminated altogether. Sulfur dioxide and particulate emission, however, may drop or rise depending on how much of the electricity comes from coal and oil power plants.

[24] "The Ultimate Incinerators," *Scientific American*, September 1995, p. 180.
[25] Electric Power Research Institute Website, *http://www.epri.com*.

Hybrid Vehicles

The much greater range offered by gasoline engines and the much cleaner operation of battery-electric and fuel cell vehicles are the major impetus for the development of hybrid vehicles. A hybrid vehicle is defined as a vehicle with at least two modes of propulsion. Today, most hybrid vehicles combine an internal combustion engine (either gasoline or diesel) and an electric motor powered by batteries. The choice is not, however, limited to these and other primary sources of power such as gas turbines and fuel cells; other types of storage devices, such as flywheels and ultracapacitors, are occasionally being used.

The idea of hybrid vehicles is not new. The French Kriéger Company, originally established to produce electric cars, introduced the first alcohol-electric hybrid car in 1902, a gasoline version in 1904, and a turbine-electric hybrid shortly before it became bankrupt in 1909. With the advent of the starter motor in 1911, much of the lure of electric and hybrid vehicles vanished and, until the oil crisis of 1973, no major research was carried out. To secure a foothold in the future automotive market, Japanese companies initiated a major investment in producing hybrid vehicles; in 1999, Honda released its two-door Insight, the first hybrid car to hit the American markets. The Insight received an EPA mileage rating of 61 mpg city and 70 mpg highway. Until recently, the Honda Insight, Honda Civic, and Toyota Prius were the only hybrid vehicles available to the public in the United States. The performance of the three major Japanese hybrids are compared and given in Table 14-8. As of 2005, almost all major automotive manufacturing companies have been introducing passenger, SUV, and pickup hybrids.

Efficiency

Hybrid vehicles offer superior efficiencies over both petrol and electric vehicles because they are operated at their most efficient modes and they can recover some energy through regenerative braking. For example, to double the fuel efficiency (2X hybrid-electric), we can use an engine only half the size and supplement the power through recovering much of the losses during braking. Let's consider two cars, one conventional and another hybrid as depicted in Figure 14-14. The conventional car (1X) consumes 100 units of fuel, losing 73 units through the exhaust, radiator, friction, and other accessories and another 11 units during standby, thereby delivering 16 units of energy to propel the car. The hybrid car (2X or two times efficiency) deploys an internal combustion engine half the size using 50 units of fuel and losing 38 units to the ambient. The standby losses are eliminated because the vehicle operates as an electric vehicle when the car stops and no motor is left running. An additional 4 units of energy can be recovered through regenerative braking as the vehicle decelerates. Like the gas-powered car, the net useful energy is 16 units.

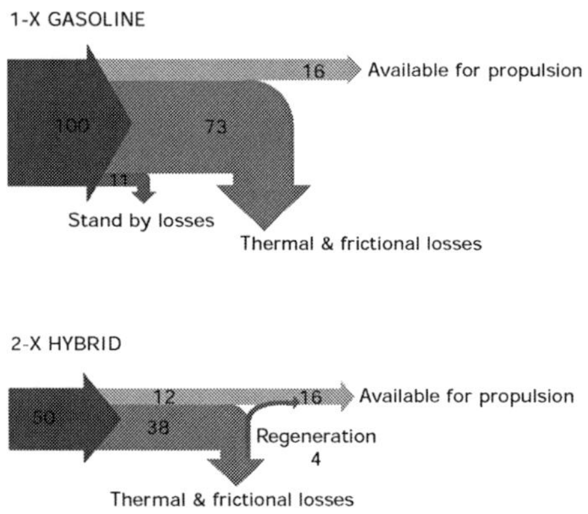

Figure 14-14
Comparison between a pure internal combustion engine and a 2x hybrid vehicle.

Table 14-8. Specifications of most popular hybrid vehicles available in the US in 2005			
	Honda Insight	*Honda Civic*	*Toyota Prius*
Configuration	Series	Series	Parallel
Engine/Electric Motor Power	65 h.p./9.7 kW	85 h.p./9.7 kW	76 h.p./50 kW
Engine compression ratio	10.8:1	10.8:1	13:1
Fuel economy (city/highway)	57/56 mpg	48/47 mpg	60/51 mpg
Emission rating	SULEV*	ULEV	SULEV*
Battery	NiMH	NiMH	NiMH
Passenger capacity	2	5	5
Weight	1975 lb	2736 lb	2890 lb
Range	528 miles	554 miles	589 miles

* Super Ultra Low Emission Vehicle
Source: Manufacturers published data.

In 1993, the US government, partnered with national laboratories and three major US automakers, Daimler-Chrysler, Ford, and General Motors, ventured the **Partnership for New Generation of Vehicles (PNGV)**, whose long term goal is to come up with a 3X car that can triple fuel economy to 80 mpg. A similar program called **FreedomCAR** is also being undertaken to develop new and alternative technologies that improve fuel efficiency and reduce the US dependence on imported foreign oil.

Configurations

Almost all hybrid vehicles operating today utilize an auxiliary power unit (APU), such as a small spark ignition engine, a diesel, or a gas turbine to complement the batteries in providing power directly, or by acting as a generator to charge the batteries. The APU can be fueled by gasoline, diesel fuel, or any number of alternative fuels and is operated at relatively steady-state optimum condition to produce very low emissions and high efficiency (Figure 14-15).

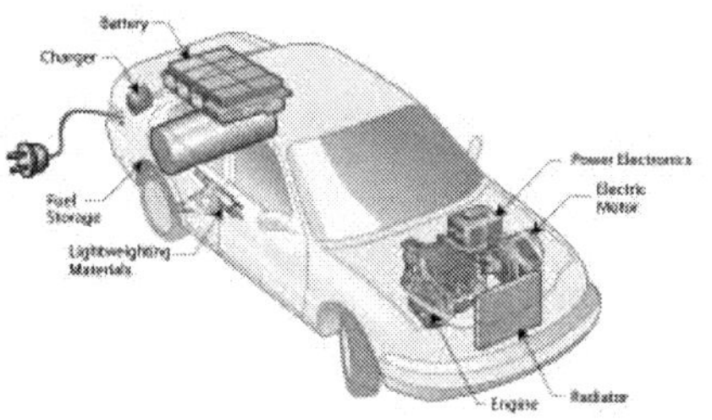

Figure 14-15
Plug-in hybrid vehicle and its components.
Image Courtesy of Argonne National Renewable Energy.

Hybrid vehicles are commonly classified as either series or parallel. The main distinctions between the two configurations are in the way the APU transfers power to the wheel and whether batteries become fully discharged (charge-depleting) or retain the charge by continuous charging (charge-sustaining).

In a series configuration, all motive power comes from the electric motor powered by batteries. The gasoline engine drives a generator to produce electricity, which either supplies power to an electric motor or charges the battery. With series hybrids, there is no mechanical connection between the engine and the wheels; power is transferred electrically to an electric motor that drives the wheels. Because electric motors generate torque that matches requirements at different speeds, series hybrids have a simpler transmissions or no transmission at all (Figure 14-16a). As a result, modern hybrid-electric drive systems are much lighter, weighing about one third to one half the weight of batteries required for a purely battery-operated electric car.

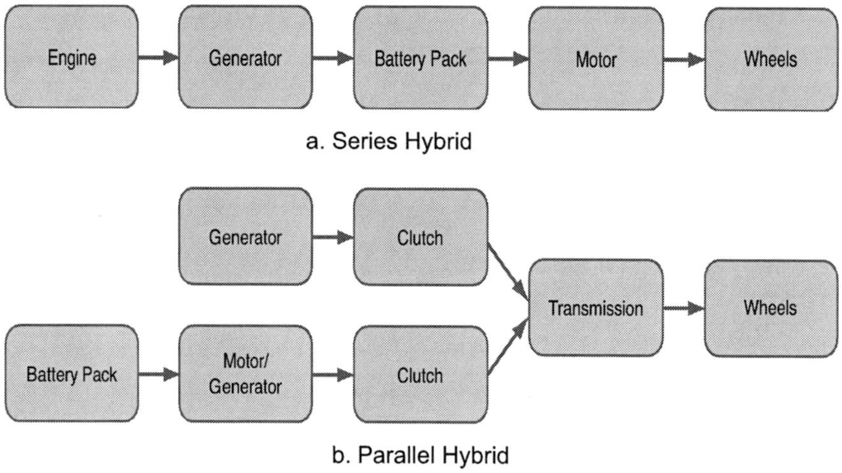

Figure 14-16
Series and parallel hybrids.

In the simplest configuration, the vehicle operates as a pure electric vehicle until batteries are depleted to their preset thresholds, at which time the internal combustion engine begins to recharge them. Since the engine does not have to meet changing power demands directly, it can be set to operate in a narrow speed range where it has the highest efficiency, the lowest emissions, or a combination of both. Alternatively, control strategies can be devised where a small IC engine operates continuously at its optimal point so as to keep the battery at or near its full charge. One major advantage of a series hybrid is that it can attain a long range with an engine only a fraction of the size of the conventional engine and with a battery pack weighing far less than those of pure electric vehicles.

In a parallel configuration, the engine, the electric motor, or both supply power to the wheels (Figure 14-16b). Parallel hybrids are primarily used in electric-only mode for short trips and city driving, whereas long trips and highway cruising are carried out by engine-only operation. The electric motor can be used to help overcome hill climbs, to accelerate quickly, and in instances when the engine cannot single handedly meet the power demand. Because the engine, electric motor, or both must deliver the power, a clutch/transmission assembly is necessary in parallel hybrids.

Air Hybrid Vehicles

Instead of storing the energy as electric energy in a battery in an electric car, air hybrid vehicles work by storing the energy needed to propel the engine in an onboard tank of compressed air. Like a conventional engine, a piston compresses air in a cylinder. As the piston reaches the top, a small amount of compressed air is released into the expansion chamber to create a low pressure, low temperature pressure wave that drives the piston to power the engine. When vehicle brakes, the engine is used as an air compressor to absorb the braking energy and store it into the air tank. The engine is essentially shut off when it stops behind a traffic light. As the car accelerates, more and more air is allowed into the cylinder until the compressed air is depleted. The tank can be refilled in a refill station or by a compressor operated by a conventional internal combustion engine.

Advantages and Disadvantages of Hybrid Vehicles

Hybrid vehicles have advantages over both electric and conventional vehicles by combining features and enhancing performance to a degree not possible using either propulsive system, and without their limitations. For example, ICE performance is a strong function of engine speed, designed to operate optimally at cruising speeds around 90-100 km/hr (roughly 50-60 miles per hour); efficiency drops rapidly at both very high and low speeds. On the other hand, electric motors have their highest efficiency at low speeds, generate no on-board emissions, have favorable torque characteristics, and can utilize regenerative braking. These characteristics make conventional engines ideal for freeway driving and constant speed operation and electric vehicles best suited for city driving and transient conditions.

Another advantage of the hybrid system is that individual components can be sized to fit different driving conditions. Since the engine is not directly coupled to the wheels, engine size can be selected to run near its optimal conditions at all times. In most designs, the engine is sized to meet only the cruising demands, whereas power for acceleration and hill climbing is offered by batteries and other storage devices. The major disadvantages of hybrid vehicles are additional complexity of dealing with two power systems and potentially higher capital cost.

Social Costs of Transportation

Many of the costs associated with driving have little to do with fuel use. Traffic congestion, lost productivity, air pollution, stress, damage to roads and other properties, accidents, and health costs are among many indirect costs attributed to driving.[26] In 2002 alone, Americans lost over 3.5 billion hours and wasted 5.7 billion gallons of gasoline , because of traffic delays. Furthermore, during the same year there were over six million accidents, resulting in about 43,000 deaths. The loss of productivity due to traffic delay alone is estimated at around 62 billion dollars. If costs associated with stress and health effects were included, the cost of owning a car would be much higher.[27]

These costs are borne by everyone. The solution suggested by car manufacturers, politicians, and policy makers has been to promote heavier and bigger cars, pack cars with more air pollution control devices, add lanes, build more roads and parking structures, and to move many industries and large corporations away from metropolitan areas and large population centers. The solution may actually be the exact opposite. For example, it may be wiser to discourage car ownership by providing a more

[26] Hawkens, Paul, Lovins, A, and Lovins, L. H., "Natural Capitalism: Creating the Next Industrial Revolution," Rocky Mountain Institute, p. 41. 1999.
[27] Schrank, D., and Lomax, T., "The 2004 Urban Mobility Report," Texas Transportation Institute, Texas A&M University, September 2004.

efficient and cheaper public transportation system and, at the same time, making personal car ownership costlier and less convenient. This may require raising automobile sales taxes, insurance fees, and maintenance costs. The revenues could then fund development of comprehensive public transportation structures. Restricting driving is also an effective way to reduce congestion and improve air quality. Some countries have enacted laws to make certain parts of large metropolitan areas off limits to passenger cars. People are to leave their cars outside the restricted zones and take public transportation to downtowns and populated city centers. Police cars, ambulances, and other emergency vehicles are exempt. In another scenario, drivers have access to certain regions during certain hours or days, and only on an as-needed basis. For example, cars with license plates that end with an odd number can travel in a city only on Sundays, Tuesdays, and Thursdays, or they have to pay an extra toll charge during rush-hours.

Office managers and business owners can encourage their employees to carpool or to use public transportation by providing free parking spaces to carpools or by paying for their bus or train fares. They may also help in paying part of the home mortgages of employees who live within a short distance of the place of employment. Some companies and organizations have successfully implemented flex-hours, allowing some employees to shift their work hours or even work part of their time at home. The local city government can participate in the effort by providing bikes to the general public that can be used free of charge in central districts where most government offices are. Architects and city planners can be especially effective in this effort by designing convenient shopping centers and large office buildings to satisfy multiple needs, thus reducing the need for frequent visits.

Summary

To cope with ever-increasing problems associated with transportation, various approaches have been proposed. Among options considered are: 1) reduce vehicle use, 2) increase the efficiency and reduce the emissions of conventional gasoline-powered vehicles, 3) switch to less noxious fuels, and 4) use cleaner alternative systems. The first choice calls for developing a large-scale mass transit system, carpooling, reducing travel by working from home, and removing the need for excessive driving by implementing more effective community designs. This will also reduce much of the social costs such as traffic congestion and delays, pollution, accidents, repair costs, and other social ills such as fatigue and stress. Although this option helps with many social and environmental problems, by itself it seems to have limited success.

The second and third approaches are to make existing transportation systems better by designing more efficient cars or vehicles that use better and less-polluting fuels. Internal combustion engines have been around for a long time and their efficiencies have been greatly improved. It seems that improving efficiencies beyond current levels only comes in small steps and no major breakthrough is eminent in the near future. Alternative fuels like methanol and natural gases have been tried by a number of investigators with marginal improvements in emissions and fuel efficiencies. Hydrogen seems to be the obvious alternative fuel. Problems associated with hydrogen storage, safety concerns, and the lack of a hydrogen distribution infrastructure have been major obstacles to commercial development of hydrogen-propelled internal combustion engines, although a number of automotive manufacturers have developed prototype units.

The fourth option is to do away with current technologies in favor of newer and cleaner alternatives. In order for electric, hybrid, or any other types of cars to find mass market acceptability, they must meet certain requirements. They must have a full-size trunk, be able to travel large distances, recharge in a short time, have all the comfort and luxury of the conventional cars, and be offered at costs comparable to conventional internal combustion engines. Unless there are major breakthroughs in advanced battery technologies, it does not appear that battery-powered electric vehicles, in their current form, can make headway in achieving a major market share.[28] Their uses would probably be limited to small service vehicles, fleet operation, and in areas where pollution, not driving range, is of primary concern.

When and if fuel cell cars become popular and infrastructure is in place, fuel cells will not only be cost effective, but may also be a source of income for their owners. As US data shows, Americans drive their cars only 4% of the time. The remaining 96% of the time cars sit idly in parking garages either at work or at home. During these times, cars can be used as tiny power plants, producing electricity which can be put on grids and sold to utility companies. In this way, a large fraction of the cost of owning and maintaining the cars may be recovered, and our reliance on foreign oil and other nonrenewable technologies reduced.

For now, hybrid vehicles consisting of a petrol engine and a battery-powered electric motor seem to be the preferred option. In the long term, fuel cells using pure hydrogen offer the best hope. In the meantime, petroleum-based reformers may be the solution as we transition to on-board hydrogen storage in the future.

A technological revolution is underway which makes future transportation systems even more efficient. New advanced polymer and carbon-fiber

[28] In hope of acquiring a great market share and to meet the California mandate of replacing 2% of the automobile sales by electric vehicles, GM introduced Impact (to be later called EV-1) to the market. The car, although well-designed technically, did not meet public acceptance, and GM had to stop its production. California also had to back down from its stated goals. Ford's Think! was somewhat more widely accepted, as it was marketed as a second (neighborhood) car to be used for city driving and only for short trips with a range of about 50 miles. Ford also had to close down its plants eventually.

composites are developed that are lighter and safer to use. Composites can be made that design frame-less "monocoques" that cost less, are many times stronger than conventional frames, are lighter than steel by two to three times, and absorb five times more energy. This helps not only in designing safer cars, but also in reducing rolling resistance and power requirements for climbing and accelerating. The reduction in required power results in smaller engines, transmissions, and other components, which in turn makes it possible to make the vehicle even lighter.

Additional Information

Books

1. Tillman, D., Fuels of Opportunity: Characteristics and Uses In Combustion Systems, Academic Press, 2004.
2. Sorensen, K., Hydrogen and Fuel Cells: Emerging Technologies and Applications, Academic Press, 2005.
3. Dhameia, S., Electric Vehicle Battery Systems, Academic Press, 2001.
4. Hussain, I., Electric and Hybrid Vehicles: Design Fundamentals, CRC Press, LLC. 2003.
5. Jefferson, C.M., and Barnard, R. H., Hybrid Vehicle Propulsion, WIT Press, 2002.
6. Spelberg, D. The Hydrogen Energy Transition: Moving Toward the Post Petroleum Age in Transportation, Academic Press, 2004.

Periodicals

1. *Fuel,* Direct Science Elsevier Publishing Company, Fuel focuses on primary research work in the science and technology of fuel and energy fuel science.
2. *Transportation Research Part C: Emerging Technologies,* Direct Science Elsevier Publishing Company; this journal focuses on scholarly research on development, application, and implications in the fields of transportation, control systems, and telecommunications, among others.
3. *Fuel Cells Bulletin,* Direct Science Elsevier Publishing Company, Fuel Cells Bulletin is the leading source of technical and business news for the fuel cells sector.
4. *International Journal of Hydrogen Energy,* Direct Science Elsevier Publishing Company, Quarterly journal covering various aspects of hydrogen energy, including production, storage, transmission, and utilization, as well as economical and environmental aspects.

Government Agencies and Websites

1. US Department of Transportation *(http://www.dot.gov).*
2. US Department of Energy *(http://www.doe.gov).*

3. US Environmental Protection Agency *(http://www.epa.gov)*.
4. National Energy Renewable Laboratory Hybrid Electric &Fuel Cell Vehicles *(http://www.nrel.gov/vehiclesandfuels/hev)*.
5. FreedomCar *(http://www.eere.energy.gov/vehiclesandfuels)*.

Exercises

I. Essay Questions

1. What are the major sources of resistance in a car? When are they dominant? How can they be reduced?

2. What are the major sources of resistance in a ship? When are they dominant? How can they be reduced?

3. What are the major sources of resistance in an airplane? When are they dominant? How can they be reduced?

4. How do spark ignition internal combustion engines work? What are their advantages and disadvantages over other types of engines?

5. How do diesel engines work? What are their advantages and disadvantages over spark ignition gasoline engines?

6. How do gas turbines work? What are their advantages and disadvantages over gasoline and diesel engines?

7. What are the major emissions from an internal engine? What are the conditions that these pollutants produce? What can we do to reduce these emissions?

8. Describe the operation of a catalytic reactor. How does it simultaneously reduce NOx, CO, and HC emissions?

9. Why are diesels more popular in Europe? What needs to be done to make diesels appropriate for the US market?

10. Name five alternatives to gasoline and diesel fuels. What are the advantages and disadvantages of each?

11. Which of the transportation modes (automobiles, trains, barges, and ships) is more energy efficient for carrying passengers? How does it differ in Europe and the United States?

12. What are problems associated with hydrogen storage, delivery, and safety? What are the sources of emissions from hydrogen combustion?

13. What are the advantages of hydrogen as a transportation fuel?

14. How is hydrogen stored in a vehicle and in bulk quantities?

15. What are the current problems with hydrogen as a transportation fuel?

16. How did hydrogen contribute to the Hindenburg explosion?

17. How does a fuel cell work? Why is it advantageous compared to batteries and other modes of propulsions?

18. What are the impediments to large-scale production of fuel-cell vehicles?

19. What is a hybrid vehicle? When are hybrid vehicles advantageous over conventional engines?

20. What is the difference between a series and a parallel hybrid?

21. What are the major manufacturers of hybrid vehicles, and what fraction of the market do they own?

II. Multiple Choice Questions

1. The main strokes of a four-stroke internal combustion engine are
   a. Intake, compression, ignition, and exhaust
   b. Compression, ignition, expansion, and exhaust
   c. Intake, compression, power, and exhaust
   d. Intake, compression, ignition, and expansion
   e. Intake, ignition, expansion, and exhaust

2. To increase the efficiency of a spark ignition engine we must

a. Increase the compression ratio
b. Use fuels with higher flame temperatures
c. Reduce exhaust temperature
d. Prevent heat losses
e. All of the above

3. In the past ten years, the average miles per gallon of gasoline used by vehicles in the US has
   a. Increased
   b. Decreased
   c. Remained about the same

4. The power required to propel a vehicle over a given distance is proportional to
   a. Speed divided by distance
   b. Power times distance
   c. Speed times distance
   d. Force times distance
   e. Force times speed

5. The main barrier to wide-scale use of electric vehicles is
   a. Finding or developing light electric motors and generators
   b. Developing faster charging systems
   c. Developing lighter and stronger chassis
   d. Reducing cost
   e. Developing batteries with higher energy and power densities

6. Most power required to propel a passenger car is for
   a. Acceleration and climbing
   b. Cruising
   c. Braking
   d. Idling
   e. Accessories

7. Aerodynamic resistance dominates all other resistance in vehicles
   a. Cruising at speeds below 40 km/hr (25 mph)
   b. Only during stop-and-go traffic
   c. During acceleration
   d. At speeds exceeding 40 km/hr (25 mph)
   e. Never; it is always less than rolling resistance

8. Internal combustion engines
   a. Are technologically fully developed with the possibility of marginal improvements in their qualities
   b. Are able to produce maximum torque only over a small speed range
   c. Do not produce any torque at start
   d. Are the biggest source of pollutant emission
   e. All of the above

9. The main advantage of diesels over spark-ignition engines is that
   a. They are heavier and therefore more durable
   b. Diesel fuel is cheaper than gasoline
   c. They can be constructed at higher compression ratios and therefore their efficiency is higher
   d. They do not require any spark for ignition
   e. All of the above

10. The main advantage of gas turbines over gasoline and diesel engines is
    a. Their relatively low emission
    b. Their high power density
    c. Their multi-fuel capability
    d. Their smooth, vibration-free power delivery
    e. All of the above

11. Nitric oxides are produced mainly as a result of
    a. Incomplete combustion
    b. Quenching by the cold surfaces
    c. The reaction of nitrogen and oxygen in the air at high temperatures
    d. Burning a very rich mixture
    e. All of the above

12. Carbon monoxides are produced mainly as a result of
    a. Incomplete combustion
    b. Quenching by the cold surfaces
    c. Poor mixing of fuel and air
    d. Burning a very rich mixture
    e. All of the above

13. The main function of catalytic converter in an automobile is
    a. To increase efficiency
    b. To increase power
    c. To reduce noise

d. To reduce pollution
e. To reduce vibration

14. Which of the following statements are not correct?
a. The major ingredients contributing to both gasoline and diesel emissions are carbon monoxide, nitric oxides, and unburned hydrocarbon.
b. Sulfur dioxide emissions are not generally high, because sulfur is removed from the gasoline and diesel fuel before they are sold.
c. Catalytic converters are effective for removal of carbon monoxide, nitric oxides, and hydrocarbons.
d. Catalytic filters can be used to remove much of the fine particulates from diesel exhausts.
e. The most efficient method to reduce nitric oxide emission is by operating the engine near its stoichiometric condition.

15. Which of the following emissions cannot be controlled by catalytic converters?
a. Carbon monoxide
b. Carbon dioxide
c. Nitric oxide
d. Hydrocarbons
e. None of the above

16. LPG is mainly
a. Natural gas with some propane and butane
b. A blend of methanol and ethanol
c. Distilled gasoline
d. Leftover natural gas after methane is removed
e. Used as a jet fuel

17. Which of the following cannot be used to power an alternate-fuel vehicle?
a. Methanol
b. Ethanol
c. LPG
d. Coal
e. Electricity

18. The major disadvantage(s) of methanol as an alternative fuel is/are that
a. It has an invisible flame
b. It is toxic
c. It is corrosive
d. It has low energy density
e. All of the above

19. What is the main advantage of hydrogen over other fuels?
a. Hydrogen can be stored at very high pressures so we can store a lot in a small volume.
b. Hydrogen is easily produced from a variety of sources.
c. Hydrogen flame is invisible so it is relatively safe.
d. When produced from renewable sources, hydrogen is the cleanest of all fuels.
e. All of the above.

20. Which of the following statements is not correct?
a. Hydrogen can be substituted for domestic energy sources.
b. Hydrogen offers better utilization of resources.
c. Hydrogen can be used as a convenient medium for storing solar, wind, and nuclear energies.
d. Hydrogen can be substituted for most fuels now in use.
e. None of the statements are incorrect.

21. Hydrogen is a clean source of energy. When burned in air the only product(s) it produces is (are)
a. Water vapor
b. Water vapor and carbon dioxide
c. Water vapor and carbon monoxide
d. Helium
e. Deuterium and tritium

22. Hydrogen is often cited as the fuel of the future. This honor is granted mainly because
a. Hydrogen is clean and available in an unlimited supply
b. Hydrogen is the main source of nuclear energy
c. Hydrogen is light and easily combustible
d. Hydrogen is the largest component of the earth's atmosphere
e. All of the above

23. The main disadvantage of battery-operated electric

vehicles is

a. That they produce the highest torque during start ups and climbing where speed is low
b. That their overall efficiency is less than that of internal combustion engines
c. That we still need fossil fuel to produce electricity and charge the battery
d. That they can travel considerably shorter distances
e. All of the above

24. Most electric vehicles operating today use
a. Lead-acid batteries
b. Iron-nickel batteries
c. Nickel-cadmium batteries
d. Nickel-metal hydride batteries
e. Zinc-air batteries

25. Which of the following can be said about fuel cells?
a. Fuel cell is a well-developed technology with most infrastructure already in place.
b. Fuel cell vehicles can go thousands of miles between each charging.
c. Fuel cells are relatively cheap to manufacture.
d. Their efficiency can be higher than those dictated by the Carnot efficiency.
e. All of the above.

26. Natural gas vehicles (NGV)
a. Are considered truly zero emission vehicles
b. Are considered highly unstable and dangerous
c. Produce no nitric oxide because there is no nitrogen in the fuel
d. Are considerably cleaner than gasoline-fueled vehicles
e. All of the above

27. Which of the following is true about hybrid vehicles?
a. They are available for purchase today.
b. They give better mileage than regular vehicles.
c. They use regular gasoline.
d. They use two modes of propulsion.
e. All of the above.

28. Fuel cells are regarded as promising technology because
a. They were used in the space program
b. They use fossil fuel
c. Their only by-product is water
d. They are cheap to produce
e. We have huge reserves of hydrogen

29. Which of the following is true about electric vehicles?
a. The main disadvantage of electric vehicles is the cost of batteries.
b. The number of battery-powered electric vehicles in California will increase to 20% by 2010.
c. Electric vehicles have limited range.
d. Most electric vehicles are powered by fuel cells.
e. Electric vehicles are superior during highway driving and at cruising speeds.

30. Which vehicle type emits no pollution whatsoever?
a. LEV
b. ZEV
c. ULEV
d. SULEV
e. ICE

31. Reformulated gasoline
a. Is lower in carbon monoxide emission because of additional oxygen dissolved into the gasoline
b. Is lower in particulate emission because of additional hydrogen dissolved into the gasoline
c. Has a higher fuel efficiency
d. Is the same as regular gasoline
e. All of the above

32. Which of the following is true of methanol as a fuel?
a. It is not corrosive, so it can be transported through existing pipeline.
b. It has a higher energy density than gasoline.
c. Because it has a higher H/C ratio than gasoline, it is relatively clean.
d. Unlike gasoline it does not produce any

formaldehyde.
e. It is relatively nontoxic.

33. You are trading your car for a hybrid car that gives you 20 more miles per gallon. If the distance you travel each year is 20,000 miles and the average price of gasoline is \$2.40 a gallon, how much do you save annually in fuel cost?
    a. \$ 416.67
    b. \$ 1,000
    c. \$ 2,400
    d. It depends on how much you paid or received in the trade.
    e. Cannot tell from the data given.

34. You have a sport utility car which gives you 20 miles per gallon. You are trading your car for a hybrid car that gives you 50 miles per gallon. If the distance you travel each year is 20,000 miles and the average price of gasoline is \$2.40 a gallon, how much do you save annually in fuel cost?
    a. \$ 960
    b. \$ 1,440
    c. \$ 2,400
    d. \$ 2,500
    e. Cannot tell from the data given.

35. An air hybrid vehicles
    a. Is another name for existing hybrid cars
    b. Is a type of internal combustion engine that operate without any fuel
    c. Is similar to a conventional engine except that compressed air from a tank provides the propulsive power
    d. Violates the first law of thermodynamics because gets energy without burning any fuels
    e. Violates the second law of thermodynamics because it compresses air without doing any work on the pistons

III. True or False?

1. Cylinders offer considerably less drag compared to spheres of the same diameter.

2. Rolling up the windows and replacing outside mirrors with inside mirrors can reduce aerodynamic drag and improve fuel efficiency.

3. To minimize aerodynamic drag, airfoils should be shaped as close as possible to tear drops.

4. To reduce rolling drag, it is best to keep tires under-inflated.

5. During cruising operation, the engine produces enough power to overcome aerodynamic resistance.

6. To reduce drag in trucks, it is best to make trucks boxier in shape.

7. For ships and boats moving at low speeds, viscous drag accounts for most of the resistance.

8. A rotary engine works similarly to a reciprocating engine, except the piston is substituted with a triangular rotor.

9. A four-stroke cycle engine goes through one revolution for each power stroke.

10. Rotary engines employ rotors instead of pistons to achieve compression.

11. Two out of every three gallons of petroleum consumed in the United States are for transportation.

12. Gas turbines have relatively lower emissions than gasoline and diesel engines.

13. The higher the C/H ratio, the cleaner the fuel.

14. Electric vehicles were more popular than petrol engines during late nineteenth and early twentieth centuries.

15. Electric motors produce maximum torque at cruising speeds.

16. Gasoline engines are most efficient at cruising speeds.

17. Electric vehicles are much quieter during operation and do not consume any power or produce any

emission when stopped.

18. Conventional vehicles produce a large amount of HC and CO during cold starts, acceleration, and braking.

19. In a series hybrid configuration, all motive power comes from the electric motor powered by batteries.

20. EV performance is a strong function of engine speed, designed to operate optimally at cruising speeds but drop rapidly at both very high and low speeds.

IV. Fill-in the Blanks

1. ______________ is the dominant drag force for automobiles at slow speeds.

2. ______________ is the result of energy lost in creating and maintaining the ship's characteristic bow and stern waves.

3. Instead of a piston, rotary engines use triangularly shaped ____________ to achieve compression.

4. Electric motors produce the highest torque at __________ speed, when it is most needed.

5. In a __________ hybrid configuration, the engine, the electric motor, or both supply power to the wheels.

V. PROJECT I - Energy Use and Pollution from Passenger Cars

In this assignment you are asked to estimate the total petroleum consumption by passenger vehicles in the United States and their contribution to the air pollution problem.

a. What kind of car do you drive? What year and model?
b. What is the power rating for your car?
c. How many miles per gallon do you get for your car in city driving? In highway driving?
d. What is the average price of gasoline in your area?
e. Besides yourself, how many other people usually ride in your car?

Now answer the following questions:

1. How many gallons of fuel do you use annually? What is the mass of fuel?
2. What is the annual cost of petroleum? Compare this with the cost of bottled water or soda you consume every year.
3. How many kilograms of air do you pollute for every mile traveled? Gasoline is a mixture of many hydrocarbons, but the chemical formula can be adequately represented as C_8H_{16}. Assume complete combustion.
4. How much carbon dioxide does your car produce? Give your results in grams/mile. What is the yearly emission of carbon dioxide in kilograms?
5. What is the total annual emission of carbon monoxide and nitric oxides for your car? The CO, NOx, and particulate emissions Federal Standards set by the EPA are 3.4, 1.0 and 0.2 grams per mile respectively, for cars less than 5 years old, and 4.2, 0.6, and 0.10 for cars older than 5 years (See Table 14-5).
6. What are the emission rates per person per year?
7. Extend your calculations to estimate the total amount of gasoline consumed by Americans every year. How does your estimates compare to published data by the Department of Transportation? How much pollution is from cars? Assuming that there are 120 million vehicles in the United States, what is the average amount of petroleum used per vehicle? Is it higher or lower than your rate of consumption?
8. Assume that you are converting your car to run on methanol (CH_3OH), a synthetic liquid produced from wood. How are the emissions affected?
9. Repeat the calculation, assuming a switch from gasoline to natural gas. Assume natural gas is primarily methane gas with chemical formula CH_4.
10. What is the impact of natural gas vehicles on global warming if there is a leakage of 1% of methane gas into the atmosphere? As a greenhouse gas, methane is about 40 times more potent than carbon dioxide.

Data and Conversion factors:

- The LHV for gasoline is 44 MJ/kg
- 1 US gallon of fuel has a mass of 2.85 kg.

Hint: Assume stoichiometric reaction:

$$C_8H_{16} + 12\,(O_2 + 3.76\,N_2) \longrightarrow 8\,CO_2 + 8H_2O + 45.1\,N_2$$

$$1\text{ kg HC} + 14.7\text{ kg air} \longrightarrow 3.1\text{ kg } CO_2 + 1.3\text{ kg } H_2O + 11.3\,N_2$$

Project II - Cars versus Bikes

In this project you will compare the amount of energy used to travel from your home to school and back by either driving your car or by riding your bike.

1. What is the round-trip travel distance between your house and the school or work?
2. How long does it take you to get to work or school if you ride a bike? Assume average speed of 20 km/h.
3. How much muscle energy do you use to make the round trip travel by bike?
4. How much food energy do you need for taking this trip?
5. How long would it take you to travel the same distance by a car?
6. What is the fuel efficiency (mpg) of your car? How many liters of gasoline do you consume? What is the total energy content of gasoline required for the trip?
7. How much money do you save annually in gasoline and other car expenses, if you bike to work or school?
8. What are the advantages and disadvantages of driving over bicycling? Discuss in terms of convenience, energy efficiency, health benefits, etc.

Economics of Energy

CHAPTER 15

We have met the enemy and he is us ~ Pogo

One of the most fundamental questions of our time is the rate at which we should consume our fossil resources and how this decision impacts our economic well-being at present and for decades into the future. Technologically, we currently have the necessary tools to exploit both the exhaustible fossil and nuclear fuels and the renewable hydro, wind, solar, and biomass resources. Other technologies such as fusion and yet unknown resources may also become available within the next few decades. In previous chapters, we covered technological issues related to the use of these resources and their effects on our environment. In this chapter we will discuss the economical issues of energy and their impact on our technological society. The economics of the environment will be covered in the next chapter.

Overview

The economic dilemma every human faces is how to satisfy unlimited wants with limited resources. The resource limitations could be technological know-how, sufficient purchasing-power, time, or materials and energy. This is why we choose (or are forced) to give up or postpone some desires to get others which are seemingly more important to us. Because of limited income, we must prioritize our desires, for example by buying a used instead of new car or by postponing the purchase of a home until we have graduated from college. At other times we give up what we already have for the sake of our children and grandchildren.

The same is true with other resources at our disposal. For example, we only have limited fossil resources, and they cannot last to meet all our energy needs for a very long time. Because we are endowed with all the resources, we have the choices of using energy more efficiently, practicing conservation, and seeking alternative sources of energy; alternatively, we may continue consuming energy with no restraints and leave future generations to resolve their own energy shortages. Because energy plays a crucial role in the economic health of all nations, understanding concepts and fundamental laws in economics is essential and is the subject of this chapter.

Competition

Perfect Competition: Efficient Market

A perfect or efficient market is a competitive market in which economic forces operate unimpeded. For a market to be perfect, the following conditions must exist:

a. There are numerous buyers and sellers. The output of the goods from each firm is small compared to the total supply of those goods so none of them are able to influence the price of the goods on their own.
b. There are no barriers to the entry of new competitions and competitors. Barriers can be legal, such as in patents and regulations; technological, such as with technologies that are too costly or available only to certain industries; or social, such as when loans or grants are available to only certain groups of people and not others.
c. There are many other firms with products that are available and practically indistinguishable from the goods of interest. Buyers could easily substitute another product if there were shortages.
d. A firm's only goal is to maximize its profits; in other words, firms have no interests (political, social, etc.) other than their economical interest.
e. All consumers and producers face the same price; alternatively, information about prices is freely available so that buyers and sellers are price takers.

In actuality, markets are not perfect and many technical, financial, and legal barriers can influence them. In a real market there are only a limited number of buyers and sellers, so it is possible that a few firms collude to dominate the industry. Because the number of buyers is finite, there is a limit to how much of a product firms can sell, no matter how low the price. Monopolies, cartels, and subsidized firms that provide specialized services, sell novelty items, or offer products protected by patents and government regulations violate one or more of the conditions set above and are therefore not considered to be perfect or efficient markets.

It should be noted that although most barriers to a perfect market (control of resources, predatory and retaliatory pricing, and excess capacity) are socially harmful, there are some advantages for a non-perfect market. For example, the economics of scale may reduce capital requirements and reduce costs to the consumers. To a certain degree patents and copy rights will foster entrepreneurship and promote innovation. Finally, non-perfect markets allow product differentiation which brings variety that spices up our daily lives.

Supply and Demand

To understand the many ramifications of modern economics, it is instructive to start with the concept of supply and demand and to explain how it determines an efficient price and quantity of a given good sold in a competitive market. In general, the theory claims that the quantity demanded for products increases as prices fall, *ceteris paribus*.[1] At the same time, sellers will be willing to offer more goods at higher prices. It is therefore reasonable to expect a demand curve to exhibit a downward slope,[2] whereas a supply curve has an upward trend as the quantity of goods offered increases. The point where these curves intersect is the **equilibrium point** (Figure 15-1). At the equilibrium point, both buyers and sellers are satisfied with the transaction and therefore there is little impetus to change. In other words, at the equilibrium, everybody is better off from the trade.[3] If the price is set too high or if supply increases, then there will not be sufficient buyers, resulting in a **surplus**. In this instance, suppliers will attempt to reduce their inventories by cutting prices, and producers, seeing the lower price, will cut production. As price falls more customers are likely to enter the market and the price and volume will increase, approaching the equilibrium.

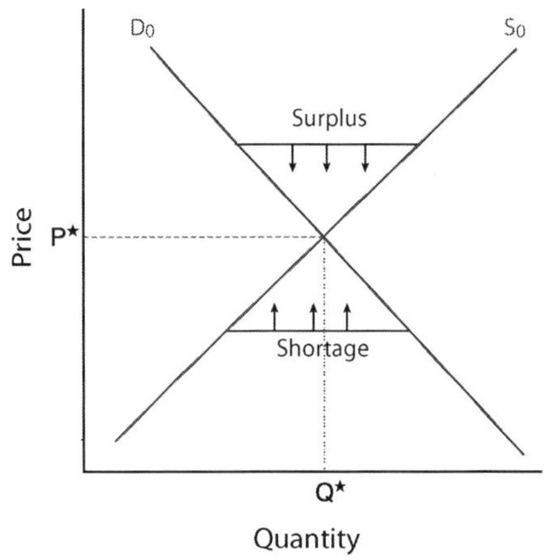

Figure 15-1
Supply and Demand. At equilibrium Q* units of the product is offered at a price of P*. When there is an imbalance between supply and demand, either a shortage or surplus results.

If the price is set too low or if demand increases, then quantity demanded exceeds the quantity supplied, resulting in a **shortage**. In this case buyers are willing to pay more (and possibly develop a black market). Seeing prices rise, more sellers enter the market, suppliers increase their production, and again we approach equilibrium.

Many factors besides price affect demand and supply. Factors that affect demand are changes in income, changes in the cost of living, personal preferences, and customers' expectations. Factors that affect supply are changes in the costs of production (such as the cost of raw materials and labor), changes in technology, and changes in the amount of taxes and subsidies.

It should be emphasized here that ***change in a quantity demanded does not mean a change in demand***. There are two ways that demand can change: a change in price or a change in overall demand. For a given market, a change in the price results in changes in quantity demanded (movement along the demand curve); the higher the price, the lower the demand, and vice versa. For example, when the price of natural gas rises, buyers move away from natural gas and use substitutes such as heating oil, kerosene, or wood. When price drops, consumers switch back to natural gas again.

Demand can also change because of changes in market conditions. For example, when there is a shift in median income, people tend to buy a different quantity of a product for the same price. Since a demand curve represents the quantity demanded at different prices, a change in demand requires a shift in the demand curve. For example the demand for natural

[1] *Ceteris paribus*, the Latin phrase meaning "other things remaining constant."
[2] It should be stressed here that, although the demand curve for the market as a whole has a downward slope, individual firms sell their product at a given price. The demand curve is a flat line for individual firms.
[3] By "better off" we only mean in terms of overall economical advantage. If the increase in a country's income comes with the destruction of cultural values, environmental degradation, and other adverse non-quantifiable social implications, then an average person might actually be "worse off" as the result of the trade.

Shortage and Hoarding

FYI ...

In 1973, following the Arab oil embargo, there was a gasoline shortage; long lines of cars formed at gas stations waiting hours to fill up their gas tanks. To prevent price gouging and black market profiteering, the government instituted a price control. Fearing increasing shortages, the average gas tank that was previously only one-quarter full was now being kept three-quarters full. With about 100 million cars on the American roads, and assuming the tanks had 20 gallon capacity, the hoarding resulted in a shortage of about one billion gallons in the first two weeks of the price control, causing a sudden increase in demand that actually exacerbated the shortage.

gas heaters increases during a cold winter. Unless supplies change, buyers are willing to pay a higher price than they would be during a mild winter, so the demand curve shifts upward and prices rise (Figure 15-2). The opposite will be true when winters turn out to be milder than expected. Supply curves can also change depending on cost. For example, as technologies improve, the cost of production of solar cells decreases, so cell manufacturers are willing to supply more cells for a given price -- the supply curve shifts to the right (Figure 15-3).

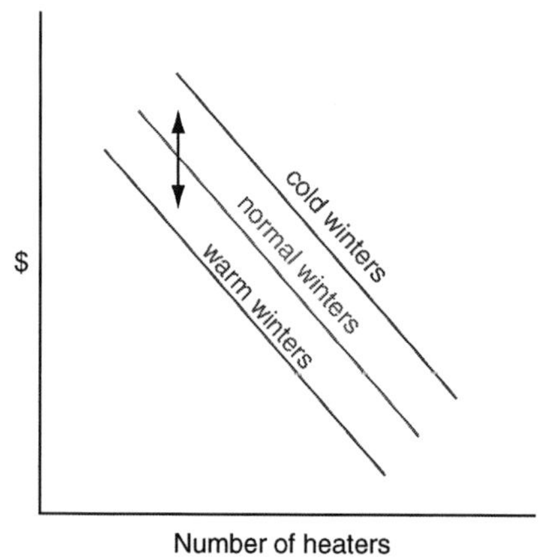

Figure 15-2
Demand shift.

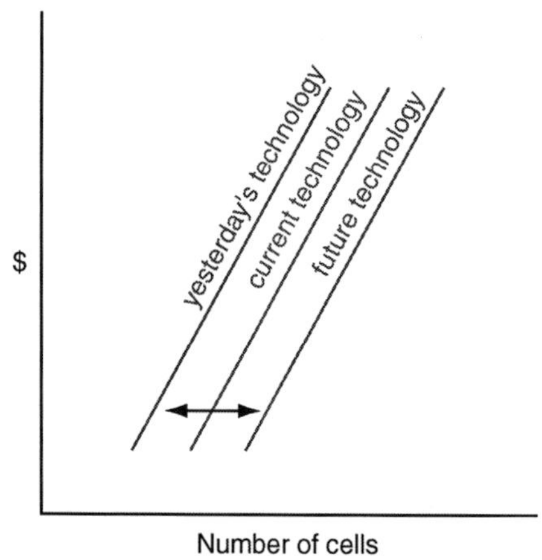

Figure 15-3
Supply shift.

Example 15-1: Depending on their incomes and needs, different customers are willing to pay different prices for gasoline, as is shown in demand curve D_0 in Exhibit 15-1. Gasoline availability is also a function of the price customers are willing to pay, as it is shown in the supply curve denoted S_0.

a. What is the equilibrium price of gasoline and the quantity supplied to the market?
b. If everything else remains the same (ceteris paribus), how would customers' behavior be affected if the price of gasoline drops 10 cents to $3.90 a gallon?
c. How much would the price of gasoline rise during summer if demand rose by 20%? Assume that the government does not interfere in controlling supply and demand.
d. Now assume that the government mandates a price freeze setting a price ceiling of $4.00 per gallon. How would customers react to the change in demand during the summer?
e. How will the market react if it anticipates future shortages as a result of a 20% cutback in production by OPEC?

Solution:

a. The market equilibrium is at point A in graph a, where 100 million gallons of gasoline are supplied by the oil companies at a price of $4.00 a gallon.
b. In this case, the demand curve is not affected. This does not mean the quantity demanded is the same. In fact as gasoline prices drop the quantity demanded will be greater (people tend to drive more or buy bigger cars), although the demand is still the same. The movement is along the demand curve to point B and 110 million gallons are sold.
c. As daily demand for gasoline climbs, the demand curve shifts to D_1, and customers are willing to buy 120 million gallons at the

price of \$4.00 (point C in graph b). In the short run, the market reacts and eliminates the initial shortage by driving prices up to point E at which 120 million gallons are sold at \$4.20 a gallon. In the long run, as prices rise, drivers will cut their driving (for example by vacationing shorter distances or by carpooling) and equilibrium moves back to point D. As a result, 110 million gallons are offered at the price of \$4.10 a gallon.

d. With the price ceiling set at \$4.00 a gallon, the oil companies are willing to sell only 100 million gallons of gasoline every day (equilibrium point A in graph c). The daily demand for gasoline, however, is 120 million gallons. To prevent shortage, customers are willing to pay more, potentially causing a black market and driving up the prices to as much as \$4.20 a gallon (point E). Because there is a price ceiling, drivers cannot legally pay more than \$4.00 a gallon. In the absence of a black market, they have to pay in non-monetary terms such as waiting in line. In a way, a new equilibrium is reached where 100 million gallons of gasoline are traded at \$4.20, of which \$4.00 is paid monetarily and \$0.20 is paid in terms of waiting in line.

e. In this case the supply curve shifts to S1 (graph c) while demand remains the same, and again we will face a daily shortage of 20 million gallons (AF), temporarily driving the price up to \$4.20 a gallon (point E). In the long run, changes in consumers driving habits and willingness to switch to alternative fuel vehicles and to electric and hybrid cars reduce the demand for gasoline. It eventually reaches a new equilibrium point G where price drops to \$4.10 again.

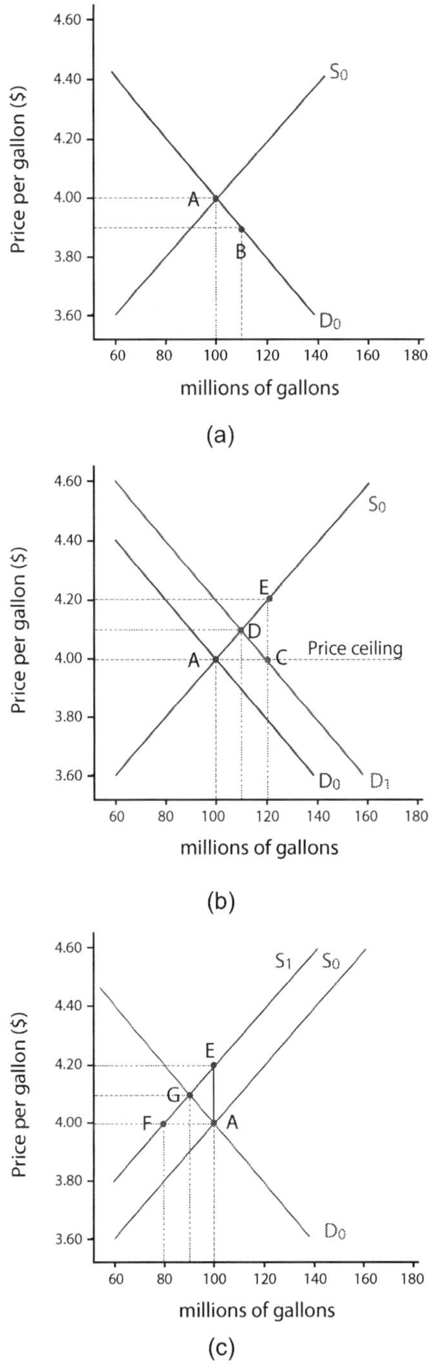

EX 15-1

It is clear from the above example that the market has a built-in mechanism that reacts to scarcity by raising prices and ultimately eliminating scarcity. The rise in price causes a decrease in the quantity demanded, and an increase in the quantity supplied by providing incentives to existing firms to up production, to invest in new technologies, and to enable new firms to enter the market.

Total and Marginal Costs

There are many kinds of costs associated with activities, but they can be generally divided into two types -- fixed and variable. **Fixed costs** exist whether or not any products are produced or services are delivered. Examples are rent, furniture, insurance, and the cost of purchasing equipment. Other costs -- called **variable costs** -- change as output changes. Examples are the costs of raw material and labor. The sum of the fixed and variable costs is the total private cost, or simply the **total cost.**

All of these costs are important to a supplier, but they are not the most important costs for the firm to decide the quantity of goods to produce. That decision depends on whether the cost of producing one additional unit pays for the revenue received for that additional unit. The additional or incremental cost of producing one extra unit of product or providing one unit of additional service is called the **marginal cost**. The marginal cost of oil production is the minimal additional cost of producing one extra barrel of oil. Because higher production levels necessitate extraction from higher-cost fields, marginal cost increases as production rises. Similarly the **marginal revenue (marginal benefit)** can be defined as the additional or incremental revenue (benefit) of obtaining one extra product or receiving one unit of additional service.[4] As long as the marginal benefit is higher than marginal cost, production of additional units is profitable. Said another way, the total profit of the firm is maximum when marginal cost of one additional unit produced is equal to marginal revenue resulting from its sale (the price it gets for its good).

Example 15-2: An agricultural firm is selling corn in a competitive market. The firm incurs a fixed cost of $20 (let's say for rent) and an additional cost which depends on the quantity of corn produced. The data is given in columns 2 and 3 in the table shown. Assuming that the market price offered by the competition is $14 a bushel, find the number of bushels that the firm must produce to maximize its total profit.

Solution: We can calculate the total cost (column 4) by summing up values of fixed and variable costs (TC = FC + VC). Marginal cost (column 5) is calculated by taking the difference between the total costs of production where one additional bushel of corn is produced. Total revenue (column 6) is simply the product of the quantity produced and the unit price of $14. Total (column 8) and marginal (column 9) profits are calculated as the difference between total and marginal revenues and cost. (TP = TR - TC and MP = MR - MC).

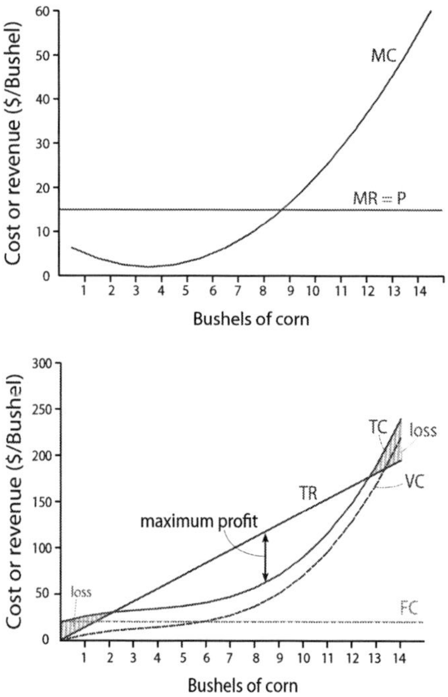

EX 15-2

As the results show, the total maximum profit the firm can make is $54.88 by selling 8 or 9 bushels of corn. This is where marginal cost and marginal revenue cross and are equal to the price of one bushel of corn ($14). The firm continues to be profitable as production increases beyond 9 bushels, although marginal profit starts to drop. Beyond 12 units of production the firm loses money at an accelerated rate. The graphical representation of the results is shown in Exhibit 15-2.

It should be noted that the analysis made above is valid only in the short run. In a perfect competition, as long as there are profits to be made, there is incentive for other firms to enter the market. As new competitors enter the market, supplies increase and prices fall. Those who experience a loss will necessarily exit the market and are replaced with new firms that enter to fill the gap. The process continues until

[4]There are instances when no marginal cost or benefit exists. For example, there is no marginal cost for an extra plate in an all-you-can-eat diner, and there is zero marginal benefit of an extra plate when you are no longer hungry.

the equilibrium is reached (P = MR = MC), at which point the firms make no profit. This is the long-run condition for perfectly competitive firms.

1	2	3	4	5	6	7	8	
Qty	Fixed Cost	Variable Cost	Total Cost	Marginal Cost	Total Revenue	Marginal Revenue	Total Profit	Marginal Profit
Q	FC	VC	TC	MC	TR	MR = P	TP	MP
0	20.00	0.00	20.00		0.00		-	
				6.32		14.00		7.68
1	20.00	6.32	26.32		14.00		-	
				3.92		14.00		10.08
2	20.00	10.24	30.24		28.00		-2.24	
				2.48		14.00		11.58
3	20.00	12.72	32.72		42.00		9.28	
				2.00		14.00		12.00
4	20.00	14.72	34.72		56.00		21.28	
				2.48		14.00		11.52
5	20.00	17.20	37.20		70.00		32.80	
				3.92		14.00		10.08
6	20.00	21.12	41.12		84.00		42.88	
				6.32		14.00		7.68
7	20.00	27.44	47.44		98.00		50.56	
				9.68		14.00		4.32
8	20.00	37.12	57.12		112.00		54.88	
				14.00		14.00		0.00
9	20.00	51.12	71.12		126.00		54.88	
				19.27		14.00		-5.28
10	20.00	70.40	90.40		140.00		49.60	
				25.52		14.00		-11.52
11	20.00	95.92	115.92		154.00		38.08	
				32.72		14.00		-18.72
12	20.00	128.64	148.64		168.00		19.36	
				40.88		14.00		-26.88
13	20.00	169.52	189.52		182.00		-7.52	
				50.00		14.00		-34.40
14	20.00	219.52	239.52		196.00		-	
				60.08		14.00		

Elasticity

Marginal quantities are closely related to the **elasticity** of demand and supply. Elasticity of supply can be viewed as the responsiveness of the market to changes in quantity supplied as prices change. The higher the number of substitutes for a given service or product, the more elastic the supply curve. When there is an infinite number of alternative products available in response to a change in price, we call the supply curve perfectly elastic (horizontal line in Figure 15-4). When there can be no change to quantity supplied, no matter the price, the supply curve is perfectly inelastic (vertical line in Figure15-4). During a very short time, when a new product enters a market or when plant operates at full capacity, supply cannot change with a change in price; i.e the supply curve

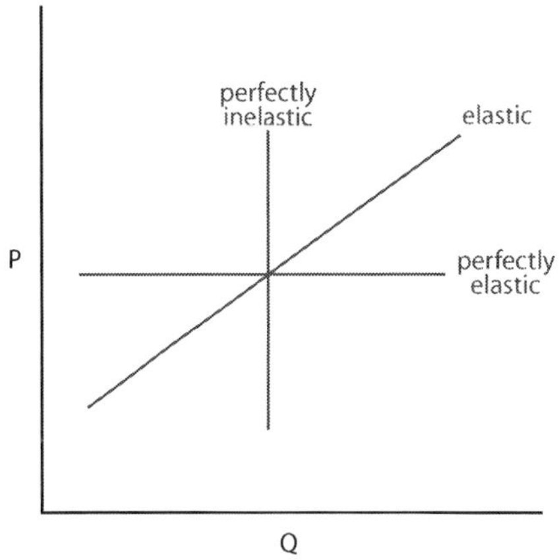

Figure 15-4
Elasticity

Price Elasticity of Gasoline and Oil

FYI ...

The sudden rise in the price of gasoline brought about by the oil crisis of 1973 and the Iranian Revolution of 1979 demonstrates the inelastic nature of these commodities in the short run. Because there are essentially no substitutes for gasoline in cars, and because of the American dependence on private cars, the price of gasoline increases suddenly following any interruption in the supply of the imported oil with practically no noticeable drop in demand. The price rises from P_0 to P_1 and the short-term demand curve is D_1 (See Shortage and Hording). Given enough time, people look for shorter and less frequent trips, switch to alternative modes of transportation such as mass transit, carpool, more economy cars, and even move their houses closer to their work, all in order to cut expenses. The long-term demand curve (D_2) becomes more elastic.

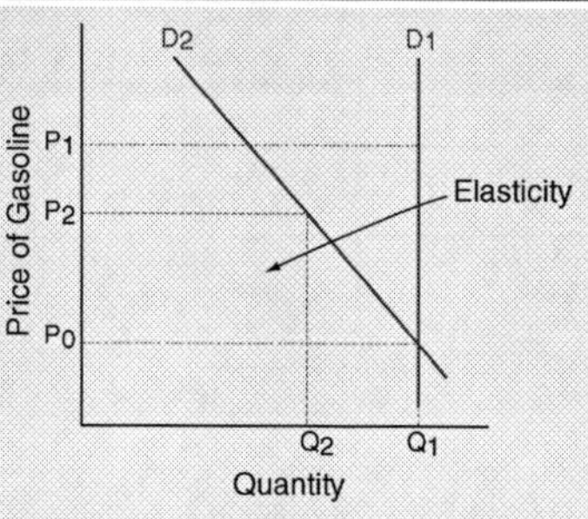

is highly inelastic. As new suppliers enter the market, the supply curve becomes more and more elastic. For a constant cost industry (such as an industry that uses inputs common to many other industries), the output is too small to affect prices of the inputs, and long run supply is perfectly elastic.

Similarly, elasticity of demand is a measure of the responsiveness of demand as prices change, i.e. the percentage change in sales caused by a percentage change in price. Demand is less elastic in the short run (there is less that we can do to moderate consumption), and more elastic in the long run (we can develop substitutes, learn to live without).

Externalities

When two people trade, the belief is that they both mutually benefit from that trade. Unless they are required by law, buyers and sellers are unlikely to consider the effect of their trades on others. Many trades and agreements have some direct and indirect effects whose costs are not borne by the buyers and sellers, and therefore prices do not reflect true costs. These costs are called **external costs** or **externalities** and include such things as the cost of health care associated with pollution and damage done to the environment, buildings, etc. Society at large must bear both the private cost of production as well as the externalities distributed among various sectors. The sum of private and external costs is called the **social cost.** Depending on whether the externalities benefit or harm the society as a whole, they can be positive or negative. Pollution is a *negative externality,* whereas cleaning the environment and research to eliminate an infectious disease are *positive externalities.* The true cost of energy is not limited to the price we pay for our utilities or at the gas station, but also includes what we eventually pay in terms of degrading the quality of life by ruining the environment and our health. Additionally, the energy sector, petroleum in particular, is highly subsidized by the government in terms of various tax incentives, low-cost access to leases, mapping, R&D on oil extraction, building pipelines, highways and other infrastructures, and finally by military expenditure to protect the flow of oil. According to one estimate, in the last 30 years, the US has spent a total of $230 trillion dollars (1998 dollars) for oil imports.[5] The government additionally spends somewhere between $100 billion and $300 billion on hidden costs like health care and lost productivity and another 6 to 60 billion dollars defending oil supplies in the Middle East every year.[6] If the costs associated with the current Iraq war are included, the last figure would be substantially higher. Many other environmental costs such as the destruction of natural habitats and wildlife and the loss of species diversity, which are not commercially traded, cannot be accurately measured and are not included in these figures.

Question: What are the external costs associated with electricity generation using solar and other renewable energy sources?
Answer: Although photovoltaic cells emit no pollutants during operation, their manufacture requires large quantities of hazardous materials, and their ultimate disposal could release toxic elements such as arsenic and cadmium into the environment. Other renewable energy sources incur similar costs; the external costs associated with them is however a lot less than with non-renewable sources.

In a new study by the European Commission, the external costs associated with various energy technologies (fossil, nuclear, hydro, solar, and wind) for generating electricity were estimated. For non-market goods such as health care costs, noise, etc., an evaluation was made on the basis of the willingness to pay for damages or willingness to accept the risks. The external cost of material was estimated from the emission produced during its manufacture. Damages to ecosystems (acid rain, ozone depletion and global warming) as a result of fossil combustion were assessed by estimating the cost of avoidance. Because of different technologies used and material and labor costs, the cost is somewhat different from country to country and from one location to another. Table 15-1 shows the marginal external costs of electricity production in Germany.[7] Data for other countries show similar trends.

Table 15-1. Marginal external costs of electricity in Germany (€ cent/kWh)

	Coal	*Natural Gas*	*Nuclear*	*Solar (PV)*	*Wind*	*Hydro*
Damage Cost						
Noise	0	0	0	0	0.005	0
Health	0.73	0.34	0.17	0.45	0.072	0.051
Material	0.015	0.007	0.002	0.012	0.002	0.001
Crops	0	0	0.0008	0	0.0007	0.0002
Total	0.75	0.35	0.17	0.46	0.08	0.05
Avoidance Cost						
Ecosystem	0.20	0.04	0.05	0.04	0.04	0.03
Global Warming	1.60	0.73	0.03	0.33	0.004	0.03

[5] Greene, D. L., and Tishchishnya, N., "Costs of Oil Dependence: A 2000 Update," Oak Ridge National Laboratory, ORNL/TM-2000/152, Oak Ridge, TN, 2000, and data updates, 2003.
[6] Hu, P.S., "Estimates of 1996 US Military Expenditures on Defending Oil Supplies from the Middle East: A Literature Review," Oak Ridge National Laboratory, Oak Ridge, TN, March 1996.
[7] Report EUR 20198, "External Costs: Research results on socio-environmental damages due to electricity and transport," European Commission Directorate-General for Research Information and Communication Unit, B-1049 Brussels, Belgium, 2003. Executive Summary at Internet: *http://europa.eu.int/comm/research/rtdinfo_en.html.*

As this data suggests, electricity generated by wind energy has a very low external cost. The cost can be reduced even further by eliminating external costs associated with noise by installing wind farms some distance away from population centers. Nuclear energy has no externality with respect to classical pollution costs such as health effects associated with pollutants like particulates and oxides of carbon, sulfur, and nitrogen. The cost of global warming is also minimal and limited to that associated with the construction of the nuclear plant. There are, however, other external costs associated with the nuclear power. When uranium ore is mined, left behind are tailings; spent fuel must also be stored. In addition, there is unaccounted risk from earthquakes, and accidental or deliberate sabotage or attack. The US government has spent billions of dollars to find ways to store the nuclear waste unsuccessfully. The cost of treatment and isolation of radioactive waste for thousands of years and damage from nuclear leaks is almost impossible to predict and therefore is not usually included in any external cost analysis. Because of the liability limits in the US, owners of nuclear plants cannot buy full liability insurance in case of major accidents or terrorist attack. The lack of an insurance market – a market failure – should be a clue that there are unaccounted externalities.

The probability of a major nuclear accident is very low, and therefore externality averaged over all nuclear plants is relatively small. Same goes for possibility of a successful nuclear sabotage. In some countries such as Switzerland, the government pays for hardening their reactors to protect against attack. That is one way to count the external cost, the cost of avoiding the consequence of attack. In the event of an accident, however, and for those directly affected by the accident, the external cost could be extremely high and greatly exceed that of all other forms of electricity generation.

Biomass has a relatively minimal impact on greenhouse gases, as the carbon dioxide emission balances with that which is used up during plant growth. The impact could, however, become more pronounced as price of biomass increases and rainforests are destroyed. Although biomass does not produce greenhouse gases, it is not clean. When burned, it still emits pollutants that are harmful to health and to the environment. The air pollution impact varies greatly depending on the type of biomass, the technology used, and the degree to which emission of pollutant gases into the atmosphere is controlled. Natural gas is relatively clean with respect to criteria pollutants. In terms of its impact on climate change, natural gas is a potent greenhouse gas. How great an impact it has, however, depends on which natural gas technologies are employed. Coal is probably the dirtiest of all fuels. Not only is coal a major producer of carbon dioxide, but it is also a major contributor to the particulates and sulfur dioxide emissions responsible for acid rain and other health effects.

Taxes

One way to internalize the external costs is to impose consumption taxes or pollution penalties. One example is waste disposal, which is taxed in the form of a lump sum fee per customer. For example, every American household is charged $10 a month for garbage collection. In some communities, these charges are included in property taxes or association fees. In either case, there are no incentives for customers to reduce waste. Unlike the US, European taxes are based on a scale starting from a low nominal value and increasing gradually as the volume of waste increases.

Energy is usually taxed indirectly in the form of gasoline and emission taxes. However, this may sometimes cause unwanted consequences. For example, taxing gasoline raises its price and so more people will switch to electric cars. To charge the batteries, electricity generated mostly from dirty coal-fired power plants is used, and the overall concentration of carbon dioxide may actually increase. It is therefore more reasonable to directly tax the pollution (for example by taxing per kilogram of carbon produced) instead of placing indirect taxes on energy usage. Several states have begun imposing pollution taxes directly. Depending on what source of energy is being used, a surplus tax per kilowatt-hour of electricity produced is imposed.

Imperfect Competition: Monopoly and Cartel

Monopoly

A **monopoly** is an industry with only a single seller or provider of goods and services. In a monopoly, there is no competition and therefore the seller can charge prices that are higher than they would be in a competitive market. Because the price being offered by the monopoly is usually high, the quantity demanded is smaller than in competitive enterprises. In monopoly firms, prices are set at the point where total profit is maximized. They can sell more at lower prices, or they can sell less but at higher prices. In either case a monopoly can either control the output or fix the product price. Once the price is set, market demand will determine the quantity that can be sold at that price. Conversely, given the output, market demand determines the price at which that quantity can be sold.

Monopolies are characterized by a lack of viable substitute goods or the existence of high barriers against the entry of potential competitors into the market. The former could be because the monopoly controls a major resource such as raw material or a specific technology necessary to produce a product. The latter could include laws that forbid competition (exclusive control over a patent on a product or on the processes needed to produce the product), laws that effectively forbid competition (through heavy regulations and subsidy), state monopoly (such as printing money and issuing stamps), or natural monopoly where production conditions and high initial cost make a sole provider more efficient (power and water distribution to private households). The term is also loosely applied to companies which have de facto control of a large share of a total market; Microsoft and AT&T (before its breakup), could, because of their shear sizes, be involved in practices that prevented others from fair competition.[8] Also classified as monopolies are firms that control the entire market through many smaller, seemingly diverse companies; in the early twentieth century, Standard Oil controlled the exploration, extraction, and transportation of petroleum.

Just like a competitive firm, a monopoly maximizes its profit when MR = MC; unlike a competitive firm, marginal revenue in monopolies is not equal to the price. In fact, marginal revenue is below price.

Example 15-3: Consider the market for heating oil in a large metropolitan area. Data of consumer behavior from previous seasons indicate that the total quantity of fuel oil sold decreases as price increases, as more and more customers switch to electricity to meet their energy needs. The cost is also a strong function of the quantity, as additional units must be imported from farther and farther distances. The total fixed cost for all firms is $4,000. The cost of production of each additional unit and the projected demand curve are given below. What is the total volume of heating oil and price offered to the market assuming that:

a. There are 100 firms who are competing for the same market?
b. The biggest firm buys out the remaining 99 smaller firms and thus becomes the sole supplier of the heating oil?

Quantity	0	100	200	300	400	500	600	700	800	900	1000
Price ($)	42.00	40.00	38.00	36.00	34.00	32.00	30.00	28.00	26.00	24.00	22.00
Marginal Cost ($)	--	10.00	8.50	8.00	8.50	10.00	12.50	16.00	19.50	26.00	32.50

Solution: Whether a firm operates as a monopoly or competes in a competitive market, it is interested in maximizing its profit. As long as the revenue of producing one additional unit (marginal revenue) is higher than its marginal cost, the firm makes a profit. In a competitive market, there are a large number of competitors, so each firm can only charge a price equal to the marginal cost. Since for a competitive firm P = MR, its profit-maximizing output is where MC = MR = P. Attempting to charge any price above its marginal cost brings in additional profit and signals competitors to enter the market. In the example given, marginal costs are reduced initially but increase as additional units are produced. At the same time, prices continue to decline with increasing output. The price becomes equal to marginal cost ($26) when 800 barrels of heating oil are sold cumulatively among the 100 retailers, bringing a total of $7,500 in profit.

Increasing production does not necessarily mean more profit. In a competitive market, each firm is only after its self-interest and tries to

[8] Some argue that a monopoly can lead to higher quality - Microsoft working hard to maintain monopoly, for example, by improving its product.

maximize its own profit, even if it does not bring maximum profit for all firms collectively. In a monopolistic firm, there is no competition so a firm can set the price as it wishes and marginal revenue need no longer be equal to the price. In this example, the monopoly's interest is in reducing the total quantity from 800 to 600 but to sell it at a higher price of $30 a barrel (A barrel of oil is exactly 42 gallons) instead of $26, increasing total profit by $750 to $8,250. As could be expected, monopolistic enterprises lead to higher prices and lower outputs.

1	2	3	4	5	6	7	8
Qty	Price	Total Cost	Marginal Cost	Total Revenue	Marginal Revenue	Total Profit	
Q	P	TC	MC	TR	MR	TP	
0	42	4,000		0.00		-4,000	
			10.00		40		
100	40	5,000		4,000		-1,000	
			8.50		36		
200	38	5,850		7,600		1,750	
			8.00		32		
300	36	6,650		10,800		4,150	
			8.50		28		
400	34	7,500		13,600		6,100	
			10.00		24		
500	32	8,500		16,000		7,500	
			12.50		20		
600	**30**	**9,750**		**18,000**		**8,250**	**Monopoly**
			16		**16**		
700	28	11,350		19,600		8,250	
			19.50		12		
800	**26**	**13,300**		**20,800**		**7,500**	**Competitive**
			26		8		
900	24	15,900		21,600		5,700	
			32.50		4		
1000	22	19,150		22,000		2,850	

Advantages and Disadvantages of Monopoly

The major disadvantages of a monopoly are lower efficiency, higher prices, and somewhat poorer quality. Because they don't have to compete in the marketplace, monopoly firms can set prices artificially high, the firm is less efficient, and innovation is restricted. As a result, the product quality compared to quality in a competitive market is poor. Some economists do not consider these actions as necessarily harmful if firms reinvest their profit into the industry to expand their internal R&D (research and development) effort, building new plants and upgrading existing equipment. According to the Wall Street Journal, Japan owes its impressive energy efficiency to big energy companies who tacitly agreed among themselves to increase the price of energy but reinvest the windfall back into energy research and development.[9]

[9] "How Japan became so Energy Efficient" *Wall Street Journal*, September 10, 1990.

The Facts: The Price of Gasoline in Caifornia

Did You Know That ...?

This is where the money we pay for each gallon of gasoline goes (May 2008):

• Cost of the crude (including profit)		$3.06	73%
• Refining		$0.28	7%
• Various taxes			
o Federal excise tax	$0.184		
o State excise tax (av)	$0.180		
o State sales tax (av)	$0.286		
o Total		$0.65	15%
• Distribution and marketing		$0.20	5%
Total cost		**$4.19**	**100%**

Source: California Energy Commission (http://www.energy.ca.gov/gasoline/margins).

Cartel

Figure 15-5
OPEC Official Bulletin

When, instead of a single seller, a combination of a group of firms control the flow of a good or a service to a market, we have a **cartel** (also called a **trust**). In a cartel, a group of industries form an alliance where they can jointly decide on strategies that effectively control production and set prices. Like monopoly enterprises, cartels' main objective is to maximize profit, except that cartels strategize to maximize collective profit of their members even when different firms have conflicting interests. In practice, one or two larger firms control cartel policies and set quota and prices. Smaller firms must exit or follow suit.

OPEC is one such entity formed in 1960 by six major oil-producing countries to control production and combat falling oil prices (Figure 15-5). In the years following the 1960s, OPEC had limited success, as the price of light crude was raised modestly from $1.80 to $2.59 in 1973. The most striking success was the Arab Oil Embargo and OPEC's cutback following the 1973 Arab-Israeli war. This resulted in a rapid rise in the price of oil to $11.65 by January 1974, which was gradually raised to $14 a barrel by 1978. The Iranian revolution in 1979 and Iran-Iraq war of the 1980s gave OPEC another opportunity to push prices to as much as $40 a barrel in 1981. Adjusted for inflation to 2007 prices, this exceeded $100 a barrel (Figure 15-6). The rapid rise in the price of petroleum forced industrial nations to consume oil more efficiently and invest heavily in alternative sources of energy such as wind and solar. Furthermore, high prices gave incentives to other nations to explore more of their own reserves and expand their production capacities. Due to these changes, along with the discovery of the North Sea and Norwegian oil deposits and the flow of Russian oil, OPEC's role as an effective cartel was greatly reduced.

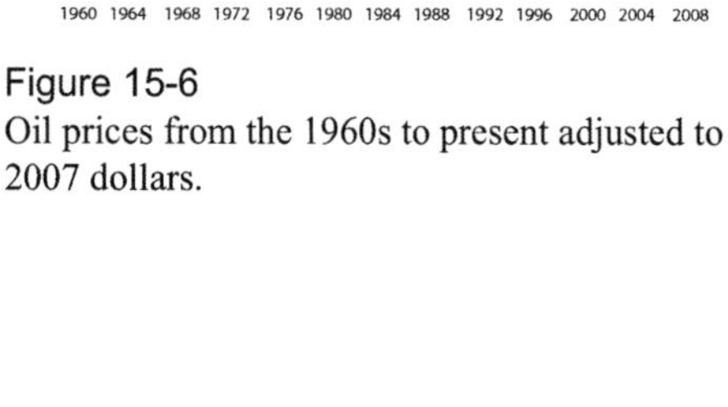

Figure 15-6
Oil prices from the 1960s to present adjusted to 2007 dollars.

Within the next few years, Saudi Arabia lost its leadership role in controlling prices, and the price of oil continued to decline – falling to $15 a barrel by February 1986. OPEC tried many times to control prices by cutting production, but the strategy remained largely ineffective, and except for a short time during the 1991 Gulf War, prices stayed under $20. As of May 2008, various economical factors and the turmoil in Iraq resulting from the US occupation have resulted in a spike in oil prices to $130 a barrel, the highest they have ever been in OPEC history.

It is important to point out that for a cartel to work efficiently, all members must agree to set aside economic differences and work in unison. In the case of OPEC, the internal economic factors and external political pressures, along with rivalry among members and emergence of non-member powerhouses, have effectively stopped OPEC from performing efficiently as a cartel.

Goods and Bads

"**Goods**" are anything that people desire, and by extension **"bads"** can be defined as all that they do not wish to have. Examples of goods are houses, cars, food, and energy. Examples of bads are pollution and noise. Garbage could be considered a good or a bad depending on whether we have to pay to get rid of our garbage, or if we are being paid by incinerator companies who burn it to produce electricity. Plastic bags can be considered to be goods when they are used to carry groceries but as bads when they no longer have any useful function and must be disposed of.

The goods that can be sold in a market or traded in a barter economy in exchange for other services and products are called **market goods**. **Non-market goods** are goods that are available to all without charge. Examples of market goods are televisions, hamburgers, oil, electricity, and in some instances even pollution. Air and water are prime examples of non-market goods.

Private and Public Goods

Private goods are goods whose consumption or use by one consumer precludes its consumption or use by someone else. For example, a car sold to you is only for your own use (or your family and friends if you wish). The gasoline that you put in your car cannot be used by any other car except yours. Private goods *are usually exclusionary*. Watching a movie in a movie theatre or riding a train requires buying a ticket, so those who do not have tickets can be excluded from watching the movie or riding the train.

Public goods are goods in common -- they can be used simultaneously by many individuals. Furthermore, once these goods are produced, they *cannot be exclusionary*, either by law or because of prohibitively high cost. Examples include the use of park services, police protection, roads, schools, and courts. The use of a public good by one person does not exclude others from using and benefiting from it as well. One person enjoying park facilities does not prevent others from enjoying the same park. Unless it is overcrowded, the use of the university library by one student does not diminish its use to others, so there is no rivalry between students over library resources.

Without exclusion, private firms have little or no incentives to provide public goods. To make a profit private firms require users to pay a fee for their products or services. For example, although TV broadcasting is a public good, cable companies can exclude non-subscribers and make a profit. For this reason, public goods are often provided by the government, which levies mandatory taxes on everyone. In certain instances, the government may hire private contractors to deliver public goods and pay them by issuing bonds or from general funds.

Two problems often associated with public goods are free riding and open access. *Free riding* is a problem when each consumer of a public good is inclined to consume the good without sharing its cost of production – to free ride on the investment of others. *Open access* refers to overexploitation of natural resources to the benefit of each consumer at the expense of the community benefit. For example, consider the decision to hire a pool man to maintain the community pool. If contributions are totally voluntary, each neighbor may not contribute their fair share, hoping to free ride on the contribution of others. As a result enough funds may not be collected. In another scenario, individual homeowners may not be as careful in keeping the community pool clean as they would have if they were the only one who used the pool. One way to correct the free riding and open access problems is to impose regulations and levy fees to each member of the community who may benefit from the public good.

Question: Is electricity a public or a private good?
Answer: Strictly speaking, electricity is a private good as it is non-rival and exclusionary. Electricity is, however, an essential part of our daily life and as such is consumed by the public at large. It may therefore be argued by some that electricity should not be treated in the same fashion as other private goods. That's to say, as a matter of economic justice or fairness, electricity must be regulated or subsidized (offered at lower prices) to low income customers.

In the United States and many capitalist countries, however, electricity has been subjected to the general laws of supply and demand and treated similarly to other conventional private goods. Because electricity generation is capital intensive, only a few firms control the electricity market. In economic terms, a production system that is controlled by limited producers is called an *oligopoly*. The problem with an oligopoly is that these firms can theoretically collude to fix prices, forcing smaller firms out of competition and thus in effect, becoming a single unregulated monopoly firm. When energy was deregulated in the United States, proponents cited "deregulation" as a measure intended to increase competition. In reality, as we saw in Chapter 13, deregulation of electricity resulted in major collusions among big energy firms who raised prices to unprecedented levels. Because the public required the service, they had no choice but to accept the highly inflated prices.

Example 15-4: In the absence of any regulatory pressure, John is dumping his garbage (a private bad) in a vacant lot adjacent to his house. Two neighbors, Sue and Bob, are particularly upset and are even willing to pay for the cost of hauling John's garbage, which is $10 a month. Sue is willing to pay $7 toward the effort; Bob has volunteered to pay $5. Would enough money be collected to clean up John's garbage?
Solution: Not necessarily! Knowing Bob's willingness to pay $5, Sue may conclude that $5 is all she needs to contribute toward the effort. By the same logic, Bob will volunteer to pay only $3, for a total of $8. As a result, each tries to free ride off the desperation of the other and not enough money is collected.

Demand curves for public and private goods
Depending on whether a good is private or public, its demand curves look different. In a market for a private good, a single price is offered, each buyer buys as many goods as desired, and total demand is simply the sum of the individual demands. For example, consider the total cigarette demand of two smokers, Mary (a regular smoker) and John (a casual smoker). Since cigarettes used by Mary cannot be used by John and vice versa, the total cigarette demand is the *horizontal sum of the demand curves* of John and Mary.

Unlike private goods, public goods must be shared by many customers. In this case, the demand curve is the *vertical sum of the demand curves* of individual customers. For example, the inhabitants of two nearby cities, A and B, may be willing to pay different taxes for developing a beachfront. The residents of city A, being farther away from beach or because of their lower median income, are willing to pay less than the more affluent residents of city B, which is also closer to the beach. Demand curves for these two cities per meter of beachfront are denoted by D_a and D_b. If the beach is willing to accommodate both cities, then total demand is found by adding the two demands vertically. That is, if the residents of city A are willing to pay $150 in taxes for developing 3 kilometers of beachfront and residents of city B are willing to pay only $80, then the total demand for 3 kilometers of beachfront is $230. Figure 15-7 shows the total demand for private and public goods.

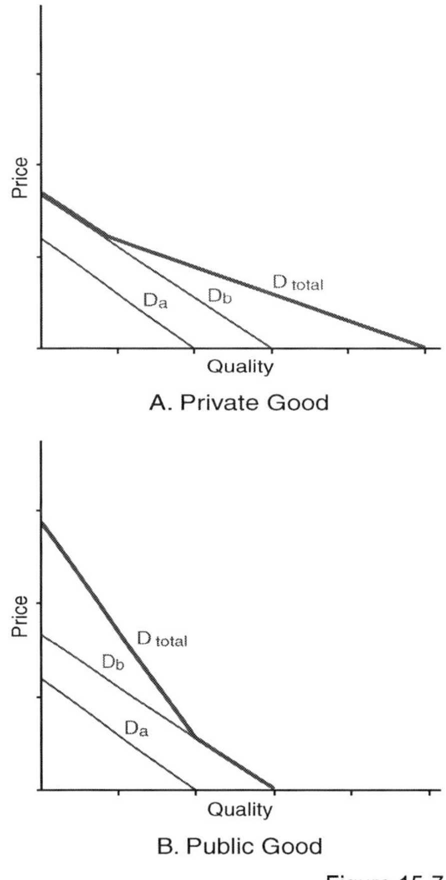

Figure 15-7
Demand curves for (a) private and (b) public goods.

a Money

Prior to the invention of money, trades were conducted by bartering – the direct exchange of goods and services for other goods and services. A shoemaker could exchange his shoes with a teacher who could tutor his children, or a tailor for a pair of pants. A housewife could borrow meat from a butcher, cook it and keep some for her family while delivering the rest to feed the butcher. In a small village with a handful of artisans, bartering was highly efficient.

Although bartering worked in feudal societies where only a limited number of products and services were available, it was largely ineffective in larger towns where a great variety of goods and services were needed. As societies grew and more goods and services were available, the number of exchanges became exceedingly large, and bartering became less attractive. With the invention of money, the need for direct exchange was eliminated, as individuals and businesses found a common unit of measure and value that could be exchanged for any number of products and services. Money also facilitated contracts for future activities and payments.

Wealth

In conventional capitalism, the quantity often cited as a measure of the economic health of a country is the **gross domestic product (GDP)**, defined as the total market value of all goods and services produced in a country in a given calendar year.[10] Another closely related indicator is the **gross national product (GNP)**, which is calculated as the GDP plus what domestic companies earn from activities abroad, minus what foreign companies make from domestic activities in that country. In other words, GDP is the sum of all goods and services produced by labor and property located in a country irrespective of who supplies them, whereas GNP is the sum of all goods and services supplied to a country's residents, irrespective of where they are physically produced.

Some economists (commonly referred to as ecological economists) object to the GDP and GNP as true indicators of wealth, because they limit wealth to only the portion of market activities where income is generated. Functions such as charity and housework are not accounted for, whereas the costs of cleaning the environment and health care are included. Furthermore, there is no distinction between desirable and undesirable outcomes, i.e. they measure only money spent, not the value received. One million dollars spent on health care research will contribute as much to GDP and GNP as one million dollars used on building a refinery in a residential community.

A better measure of human welfare may be the **net national welfare (NNW)**, defined as the total annual output of both market and non-market goods and services, minus the cost of externalities, minus the cost of depreciation of natural and human-made capitals used up in production.

Future and Present Worth

In Chapter 1, we gave a few examples of how a quantity such as savings in a bank, population, or energy consumption grows exponentially. With no penalty for early withdrawal, a deposit of $100 in a 10-year CD (certificate of deposit) receiving a 10% interest, grows to 100 (1.10) = 110 after a year, 100 $(1.10)^2 = 121$ after the end of second year, and $100(1.1)^{10} = 259$ dollars at the maturity date. Mathematically speaking, if the present value of a quantity is P and the rate of growth is r, then the future worth of a present value after time n years have passed is (F/P) given by:

$$F = P(1+r)^n \tag{15-1}$$

Rearranging this equation, we can find the present value of a quantity if we know its future worth, i.e. the *present worth of a future value (P/F)* is:

$$P = F(1+r)^{-n} \tag{15-2}$$

For convenience, values of *F/P* and *P/F* are tabulated for various n and r values and are given in Tables 15-2 and 15-3.

Table 15-2. Future Worth of a Present Value

$F/P=(1+r)^n$

r/n	1	2	3	4	5	10	15	20	25	30
0.01	1.01	1.02	1.03	1.04	1.05	1.10	1.16	1.22	1.28	1.35
0.02	1.02	1.04	1.06	1.08	1.10	1.32	1.35	1.49	1.64	1.81
0.03	1.03	1.06	1.09	1.13	1.16	1.44	1.56	1.81	2.09	2.43
0.04	1.04	1.08	1.12	1.17	1.22	1.48	1.80	2.19	2.67	3.24
0.05	1.05	1.10	1.16	1.22	1.28	1.63	2.08	2.65	3.39	4.32
0.06	1.06	1.12	1.19	1.26	1.34	1.79	2.40	3.21	4.29	5.74
0.07	1.07	1.14	1.23	1.31	1.40	1.97	2.76	3.87	5.43	7.61
0.08	1.08	1.17	1.26	1.36	1.47	2.16	3.17	4.66	6.85	10.06
0.09	1.09	1.19	1.30	1.41	1.54	2.37	3.64	5.60	8.62	13.27
0.10	1.10	1.21	1.33	1.46	1.61	2.59	4.18	6.73	1.083	17.45

Table 15-3. Present Worth of a Future Value

$P/F=(1+r)^{-n}$

r/n	1	2	3	4	5	10	15	20	25	30
0.01	0.99	0.98	0.97	0.96	0.95	0.91	0.86	0.82	0.78	0.74
0.02	0.98	0.96	0.94	0.92	0.91	0.82	0.74	0.67	0.61	0.55
0.03	0.97	0.94	0.92	0.89	0.86	0.74	0.64	0.55	0.48	0.41
0.04	0.96	0.92	0.89	0.85	0.82	0.68	0.56	0.46	0.38	0.31
0.05	0.95	0.91	0.86	0.82	0.78	0.61	0.48	0.38	0.30	0.23
0.06	0.94	0.89	0.84	0.79	0.75	0.56	0.42	0.31	0.23	0.17
0.07	0.93	0.87	0.82	0.76	0.71	0.51	0.36	0.26	0.18	0.13
0.08	0.93	0.86	0.79	0.74	0.68	0.46	0.32	0.21	0.15	0.10
0.09	0.92	0.84	0.77	0.71	0.65	0.42	0.27	0.18	0.12	0.08
0.10	0.91	0.83	0.75	0.68	0.62	0.39	0.24	0.15	0.09	0.06

Example 15-5: According the US Department of Energy data, the world production of crude oil increased from 60 million barrels per day in 1981 to 74 million barrels per day in 2001. What is the average rate of increase of the world's crude over this period? Calculate the expected volume of crude in 2020 if the rate of production continues to increase at 1% per year.
Solution: We can think of the value in 2001 as the future worth of the value in 1981, i.e. F/P = 74/60 = 1.23. Consulting Table 15-2, we can see this corresponds to a rate of increase of about 1% per year over the past 20 years. The same table can be used to extrapolate future production from 2001 data. Interpolating for n = 19, we can find F/P = 1.21. The 2020 consumption of crude can be estimated as 74x1.21 = 90 million barrels. It is left to the student to verify these results by applying equations 15-1 or 15-2 instead of using tables.

Example 15-6: A new substitute to Middle Eastern oil is expected to be found in the next 20 years. The cost of production of energy equivalent to one barrel of oil is estimated at $100. What should the price be of a barrel of oil sold today in a competitive market? In a market with substantial monopoly owner?
Solution: Assuming the market is efficient that no cost is associated with the extraction of oil from the Middle East, and that the cost of production remains the same, the oil should be priced at a value equal to its future price discounted to present time:

$$100x\{(P/F)_{n=20,\, r=0.05}\} = 100x0.38 = \$38.$$

It is very difficult to predict the cost of future technologies with certainty. Monopolies usually price their commodities at the cost of substitutes today, which could be substantially higher than $38 at today's prices.

Example 15-7: The US consumed 7.3 billion barrels of oil (bbo) in 2003, 2% more than it consumed in 2002.[11] Assuming the consumption continued to increase at the same 2% per year, how much petroleum would be left in US reserves at the beginning of 2006? The total US petroleum reserves were estimated to be 100 billion barrels at the start of 2003.
Solution: In 2004, we would be consuming 7.3x1.02 = 7.5 billion barrels of oil (bbo). The remaining reserves would be 92.5 bbo. In 2005, we would be consuming 7.5x1.02 = 7.6 bbo. The amount of oil that would remain at the end of 2005 is 84.9 bbo.

Doubling Time

Assuming a quantity grows in an exponential manner at a rate of r per year, the quantity grows after n years as given by equation 15-1. Of interest is

[11] US DoT, http://www.bts.gov/publications/national_transportation_statistics/2003/html/table_04_01.html

the amount of time it takes for a present value to double for a given rate of growth, i.e. F/P = 2. The solution can be found by searching in the table of F/P factors for a given r values. For example, for a rate of 8% per year, the doubling time is 9 years. In general, it can be shown that doubling time is approximately calculated by dividing 72 by the percent interest rate (annual percentage growth rate). This approach is called the **Rule of 72** and is expressed algebraically as:

$$T_2 = \frac{72}{R} \qquad (15\text{-}3)$$

where T_2 is the doubling time in years, and R is the annual percentage growth (R = 100 r).[12]

> **Example 15-8:** How long does it take for a deposit of $5,000 in a saving account to yield a total of $20,000 in principal and interest? The bank pays 6% interest.
> **Solution:** At the rate of 6% (r = 0.06), it takes 72/6 = 12 years to double the deposit. So the investment doubles to $10,000 after 12 years and doubles again to $20,000 after 24 years.

Cumulative Value of a Series of Future Payments

Also of interest is the cumulative value of a series of installments paid in equal intervals. For example, the total principal and interest paid on a home mortgage is the sum of all monthly payments over the course of a 30-year loan. Similarly, the cumulative value of energy consumption is the sum of energy used annually from the original reserves.

The analysis presented in the previous section can be used to forecast energy consumption from the present rate of consumption assuming a certain growth rate. Of greater consequence is the amount of energy remaining at a future point in time. We calculated the remaining US oil for the years prior to 2006. That procedure can be used until the oil reserves are depleted. A better way is to find a formal relationship for the cumulative consumption is by adding all future consumption by summing up the series:

$$C = P[1+(1+r)+(1+r)^2)+(1+r)^3)+\ldots.+(1+r)^{n-1}]$$

The mathematical series converges to give:

$$C/P = \frac{(1+r)^n - 1}{r} \qquad (15\text{-}4)$$

---

[12] In this example, we have assumed that the interest rate is compounded yearly. If the interest rates were compounded monthly, or even better, daily, then our money would have doubled in a shorter time. Taken to the extreme, exponential growth would occur on a continuous basis as it does naturally in population, energy consumption, etc., further shortening the doubling time, i.e. $N = N_0 e^{rt}$ (See equation 1, Appendix B). The doubling time is found by substituting $N/N_0 = 2$ to give:

$$T_2 = \frac{\ln 2}{r} = \frac{0.693}{r} = \frac{70}{R}$$

For the example above, the doubling time is calculated as 0.693/0.08 = 8.66 years instead of 9 years calculated with equation 15-3.

Equation 15-4 can also be interpreted as the future value of the stream of annual payments, accumulating interest compounded over time. The C/P factors are tabulated in Table 15-4.

Table15- 4. Cumulative Value of a Series of Future Payments										
$C/P=[(1+r)^n-1]/r$										
r/n	***1***	***2***	***3***	***4***	***5***	***10***	***15***	***20***	***25***	***30***
0.01	1.00	2.01	3.03	4.06	5.10	10	16	22	28	35
0.02	1.00	2.02	3.06	4.12	5.20	11	17	24	32	41
0.03	1.00	2.03	3.09	4.18	5.31	11	19	27	36	48
0.04	1.00	2.04	3.12	4.25	5.42	12	20	30	42	56
0.05	1.00	2.05	3.15	4.31	5.53	13	22	33	48	66
0.06	1.00	2.06	3.18	4.37	5.64	13	23	37	55	79
0.07	1.00	2.07	3.21	4.44	5.75	14	25	41	63	94
0.08	1.00	2.08	3.25	4.51	5.87	14	27	46	73	113
0.09	1.00	2.09	3.28	4.57	5.98	15	29	51	85	136
0.10	1.00	2.10	3.31	4.64	6.11	16	32	57	98	164
0.15	1.00	2.15	3.47	4.99	6.74	20	48	102	213	435
0.20	1.00	2.20	3.64	5.37	7.44	26	72	187	472	1,182
0.25	1.00	2.25	3.81	5.77	8.21	33	110	343	1,055	3,227
0.30	1.00	2.30	3.99	6.19	9.04	43	167	630	2,349	8,730

Example 15-9: Given the data in example 15-7, calculate how long US petroleum reserves will last. Repeat the calculation for the total world petroleum reserves. The remaining world oil reserves are estimated at approximately 2,000 billion barrels; the current rate of consumption is 30 billion barrels annually and is expected to rise at a rate of 4% per year.

Solution: In this example, C is cumulative consumption equal to total US oil reserves in 2003 ($C = 100$ bbo) and P is the total oil consumed in 2003 ($P = 7.3$ bbo), so $C/P = 100/7.3 = 113.7$. Assuming that the annual rate of consumption continues to increase at 2% a year, we can refer to Table 15-4 to find $10 < n < 15$ years. Interpolating, a better estimate will give a value of $n \approx 12.3$ years. Similarly, for the world, $C/P = 2{,}000/30 = 66.6$. For $r = 0.04$, we can estimate we will have petroleum for another 40 years.

Annuities (Net Present Value)

Investors are interested in the return that their capital investment brings to them. It is also understood that the greater the risk they take, the higher their expectations for returns will be. The minimum that any investor expects as their **rate of return** is the interest that they could collect if they had deposited their investment into a bank or other financial institution. Depending on the perceived risk that different projects have and the tolerance of the potential investor to the loss of capital, different investors may expect different rates of return.

Investors do not necessarily put all the capital up front. Rather, it is common that they make a series of payments as the project proceeds. To make a wise decision on the profitability of the investment, all future revenues and costs must be discounted to the present. The discount rate is commonly chosen to be the expected rate of return.

The present value of a stream of annual payments or annuities, A, is calculated by summing Equation 15-2 over all future payments:

$$P = A\left[1 + \frac{1}{(1+r)} + \frac{1}{(1+r)^2} + \ldots\ldots + \frac{1}{(1+r)^n}\right]$$

The series converges to give:

$$P/A = \frac{(1+r)^n - 1}{r(1+r)^n} \qquad (15\text{-}5)$$

The P/A factors are calculated and summarized in Table 15-5.

Example 15-10: A $10,000,000 lottery win can be awarded either over a 20 year period, by installments of $500,000 each, or as $5,000,000 in cash right away. Assuming the winner plans to deposit his money in a 20-year CD that pays 8% in interest, which option makes more sense?

Solution: The present value of the annual installments is $\{(P/A)_{n=20,\, r=0.08}\} = 10.59$. The present worth of all future installments is $500,000x10.59 = $5,295,000 and is somewhat better than $5,000,000 in cash paid out today.

Example 15-11: The population of a large metropolitan area is expected to increase by 200,000 people within the next 5 years. To meet the future demand, a 200 MW wind farm is proposed for construction. The cost of construction is expected to be $5M in the first two years and an additional $2M for the following three years. The plant is expected to generate a net income (revenue minus maintenance costs) of $2M for 20 years after it is completed. Does this investment make economic sense? Assume an interest rate of 5%.

Solution: In the first five years that the plant is under construction, $16M are spent on construction. Starting year 6, the plant will generate revenues of $2M in annual income for the next 20 years after maintenance costs are deducted. A summary of the cost and income schedule is given below:

Year	1	2	3	4	5	6	7	...	25
Cost (M)	5	5	2	2	2	0	0	0	0
Income (M)	0	0	0	0	0	2	2	2	2

Please note that P/A factors assume that all payments are equal and start from year 1. As mentioned, there is no income for the first 5 years. To correct for this we subtract the net present value of income that would be generated for 25 years and subtract from it the lack of income opportunity for the first five years. Similarly, we can break down costs as annual cost of $2M for five years, in addition to annual costs of $3M a year for the first two years. This is best understood if the timeline table is rearranged as:

Year	1	2	3	4	5	6	7	...	25
Cost (M)	2	2	2	2	2				
Cost (M)	3	3							
Income (M)	2	2	2	2	2	2	2	2	2
Income (M)	-2	-2	-2	-2	-2				

The approach is to calculate P/A factor for a series of annual payments starting in year 1. Consulting Table 15-5, the net present values of income and expenses are:

NPV(income) = [2(P/A, 0.05, 25) - 2(P/A,0.05,5)] =2(14.09)–2(4.33)=9.52 M

NPV (cost) = [3(P/A,0.05,2) + 2 (P/A,0.05,5)] = 3(1.86) + 2(4.33) = 14.24 M

NPV (income) / NPV (costs) = 19.52/14.24 = 1.37 ; thus the investment makes sense!

Capital Recovery Factor

The capital recovery factor (annualized cost) represents the series of annual payments to pay off a loan. This is the reverse of the previous problem -- we know the present worth of all future payments (i.e. the loan principal). The result is:

$$A / P = \frac{r(1+r)^n}{(1+r)^n - 1} \qquad (15\text{-}6)$$

A/P factors are tabulated in Table 15-6.

Example 15-12: What is the annual payment required to pay a $250,000 loan at 7% over 15 years?
Solution: Annual cost = loan amount x $[(A/P)_{r=0.07, n=15}]$ = ($250,000) x 0.11 = $27,500

Internal Rate of Return

A convenient way to evaluate the economic benefit of a project is to calculate the effective rate of return of all future transactions. Since payments and revenues take place at different times, a proper method of evaluation is to discount them to present value. The internal rate of return is computed simply by equating the NPVs for all future costs and incomes.

NPV (cost) = NPV (income)

Table 15-5. Net Present Value of a Series of Future Installments (Annuities)										
$P/A=[(1+r)^n-1]/[r(1+r)^n]$										
r/n	***1***	***2***	***3***	***4***	***5***	***10***	***15***	***20***	***25***	***30***
0.01	0.99	1.97	2.94	3.90	4.85	9.47	13.87	18.05	22.02	25.81
0.02	0.98	1.94	2.88	3.81	4.71	8.98	12.85	16.35	19.52	22.40
0.03	0.97	1.91	2.83	3.72	4.58	8.53	11.94	14.88	17.41	19.60
0.04	0.96	1.89	2.78	3.63	4.45	8.11	11.12	13.59	15.62	17.29
0.05	0.95	1.86	2.72	3.55	4.33	7.72	10.38	12.46	14.09	15.37
0.06	0.94	1.83	2.67	3.47	4.21	7.36	9.71	11.47	12.78	13.76
0.07	0.93	1.81	2.62	3.39	4.10	7.02	9.11	10.59	11.65	12.41
0.08	0.93	1.78	2.58	3.31	3.99	6.71	8.56	9.82	10.67	11.26
0.09	0.92	1.76	2.53	3.24	3.89	6.42	8.06	9.13	9.82	10.27
0.10	0.91	1.74	2.49	3.17	3.79	6.14	7.61	8.51	9.08	9.43
0.15	0.87	1.63	2.28	2.85	3.35	5.02	5.85	6.26	6.46	6.57
0.20	0.83	1.53	2.11	2.59	2.99	4.19	4.68	4.87	4.95	4.98
0.25	0.80	1.44	1.95	2.36	2.69	3.57	3.86	3.95	3.98	4.00

Table 15-6. Capital recovery factor										
$A/P=[r(1+r)^n]/[(1+r)^n-1]$										
r/n	***1***	***2***	***3***	***4***	***5***	***10***	***15***	***20***	***25***	***30***
0.01	1.01	0.508	0.340	0.256	0.206	0.106	0.072	0.055	0.045	0.039
0.02	1.02	0.515	0.347	0.262	0.212	0.111	0.078	0.061	0.051	0.045
0.03	1.03	0.524	0.353	0.269	0.218	0.117	0.084	0.067	0.990	0.051
0.04	1.04	0.529	0.360	0.275	0.225	0.123	0.090	0.074	0.064	0.058
0.05	1.05	0.538	0.368	0.282	0.231	0.130	0.096	0.080	0.071	0.065
0.06	1.06	0.546	0.375	0.288	0.238	0.136	0.103	0.087	0.078	0.073
0.07	1.08	0.552	0.382	0.295	0.244	0.142	0.110	0.094	0.086	0.081
0.08	1.08	0.562	0.388	0.302	0.251	0.149	0.117	0.102	0.094	0.089
0.09	1.09	0.568	0.395	0.309	0.257	0.156	0.124	0.110	0.102	0.097
0.10	1.10	0.575	0.402	0.315	0.264	0.163	0.131	0.118	0.110	0.106
0.15	1.15	0.613	0.439	0.351	0.299	0.199	0.171	0.160	0.155	0.152
0.20	1.20	0.654	0.474	0.386	0.334	0.239	0.214	0.205	0.202	0.201
0.25	1.25	0.694	0.513	0.424	0.372	0.280	0.259	0.253	0.251	0.250
0.30	1.30	0.735	0.549	0.461	0.410	0.324	0.306	0.301	0.300	0.300

Example 15-13: An investor is considering investing in the wind farm project described above. The investor also has an opportunity to purchase a government bond which is expected to earn a 10% return. Which option would be advantageous?

Solution: To have a fair basis for comparison, the investor should compare the internal rate of return for the project and the rate of return from the bond. The internal rate of return can be calculated by equating the NPVs for income and cost.

NPV (income) = NPV (cost)

$$2(P/A, r, 25) - 2(P/A, r, 5) = 3(P/A, r, 2) + 2(P/A, r, 5),$$

or

$$\Delta\,(NPV) = 2(P/A, r, 25)-4(P/A, r, 5) -3(P/A, r, 2) = 0$$

The problem must be solved by trial and error. Assume a rate of return and evaluate the equation above until it converges.

r	Δ (NPV)
0.07	2(11.65) -4(4.10) -3(1.81) = 1.47
0.08	2(10.67) -4(3.99) -3(1.78) = 0.04
0.09	2(9.82) -4(3.89) -3(1.76) = -1.20

The effective rate of return for investing in the wind farm project is about 8%, which is less than the 10% that the bond would earn. Although the investment in the bond is a better choice financially, an environmentally contentious investor may still consider the investment in the wind farm.

Cost of Living Adjustment (COLA)

How can we compare the cost of acquiring a good or service at two different times? Our salary today is probably higher than it was 10 years ago, but does that mean that we are richer today? The answer depends on whether we can buy less or more with the money we earn today than with our income of 10 years ago. This means our salaries today must be discounted to account for inflation.

Table 15-7. The 2008 CPI Conversion Factor

Year	*CF*	*Year*	*CF*	*Year*	*CF*	*Year*	*CF*
1810	0.066	**1935**	0.067	**1979**	0.352	**1995**	0.740
1820	0.063	**1940**	0.068	**1980**	0.400	**1996**	0.762
1830	0.05	**1945**	0.087	**1981**	0.441	**1997**	0.779
1840	0.047	**1950**	0.117	**1982**	0.469	**1998**	0.792
1850	0.042	**1955**	0.131	**1983**	0.484	**1999**	0.808
1860	0.045	**1960**	0.144	**1984**	0.504	**2000**	0.836
1870	0.07	**1965**	0.153	**1985**	0.522	**2001**	0.860
1880	0.055	**1970**	0.189	**1986**	0.532	**2002**	0.873
1890	0.049	**1971**	0.196	**1987**	0.551	**2003**	0.893
1900	0.046	**1972**	0.203	**1988**	0.575	**2004**	0.917
1905	0.048	**1973**	0.216	**1989**	0.602	**2005**	0.939
1910	0.051	**1974**	0.239	**1990**	0.635	**2006**	0.957
1915	0.052	**1975**	0.261	**1991**	0.661	**2007**	0.978
1920	0.103	**1976**	0.276	**1992**	0.681	**2008**	1.000
1925	0.106	**1977**	0.301	**1993**	0.701		
1930	0.096	**1978**	0.326	**1994**	0.720		

To convert dollars of any year to 2008 dollars, divide the dollar value of that year by its conversion factor. For example, $1,000 in 1950 dollars is equivalent to 1,000/0.117=$8,547 in 2008 dollars.

The **Consumer Price Index (CPI)** was designed in 1983 to measure the average cost of living of average consumers. CPI calculated the cost for a basket of products and services (food, housing, transportation, clothing, health, etc.). Although CPI was a good indicator of inflationary pressure for a few years, its value became less and less indicative of the cost of living. The main reason is because CPI was based on a fixed set of commodities and ignored behavioral changes as technologies matured and peoples' taste for goods and services changed. Furthermore, CPI ignored the impact of consumption on declines in the price of goods. In 1983, an average person traveled fewer miles and paid less for gasoline. The cost of a computer was much higher, however. As prices of some commodities rose sharply, consumers found other alternatives. As a result, the consumer behavior today is widely different from that of twenty years ago, and the "basket" of goods and services used in 1983 does not give a true representation of the change in purchasing power of the average consumer.

The price of any item in dollar-values of year X can be calculated in terms of dollar-value of the price of any item in dollar-values of year Y from the equation below:

$$\textit{Dollar value in year } X = \frac{\textit{CF in year } X}{\textit{CF in year } Y} \cdot (\textit{Dollar value in year } Y)$$

Example 15-14: Following the Iran-Iraq war in 1981, the price of Middle Eastern oil reached an all-time high of $40 a barrel. What was the price in 2005 dollars? Adjusting for inflation, are we paying more or less for gasoline today compared to 1981 prices?
Solution: Using the CPI indicator (Table 15-7) to adjust for inflation gives us a conversion factor of 0.47; every 2005 dollar was worth only 47 cents in 1981. The price of 1981 gasoline in 2005 dollars can be calculated by dividing the 1981 price by 0.47 (40/0.47 = $85 in 2005 dollars). Compared to the $55 price of gasoline today, we are buying oil at a relatively cheaper price than in 1981.

Example 15-15: The average of price of gasoline in the US was $2.60 a gallon at the gas pump in 2004. What was the price in 1950 dollars?
Solution: Consulting Table 15-7 for CF values, the price of gasoline in 1950 dollars can be calculated as:
0.125/0.977 x $2.60 = $0.33.

International Trade

International trade has always been a key component of relations among nations and is critical for the economical development of all nations. As we discussed earlier, trades take place because both sides are better off as a result of the transaction. International trade is particularly important because different countries have different resources and therefore, different opportunity costs. In other words, some resources have a comparative advantage over other resources; different countries may be better off putting their resources towards production of one product than another and using trade as a means to acquire their needs—in the process both countries benefit. To make any trade possible, there are middlemen or entrepreneurs that facilitate the transactions, generating profits and at the same time allowing both countries to benefit economically.

Example 15-16: Saudi Arabia is a large supplier of oil, but its desert climate is not particularly suitable for agricultural products such as wheat. It costs the Saudi government $5 for every barrel of oil it produces and $50 for a ton of wheat. The United States, on the other hand, does not have large oil resources, but has a high opportunity cost of producing oil ($25 a barrel). US temperate climate and fertile soil is ideal for growing wheat and can produce it at relatively lower cost ($15 per ton). Assuming each country's production satisfies its own national needs, is there anything gained from trades between the two?

Solution: The volumes of production for oil and wheat and the total cost of production are given in column A for each country. For Saudi Arabia, the cost of production of one ton of wheat is higher than for one barrel of oil. It therefore makes sense for Saudi Arabia to dedicate its resources to the production of additional oil; it can invest all its resources to increase production of oil to 3,000 barrels and buy wheat from the US (Column B), keeping its total production cost the same. By the same token, it makes sense for the US to divert its resources to production of wheat, increasing its volume to 1,250 tons, and to import its oil from Saudi Arabia. Now Saudis can export 150 barrels of their oil to the United States and keep 2,850 barrels for domestic use or sell them to other countries. Without trades it would have cost Saudis $24,250 to produce this amount of these two products (Column C), resulting in a net savings of $9,250 directly resulting from the trade. The United States, on the other hand, buys 150 barrels of oil while selling 200 tons of its wheat to Saudi Arabia, meeting its domestic oil needs and keeping 50 tons of wheat in excess of what it would have without these trades, saving $750. It is clear that both countries have benefited from this trade.

	Saudi Arabia			US		
	A	B	C	A	B	C
Production, oil (barrels)	1,000	3,000	2,850	150	0	150
Production, wheat (tons)	200	0	200	1,000	1,250	1,050
Total production cost ($)	15,000	15,000	24,250	18,750	18,750	19,500
Net saving ($)			9,250			750

The example above shows that both countries benefited from free trade with a net benefit that is spread across the society. Free trade, however, is not free of cost. For example, US oil workers may be laid off or must switch to farm work. The reverse is true for Saudi Arabian nationals. As a result, some sectors of the society will be unsatisfied. To counteract the adverse effects of free trade, these groups, through their trade unions, hire lobbyists encouraging lawmakers to enact trade barriers, organize strikes, or boycott a particular good imported from a foreign country. Usually, the cost incurred to these individuals is small compared to the benefit gained by the populace at large, however.

> **Question:** The European Union places a high tariff on food imported from non-EU countries. What would the effect on EU population be?
> **Answer:** Although some jobs are saved as a result of lower imports, the general public will suffer from higher prices of food in EU member countries. With no restrictions on free trades, the EU food prices drop and everybody benefits. The farmers, through political campaigning and advertisement, will try to inform the public of the costs, not the benefits, of free trade. One approach to neutralize the opposition to free trade is by passing on some of the benefits to farmers and those who are directly affected and still making the whole population better off overall through free trade.

Sources and Types of Barriers and their Impacts

There are several reasons that countries impose trade barriers. The most common ones are: a) learning by doing, b) economics of scale, and c) national security.[13]

a. **Learning by doing** – Trade barriers are imposed when the government believes that they can make a particular good more internationally competitive only if an infant industry is given enough time to gain experience to improve the product and reduce manufacturing costs. For example, a country may restrict the import of plastics to give local manufacturers time to improve their inferior product and eventually compete with a superior foreign product.
b. **Economics of scale** – Trade barriers are intended to allow a higher output of a certain good by local manufacturers. This reduces the cost per unit, making the good more attractive to foreign competition. For example, a country may be able to produce cotton at below international market price only if the volume of production exceeds a certain level.
c. **National security** – Trades are restricted by a country on the grounds that their national security may be threatened. This restriction may be imposed on both imports and exports. For example, a country may put a restriction on volume of imported oil because they don't want to be dependent in a time of war. At the same time, the export of certain military equipment or specific chemicals may be banned to assure they do not fall into unfriendly hands. Trades may also be restricted by a country as a means of putting pressure on or punishing another country to achieve a certain political goal. For example, the

[13] Colander, D. C., Macroeconomics, 3rd Ed., Irwin McGraw-Hill, 1998, pp. 489.

US restricts trades with Cuba and North Korea to pressure these governments to change their policies on immigration or developing certain nuclear technology. Trade barriers can also be imposed to protect against transmission of insects, or to protect human health.

Depending on the impact they are intended to make, many kinds of trade barriers may be imposed. These include tariffs, quotas, embargos, regulatory trade restrictions, and nationalistic appeals. **Tariffs** or custom duties are taxes imposed by governments on imported goods. They are mainly to increase revenues and encourage consumption of domestically produced goods. **Quotas** refer to quantitative limitations placed on imports. An **embargo** is a total restriction on the import or export of a good. **Regulatory trade restrictions** are designed by governments to set certain standards and requirements to limit imports. **Nationalistic appeal** refers to a direct pleading to citizens to buy domestic products rather than products manufactured and imported from other countries.

No matter which type is used, trade barriers can only work if the affected country does not retaliate in kind or by other means. This is highly unlikely and, except in rare circumstances, trade barriers will not work in the long run. As trades become more and more global in nature, countries find other means to counteract by imposing their own barriers or finding alternative markets in which to sell and buy their necessities.

Mixed (Dual) Economies

The free market economy is practiced in developed countries and international trades and is based on established laws of supply and demand. Many problems inherent in developing countries, however, hinder the full implementation of this system. These include political instability, corruption, overpopulation, and lack of financial infrastructure. Political instability results from social inequality, excessive poverty, lack of political freedom, and arbitrary decisions by authoritarian regimes. As a result, the transition of government is unlikely to be orderly and is often associated with coups, bloodshed, and political apathy among the people. Political instability discourages foreign and internal investment and encourages the flight of qualified engineers and technocrats. Because there are no checks and balances on government leaders, citizens have a deep mistrust of their government, and corruption is usually rampant and a way of life in many developing nations.

Generally, low-income developing countries have a dual economy, one being the traditional, often non-market economy, and the other a western-style free market economy. The traditional economy sector financing is limited, and trades are largely in local currency and in cash, even at times by bartering. These systems do not have an institutional apparatus to collect taxes. Even if such a system is in place, it is often rigged with fraud and bribery. The free market economy sector is used in large financial transactions and when the government or private enterprise engages in international trading. Financing is usually possible and often the norm.

Globalization

Globalization, in the broadest terms, refers to the integration of world economies through increased trade and knowledge and the free movement of material, labor, and capital across international borders in search of new markets and investment opportunities. Although globalization is not new and existed at some level centuries ago, the technological advances of the past few decades have made it easier and quicker to complete transactions and the trend toward globalization has accelerated. Whether globalization benefits or harms a particular country has been the topic of many debates. Some consider globalization and free market economy to be the keys to future world economic development. They foster competition and improve efficiency by making the flow of technological innovations, skilled labor, and free flow of capital easier. They also help developing and underdeveloped countries make progress towards democracy and a cleaner environment as their living standards rise.

Critics of globalization argue that markets do not necessarily ensure that the benefits of increased efficiency are shared by all. Furthermore, these policies, written primarily by large corporations and western governments, require such agencies as the **International Monetary Fund (IMF)**, the **World Bank**, and the **World Trade Organization (WTO)** to oversee trade and investment agreements negotiated between member states. These organizations often impose regulations that put industrialized nations at an unfair advantage and at the same time increase their access to natural resources of the developing countries. According to these rules, all resources, no matter where they lie, must be available for sale to the highest bidder, but the technologies used to discover, extract, and process these resources are regarded as proprietary, or "intellectual property" of the developers.

The role of energy is particularly important in shaping this debate. Certain types of energy such as wind and solar are inherently decentralized. Others, such as fossil and especially nuclear, are controlled by large corporations and governments; therefore the control over these resources and supply routes brings about economic as well as political power. As we have seen by now, the world is divided into two groups: those who have vast fossil resources and those who have not. The countries that are endowed with these resources are mainly developing countries, mostly in the Middle East. Western and other industrialized countries are by far the greatest consumers of energy. To secure the flow of petroleum and other natural resources to industrialized countries, through a twin strategy of military intervention and economic globalization, governments are required to open their borders to international trade and investment without the imposition of taxes, tariffs, and other regulatory barriers.

So far, the result of globalization has been mixed as best. Rich countries have in large benefited from international trades, although many of their citizens (especially those hired in low-tech industries such as textile, agriculture, and some electronic products) have suffered from losing jobs to lower-paid foreign workers. During the past century, the average per capita GDP of many industrial countries increased by as much as six-fold, although the income gap between rich and poor countries has been widening.[14] In the last three decades, some developing countries, especially in Asia, have been able to benefit from globalization by increasing their share of world trade. Other countries, notably in Africa and in South America, have not been able to integrate with the world economy and their per capita income has actually declined. What is clear is that globalization is here to stay and if they are to benefit, individual countries must embrace globalization on their own terms, taking into account their own history, culture, and traditions.

The two most populous countries in the world, China and India, are growing rapidly to become world's new economic powerhouses. This means that not only their demand for energy and fossil fuel will accelerate, but also their ecological footprints, the strain on natural resources, release of carbon into the atmosphere, and higher level of environmental pollution. Unless drastic effort in finding new alternative resources and implementing innovative conservational measures are made, the competition for control of these resources and potential for additional regional conflicts will increase.[15]

Summary

Higher energy prices have direct and dramatic effects on the global economy. Because fossil fuels, especially petroleum, comprise the majority of our energy consumption, rises in oil prices have correlated closely with inflationary pressures. Though the US has successfully reduced its dependence on imported oil in the last decade (measured as energy consumption per dollar of the GDP), other countries have not been able to follow suit. Moreover, over half of the US total oil consumption is now imported, and this number continues to increase steadily into the future. As the price of oil climbs, other commodity prices will also increase, likely affecting prices of non-energy exported and imported goods, which in turn affects the global economy and may result in political instability in developing countries.

In the United States, government and business focus on the short-term profit motives and quarterly earnings and not much on long-term interests. Banks and other sources of funding will seek to receive a return on their investments in as early as a 3-5 year period, rarely exceeding 10

[14] "World Economic Outlook," International Monetary Fund, Washington D.C., May 2000, *(http://www.imf.org/external/pubs/ft/weo/2000/01/index.htm).*

[15] Stiglitz, J.E., Globalization and its Discontent, Norton Publishers, New York, 2002.

years. As a result, in many instances lack of clear immediate revenues will stifle innovation or give it away easily to foreign competitors. For example, during the 1970s when energy sufficiency and environment became a national security issue, the US spent a tremendous amount of resources to support research and development of renewable energy technologies such as solar photovoltaic, solar thermal power, and wind. In the 1980s, soon after the energy crisis subsided and oil prices dropped, so did the support for alternative sources of energy and environmental protection. The Reagan administration reversed the energy independence policy of the former Carter administration by slashing the federal research funds from $150 M to zero and much of their tax credits and subsidies as well. At the time, the United States was the leading manufacturer of solar cells and wind turbines. Japan and much of Europe capitalized on these inventions by buying American companies and at the same time developed long-term energy policies that reduced the imports of foreign oil and instead relied on the development of cleaner technologies and alternative sources of energy.

Today, Japan is the leading producer of solar cells and fuel cells, and over one-half of all hybrid and electric vehicles sold in the United States. Germany is the largest manufacturer of wind turbines. Spain and several other European countries are developing the next generation of solar thermal power plants. Unless the United States reverses its "cheap oil policy," not only will its reliance on Persian Gulf oil increase, but much of its engineering ingenuity and know-how will be relegated to foreign competitors.

Additional Information

Books

1. Colander, D. C., Economics, 3rd E., Irwin-McGraw-Hill, 1998.
2. Bosselman, F., Energy, Economics and the Environment, Second Edition, Foundation Press, 2005.

Periodicals

1. *Energy Economics*, Science Direct Elsevier Publishing Company. Publishes research papers concerned with the economic and econometric modeling and analysis of energy systems and issues.

Non-Government Organizations and Websites

1. United States Association for Energy Economics *(http://www.usaee.org)*.
2. International Monetary Fund *(http://www.imf.org)*.
3. The World Bank *(http://www.worldbank.com)*.

Exercises

I. Problems:

1. The worldwide consumption of petroleum was 150.8 quads in 2002. Assuming the consumption continues to increase at 2% per year, what will the world consumption of petroleum be in 2025?

2. A bacterial colony in a jar grows at the rate of 10% every hour. Find:
   a. Doubling time
   b. If the jar were initially 1/8 full, how long would it take before the jar is completely full?
   c. Assume that a person is accidentally infected with only one bacterium. What is the bacterial population after one week?

3. If an employer were to offer you a) $10,000 a year in initial salary and a $2,000 raise every year thereafter, or b) $1 a year in initial salary, but would double it every year for the next twenty years, which one would you choose? Why? Assume that the prevailing interest rate remains at around 5% for the 20 year period.

4. A loan agency has offered a customer to pay off his $ 100,000 existing loans using one of the following three options:
   a. By paying balance immediately in cash
   b. By paying 10 equal payments of $11,000 each annually.
   c. By paying $50,000 now and a balloon payment of $60,000 in 10 years.

   Which option makes more sense to the customer? Assume annual interest rate of 5%.

5. The national average price of regular gasoline in the United States is given by the *DoE Energy Information Energy* (all data are for August) as following:

1992	$1.03	a gallon
1994	$1.13	"
1996	$1.17	"
1998	$1.00	"
2000	$1.45	"
2002	$1.36	"
2004	$1.85	"
2006	$3.00	"

   a. Adjust the prices for inflation, which year offered the cheapest gasoline? most expensive gasoline?
   b. What is the price of gasoline in today's dollar? in 2000's dollar? In 1950's dollar? In 1900's dollar?

II. Multiple Choice Questions

1. In an efficient market
   a. There are numerous buyers and sellers
   b. There are no barriers to entry
   c. There are numerous products that are essentially indistinguishable from the good of interest
   d. Firms' only goal is to maximize their profits
   e. All of the above

2. The supply curve is upward sloping because
   a. As prices increase, so do costs
   b. As prices increase, demand for a product declines
   c. As prices increase, demand for a product also rises
   d. As prices increase, suppliers are willing to produce more
   e. As prices increase, suppliers are willing to produce less

3. As the cost of production of cell phones falls,
   a. The demand curve for cell phones shifts to the left
   b. The demand curve for cell phones shifts to the right
   c. The supply curve for cell phones shifts to the left
   d. The supply curve for cell phones shifts to the right
   e. Neither the supply curve nor the demand curve shift

4. In a perfectly efficient market, a firm's long-term profit will be maximized when
   a. Price equals fixed cost
   b. Price equals total cost
   c. Fixed cost equals total revenue
   d. Total cost equals total revenue
   e. Marginal cost equals marginal revenue

5. Which of the following is not consistent with the existence of a perfect competitive market?
   a. Absence of low transaction costs
   b. Limited competitors
   c. Significant barriers for entry into the market
   d. Absence of a sufficient number of indistinguishable substitutes
   e. All of the above

6. When a newer technology replaces an older technology
   a. Both demand and supply curves shift to the right
   b. Both demand and supply curves shift to the left
   c. Demand curve does not change, but the supply curve shifts to the right
   d. Supply curve does not change, but the demand curve shifts to the right
   e. Neither the demand curve, nor the supply curve are affected

7. When median income increases,
   a. Both demand and supply curves shift to the right
   b. Both demand and supply curves shift to the left
   c. Demand curve does not change, but the supply curve shifts upward
   d. Supply curve does not change, but the demand curve shifts upward
   e. Neither the demand curve, nor the supply curve are affected

8. Which of the following is considered to be a positive externality?
   a. A power plant emits pollution into the air thus reducing its cleanup cost.
   b. A man blasts his music and gives his neighbor a headache.
   c. A little girl buys a pretty doll from a toy store.
   d. A power plant switches from coal to natural gas.
   e. None of the above.

9. As a result of higher demand for a product
   a. The demand curve does not change, but price rises
   b. The demand curve does not change, but price falls
   c. The demand curve shifts to the right, but price rises
   d. The demand curve shifts to the right, but price falls
   e. Both price and quantity supplied increase

10. As a result of a rise in the cost of living, we expect
    a. Both price and quantity of goods to rise
    b. Price increases as quantity of goods falls
    c. Price does not change as quantity of goods falls
    d. Quantity of goods does not change as price increases
    e. Both price and quantity of goods increase

11. Exponential growth may be characterized by
    a. The absolute amount of growth every year remaining constant
    b. The percentage growth every year being constant
    c. The percentage growth every year being exponential
    d. The growth rate doubling every year
    e. None of the above

12. Linear growth may be characterized by
    a. The absolute amount of growth every year being constant
    b. The percentage growth every year being constant
    c. The percentage growth every year being exponential
    d. The growth rate doubling every year
    e. None of the above

13. According to the latest (2000) census results, Nevada showed the fastest population growth (5.4%) in the United States. In what year would the population of Nevada double?
    a. 2005
    b. 2013
    c. 2018
    d. There is not enough information to predict the year.

e. None of the above.

14. In the US, the doubling time of our exponentially growing electricity consumption rate was 10 years from 1900 to 1970. How did the electricity consumption from 1960 to 1970 compare to all US electricity consumption prior to 1960?
    a. It was twice as much.
    b. It was half as much.
    c. It was an equal amount.
    d. It was 1/6 as much.
    e. It cannot be determined from given information.

15. Since the 2000 census, the US population has grown at a rate of 1.2% per year. In what year will the US population double?
    a. 2058
    b. 2083
    c. 2120
    d. There is not enough information to predict the year.
    e. None of the above.

16. In 75 years, a small community of 10,000 people that enjoys a constant 2.8% annual growth rate will attain a population of about
    a. 28,000
    b. 30,000
    c. 80,000
    d. 2,100,000
    e. None of the above

17. An employer guarantees a salary increase of 7.2% a year for all employees. If a particular employee is hired at an initial annual salary of $30,000, how long would it take for their salary to reach $60,000?
    a. About 7 years
    b. A bit sooner than 10 years
    c. 10 years
    d. A bit longer than 10 years
    e. About 14 years

18. Statistics show that for the last few decades, the US consumption of petroleum has doubled every decade. The corresponding annual percent growth rate is
    a. 2%
    b. 7%
    c. 10%
    d. 13%
    e. None of the above

19. $200 invested at 5% compound interest will grow to $800 in about
    a. 4 years
    b. 28 years
    c. 40 years
    d. 42 years
    e. 56 years

20. What is the present worth of $1,000 payable in 5 years assuming an average discount rate of 4%?
    a. $ 820
    b. $ 850
    c. $ 1,000
    d. $ 1,170
    e. $ 1,220

21. How much interest does $1,000 placed in a 5-year CD paying 4% accumulate?
    a. $ 2,220
    b. $ 1,220
    c. $ 820
    d. $ 220
    e. None of the above

22. If a sin tax is imposed on tobacco and alcohol
    a. The demand curve shifts to the left
    b. The demand curve shifts to the right
    c. The supply curve shifts to the left
    d. The supply curve shifts to the right
    e. Neither the supply curve nor the demand curve shift

23. What happens to the price of air travel when the price of travel by rail increases?
    a. The demand curve shifts to the left.
    b. The demand curve shifts to the right.
    c. The supply curve shifts to the left.
    d. The supply curve shifts to the right.
    e. Neither the supply curve nor the demand curve shifts.

24. The social cost of a product or activity is the sum of
    a. Fixed and variable costs
    b. Private and variable costs
    c. Private and fixed costs
    d. Private and external costs
    e. Private and public costs

25. Firms often collude to form a cartel because
    a. They can function like a monopoly
    b. It allows them to spend more money on innovation
    c. It furthers competition and increases their market shares
    d. Cartels are more efficient
    e. All of the above

26. Net National Welfare (NNW) is defined as
    a. The total value of all goods and services produced in a country in a given calendar year
    b. The sum of all goods and services produced by labor and property located in a country irrespective of who supplies them
    c. The sum of all goods and services supplied by a country's residents irrespective of where they are physically produced
    d. The total annual output of both market and non-market goods and services, and minus the cost of externalities and depreciation of natural and human-made capitals used up in production
    e. The total money spent on welfare agencies

27. The United States has a GDP in the order of
    a. 10 billion dollars
    b. 100 billion dollars
    c. 1 trillion dollars
    d. 10 trillion dollars
    e. 100 trillion dollars

28. The Gross National product (GNP) is defined as
    a. The total value of all goods and services produced in a country in a given calendar year
    b. GDP plus what domestic companies earn from activities abroad plus what foreign companies make from domestic activities in that country
    c. The sum of all market and non-market goods and services supplied by a country's residents
    d. The sum of all goods and services produced by labor and property located in a country irrespective of who supplies them
    e. Another name for Gross Domestic Product

29. One example of an objection raised against globalization and free-trade is
    a. Richer countries can use their economic muscles to force poorer countries to accept their wastes
    b. Resource depletion is accelerated because richer countries can have unrestrained access to natural resources of poorer countries
    c. Richer countries can impose regulations that put them at an unfair advantage over underdeveloped and developing nations
    d. All of the above
    e. None of the above

30. The quantitative limitation placed by a government on an import is called a(n)
    a. Quota
    b. Tariff
    c. Embargo
    d. Import tax
    e. Excise tax

III. True or False

1. There are two ways that demand can change: in price or in overall demand.

2. As price increases due to a shift in the demand curve, suppliers respond by reducing the supply.

3. As the cost of production of a product goes down, the price declines, causing the demand curve to shift to the right.

4. As the price of gasoline rises, the demand for automobiles shifts to the right

5. Along a standard demand curve, price and quantity are inversely related—when price rises, quantity demanded decreases and vice versa.

6. When there can be no change to quantity supplied

no matter the price, the supply curve is perfectly elastic.

7. When the price of air travel rises, the price of travel by rail increases also. This is because airline and rail travel are substitutes.

8. Externalities arise when private costs and social costs or benefits do not match.

9. Depending on their implications, some goods could have both positive and negative externalities.

10. Police protection is a prime example of a private good.

11. In the long run, no profit can be extracted in a perfectly competitive market.

12. When MB= MC, total net benefits are maximized.

13. The social benefit of a trade is the sum of the private benefit to buyer and seller and the external benefit to all other members of society.

14. International trade facilitates the transfer of technology from industrial to developing nations.

15. During off-tourist seasons there will be an increase in equilibrium price along with a decrease in quantity sold.

IV. Fill in the Blanks

1. As the cost of production of a product goes down, the price to the customer declines, making it more affordable. This results in a shift along the ____________ curve.

2. ____________ is the money that must be paid to buy a product. ____________, on the other hand, refers to the foregone value of resources that are used to make a product.

3. When a price ceiling is imposed below the equilibrium price, a ____________ results.

4. In a competitive market, surplus occurs if the price is set ____________ the equilibrium price.

5. A monopoly is inherently ____________ efficient than a competitive market.

V. Project - Power Plant Finances

It is estimated that a nuclear power plant will cost $3 billion to build. The utility needs to raise $2.4 B for the plant construction and equipment. The money is needed over a 5 year period ($480 M each year for 5 years) which is how long it takes the plant to be built. There are three options available:

a. Borrow the total sum at the beginning of year 1 at a rate of 8%. The excess money, will be invested in the stock market, where an historic average of 5% has been returned.
b. Borrow money from the bank as needed, i.e. $480 million at the beginning of year 1, 2, 3, 4, and 5. The bank charges 7% per year interest.
c. Borrow the money monthly at the prime rate, i.e. borrow $8 M for 60 months. The prime rate at the beginning of the year 1 is 4%, but is expected to raise by 1.5% every year for the next 5 years.

At the end of year 5, the plant is operational and the utility will be able to make a monthly payment of $5 million until all of the loan is paid off. Find:

1. What is the net present worth of the three options proposed above? Which option is economically more viable?
2. What is the effective interest rate of all future payments?
3. Redo the problem if a 3% cost of living adjustment is considered.
4. How many months does it take for all loans to be paid off?
5. What is the cumulative value of all payments for the option considered?

Economics of the Environment

CHAPTER 16

Not blind opposition to progress, but opposition to blind progress
~ Unknown (Sierra Club's slogan)

The view of the average human as builder conflicts with the view of the average human as destroyer which underlies the thought of many doom sayers
~ Julian Simon (1932-1998)

The environment is not only the source of raw materials, but is also a sink for the wastes generated as a result of exploitation of those same resources. As the economy continues to expand and industrial output grows, we are generating more and more hazardous wastes and toxic chemicals which will eventually end up in the air, in the water supply, and on land. Because society requires many services and products, it is understandable that a certain amount of pollution must be tolerated. The question is where the balance lies and how much longer we can exploit our natural resources before the costs associated with environmental degradation outweigh the benefits of increasing production and material wealth. Economists differ in their approach to reducing environmental pollution but, at the same time, assure that society as a whole benefits from proposed cleanup strategies. Most economists measure the cost/benefit ratio of the environmental protection in monetary terms; there are some, however, who are not willing to put a figure on the natural capital and the ecological systems that support life. The dilemma is that, although sources are owned privately (either by state or private enterprises), the same entities have free access to the sink (atmosphere) which is owned by the public as a whole.

In this chapter we will discuss various methodologies that are used by traditional and ecological economists to measure the cost of polluting the environment in both monetary and non-monetary terms and the role the government plays in regulating pollution.[1] In particular we will address pollution generated from fossil combustion and the costs and benefits of environmental cleanup.

Open Access and the Tragedy of the Commons

One of the major problems with environmental cleanup is that most natural capitals are not (and cannot be) privately owned. In economic terms, this is called *the open access* problem, or the *tragedy of commons.*[2] When a resource (such as a grazing land or a lake) is accessible to all, there is little incentive for individuals to protect it. A cattle herder or a fisherman will *rationally* argue that if they do not exploit the land for

[1] For more detailed information, please refer to the list of books given at the end of this chapter.
[2] Hardin, G., "The Tragedy of Commons," *Science*, vol. 168, December 13, 1968.

more grazing, or use the lake for more fishing, somebody else will. It is therefore logical to exploit the land or to overfish the lake to its maximum limit and reap the benefits before somebody else does. Same can be said of natural sinks like the atmosphere. A factory owner or a coal company have little incentive to bear the high cost of cleaning the air even though their children and families living nearby may have to breathe the same air. In a way, in absence of environmental regulations, they reap the full benefit of avoiding the cleanup cost, although they have to pay only a fraction of the cost that air pollution will bring to the entire community.

Is there a right amount of pollution?

Although having a clean environment is desirable to almost anyone, there is a limit to how clean it can get and how much it can cost. There are two schools of thought on what constitutes an efficient amount of pollution. Traditional economics treats pollution as a good that follows the same law of supply and demand that governs all other goods; it is most efficient at a level where the marginal cost of pollution reduction (supply) is equal to the marginal benefit to the society, or what society is willing to pay (demand) to avoid it. Any measure that reduces or increases pollution beyond this level results in the reduction in total benefits to producers and consumers as a whole.

The second group of economists, called ecological economists, rejects this view as inherently unethical and unfair. They point out that personal safety is the innate right of every individual, and, independent of how much it costs, pollution must be brought below a certain level of risk to its victims. The former group advocates an **efficiency standard**, whereas the latter group believes a **safety standard** must be the basis of pollution control. Not surprisingly, efficiency advocates consider the safety standard as inefficient, whereas safety advocates consider efficiency standards to be inadequate in safeguarding the public and the environment.

Efficiency Standard

As we saw in the previous chapter, an efficient market is a market in which demand and supply are in balance. The market demand curve represents the amount people are willing to pay for one additional unit of output at each price, i.e. what people pay for the pleasure of having one extra unit of the good (marginal benefit). Similarly, the market supply curve represents cost incurred by the firm to produce each additional unit of good (marginal cost). If the price rises above this equilibrium point, the quantity demanded for the good drops, but suppliers would be willing to increase their production and new suppliers are willing to enter the market. The imbalance between the supply and demand pushes the price down and equilibrium is achieved again. The opposite will be true when price drops, but the final result will be the same. Applying this concept to pollution control, we note that the marginal costs of reduction will be increasingly higher per unit reduction in pollution (easier cleanups are done first) and marginal benefits of reduction are lessened as air gets cleaner and cleaner. For example, it is costlier to reduce nitric oxide concentrations from 100 to 99 ppm than it would be to reduce it from 1000 to 999 ppm. At the same time, the marginal benefit the society receives (in terms of lower health risks, for instance) with reduction in nitric oxide's concentration drops (Figure 16-1). Figure 16-2 shows the estimated marginal costs of nitric oxide abatement (in Euros/ton) for OECD and the former Soviet Union in 2010. As the data shows the graph is highly nonlinear. With the current technology, the cost of emission reduction is fairly small for removal of up to 50% of the pollutants, rises rapidly up to 80% and becomes prohibitively expensive if the last 20% of the emission is to be cleaned up.[3] That is, as the level of pollution in the atmosphere decreases, we reach a point where further cleanup effort requires costs too excessive for the society to bear.

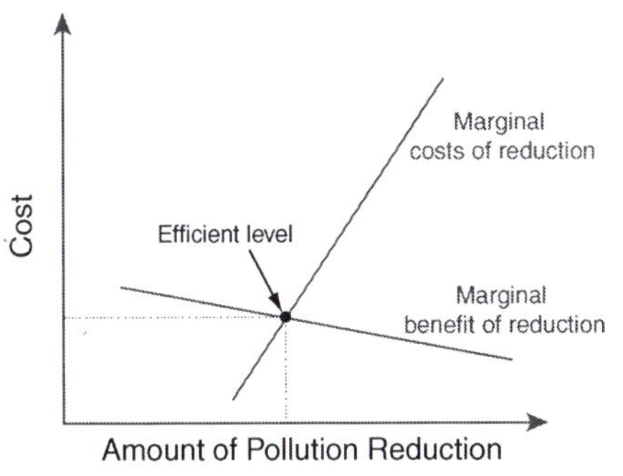

Figure 16-1
Efficient level of pollution control

a Coase Theorem

There is no question that a clean environment will benefit public health and reduce medical costs. It seems to be a no-brainer that whoever pollutes the environment has the moral responsibility to pay for cleaning the environment. From an economical standpoint, however, society as a whole does not care whether the polluter or the victim pays for the cleanup. As long as victims and polluters reach a mutual agreement, in the absence of any regulation and in a truly competitive market, it makes no difference whether the victim or the polluter pays for the cleanup. In other words, it does not require victims to necessarily be compensated. That is the equilibrium point is where bargaining between the two parties leads to efficient levels of pollution.

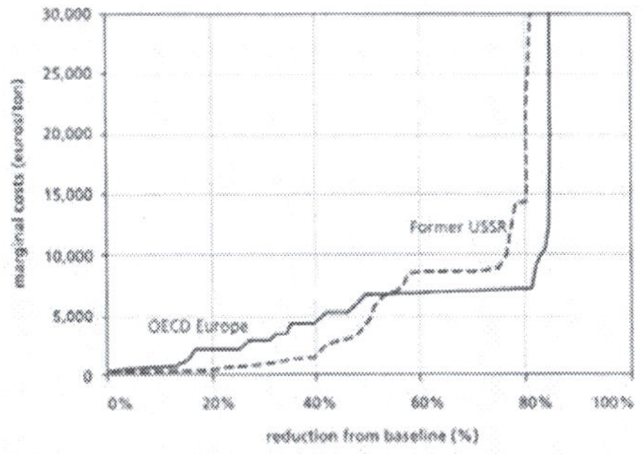

Figure 16-2
Costs of Pollution Abatement *[Ref. 2]*.

Question: To protect the public, environmentalists propose strict environmental regulations that restrict the dumping of pollutants into the atmosphere. How does the efficiency standard address the concern for public safety?
Answer: Victims always have the option of suing the polluters to recover the cost of cleanup and other damages. By doing so, the cost of production increases, which forces polluters to invest in cleaner technologies or acquire effective pollution control equipments.

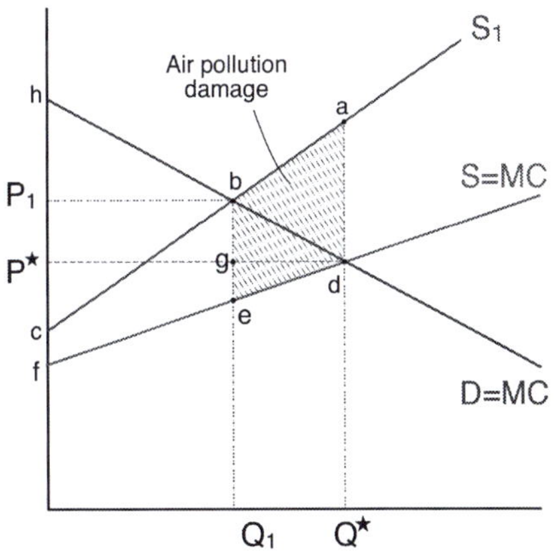

Figure 16-3
Effect of pollution on equilibrium

Consider, for example, the transportation of crude oil from Alaska to refineries in California.[4] The suppliers ship Q* barrels of oil at a cost of P* that includes the price of the good and ordinary transportation expenses such as wages for the crew, the cost of the ship, port taxes, etc. This is shown along the supply line S (Figure 16-3). In addition to these expenses there are costs associated with the spillage of oil along the route used by the oil tankers. If these costs are internalized, the supply curve shifts from

[3] Harmelen, T., et al., "An analysis of the costs and benefits of joint policies to mitigate climate change and regional air pollution in Europe." *Soil and Water Pollution,* 5, pp. 349-365, 2002.
[4] This example is borrowed from Economics Today, by Roger LeRoy Miller and Daniel K. Benjamin, Addison Wesley Educational Publishers, Inc, 2001-2002.

S to S_1 and the quantity shipped is reduced from Q^* to Q_1.

Let's consider two situations: a) the tanker operators have the right to spill without having to pay any penalties for environmental damage; and b) the fishing industries have the right to demand that tanker operators reduce the shipment or even stop it altogether.

If tanker operators own the rights to shipping, then they would continue to ship Q^* barrels instead of the optimum value Q_1. In this case fishers have to convince the operators to reduce their shipment from Q^* to Q_1. If they succeed, damages would decline by the area shown as trapezoid $abed$. This requires fishermen to make two side payments, one to the tanker operators for forfeiting a profit equal to the triangular area bdg and another to the operators' customers, the refineries, in an amount equal to the triangular area gde for compensating their losses from lower sales. The total payment to operators and refineries is therefore $bde = bdg + gde$. Fishers have an incentive to pay for these costs because they are still left with a net gain of $abd = abed - bde$. Overall, fishers are better off reducing environmental damage as a result of shipments being reduced from Q^* to Q_1, while there is no monetary impact on operators or refineries. We can therefore conclude that a side payment will be paid by fishers to operators and refineries, and only Q_1 units of oil will be transported.

If the fishers own the rights to clean water, then they can opt to insist on reducing the shipment or stop it altogether. In this case, the operators and refineries may choose to offer side payments to fishers for obtaining the rights. But how much are they willing to offer? If they manage to obtain the rights of shipping Q_1 barrels of oil from fishers, they are to benefit with a profit as high as $hbef$, the difference between their revenues and costs. At the same time, fishers are willing to accept as low as $cbef$, an amount equal to the environmental damage to them. For any amount above Q_1, fishers demand an amount exceeding the additional revenues by the tanker operators which therefore cannot be collected. We can conclude that a side payment will be paid by operators and refineries to fishers, and again, only Q_1 units of oil will be transported.

This example shows that, irrespective of whether the polluter or victim pays for the pollution, the result is exactly the same. This is named **Coase Theorem** after Ronald Coase, the 1991 Nobel Laureate in Economics.

> **Question:** What do advocates of the efficiency standard say about second-hand smoking?
> **Answer:** Smokers have as much right to enjoy smoking as non-smokers have to enjoy clean air. We cannot discriminate against anybody on the basis of whose rights are more valuable.

Economic Efficiency

The term "efficiency" means different things to different people. Engineers talk about efficiency as the ratio of the desired output from a device to necessary input. An energy-efficient refrigerator uses less electricity for the same amount of cooling than a less efficient refrigerator uses (See Chapter 5). An efficient worker performs more work in a shorter time, and an efficient programmer writes the same program with fewer instructions.

Social scientists refer to efficiency as the cost that society as a whole will pay compared to the cost if certain actions were or were not undertaken. For example, the social cost if there are no energy conservation policies could include a rise in pollution, degrading health, social unrest, hunger, job loss, and a decline in the economy overall.

Economists refer to efficiency in terms of the maximum profit that can be gained from a given transaction. The most widely accepted definition of economic efficiency is **Pareto efficiency**. A market is most efficient when it makes the public as a whole "better off." As long as the market has not reached its most efficient point, the economic pie enlarges and, in principal, everyone can benefit by getting a bigger slice.

> **Question:** When does the market for a car sales reach its peak economic efficiency?
> **Answer:** As long as there are buyers who are willing to pay for the cost of owning a car and there are sellers who are willing to satisfy customers' demands, the market has room for additional cars. That is, the market is Pareto efficient when demand balances supply.
>
> **Question:** Is a "zero pollution" solution compatible with the concept of economical efficiency?
> **Answer:** The zero pollution option will not make the outcome economically efficient. Although the public is willing to pay for certain amount of cleanup cost, at a certain point the increase in the marginal cost of cleanup does not justify the marginal benefits and brings enough economic hardship to a large number of people that the happiness of some would not make up for the unhappiness of the rest.

Safety Standard

Can one put a price tag on safety? Efficiency advocates, as we saw in the previous section, set price at the point where the market is at its Pareto efficiency. Safety advocates, however, believe that people have a right to protect themselves against unsolicited harm to their immediate environment and see safety as an essential human right that must be assured at an acceptable level no matter the cost.

Proponents of safety (or fairness) standards disagree with the Coase hypothesis and argue that polluters are the ones who must pay for the cost of cleaning the environment -- the public should not be victimized at the expense of profiteering by the polluting firms. In addition to being fair to the victim, the polluter-pays principle is superior on two main economic grounds. First, it removes the incentive for polluters to pollute (free-ride), as they, rather than the victim, would have to pay for the cleanup. Second, if polluters do not pay for the pollution, their production costs drop. This encourages even more polluters to enter the market, produce more goods, and create even more pollution. Furthermore, safety advocates refute the claims by efficiency advocates that suing the polluters can remedy the problem of environmental pollution by pointing out that victims often do not have the resources that large corporations have. Litigation is often quite costly and can take many years. Therefore it is very likely that victims tolerate the pollution damage, giving polluters a free hand to pollute with little fear of retribution. Another concern raised with this option is that there are always those who try to free ride on the outcome of the lawsuits of a few without exposing themselves to excessive costs associated with the lawsuits. For example, a noisy tenant blasting his stereo in the middle of the night might inconvenience more than a few neighbors. Although one or more victims may be inclined to sue and evict this neighbor, others may refuse to contribute to the effort and hope others carry the burden.

Proponents of efficiency standards, for their part, point to a number of drawbacks in safety standards. The first criticism is that safety standards are by definition inefficient. Voters often tend to walk on the side of caution and overestimate the risks; thus the cleanup costs will be excessive. Furthermore, with budgetary constraints, the public will be best served if the money is spent on environmental projects that save the largest number of lives. Because we have limited financial resources for

Safety vs. Efficiency Standards

Point/Counter

P: Everybody has the right to a clean and healthy environment. Polluters have the responsibility to clean the environment, no matter the cost.

CP: Of course, every human being likes to breathe clean air, drink clean water and live in an environment free of toxic chemicals. The question is: at what cost, and does the price justify the outcome?

P: We owe it to our children and grandchildren to protect our environment and preserve our natural resources.

CP: The resources are used for our wellbeing and that of future generations. The cleanup cost is justified as long as it does not impede progress. Our children will be collectively better off if the economy expands at a faster rate than resources deplete.

P: Deciding on safety issues based purely on cost-benefit analysis puts a higher value on the health and comfort of the wealthy, and is therefore immoral.

CP: The government has a limited budget to spend on safety and cleanup measures. The cost of cleanup might be too draining on the economy of low-income neighborhoods and may adversely affect them by incidentally increasing the violence, noise, drugs and pornography, rampant among poor and economically disadvantaged groups. More money spent on the environment leaves less money available to spend on other safety measures such as gun control, accident prevention, and police protection.

environmental protection, accepting a small increase in a particular risk may release money to carry out other safety measures. For example, by allowing a higher risk of cancer from pesticide use and diverting funds to reduce cancer from tobacco smoking, investing in accident prevention measures, and educating the public about drug abuse, gun safety, and gang violence might in fact saves more lives overall. Ultimately, how strict the safety standards should be requires a cost-benefit analysis.

> **Question:** Who should pay for hauling garbage and constructing new landfills, the government or private homeowners?
> **Answer:** When the government pays for the cost, individual households have little incentive to reduce waste, and landfills fill sooner. At any rate, in the long run, the cost will be distributed through all households by requiring them to pay higher taxes.

The Benefits of Environmental Protection

Depending on how one views the environment, different values can be placed on protecting it. The utilitarian considers the value of the environment only as long as it brings him happiness. Anything that does not affect him directly has little immediate value to him. Therefore, nobody can be easily persuaded to pay for expenses today even though he might eventually be rewarded by cheaper products, better health, and a more pleasant environment. The **environmental ethicist**, on the other hand, sees protecting the environment as a moral issue; the environment belongs not only to us, but also to all living organisms and to all future generations. Society should seek to protect it whether it directly benefits today's humans or not.[5]

Environmental protection has both market benefits and non-market benefits and may be broken down into values for generating revenues, providing non-monetary services to the community (**use value**), presenting unknown potential benefits (**option value**), and preserving the environment even if it does not benefit us directly (**existence value**). An example of the use value of a lake is an increase in the number of water sport enthusiasts, swimmers, and fishermen who use the lake. An example of option value is its value to future generations and the possibility that saving the environment and its ecological systems may one day provide the key to a currently non-curable disease. The existence value of the lake is the value individuals place on simply saving the ecosystem because of our belief that other creatures -- animals and plants -- have the same inherent right to these resources as humans.

Cost-Benefit Analysis

The economical benefits of environmental protection can be divided

[5] The differentiation between utilitarians and environmentalists only makes sense when basic human needs are satisfied. The poor have not seen any point in protecting the environment, as they do not receive much of the benefits. When poor countries are given sufficient incentives, their attitude will necessarily change and they will see the value of environmental protection and conservation.

into its market and non-market benefits. **Market benefits** are measured in terms of increased efficiency and production, savings in the cost of energy expenditure, reduction in cost of health care, and in the creation of new jobs. For example, cleaning the water in a polluted lake can increase commercial fishing and tourism, reduce the rate of infectious disease, and appreciate the price of housing adjacent to the lake, all of which can be measured in dollars. One approach is to use a **hedonic regression** to estimate the utility or pleasure associated with an improved environment. This technique assumes the price of an item can be calculated by assigning certain characteristics to its constituents and obtains estimates of the value of each characteristic. In essence, it assumes that there is a separate market for each characteristic. For example, hedonic regression can be used to assess the value of real estate in the absence of specific market transaction data. Because one building is different from another, it is often difficult to find an identical building that was sold recently. Instead, it is assumed that a house can be decomposed into several characteristics such as location, number of bedrooms, size of a lot, quality of neighborhood schools, etc. Hedonic regression treats these attributes separately and estimates prices for each of them. This information can be used to construct a price index that can in turn be used to compare the price of housing in different cities or at different times.

The **non-market benefits** of cleaner lakes are the enhanced diversity of marine life, more enjoyable swimming, recreational fishing and boating, and improved air quality. Non-market benefits cannot be measured directly but can be estimated by polling people on their hypothetical *willingness to pay*, to avoid incurring specific damage and for the benefit of enjoying a cleaner environment, or *willingness to accept* compensation for a certain amount of pollution and its associated actual or perceived risks (**contingent valuation**). Since different people value the benefits differently, the result is contingent upon the questions asked.

Actual risk is often estimated as the probability that a certain pollutant causes a certain number of deaths every year, spread over the affected population. This is usually done by examining data from past cases of human exposure to the pollutant or by extrapolating laboratory data from animals exposed to various doses of pollutants to humans. Many times, however, actual risks do not represent the perceived risks of exposure to pollution. For example, people may perceive that the risk of radiation from nuclear reactors is much higher than the risk from coal reactors, even though data shows the actual health risks from exposure to radiation are far less than those of exposure to air toxins.

Once the environmental benefits of environmental protection are evaluated, we need to evaluate costs of noncompliance by determining losses to property and to humans. Although placing a monetary value to human life seems highly insensitive, unfortunately it is the only way the cost of environmental degradation and damage to humans can be measured. Perhaps a better way to view this approach is to look instead into the benefits of saving lives by investing in various pollution abatement strategies. Depending on age, education, and income, the US Environmental Protection Agency puts a number ranging from $475,000 to $8.3 million for the value of human lives.[6] In contrast, the value of life in many poor countries has been estimated to be in tens of thousands of dollars.

Question: Some propose using the discounted future earnings of an individual to estimate the damage resultant from a loss of life. Others suggest the best way to measure the value of a life is by the risk various people are willing to take to accept similar jobs. Comment!
Answer: The first method is often used by the US court system to determine monetary damage for a loss of life. The main objection to this method is that lives of retired people or volunteers are greatly underestimated. The second method looks into how much people value their own lives when they participate in dangerous sports, volunteer for military service, or apply for risky jobs such as police officers and firefighters. The main objection to this approach is that, to support their families, many people have no choice but to accept high risk jobs. Secondly, very few people are adequately informed on the actual risks they are undertaking when they accept a new job or participate in a dangerous endeavor, and therefore their choice of the job or activity is not a true indication of the actual risk they may face. Finally, this method does not take into account the preference of many participants whom are inherently more risk-taking with no relevance to the proceeds they receive. One thing is certain; no matter which approach we use, we must face the ethical dilemma of valuing some lives more than others.

Economic Impacts

The economic impact of environmental protection can be studied in terms of the direct cost of cleanup, the potential loss of jobs, and the costs associated with monopoly power. Engineering cost estimates of environmental protection in the United States have been steadily increasing from 1% of the GNP in the early 1970s to 2.8% of the GNP ($224 billion) in 2000, much of which is for improving water quality and cleanup of old hazardous solid waste sites. The impact of environmental protection on employment has been minimal at best. Despite what many "pro-business" activists claim, environmental regulations have not resulted in a large number of plant shutdowns and massive job losses. In fact, labor statistics show that although some workers were forced to retrain and change jobs, environmental protection has probably led to a small net gain in US employment.[7] In fact, in the past quarter of a century,

[6] "Putting a Price Tag on Life," *Newsweek*, January 11, 1988, p. 40.
[7] Goodstein, E., "The Trade-Off Myth: Fact and Fiction about Jobs and the Environment," Washington, DC: Island Press, 1999.

the majority of US jobs lost have been due to increased competition from newly emerged industrial powerhouses like China, Korea, India, and Brazil, or due to the transfer of much of the US manufacturing capability offshore, where labor is cheaper and taxes are lower. In the few cases that there has been some loss of income due to environmental regulation, the utility derived from a cleaner environment compensated for the drop in utility from additional material goods.

Probably the most direct impact of environmental protection has been the consolidation of major environmental firms and an increase in the monopoly power that these firms exercise. The main reason for these mergers is that the cost of equipment required for cleanup is quite high and cannot be afforded by many small businesses. As we discussed in the previous chapter, monopolies may not be as efficient as a competitive market and prices charged are somewhat higher.

> **Question:** The costs for many of new and clean energy technologies such as photovoltaic and wind have already dropped below the social cost of older and polluting technologies such as coal and oil. Why haven't these technologies been promoted more aggressively or been adopted by the public in great volume?
> **Answer:** It is true that renewable energy technologies are the socially preferred option; however, their private costs are still above those of the non-renewable energy sources. Unless they have a substantial profit advantage, no government promotion or subsidy can make clean technologies succeed. It should be noted that being competitive in the market is not enough for public transition to cleaner technologies. For example, in many instances it can be shown that it makes economic sense to switch from electric and gas heating to solar heating and geothermal heat pumps. Nonetheless, many people are reluctant to invest in upfront costs to install new technologies because they lack knowledge about advantages of clean technologies and have a strong perception that oil and coal are well-established, mature, and proven technologies. Clean technologies have another major barrier to overcome. That is, they are offering consumers a service that they already have. Furthermore, unless services are offered by financially secure, well-established firms, they often do not have necessary resources to counteract advertising campaigns by existing firms. In the absence of such incentives it is reasonable to assume that, without government action (regulation and elimination of subsidies to conventional technologies), clean technologies will, for now, serve only niche markets.

Environmental Legislation

Many economists consider the public ownership of resources (land, water, air) and free access to the environment as the main causes of environmental pollution; therefore, the only remedy for the abuse is through environmental regulations, pollution taxes, and fines. Regulations are usually mandated by Congress and implemented by the government. Depending on their political affiliation, different people see the government's role differently.[8]

The regulative systems of most governments involve complicated and lengthy processes that leave doors open to abuse, political maneuvering, and industry influence. In the United States, after a particular problem is identified, Congress acts to pass a bill that requires the government to study the problem and take necessary actions. The task of risk assessment and safety regulations is delegated to the EPA who may then study the problem in-house or hire consultants. During this process, the EPA seeks input from the industry, the public, and environmental groups. The agency then drafts a regulation which it refers to the state governments for enforcement. It is taken for granted that no matter how insignificant the ruling is, numerous lawsuits by various interest groups will stall the process. Many of the dirtiest industries such as petroleum, chemical, and manufacturing firms have huge financial and technical resources that they spend on advertising, promotion of the election of friendly candidates, and utilization of a large lobbying staff and legal firms that can represent them in every step of the way. Even the EPA's top managers are influenced by high-ranking government officials who may hold the key to their promotions and job security and by industries who may, at some point in the future, hire them with very generous benefit packages.

The best solution to reduce conflicts-of-interest is to eliminate or at least greatly reduce the role of money in legislative and regulatory processes. This may require banning or limiting political advertising, restricting or limiting contributions of industry lobbying groups to political campaigns, eliminating tax deductible benefits to lobbying expenses, and disallowing government officials to accept employment at firms whom may have benefited from rulings during the tenure of those officials for a certain number of years (let's say 3-5 years after they leave their government posts). The same can be said about top-level industry managers who may be appointed to sensitive government posts.

Finally, to implement political reform we must assure that the principles of democratic government are taking hold. The collapse of the Soviet Union demonstrated that the centrally planned economy in socialist countries was even less effective in preventing environmental disaster than market-based economies. The point is that democracy is a necessary ingredient to assure the free flow of information and to empower citizens

---

[8] Depending on their political affiliation, different people see the government's role differently. Conservatives want to limit the government's role as much as possible, accusing the government of being run by politicians and "special interests." The environmental pollution, they insist, will be resolved only when industries agree to reduce emissions voluntarily. The public as a whole will decide what steps to take to force polluters to cut their emissions by boycotting, suing, or switching to more environmentally friendly competitors. Progressives see the "good" government as necessary to serve the interest of the public by providing health care and assuring their welfare. They object, however, to the role of the big businesses that – through their vast financial and intellectual assets – can influence the government and rally the public to support only those environmental measures that benefit them directly no matter what the short- and long-term costs to consumers.

and institutions to challenge the policies of the government when it is practicing erroneous environmental initiatives.

There is no single approach to reduce the harmful effects of environmental pollution. Three approaches -- command-and-control regulations, incentive-based regulations, and process-based regulations (or a combination of them) -- are often used to address a variety of environmental problems.

Control-Based Legislation

Control-based legislation (also called **command-and-control legislation**) relies on directives (commands) on how polluters should reduce their pollution and by how much. These policies are enforced by imposing fines, penalties, and taxes for non-compliance. Under command-and-control (CAC) plans, firms are required to deploy the **Best Available Control Technology** (BACT) to control emissions. Although this approach dictates a fair and balanced regulation to all players irrespective of their size, it is not necessarily an efficient one. For example, consider two firms A and B producing essentially the same product and residing in the same geographical area. Because of its size and the cleaner technology it employs, A has a lower cost per unit of pollution reduced than B. From a fairness standpoint, A and B must reduce emission per unit output by the same amount. From an efficiency standpoint, it is best if A reduces emission by a greater amount than B, keeping the combined emission the same. The combined cost of emission abatement by the two firms will be lower.

In the United States, environmental regulations are largely based on the CAC approach for most pollutants and are administered by the EPA through such statutes as the 1990 US Clean Air Act, which imposes penalties for those entities that violate the federal and state guidelines on the type and amount of toxic substances, handling, use, and disposal. The rules are imposed uniformly for all polluters without considering costs and circumstances. The problem with this approach is that it is, by definition, "not cost efficient" and can sometimes be outright ridiculous. For example, EPA regulation requires that 30 percent of the organic matter must be removed from inflows to sewage treatment plants. Although this regulation makes sense most of the time, it does not make sense in Alaska, which has some of the purest waters in the world. To comply with regulation, Alaska had to build a new $135 million treatment plant to remove 30% of almost nothing. Alaskan officials requested a waiver but were turned down. As a result, Alaskans had fish guts dumped into the water just to be removed again downstream before it entered the treatment plant. Although the EPA mandate was satisfied, the water was more polluted than it would have been if nothing had been done at all.[9]

The one exception to CAC legislation is the distinction it makes between

[9] Dilorenzo, T. J., "Unfunded Federal Mandates: Environmentalism's Achilles Heel?" Contemporary Issues Series 62, Center for Study of American Business, December 1993.

new and old sources. Under this ruling, older sources (such as cars) do not have to meet the same stringent standards as newer equipment. The reasoning behind this decision is both technical (higher costs are associated with cleaning the older sources) and political (it is much easier to pass regulations restricting firms that are not yet around or are in infancy). The problem with this approach is that it promotes the use of older and dirtier equipment for much longer, reduces the incentive for putting additional resources into research and development (R&D) work, and slows down investments in costlier, newer, and cleaner technologies. This is a serious objection that has been raised against any technology-based regulations for both new and old sources. The fear of developing better technologies is that, under BACT, firms may then be legally obligated to upgrade their pollution control equipment at existing and newer facilities. Over the long run, this approach becomes less cost-effective as the pace of technological innovation slows, and the cost of pollution abatement becomes higher.

Question: Examples of the CAC approach are the imposition of gasoline taxes and CAFÉ standards discussed in Chapter 14, both of which were designed to increase fuel efficiency. Comment on the viability of these approaches.
Answer: The main problems with gasoline tax are that both clean and dirty cars are taxed at the same rate and taxes must be quite high to be effective. The main concern raised against CAFÉ standards is that higher gasoline efficiencies promote more driving, and as a result total vehicle-miles traveled may increase.

Incentive-Based Legislation

To counteract many problems associated with control-based legislation, many economists propose developing incentive-based (IB) regulations in the form of emission credits, allowable tax deductions, and other incentives. Likewise, those who are not complying will be penalized by paying higher taxes with fewer emission credits. Two schemes have been proposed: 1) to enact **pollution tax** proportional to the amount of pollutants discharged into the air; and 2) to issue **pollution permits** (emission credits) that can be auctioned off or freely traded in a pollution market.

Both systems are designed to put a price on pollution and provide incentives to firms and others to seek less polluting alternatives or better cleanup strategies. With taxing, the cost is direct – more pollution means more taxes. As an incentive, some tax breaks may be given to those industries that have met and exceeded the requirements set by law or have invested in research and development of cleaner technologies. With pollution permits (also called ***emission trading*** or ***cap and trade*** system), a regulatory body--usually the government-- sets an overall cap on the country's carbon emission that drops annually; less pollution offers the opportunity to free permits for sale, whereas more pollution requires additional permits to be bought from the government or from less polluting competitors. This encourages companies to invest in cleaner technologies or find efficient ways to cut emissions and sell the unused portions of their permits to others. There are, however, sharp differences on how the pollution permits should be allocated. Some economist favor allocating higher number of permits to industries which produce more pollution. The rationale is that as the overall cap tightens each year, the biggest polluters face the largest challenges in cutting emissions. Other economists, propose allocating permits through an auction. Under this system, every company – large or small - would have to buy rights to pollute. As a result, the biggest polluters would have to pay the most - thereby providing them with the greatest incentive to cut emissions right from the start. [10]

IB regulations applied to automotive emissions can take several forms that encourage both the fuel efficiency and, at the same time, lower the vehicle-miles traveled. One possibility is an auto emission tax that is adjusted based on the total emission the car has produced over the period of one year since the last smog check. Congestion tax (higher toll-charges for single-occupant cars or during rush hours) discourages traveling alone or during peak traffic. Another solution is to charge auto insurance on a *pay-by-the-mile* system; less would be charged to smaller cars and to people who drive less. Insurance rates will be higher for bigger cars and individuals that drive higher mileages. The final solution is the imposition of a revenue neutral scheme referred to as *feebate*; this system penalizes gas-guzzling cars by charging higher tax rates while rewarding high fuel-efficient cars by giving them rebates.

> **Example 16-1:** Suppose a city ordinance requires a reduction in total particulate emissions in an industrial area by 100 tons. There are only two polluting firms, A and B, each responsible for producing 150 tons of particulates or one-half of all of the particulate emission in that area. What is the best strategy to achieve the required results in two scenarios in which firms can and cannot trade pollution permits? The cost of pollution abatement is \$50 per ton for firm A and \$200 per ton for firm B. What is the total cost of reduction in each case?
> **Solution:** Firm B has a much higher cleanup cost than A. If emission trading is permissible, it makes sense for firm B to buy 100 tons of emission by paying firm A \$50 per ton for a total cost of \$5,000. If emission trading is not allowed or firms do not reach an agreement, then firms must cut emissions by 50 tons each. The cost of pollution abatement is 50x\$50 = \$2,500 for A, and 50x\$200 = \$10,000 for B resulting in a total cost of \$12,500.

The main objections raised against incentive-based regulations are hot

[10] Robert Reich online blog, "Why McCain's "Cap-and-Trade" Won't Work Nearly as Well as Obama's,", May 27, 2008 *(http://robertreich.blogspot.com)*.

spots, monitoring, and enforcement. **Hot spots** refer to the elevated concentration of pollution near firms who find it economically advantageous to avoid or defer cleanup by paying a tax penalty or purchasing additional permits. This is particularly important for concentrated pollutants that are not uniformly mixed. Examples are nuclear wastes, toxic wastes, and sulfur dioxide or particulate emissions that can easily buildup in the vicinity of power plant smokestacks. The IB strategies on pollutants like carbon dioxide and chlorofluorocarbons are less important because their impact on the overall air quality depends on the total volume of their emission, not on local concentration of these pollutants. To correct for this problem, it is often suggested that penalties and permits are issued such that the maximum concentration of different pollutants within a given geographical area is not to exceed the state and federal ambient air and water quality standards.

Another problem raised by IB regulation is **monitoring and enforcement**. Unlike a CAC system where firms are required to install and maintain certain air quality abatement equipment, in incentive-based approaches, no such requirement exists so constant monitoring and precise measurement of concentrations of pollutants in multiple monitoring locations is required to assure compliance. This can be quite complicated and the cost, except under special circumstances, may not be warranted.

Process-Based Legislation

Unlike control-based and incentive-based statutes that regulate the amount of pollutants that an industry is allowed to release into the atmosphere, process-based legislation looks at the process or technology itself. Depending on technology, different regulations can be enacted during different stages of development from R&D to prototyping, field testing, and commercialization. Process-based legislations are usually devised by various federal and state agencies that are charged with the administration of environmental policies.

When Does Noncompliance Pay Off?

Not all polluting industries choose to comply with regulations. From a purely economical standpoint, if polluters' marginal cost of compliance outweighs its marginal benefit, polluters are better off ignoring regulations. They may instead opt to pay fines and accept other punishments such as adverse public relations, potential losses in sales, and possibly even jail time. Costs of compliance are related to the cost of new pollution reduction equipment and the additional resources that are ultimately diverted to pollution control efforts. Benefits associated with compliance can be measured as the money saved due to reductions in the amount of fines and other penalties, improved public relations, and a potential for higher sales volume. The marginal cost of compliance increases as a firm decides to comply more and more with the regulations. The marginal benefit of compliance is highest when a firm does not comply with any

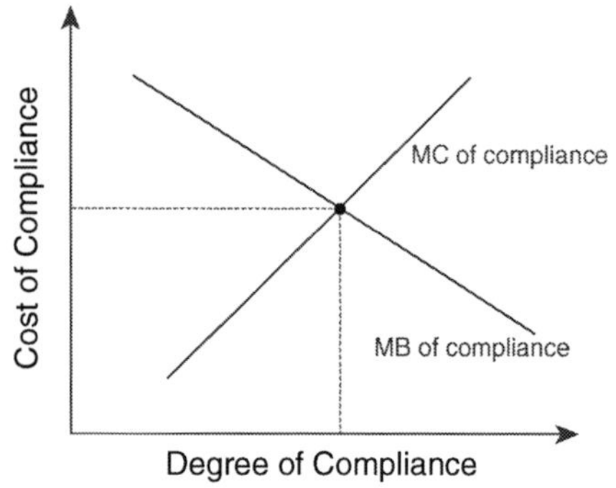

Figure 16-4
Efficient level of pollution control

regulations and thus is most vulnerable to being fined. As a firm complies more, often its chances of being caught and so its marginal benefit falls (Figure 16-4). The point where MB and MC of noncompliance cross is where the firm can afford to avoid compliance. For any additional incidence of non-compliance, the firm will have to pay fines that exceed its marginal benefits. From the discussion above, it can be concluded that to reduce non-compliance, either the marginal cost of noncompliance must be increased (such as stiffer fines or longer jail time) or its marginal benefits reduced.

Example 16-2: A power generating station is considering reducing the total particulate emission from its smoke stacks. Reducing the particulate emission is easiest for the larger particles with the biggest mass, and therefore a large portion of particulate mass can be removed with filters at a relatively low cost. The medium-sized particles can be removed by bag houses at somewhat higher costs, whereas the smallest particles require very expensive electrostatic precipitators. A cost analysis is made to calculate the marginal cost (cost for an additional unit of reduction) of particulate cleanup. The result is tabulated below:

% reduction in total mass	Total costs in millions of dollars
10	1
50	11
90	33
95	61
99	96

To promote a cleaner environment, the government is offering polluters a tax credit that varies depending on the degree of reduction that is made. The most benefit can be reaped if the largest, more massive particles are eliminated because these particles most effectively reduce visibility, form fog, and soil buildings. The amount of tax credit reduces as particle sizes become smaller. The table of tax credit is shown below:

% reduction in total mass	Tax Savings (Revenues) in millions of dollars
20	30
40	52
60	68
80	80
100	9

What is the Pareto efficient point for pollution reduction?

Crime and Punishment

Digging Deeper ...

In every society, there are always those who become involved in illegal or criminal activities for various reasons: to reduce costs (avoiding buying and installing pollution-control devices), to increase revenue (illegal commercial fishing, and logging), or to increase utilities (illegal hunting or swimming). An economic analysis of the situation shows however, that there is a point where the risk of detection and punishment does not justify the benefit reaped from such activities. The main task of regulatory and law-enforcement agencies is to impose a high level of monitoring and enforcement, or create enough deterrence (for example, large fines or long jail times) to convince the criminals (the environmental polluters in our discussion) that crime does not pay.

Example: Suppose you are given the option of receiving a guaranteed $1,000 or have a 50% chance to receive double or nothing. Which one do you take?
Solution: Mathematically, both options have the same expected value of $1,000. The conservative takes the $1,000 payoff; the risk taker will gamble hoping for the $2,000. Both options are exactly the same for a risk-neutral person. The majority of firms are risk neutral by definition.

Example: Suppose a firm determines that there is only a 20% chance of being caught dumping its hazardous waste into a nearby river and if caught there is a 50% chance of being punished. Furthermore, the firm estimates a savings of $100,000 from noncompliance. How large a fine would create sufficient deterrence against noncompliance by the firm?
Solution: The penalty, X, should be large enough to washout the potential (expected) gain from noncompliance. i.e., 0.2*0.5*X = $100,000, or X = $1 million dollars.

Solution: The optimum pollution reduction level is at the point where MC and MB of pollution abatement are equal. The marginal benefits and costs are calculated and shown graphically. The two curves cross at a percent reduction where MC = MB = $14 million, for a reduction of 68% in total particulate mass.

Global Issues

During the past few decades, our environment has undergone changes unforeseen by our predecessors. Global population has nearly doubled and the gap between rich and poor nations has widened. International conflicts have escalated, many of them because of competition for control of limited natural resources. We discovered that the sky over Antarctica (and to a smaller degree all over the world) has been losing ozone rapidly. Acid rain has damaged millions of acres of land and has adversely affected marine life and other species. More than half of the tropical forests have been destroyed or greatly degraded. The damage incurred by properties alone is estimated at billions of dollars. Many valuable species of plants and animals are now extinct or endangered. Our ever-increasing dependence on petroleum has not only affected our health, but has also resulted in global warming, the full consequences of which are yet to be determined. The environmental issues associated with fossil combustion were discussed in some detail in Chapter 8.

Until relatively recent times, most of our activities that affected the environment were local in scope. Cleanup required taking small steps over small geographical areas and short times. Today, the rapid rise in consumption of both materials and energy has made many environmental issues global in nature. Nuclear testing by one nation can have environmental repercussions thousands of miles away. The burning of fossil fuels, use of non-biodegradable products, and dumping of toxic wastes in rivers can impact the environment globally.

One feature that separates global and local issues is their scale, both in magnitude and duration. Global issues are persistent; they cannot be solved overnight by one nation or one country alone. Resolving these issues may require long-term planning. Global issues are interconnected; solving one may in fact solve many other problems, or we may need to solve a multitude of other problems simultaneously. For example, reducing greenhouse gases will reduce both global warming and ozone depletion, while narrowing the gap between rich and poor nations can also solve many health, hunger, employment, and security issues. Global warming, ozone depletion, and acid rain are among the most severe problems we are facing worldwide.

Global warming has rightly received the most attention, as we may have to face consequences that last hundreds of years as a result. The Kyoto Treaty was a positive step that required cooperation between industrial and poor countries. Under this plan, rich countries would reduce their emissions of greenhouse gases and at the same time provide incentives for poor countries to follow suit. Unfortunately, the United States and Australia, two of the most industrialized countries and the largest contributors of greenhouse gases, cited economic hardship and pulled out of the treaty, weakening the agreement and reducing its likelihood of success. Despite this disappointment, the treaty was finally ratified in 2005. The impact of the treaty on curbing overall carbon emission is yet to be determined.

Ozone depletion was recognized as a major environmental issue at the 1987 Montréal Protocol and later in the Rio Summit on biodiversity in 1992. Enforcement of the treaty required signatories to ban the import and export of products containing CFCs to and from nonmember countries. This provision not only provided incentives for member countries to find ozone-friendly substitutes, but the reduced trades with nonmembers also provided impetus for others to join the agreement. The two common substitutes for CFCs are hydrofluorocarbon (HFC), hydrochlorofluorocarbon (HCFC). The former is not an ozone depleting substance but neither is as efficient a coolant; the latter is a mildly ozone depleting substance slated to be phased out by 2030.

Acid rain is formed when the sulfur and nitric oxides emitted from the stacks of coal and oil power plants react with the humid atmospheric air. The "UN Convention on Transboundary Air Pollution" drafted the "Protocol to Abate Acidification, Eutrophication, and Ground-level

[10] UN Environmental Commission for Europe (*http://www.unece.org/env/lrtap/multi_h1.htm*).

Ozone."[10] This protocol sets ceilings for emissions of four pollutants, sulfur dioxide (SO_2), nitrogen oxides (NOx), volatile organic compounds (VOCs), and ammonia (NH_3), to be reached by 2010. The protocol also sets limits for specific emission sources (power plants, refineries, farms, etc.) requiring the best available techniques to keep emissions down.

Globalization and the Environment

As discussed in the previous chapter, globalization has removed many of the trade barriers to allow freer movement of capital. Proponents of globalization argue that an increase in trades stimulates investment, reduces poverty, allows transfer of cleaner technology, improves health, creates jobs, and provides environmental benefits in the form of greater investment in air and water quality. Critics contend that globalization encourages ever more economic growth and production with no real concern about unequal and unsustainable pattern of consumption.[11] To promote globalization, international agreements are made that removes tariffs and other barriers to trade. As a result of such agreements, subsidies for many essential goods are eliminated and, because they cannot effectively compete with well-developed highly automated agricultural and manufacturing practices of developing countries, many thousands of small farmers, artisans, and independent workers lose their livelihoods. Another charge against globalization is that, since environmental laws are less strict and corruption is more rampant in some of the developing countries, many developed countries have moved their manufacturing sectors to these countries. Weaker environmental laws in the developing countries have given governments and large corporations an effective argument to lower their own environmental laws, thus leveling the playing field to the detriment of the global environment.

Environmental Issues in Developing Countries

Many of the environmental problems in developing countries are directly related to poverty (See box "Poverty and Wealth"). For people who lack safe drinking water, have inadequate sewage systems, and live in unsanitary conditions, conservation measures and environmental issues are one of the last things on their minds. Only when they achieve simple means of subsistence and have sufficient resources to afford thinking about other problems do they become interested in long-term health issues and consequences of environmental pollution. Most of these countries are heavily in debt and therefore any attempt at conservation and environmental protection must be linked to economic development and debt relief. Unless richer countries are willing to subsidize poorer countries to carry out environmental programs, the situation will deteriorate even further and the environment will suffer. Debt relief does not have to come as a handout, however, and could require the debtor country to spend equivalent money in their local currency on environmental programs

[11] Dauvergne, P., Handbook of Global Environmental Politics, Edward Elgar Publishing, 2005.

Poverty and Wealth: The Facts[i]

Did You Know That ...?

- The world's 200 richest individuals put together own as much as 41% of the world population.
- 1.2 billion people in the world, roughly one in five, live on less than US$1 a day. Another 2.8 billion on less than $2 a day. Over a billion do not have access to clean water.
- 17-40% of the world's population are hungry or suffer from malnutrition. 40 million people die each year as a result.
- In India, only 4% of people reach the age of 65, and nearly 4 out of every 10 people are less than 15 years old.
- The richest fifth of the countries account for 86% of total consumption, 45% of all meat and poultry, 58% of all energy, and own 87% of all vehicles and 74% of all telephone lines. The poorest fifth of the countries account for miniscule 1.3% of total consumption, 5% of all meat and poultry, 4% of all energy, and own less than 1% of all vehicles and 1.5% of all telephone lines.

[i] *Singh, D. N., "Population Growth and Development Relationship," in Strategies and Development Planning, Edited by Ashok Kumar Rai, 1977, pp. 368-388. ISBN: 81-7100-905-0.*

such as family planning, water sanitation, land remediation, air quality improvements, etc. This is particularly important, in light of the fact that, as developing countries become more affluent and undoubtedly emulate the living standards of the developed countries, their environmental footprints becomes larger.

Another powerful tool to promote environmental protection is international marketable permits. Under this plan, once the target for reduction of a certain pollutant is set, a fixed number of permits are issued and distributed among developing and developed countries. Developing countries have incentive to maintain their environment and, by doing so, they will have additional permits that they can exchange for cleanup equipment or upgrade their old and dirty technologies with newer clean technologies. Developed countries can buy the permits to avoid penalties and find new markets for shipping clean technologies. For example, a country can trade one million tons of carbon dioxide (for example by switching from coal to natural gas) for a shipment of wind turbines, photovoltaic cells, or consulting expertise for increasing crop yields without the need for additional fertilizers.

The same goes for population control. In many poor countries, larger families provide a safety net that can help at times of sickness and old age. Children can provide labor and bring in additional income. Because the infant mortality rate is higher, parents seek to have more children – especially males – to assure enough of them survive to maintain minimal living standards and to carry on their genes. It might be argued by some that a reduction in infant mortality rate will increase population which will, in turn, exacerbate the problem. This is what happened in 1950s and 60s when improved health conditions, better nutrition, and vaccinations saved the lives of millions of babies, resulting in a population boom in many poor countries. The data also shows that after the initial surge in population the growth rate declined. This is attributed to better access to education, less pressure to have larger families, and an increasing allocation of family resources to improving the quality of life of fewer children. It now appears that proper family planning is one of the best strategies and is an economically sensible way to address a wide range of global environmental problems in poorer countries.

Due to the absence of popular and democratic governments in many developing countries, especially in Africa, Asia, and the Middle East, environmental problems continue to become increasingly severe. International businesses have great incentives to collaborate with corrupt officials to pass laws that are inherently in conflict with the principal of conservation of these countries' natural capital. It is therefore fair to conclude that protection programs are possible only if they involve the full partnership of the host countries monitored by international teams of economic ecologists and consultants. Only those technologies that result in a more equitable distribution of wealth, eliminate or reduce poverty, make education and health services accessible, result in the least damage to the environment, and produce work can benefit these countries and are sustainable.

Summary

Benefits of environmental protection and optimum amounts of pollution are difficult to measure. Economists are divided over the economic merit of pollution abatement, although they generally agree that some degree of environmental protection will benefit overall human wellbeing.

Because the environment is a public good, it is subject to the same open access problem as other goods. Individuals have less inclination to engage or invest in long-lived environmental protection measures than if they had ownership of the good. Poor countries are also banking on receiving the benefits of environmental protection actions taken by more developed countries and, in that sense, on getting a free ride. Greenhouse gases, ozone, and acid rain are public goods because measures taken to reduce them by one country benefit all the others. Within a country or municipality, the government can overcome the free riding problem by imposing taxes and fines. At the international level, imposition of environmental laws requires international cooperation, ratification of treaties, and the financial commitment of rich and industrial nations. Since the United Nations does not have the power or authority to force any member states to comply with the laws, the terms for enforcing the laws must be clearly spelled out in the treaty itself.

There is no clear strategy to ensure public health while at the same time maintaining optimal efficiency. When a command-and-control system is used, firms are required to install cleanup equipment. However, the government must ensure that the cost of abatement is not prohibitive and the economic welfare of the firm is also considered. Two incentive-based approaches were discussed. Permits can be issued by the government, auctioned off, or traded between firms to achieve an efficient level of pollution. Alternatively, fines and taxes can be imposed to reach the same results. Unfortunately, both approaches also provide incentives for polluters to lie. In a tax system, firms are encouraged to underestimate pollution as taxes will be imposed per unit volume or weight of emission. With permits, firms have an incentive to overestimate emission, because they can get more permits. Process-based regulations are effective in assessing the total amount of waste in a given process, differentiating among various manufacturing technologies and helping firms in their initial planning stages of development. What is clear is that neither safety nor efficiency alone can be used to set standards, and some risk/benefit analysis is needed to decide on the most effective approach.

Additional Information

Books

1. Chapman, D., Environmental Economics: Theory, Application, and Policy," Addison-Wiley, 2000.
2. Goodstein, E. S., Economics and the Environment, 4th Ed., John Wiley & Sons, 2002.
3. Siebert, H., Economics of the Environment: Theory and Policy, Springer Verlog, 2004.
4. Dauvergne, P., Handbook of Global Environmental Politics, Edward Elgar Publishing, 2005.

Periodicals

1. *Journal of Environmental Economics and Management* (JEEM), the journal of Association of Environmental and Resource Economics.
2. *Ecological Economics* – Direct Science Elsevier Publishing Company, the journal of the International Society for Ecological Economics (ISEE).
3. *Environmental Economics and Policy Studies* – Published by Springer-Verlog, New York is the official journal of the Society for Environmental Economics and Policy Studies.

Government Agencies and Websites

1. US Agency for International Development *(http://www.usaid.gov/)*
2. National Center for Environmental Economics (*http://yosemite.epa.gov/ee/epa/eed.nsf/pages/homepage).*

Non-Government Organizations and Websites

1. United Nations Development Program *(http://www.undp.org).*
2. United Nations Environment Programme *(http://www.unep.org).*
3. Intergovernmental Panel on Climate Change *(http://www.ipcc.ch).*
4. World Resource Institute *(http://www.wri.org)*
5. Union of Concerned Scientists *(http://www.ucsusa.org).*

Exercises

I. Essay Questions

1. Describe what economists mean by
   a. Existence value
   b. Safety standard
   c. Contingent evaluation
   d. Command-and-control legislation
   e. Economic efficiency

2. What are the differences between existence, use, and option values? Give examples for each.

3. According to one study, the maximum price elasticity for mpg is 0.21, meaning that a 1% increase in price leads to 0.21% increase in fuel economy. How much should the price of gasoline be increased if we desire to achieve a 10% increase in automobile fuel efficiency?

4. Jill and Jack are sharing an office in a manufacturing firm. Jack is a smoker and claims much more relaxed and productive if he smokes three cigarettes a day. Jill hates smoking and claims that she cannot handle the smell of even a single smoke. Assume that there is no regulation that bans smoking in the office, who is right? Is there a point where a compromise can be reached?

5. Should smoking be banned in public places? If yes, should imposing the ban be the responsibility of a government agency? Should it be decided through voting?

6. A new landfill in your neighborhood is expected to raise the cancer rate by 1 in 1000 during your life time. To reduce the risk to 1 in 10,000, new monitoring and safety equipments have to be installed which will cost $100 million. To pay for the cost, your property tax must be increased by $1 a month. Would you pay for it? What if higher costs necessitated an increase in property tax of $10 a month? $50 a month? $100 a month?

7. A wireless telephone company needs to install their high-frequency, high-voltage transformers on the rooftop of your building. They offer you a flat fee of $10,000 for the right to install their equipment. Would you take it? There is some research that indicates a 1% increase in the rate of cancer for people living nearby these transformers. Would your neighbors have the right to sue you if you accepted the company's offer?

8. The same company wants to install a second unit in a poor neighborhood of the city. The company offers $2,000 for obtaining the right. Is this fair? What if the neighborhood voted to accept the offer?

9. Discuss three methods of encouraging automotive fuel efficiency. What are the advantages and disadvantages of each?

10. What is the globalization and its effect on the environment? Is the overall trend positive or negative? How has globalization affected your community?

II. Multiple Choice Questions

1. On the basis of the traditional economic theory
   a. The cleaner the environment is, the more sustainable economy will be
   b. Cost of environmental cleanup is a major drag on economy
   c. Society must bear a certain amount of pollution
   d. Pollution is a good that follows the law of supply and demand
   e. Both c and d

2. When it comes to the environmental protection, efficiency standard advocates believe that
   a. Personal safety is the innate right of every individual, and no matter the cost, must be guaranteed
   b. The marginal cost of reduction drops per unit of reduction in pollution
   c. The marginal benefit of reduction increases as air gets cleaner and cleaner
   d. An efficient market is a market in which demand and supply are in balance
   e. All of the above

3. The unknown potential benefits of the

environmental protection are called

a. Use value
b. Option value
c. Existence value
d. Hidden value
e. Contingent value

4. Which of the statements below is consistent with the operation of a profit-maximizing firm in a competitive market that does not penalize polluters no matter how much they pollute?
   a. Consumers pay the full social cost of production.
   b. Consumers pay only the external cost of production.
   c. Consumers pay only the external marginal cost of production.
   d. Consumers pay less than the socially efficient price since producers are subsidized by society.
   e. None of the above.

5. Which of the following is consistent with the command-and-control regulation?
   a. Taxing producers equal to the external cost created by their pollution.
   b. Requiring firms to lower their pollution level according to best available control technology.
   c. Requiring firms to reduce their pollution level to zero.
   d. Requiring firms to buy pollution permits to offset their pollution.
   e. Requiring firms to pay a penalty equal to the external cost of pollution.

6. Which of the following statements <u>is not</u> correct?
   a. The economical benefits of environmental protection can be measured only by its monetary benefits.
   b. Hedonic regression implies the pleasure one gets from an improved environment.
   c. Non-market benefits of environmental protection can be estimated by people's willingness to pay for those benefits.
   d. The willingness to accept compensation for some level of pollution is called contingent evaluation.

7. Suppose two firms A and B each produce 10 tons of emission every month. The local city council passes a bill that requires total monthly emission to be reduced by 8 tons. It costs firm A $5 and firm B $20 to cut emissions by one ton. The most efficient way to comply with the ordinance is for emission trading that results in
   a. Firms A and B negotiating to trade emission permits
   b. Firm A cutting emission by 8 tons
   c. Firm B cutting emission by 8 tons
   d. Each of the firms A and B cutting emissions by 4 tons
   e. Both a and b

8. In the example above, assume that the law sets a maximum particulate emission of 12 tons per each firm. What is the optimally efficient cost of pollution abatement?
   a. $ 60.00
   b. $ 90.00
   c. $ 130.00
   d. $ 240.00
   e. Not enough data is given.

9. The main advantage of control-based regulations is that
   a. Because it excludes or relaxes fines on old sources, the cleanup costs are less
   b. Because regulation is based on imposing the best available control technology, there is great incentive for firms to spend money on R&D activities
   c. It allows old industries to survive and thus hire new employees
   d. It provides incentives to those who help improve air quality and reduce emissions
   e. All of the above

10. Which of the following statements is correct?
   a. Both control-based and incentive-based regulations provide incentives to firms to reduce emissions.
   b. Tradable pollution permits allow same amount of reduction in pollutant levels, but at a lower cost.

c. The main problem with incentive-based regulation is the creation of hot spots.
d. Tradable pollution permits encourage firms to invest in control abatement technology to make additional money by selling their permits to dirtier firms.
e. All of the above.

11. The main objection raised against incentive-based regulations is that
    a. Penalties cannot discourage firms from polluting
    b. Localized hot spots may appear
    c. It is always possible to buy permits at a lower cost than it takes to install new pollution control equipment
    d. It offers the opportunity to sell permits given out free
    e. All of the above

12. Which of the following statement(s) is/are correct?
    a. Proponents of safety standards believe that polluters must pay for the pollution.
    b. Safety standards remove incentives for polluters to free ride.
    c. Safety standards are inefficient.
    d. All of the above.
    e. a and b, but not c.

13. Non-market value of the yet-unknown potential benefits of a resource is called
    a. Use value
    b. Option value
    c. Existence value
    d. Utility value
    e. A charity

14. ________________________ see protecting the environment as a moral issue.
    a. Hedonists
    b. Utilitarians
    c. Environmental ethicists
    d. Existentialists
    e. Environmentalists

15 On the basis of efficiency standards, which of the following can justify subsidizing the development of a hydrogen infrastructure?
    a. The funding for hydrogen research has been increasing in the last few years.
    b. The hydrogen sector has a strong lobbying presence in Washington D.C.
    c. Switching to hydrogen reduces our dependence to foreign oil and is therefore a positive externality.
    d. Hydrogen is a necessary component for fuel cell operation.
    e. All of the above.

III. True or False

1. A competitive market is Pareto efficient.

2. The main drawback of traditional command-and-control regulation is that it is not economically efficient.

3. Incentive-based regulation is superior over command-and-control regulation because it benefits society by generating the same level of pollution at a lower cost.

4. A Pareto efficient market is one that assures the seller is better off than the buyer as a result of a trade.

5. Zero pollution is fundamentally in-compatible with an efficient market.

6. Environmental protection necessarily brings about a certain number of job losses.

7. In the last few years, the majority of US job losses have been due to increased competition from low wage foreign workers.

8. An environmental ethicist considers the environment valuable because it is the source of happiness to him and others.

9. An example of socially efficient practices is a subsidy to educational systems, because it promotes literacy which is a positive externality.

10. The environment of the Former Soviet Union was considerably cleaner than that of western countries because firms had no profit motives to produce pollution.

IV. Fill-in the Blanks

1. According to the __________ theorem, it does not make any economic difference who pays for the pollution, the polluters or the victims.

2. ____________ advocates disagree with the Coase hypothesis and argue that polluters, not victims, should pay for the cleanup.

3. The____________see the value of the environment only as long as it brings them happiness.

4. Polling people to find their willingness to pay for a product is called _________________.

5. The environment is a _________ good and is therefore subject to the same law of ________ ______________ as other goods.

6. The enjoyment one gets from a cleaner beach, enjoying water sports, and a greener park is examples of the ______________ benefits of environmental cleanup.

7. In the United States environmental regulations are usually mandated by _______________.

8. The agency responsible for implementing US laws on environmental protection is ______________.

9. Under command-and-control strategy, to control emissions, firms are required to deploy the _____________________.

10. An area near a source where there is an elevated concentration of pollution is called a __________.

A Blueprint for a Sustainable Future

CHAPTER 17

Sustainable development is development that meets the needs of the present without compromising the ability of future generations to meet their own needs ~ The World Commission on Environment and Development (1987)

We started this book with the views expressed in The Limit to Growth, which painted a pessimistic picture of the world if population and energy consumption were to continue their exponential growth. We would like to conclude the book by giving a sharply contrasting, optimistic view of how the future could be if we take the necessary actions to preserve the environment and our natural resources. This does not necessarily have to be a painful process, but can be accomplished utilizing small steps that enrich our lives as well as the lives of our children and grandchildren.

We will start this chapter by discussing what sustainability is, then offer recommendations on how to achieve it on personal, national, and global levels. All it takes to achieve these goals is an open mind, a little sacrifice, and a well thought-out pathway.

Sustainability

The technological advances of the last few decades have provided unprecedented opportunities for greater material wealth, better health care, and a more comfortable life for the great majority of people, but have also resulted in the accumulation of a tremendous amount of waste, depletion of a large fraction of our natural resources, and the acceleration of environmental degradation. This naturally leads to the question of how to continue with economic progress while keeping environmental damage to a minimum. Two approaches to sustainability are proposed by economists.

Neoclassical economists consider a sustainable environment to be that which provides an average person a future with a living standard that is better, or at least as good, as that of an average person today.[1] They maintain that our children will be collectively better off if the economy expands at a faster rate than resources are depleted. After all there are no guarantees that we will ever run out of natural resources. Neoclassical economists are *technological optimists*; they believe that a free-market economy has the built-in mechanisms that would prevent ecological collapse, by promoting technologies that will help overcome resource scarcity and environmental damage. They cite increases in food production, discoveries of new energy reserves, and better and more efficient manufacturing technologies of the last century as proof that the overall standard of living around the globe

[1]Neoclassical economics is a school of thought in which maximizing profit is the ultimate goal of any organization.

has improved. They also claim that the living standard will continue to improve in the future and conclude that we are living in a sustainable environment.

Ecological economists (or ecologists), on the other hand, consider an environment sustainable only if future generations can enjoy essentially the same resources and a similar environment as we do today.[2] Although technological advances can help prevent resource scarcity and environmental damages in the short term, they will ultimately be used to accelerate the ecological abuses. Unlike the neoclassical economists, ecologists see material goods as only useful in serving people to a point; therefore, economic well-being alone cannot compensate for a loss of natural resources or environmental degradation.

Question: Can we maintain a sustainable environment and still use fossil fuels and pollute the air?
Answer: Neoclassical economists do not see continued use of fossil fuels as necessarily bad, as increases in the rate of energy consumption increase economic output and help improve living standards. By allocating sufficient resources we can develop technologies that combat environmental damage and maintain a reasonable degree of sustainability. Ecological economists believe that sustainability is possible if benefits of fossil consumption outweighs the damages it causes to the environment. To be sustainable, the noted economist, Herman Daly, set three citerias[3]:

1. *For renewable resources,* the sustainable rate of use can be no greater than the rate of regeneration. For example, geothermal hot water and steam resources are used at a rate that is being regenerated by nature.
2. *For nonrenewable resources,* the sustainable rate of use can be no greater than the rate at which a renewable resource can be substituted for it. For example, photovoltaic and wind technologies allow equivalent amount of energy and at comparable prices to become available, once the nonrenewable resources are exhausted.
3. *For environmental pollution,* the sustainable rate of emissions can be no greater than the rate at which that pollution can be rendered harmless. For example, the volume of carbon dioxide emission balances the removal of other substances that there is no net contribution of green house gases to the environment.

Question: According to a study, from 1930 to 2000, the money value of the world industrial output grew by a factor of 14.[4] During the same period -- because of the population growth -- the average per

[2] Ecological economics is a branch of economists who see a direct link between the health of the ecosystem and that of human beings. Ecological economics is sometimes referred to as "Green Economics."
[3] Daly, H., "Toward Some Operational Principles of Sustainable Development," *Ecological Economics* 2, 1990.
[4] Meadows, D., et al., Limit to Growth: 30-Year Update, Chelsea Green Publishing, 2004.

capita output increased by a factor of 5. Has society's progress been sustainable?

Answer: Neoclassical economists see this data as an indication of an improved standard of living during that period and thus conclude that the progress was sustainable. The ecological economist, argues however, that if the population had been stable, the material standard of living would have grown by a factor of 14 also. Furthermore, although the cumulative wealth has been greater, the gap between the rich and poor nations has increased, and therefore we experienced an unsustained growth.

Among the first to quantify the impact of increasing consumption on the environment were Ehrlich and Holdren, who proposed a simple model commonly known as the IPAT equation.[5] According to this model, total environmental damage *(ecological footprint)* is determined by three factors: the number of people, consumption per capita, and the state of technology (damage done per unit of consumption). This is represented as:

$$I = P \cdot A \cdot T \qquad (17\text{-}1)$$

Impact = **P**opulation X **A**ffluence X **T**echnology

where:

- I is the impact on the environment resulting from consumption,
- P is the population,
- A is the degree of affluence measured as consumption per capita, and
- T is the technological factor that accounts for both the positive and negative impacts of technology on the environment.

Example 17-1: Using the IPAT model estimate what impact automobiles have on global warming.

Solution: Assuming 6.3 billion people live on earth, that there is roughly one car for every 7 persons, and that with the current state of technology, each car produces 5.4 tons of CO_2 per year, we can evaluate automobiles' impact as:

$$I = P \cdot A \cdot T$$

$$\frac{CO_2\,Emission}{year} = (6.3\ billion\ people)\left(\frac{1\ car}{7\ persons}\right)\left(\frac{5.4\ tons\ of\ CO_2}{car \cdot year}\right) = 4.86\ billion\ tons\ of\ CO_2/\,year$$

This accounts for roughly 22% of total carbon dioxide emission (and thus global warming).[6]

[5] Ehrlich, P. R., and Holdren, J., "Impact of Population Growth," *Science* 171, no. 3977, pp. 1212-1219, 1971.

[6] It is assumed that the world's total carbon dioxide emissions will remain constant at their 2002 level of 24.5 billion tons *(http://www.eia.doe.gov/pub/international/iealf/tableh1co2.xls)*.

Population

The first factor in the IPAT equation is population. The neoclassical economists see population as an asset and believe its growth can actually be a positive factor for development. Their argument is that, with the right environment and proper training, people can be equipped with the skills to produce more than they consume. In addition, having more children lowers the median age and helps to remedy the shortage of manpower which would otherwise be a major problem in sustaining economic and environmental health.

The world suffers from excessive use of resources by the West as much as the excessive population in India and elsewhere. -- Indira Gandhi, the Indian Prime Minister in her 1972 address to the United Nations.

Contrary to the neoclassical view, the ecologists consider population as the greatest drain on resources and the main cause of ecological degradation. This is no more evident when we consider that countries with the most population, China, India, Indonesia, and Brazil are among some of the poorest; indeed the same countries have some of the highest birth rates. Contrasting these with sparsely-populated countries such as Switzerland, Luxemborg, and Finland with the highest per capita in the world, shows that more population - means less, not more - wealth. [7] More people mean that the already scarce natural resources and social services must spread over a larger population.

This view was originally expressed by Thomas Malthus (1766-1834) who, in "*An Essay on the Principle of Population,*" warned that the exponential increase in population along with the linear increase in availability (consumption) of resources would inevitably lead to a massive number of poor dying from famine, disease, or war.[8] Based on theory, famine and war were two necessary ingredients of any development process, a view attacked vigorously by Marx.[9]

In the last few years, social scientists, ecologists, and others have proposed new approaches to slowing the rate of population growth. Some, like ecologist Raymond Cowles, suggest economic motivations such as taxing large families and giving cash incentives to those who choose to have fewer children.[10] Others, like Ken Boulding, take a more radical approach by proposing to introduce a "child credit," similar to the concept of a "pollution credit" introduced in the previous chapter.[11] According to this plan, every woman is allowed to have a set number of children. Those who opt for fewer children can sell their credit to those who wish to have more. Family planners can adjust the number of permits to assure that sustainability is maintained.

---

[7] Diamond, J., Collapse: How Societies Choose to Fail or Succeed, Penguin Group, USA, 2004.

[8] Malthus was unable to take into consideration technological advancements such as the improvement in the quality of seeds, pesticides, and fertilizers or the effects of mechanized production and more effective irrigation methods. In addition, the recent development of biotechnology promises the possibility of genetically modified crops that will increase production and decrease infestations. He also overlooked advancements in education and the impact of women entering the workforce, two factors that reduce population growth. Furthermore, population growth in several industrialized countries has ceased and still others have achieved negative growth. From the 1950's to 1980's, the Green Revolution era, food production increased more rapidly than population, however it has become apparent that the use of land was excessive and the rate of crop yield has dropped significantly in the last twenty years, falling below the rate of population growth. A copy of the original paper published in 1798 can be found at http://www.esp.org/books/malthus/population/malthus.pdf.

[9] Meek, R. L., . Marx and Engels on Malthus. London: Lawrence & Wishart, 1953.

[9] Garrett Hardin, Living without Limits: Ecology, Economics, and Population Taboos, Oxford University Press, 1993.

[10] Boulding, K., Collected Papers, Vol II., Foreword by T. R. Malthus, Population First Essay, Colorado Associated University Press, 1971.

Betting on Sustainability

FYI ...

In 1980, the famous neoclassical economist Julian Simon challenged his colleague Paul Ehrlich, a noted ecological scientist, to put his money where his mouth was by betting that the prices of five metals would decrease in the next decade. According to Simon, the world will never run out of any of its natural resources, including energy. By 1990, the price of the basket of metals had gone down, partly because of recession and partly because of the rapid rise in the price of oil. The lower price of metals was meant to indicate that there are no shortages of these metals.

Having won the earlier bet, in 1995, Simon issued a new challenge in which he predicted that, in contrast to the doomsayers' scenario, every trend pertaining to human welfare would get better. Rather than accepting the new challenge, Paul Ehrlich and another colleague came with their own list of 15 global indicators that, they claimed, would get worse in the decade that followed.[i] The list included such items as the increase in concentration of several global warming gases in the atmosphere, the decrease in area of fertile cropland per person, fisheries, plants and animals, an increase in deaths due to AIDS, and an increase in the gap between the richest 10% of population and the poorest 10% of the population. Simon refused to wager citing that he would bet only on direct measures of human welfare, such as median income and life expectancy. Ehrlich and his colleague did not agree to the change, arguing that income and life expectancy are not good indicators of environmental sustainability, and therefore only those negative environmental or social trends can truly represent whether the environment is sustainable or not.

Although nobody agreed on the criteria most suitable to measure sustainability, the offer brought about public awareness of environmental issues and countered the claims made by Simon.

[i] *Ehrlich, P, and Ehrlich, A., The Betrayal of Science and Reason, Island Press, pp 100-104, 1996.*

Depending on how they see the population threat, different countries have opted for different coping strategies. The Vatican and a few Latin American countries do not consider over-population to be a problem and are openly opposed to abortion and any type of birth control. In contrast, Europeans, through education and by making birth control (including abortion) readily available, have succeeded in achieving a near zero population growth. Americans are divided on the question of abortion and women's right to choose. China encourages late marriages and late child bearing by providing financial subsidies, longer maternity leaves, and better housing to couples with fewer children.[12] Several developing countries have followed China's lead by providing public subsidies for housing, health care, and insurance to families with fewer children. During the last decade, Russia actually had a negative growth rate, mainly because of a large number of emigrants leaving for other countries.

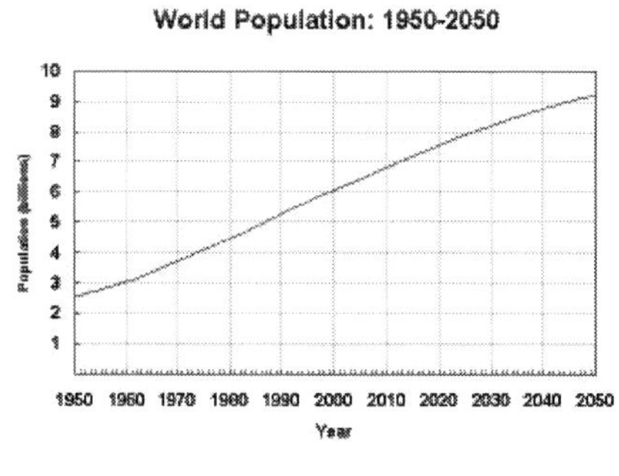

Figure 17-1
Past, present, and projected population in developed and developing countries.
Source: U.S. Census Bureau, International Data Base, April 2005.

As it stands, many developed countries have succeeded in stopping, or at least slowing, the growth rate of their populations. Unfortunately, population continues to increase in many poor and developing countries, and is likely to continue until it reaches equilibrium by mid-century (See Figure 17-1). As data indicates, the world population has steadily increased throughout history in a manner that exceeds simple exponential growth, as noted by shorter and shorter doubling times.[13] It is expected that the world population will continue to increase to 7.2 billion by 2015

[12] Carnell, B., "China's One-Child Policy," 1977 *(http://www.overpopulation.com/one_child.html).*

[13] Although the population has not always increased, when viewed over a period of many generations, it has shown steady growth. For example, the bubonic plague ("black death") pandemic of 1347-1351 killed 25 to 50 million, almost one in every three people in Europe. Nevertheless, the population recovered shortly thereafter and has been increasing ever since. The same can be said of the temporary decreases in population resulting from major wars, famine, or even a nuclear holocaust.

– with 95% of its increase in developing countries – before it stabilizes to 10 billion at around 2050.[14] Table 17-1 shows the expected number of people living in different parts of the world by 2050.

We cannot talk about population control without looking at immigration. It has been practiced throughout the centuries, sometimes by those in search of better economic opportunities, sometimes to escape political persecution, and at times to satisfy curiosity. In any instance, immigration has promoted diversity, has caused better mutual understanding among people, and has made the world seem ever smaller. In the twentieth century, faster and more convenient means of transportation, mainly airplanes and ships, have accelerated this pattern. Oftentimes, when cheaper labor is sought, immigration has been promoted, whereas at other times economic problems at home have made foreign workers and expatriates unwelcome. What is clear is that, unless there is a more equitable distribution of wealth, a greater effort in educating the public, and better ecological policies in place, immigration from developing to developed countries will continue to be a problem.

Table 17-1. Population growth rate and doubling time in 2001

Region	*Population in 2001*	*Growth %*	*Doubling time in years*	*Projected Population in 2050*
Africa	818 million	2.4	29	1.8 billion
Asia	3.72 billion	1.4	50	5.2 billion
North America	316 million	0.5	140	452 million
Latin America	525 million	1.7	41	814 million
Europe	727 million	-0.1	-	662 million
Oceania	31 million	1.1	64	46 million
World	6.14 billion	1.3	54	9.0 billion

Source: World Resource Institute (http://www.wri.org).

Affluence

The second factor appearing in the IPAT model is affluence. Affluence can be regarded as the increase in material gain and is usually measured as the per capita consumption of a certain good or a market activity. Neoclassical economists see more consumption and higher GDPs (gross domestic products) as indications of greater wealth, a healthier economy, and a better standard of living. The ecological footprint as a result of higher consumption is largely ignored. Consequently, these economists object to reducing consumption on the grounds that it slows down economic expansion and results in loss of jobs. Ecologists, however, disagree with this assertion on two accounts. First, material gain can bring happiness only up to a point, beyond which there is no correlation between the two. Second, higher GDP per capita is not a sign of wealth, as it includes costs associated with pollution and other environmental damage. If it were, a country could exhaust all its mineral

[14] *"Global Trends 2015: A Dialogue About the Future With Non-government Experts"*, Centeral Intelligence Agency, Report *GT-2015, 2000 (http://www.cia.gov/cia/reports/globaltrends2015).*

resources, cut down all its trees, erode its soil, pollute its aquifers, and kill all its wildlife and fisheries, while its economy continued to grow.[15]

Figure 17-2 shows the per capita energy consumption versus the GDP for several countries. Although correlation between GDP and energy consumption per capita was high in the past, the higher efficiencies achieved by newer technologies along with some conservation measures and the shift toward e-commerce has decoupled GDP from energy consumption for many industrial countries.

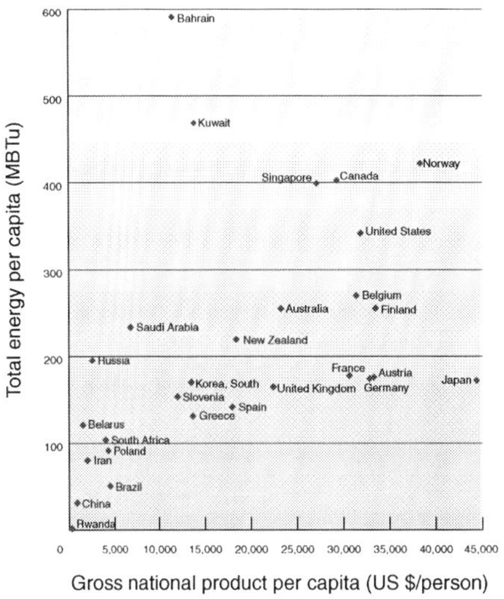

Figure 17-2
Energy consumption vs. GDP for several countries.
Data compiled from the World Bank.

The data from the General Accounting Office (GAO) indicates that the US GDP has continuously grown by 2.5% per year since 1973, reaching 11.7 trillion dollars by 2004.[16] It is also estimated that up to a quarter of the US GDP is wasted on nonproductive human activities (lost productivity due to traffic congestion, absenteeism, and costs associated with highway accidents, property damage, environmental cleanup, and maintaining law and order) where no value is generated. Table 17-2 gives the breakdown of various nonproductive expenses in the United States. In addition, many of the social costs associated with climate change, distress, illnesses, and substance abuse have not been included, but may have to be borne out by future generations in terms of increases in crime rate, mental depression, unemployment, and other social ills. If we subtract the cost of wasteful activities from the GDP, we might be surprised to find that the economy may not have grown at all!

Table 17-2. Nonproductive expenses in the US

Activity	*How?*	*Billion $*
Highway accidents	Health care, repairs, loss of productivity, police, and judicial systems	150 100
Highway congestion	Loss of productivity	1,000
Driving (other)	Road repairs, pollution	400
Diet	Obesity, heart disease, substance abuse	450
Crime	Health, police, and prison	500
Cleanup of nuclear material	Extraction, processing, and disposal, nuclear weapons	300

Source: Hawken, P., Lovins, A., and Lovins, H., Natural Capitalism: Creating the Next Industrial Revolution, Rocky Mountain Institute, pp. 57, 1999.

Question: In Chapter 15, in addition to the GDP, we defined another indicator - the net national welfare (NNW) - to gauge the state of the economy of a given country. How can we use these indicators to measure the degree of sustainability?

Answer: If both the GDP and NNW grow from one year to the next, no matter whether you consider yourself a neoclassical or an ecologist,

[14] Even if monetary gains were the only indicator of affluence, higher income (GDP per capita) does not signify wealth because these values are calculated based on a single currency, usually $US, and cannot be translated to the real purchasing power of a citizen in his own country. In response to these problems economists have devised an alternative measure known as Purchasing Power Parity (PPP), which tracks the cost of a basket of traded and non-traded goods and services across countries.

[16] Bureau of Economic Accounting, US Department of Commerce (*http://www.bea.gov*).

you will see the environment as sustainable. If GDP continues to grow while NNW falls, the environment is sustainable according to views held by neoclassicals, but not to ecologists. If both the GDP and NNW decline, then we have a non-sustainable environment. The major problem with the NNW indicator is that it is not an easy way to determine the external costs and gains resulting from non-market goods, so this type of calculation is subject to interpretation and therefore highly subjective.

Effects of Affluence on Waste and Environmental Damage

The affluent lifestyle enjoyed in many industrial countries, in particular the United States, comes with a price: an ever-increasing volume of waste and a rapidly degrading environment. Waste is not only a result of inefficient material and energy use, but also a consequence of our unquenchable appetite for over-consumption, due to consuming products that are either unnecessary (such as disposable products, electric toothbrushes, motorized lawn mowers, etc.) or excessive (such as large cars, excess packaging, oversized air conditioners, super-sized food portions, etc.).

Waste problems in the United States became particularly severe following the Second World War, where much of its technological capability was redirected towards manufacturing consumer goods, newer drugs, higher quality cosmetics, and more potent pesticides, fertilizers, solvents, and lubricants. These products made life easier for most, but also resulted in numerous health problems and deterioration of the environment. As the economy became stronger and the average purchasing power of American families grew, so did the rate of consumption and the volume of waste. Today, according to EPA estimates, over 250 million tons of toxic waste are produced every year spread over about 50,000 toxic waste sites in the United States. Another 12 million tons of pesticides are used on crops, forests, and lakes.

The cleanup and safe disposal of toxic waste has been debated between those who promote building bigger and bigger landfills and those who prefer incineration. Unfortunately, each method has certain drawbacks; landfills, even those lined with multiple layers of clay and nonporous plastics, are largely ineffective at keeping toxins out of the surrounding soil. Incineration produces significant amounts of air pollutants and is suitable only for organic matters.

The best way to combat waste is to remove incentives for being wasteful and increase material and energy efficiency through better designs, more efficient utilization of materials, increasing product durability, and by recycling and reusing the scraps. Unfortunately, our economic system favors waste while discouraging efficiency. For example, commission fees charged by building contractors, architects, and engineering firms are based on the total cost of the project, and therefore the tendency is to oversize everything. The same can be said of the OEMs who furnish equipment and machinery. The sensitive alternative is to link prices and fees to efficiency rather than to a percentage of total cost. In other words, designers and engineers should be rewarded for what they save, not what they spend.

Technology

The third factor impacting the environmental damage is technology.[17] Technology can influence the environment both positively and negatively. On one hand, technology makes it possible to design machines that have better efficiencies and make elaborate tasks much easier to perform. On the other hand, technology promotes automation, usually at the expense of faster resource depletion and additional waste. Consider, as an example, the impact of technology on automobile use. Automobiles have given us mobility by allowing us to move around faster but have also limited our physical activity, which affects our health and general wellbeing. Increased efficiency, measured as miles traveled per gallon of fuel, has not helped either, as it has only caused people to buy larger, heavier cars and travel longer distances. Fortunately, technological advances are on the horizon which, if used wisely, give us the promise of a sustainable future, affecting everything we build and consume from microchips to potato chips. The most promising innovations are in the fields of manufacturing, transportation, agriculture, and energy.

Consumerism and the Rat Race

FYI ...

The increased pattern of consumerism in western cultures is mainly the result of inherent pressure to adapt our lifestyles to a particular social norm. The motive for this behavior might be the desire of a particular person to be similar to others (bandwagon effect), or to distinguish himself from others (snob effect), or appear to be of a higher social status than others (the Veblen effect). An example of the **bandwagon effect** is adhering to a brand name which has become popular with a certain age group. An example of the **snob effect** is a punk hairstyle, only to stress a distinction from the average person. The **Veblen effect**[i] refers to consumption of an expensive item like a Lexus (instead of a Toyota). Although each of these social behaviors can be satisfying in the short term for certain groups of consumers, in the long term and for the society as a whole, more material wealth will eventually fail to translate into happiness. If everyone races to get ahead of others by working harder, making more money, or assimilating to the social norm, then the differences in social status will eventually disappear. As a result, everybody must run faster to maintain the edge and the society as a whole becomes less satisfied. For example, since there is a limited amount of space for houses overlooking oceans, then the prices of these houses increase and one has to work harder to be able to achieve similar social status or comfort they had with much less work and lower income.

Even by age 5, my daughter had already chosen her favorite store and brand name.

[i] Named after the American sociologist and economist, Thorstein Veblen (1857-1929).

[17] Much of the material in this section has been adapted from Natural Capitalism: Creating the Next Industrial Revolution, Rocky Mountain Institute, 1999. (See list of additional reading at the end of this chapter).

Manufacturing

Manufacturing is undergoing major changes in reducing both material and energy consumption. For example, optical fibers have forty times the carrying capacity of copper, which can potentially save an enormous amount of materials and energy during extraction, manufacturing, and implementation of conventional copper cables. The same can be said of new composites that can replace heavy steels. Molded plastic parts produce far less waste than metals, potentially saving a great amount of material.

More versatile design tools and better manufacturing techniques allow for the design of better and more efficient devices without sacrificing their functionality. The manufacturing practices prevalent in the past have been based mostly on the one-time use of materials, creating a large amount of waste that must be disposed. *Just-in-time manufacturing* and *made-to-order manufacturing* are two approaches being popularized in many manufacturing sectors; they use the exact amount of materials that are being delivered at the exact time they are needed. This reduces the volume of the scrap materials, eliminates the need for on-site storage, and minimizes the risk of spoilage and environmental damage.

New CAD/CAM (*computer-aided design* and *computer-aided manufacturing*) tools are available that help design products that weigh less and are stronger, while consuming less energy and material. This is achieved primarily by making them less bulky while assuring thermal and mechanical stresses in critical areas are kept below their safe limits. Rapid prototyping is a relatively new concept that uses CAD software to manufacture three-dimensional prototypes of a device at only a fraction of the cost, energy, and material. The *design for manufacturing* is based on minimizing material and energy use during the entire chain of product development from material extraction, to processing, to manufacturing, to use, to reuse, and to disposal. Making products lighter and more durable is obviously the best choice. Proper reuse of material and recycling waste will play a large part in creating a sustainable society. For example, we can re-pulp paper products into cardboard, shred them and use them as insulation, or incinerate them to generate heat. Clothing and furniture can be reused by more needy individuals, recycled into new products, or burned to generate energy. Glasses can be refilled over and over again or be crushed and made into new products. Metals can be re-melted and reused to manufacture new products.

> **Question:** A beer canning company intends to reduce material waste by redesigning the shape of their beer cans. What do you suggest?
> **Answer:** For the same volume, beer cans that are fatter and shorter have the lowest total surface area and a smaller mass. It is left as an exercise to show that minimum surface area is achieved when a can's height is equal to its diameter. Another factor that can affect the choice

is the practicality of the design.

Transportation

As we discussed in Chapter 14, transportation systems are changing with major impacts on the world energy supplies. These include a complete redesign of cars using lighter and stronger materials, more efficient hybrid and fuel cell cars, maglev rail, and more aerodynamic sea and air vehicles. Development of low-cost public transportation systems, faster communications that allows efficient home offices, and new architectural concepts to build modern residential communities and financial centers will greatly reduce the driving needs.

Agriculture

Scientists have already developed bacteria that can absorb nutrients more effectively and a natural antifungal agent that protects wheat and other seeds staves off frost development in plants. It is predicted that by the end of this century making better farming and irrigation systems, developing hybrid seeds, producing food from municipal waste, planting trees and growing food in unfriendly climates and salty soils, and hydroponics (a method of growing plants without soil by providing food and water directly to the roots of the plant) will alleviate the food shortage in many poor countries[18], eliminating the need to transport food across continents.[19] Genetic engineering will also play an important role in accelerating the growth of livestock and modification of crops. Genetic engineering enables scientists to create plants, animals, and micro-organisms by manipulating genes in ways that do not occur naturally. The approach, however, is highly controversial as there is not adequate scientific understanding of the impact on the environment and human health (Figure 17-3).[20]

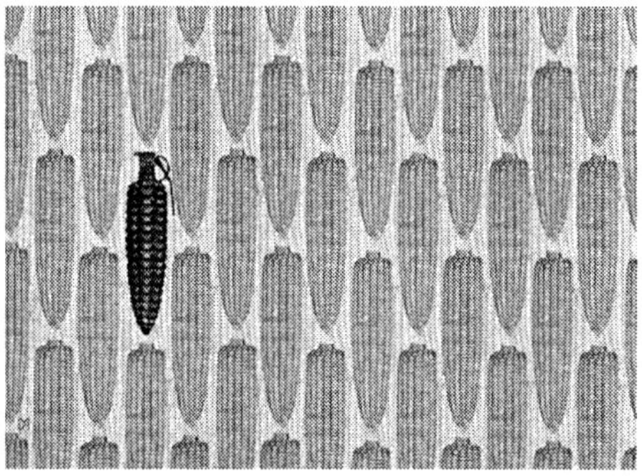

Figure 17-3
Corn grenade: the winning image from the Greenpeace Seeds of Trouble competition, 2002.
Image courtesy: Greenpeace International.

Energy

The energy industry will be affected in two fundamental ways: by finding new and alternative sources of energy and by reducing energy need through more energy-efficient designs and practices.

Within the next couple of decades, innovative processes that convert coal, oil shale, and tar sands to gaseous and liquid fuels, and possibly, breeder technology will help supply our immediate energy needs. Advances in the field of synthetic biology will enable scientists to produce an artificial genome and re-engineer new organisms that efficiently, and at much-reduced cost, turn plant fiber into ethanol and other biofuels. Scientists at the University of California, Riverside have discovered new strains of petroleum-degrading bacteria that thrive with no light or water, breathe

[18] Kahn, H., et al., The next 200 years: A scenario for America and the World, Morrow Publishers, New York, 1976.
[19] Today in the United States, food travels an average of 1300 miles from farm to plate; often the energy needed to produce it is many times the energy contained in the food itself.
[20] Greenpeace International (*http://www.greenpeace.org/international/campaigns/genetic-engineering*).

carbon dioxide instead of oxygen, and break down toxic petroleum compounds to give off methane gas.[21] These properties can pave the way to develop new and cheap class of biofuels, not to mention cleaning up oil and other hydrocarbon spills.

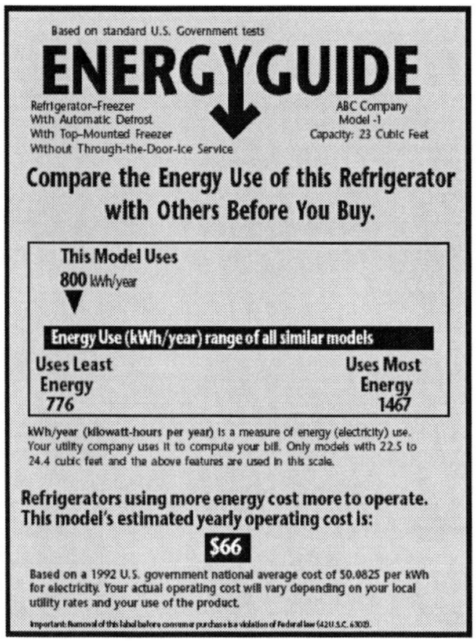

Figure 17-4
Efficient appliances are authorized by the DoE with special Energy (Green Star) labels. These models can save as much as 30% in energy.

In the near future we will witness fuel cells become ready for commercialization for providing premium electricity at efficiencies of 50-60% for building applications. The waste heat provided by high-temperature fuel cells can be recovered to produce steam that can be used directly in industrial applications or expanded in a gas turbine to produce more electricity. The waste heat from low-temperature fuel cells can be used for cooking and is ideal for hot water heating and space air conditioning. Further advances in solar, wind, wave, and ocean thermal technologies are expected to bring down prices, making renewable energy competitive or cheaper than fossil fuels or nuclear fission, alleviating many of our concerns about energy and its effects on the environment. Fusion technology, although currently not available, may prove to be our ultimate source of clean and inexpensive energy for many centuries.

In the relatively near future, more energy-efficient devices are expected to flood the market (Figure 17-4). These devices operate by automatically adjusting to consume the least amount of energy required for the task they are designed to perform. Washing machines will be equipped with fuzzy controllers[22] that adjust the amount of detergents, water, etc., depending on how dirty clothes are. Dryers, microwave ovens, air conditioners, refrigerators, and vacuum cleaners are being designed that work based on similar principles.

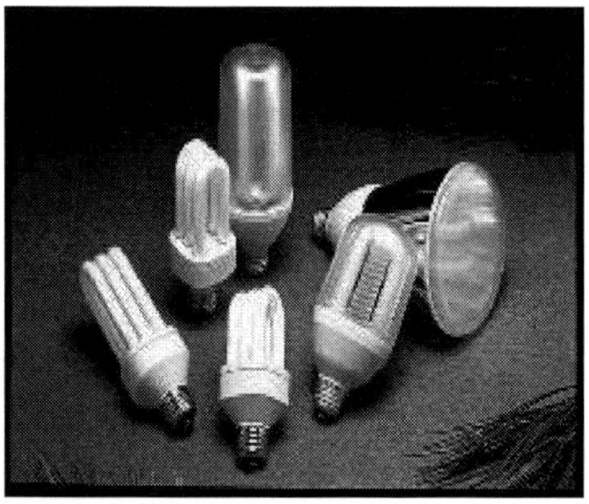

Figure 17-5
Compact fluorescent lamps.
(National Renewable Energy Laboratory, US Department of Energy).

Lighting and illumination is making great progress toward energy efficiency. Incandescent lamps can be replaced with compact fluorescent and tri-chrome lamps with significantly higher efficiencies. An **incandescent** bulb contains a tungsten coil filament that turns red when electricity flows in a vacuum. Only 5% of the electricity is converted to light. The remaining 95% is used for heating the element. On the other hand, **fluorescent** lights operate by exciting the mercury gas through electric discharge to emit ultraviolet light, which in turn converts to visible light by phosphorous coating of the tube. Fluorescent lights are four times more efficient than incandescent lights. Compact fluorescent lights and **tri-chrome** lamps with rare-earth elements emitting in the visible range that increase efficiency even further are becoming available.[23]

New architectural concepts allow construction of "green" or ecologically-friendly smart buildings that incorporate many energy-saving measures at lower costs without sacrificing convenience or safety. For example,

[21] Kim, J. and Crowley, D., "Microbial Diversity in Natural Asphalts of the Rancho La Brea Tar Pits," *Applied and Environmental Microbiology*, April 6, 2007.
[22] Unlike traditional on/off controllers, fuzzy controllers work by assigning a weighting function that can vary by any number between zero and one.
[23] Schuman, J., et al., "Technology Reviews: Lighting Systems," LBL-33200, September 1992.

passive heating and cooling systems, solar walls and roofs, solar collectors deploying newer phase change materials, window overhangs, improved insulation, double-glazed air-tight glass, low-emissivity window coatings, windows with aerogel glazing, and better sensors are some of the technologies that are finding their way into average households.[24] Photovoltaic cells can eventually be shaped to directly replace walls and windows, or rolled out like a blanket, or layered like paint onto curved surfaces. Wind turbines can be designed as an integral part of future buildings, making these buildings a net exporter of energy (See Figure 3-15). Future communities can be designed to eliminate repeat trips and reduce commuting, both of which are essential for designing a sustainable future.

A Blueprint for a Sustainable Future

Throughout the ages civilizations have risen and crumbled, many times as a result of their own mistakes and those of their forefathers. Many of these mistakes were, of course, inevitable, because they lacked the necessary technological know-how, and their immediate survival depended on the unsustainable exploitation of their natural resources. In addition, the population was relatively small and natural resources were abundant, so they moved to more fertile lands and friendlier environments and set up new centers of civilization. Because of the availability of a vast amount of resources and the relatively small population, the ecology remained to a great extent sustainable. It has only been in the last century that accelerated growth in technological innovation, along with exponential growth in population, the rapid pace of economic expansion, and a lack of respect for the environment, have brought about conditions that could put us at risk for ecological disaster. As Diamond asserts, the main problem is not the world population itself - although it has been named most often - but the total population impact. If all people living in the third world were to raise their standard of living and adopt the life styles of inhabitants in the first world, we would need twelve times the resources we are consuming and inflict twelve times the environmental damage we are causing today, something the earth can certainly not sustain.[7]

In an excellent book called Natural Capitalism,[25] published by the Rocky Mountain Institute (a non-profit organization dedicated to promoting sustainable alternatives) Hawken and Lovins propose a new system in which **resource productivity** rather than human productivity drives the new economy. By increasing resource productivity, the same output is achieved but less material and energy are consumed. The authors question the widely accepted assumption that what is good for us must also be good for the world, proposing instead a radically new and contrasting paradigm which proclaims what is good for the world will be good for us. To make our lives better we must strive to make the world better, and to make the

[24] Rosenfeld, H., et al. "Technologies to Reduce Carbon Dioxide Emissions in the Next Decade," *Physics Today*, Nov. 2000, pp 29-34.
[25] Hawkens, Paul, Lovins, A, and Lovins, L. H., Natural Capitalism: Creating the Next Industrial Revolution, Rocky Mountain Institute, 1999.

world better we must first understand it more. The new economy will contrast modern capitalism in that economic prosperity is not measured by gains in material well-being, but by natural capital that includes all the familiar resources used by humans: water, minerals, oil, trees, fish, air, soil, etc., and encompasses all living organisms, humans, animals, plants, wetlands, oceans, and marshlands.[26] Furthermore, unlike "industrial capitalism" that considers only human financial and manufactured capitals as important in creating wealth, "natural capitalism" also keeps stock of **natural capital** (natural resources, living systems, and ecosystems) and includes them in determining wealth. Whether we follow a sustainable or a non-sustainable future depends largely on how we preserve our biological diversity, how much we increase resource productivity, how much of the natural resources we exploit, and how much we leave for our children. According to many environmentalists and ecological economists we have been, and unfortunately still are, continuing many practices that are unsustainable.

To assure sustainability, policy makers, economists, and educators have much work to do. Today's economy is driven by maximizing profit with little regard to natural capital. Many policies are in place that tax innovation and resource productivity, but encourage waste and inefficiency. For example, we continue to support fishing fleets, even though existing fishing capacity far exceeds the sustainable yield of oceanic fisheries. Subsidizing forest industries by building roads for transporting wood has increased logging. Similarly, subsidizing agricultural farms to cover irrigation and pesticide costs has resulted in large-scale land abuse and salinization.

As we discussed earlier, sustainability cannot be achieved unless poverty

Sustainability: The Facts

Did You Know That ...?

- To travel one kilometer by bike requires approximately 5-15 watt-hours (Wh) of energy, while the same distance requires 15-20 Wh by foot, 30-40 Wh by train, and over 400 Wh in a singly occupied car.[i]
- Air-, water-, and food-borne diseases are spreading.
- Almost one half of the planet's tropical forests have been destroyed or severely degraded.
- Most of our medicines are extracted from plants, many of which are now endangered.
- Ninety percent of all large fish have disappeared in the past half century.
- 70% of all people do not have access to fresh water and 20,000 die each day as a result of contaminated water.
- Because of the climate change, it is expected that over one million species will be lost in the coming 50 years.
- At the current rate, world population is increasing by 80 million people every year.
- Since the mid-twentieth century, the population has doubled, while at the same time grain consumption has tripled, seafood harvest has increased four times, paper use has increased six-fold, and consumption of fossil fuels has increased by a factor of four.
- 16% of the world's population is consuming some 80% of its natural resources.
- The world's 500 largest corporations control 25% of the world's economic output.

[i] *Bouwman, M. E, An environmental assessment of the bicycle and other transport systems, University of Groningen, Netherlands, 2000 (http://www.velomondial.net/velomondiall2000/pdf/bouwman.pdf)*

[26] Swamy, M, and Kumara, R., "Does Non--Inclusion of Intangible Asset Values Lead to Distortion of Financial Statements and Mislead Judicious Financial Decision Making?," *Journal Of Financial Management & Analysis* Vol 17, Iss 1, Jan-Jun 2004.

is eliminated. Many developing countries have been borrowing heavily to subsidize basic necessities and much of their economic development needs. The total external debt of these countries has grown tenfold during the past three decades, from $245 billion in 1970 to more than $2.4 trillion in 2005.[27] These countries' poverty cannot be eliminated unless there is a sincere effort and a substantial commitment from richer countries in the form of debt relief, transfer of clean technology, international development aids, and reduction of trade barriers on goods imported from poorer countries.

Educators also have a great degree of responsibility for introducing their students and the public to issues related to energy and the environment, and for advising them on policies and practices that discourage wastes and promote sustainability. Unfortunately, even many consumers who claim to favor environmentalism are not willing to pay more, accept the loss of some comfort, or forfeit minor customary features to buy more environmentally friendly products. The number is growing, however, as the sales of consumer products that carry labels identifying them as environmentally-friendly products by such internationally-recognized organizations such as *Forest Stewardship Council* (FSC), *Marine Stewardship Council* (MSC), and *Leadership in Energy and Environmental Design* (LEED) is gradually increasing.[7] We believe consumer attitudes will significantly change as health effects associated with increasing pollution, global warming, and depletion of natural resources become more apparent.

Simplicity, simplicity, simplicity ~ Henry David Thoreau (1817-1862)

Below are some guidelines for what we can do as small steps to move toward a sustainable future or at least reduce the environmental footprint and slow down the damage we are already causing to the environment.

On the Personal Level

Simplify your life – Use simpler machines. Use tools that require more of your muscle power, like bicycles, manual hacksaws, etc., which are also good for your health.

Find a hobby – Participate in leisurely activities like music, sports, arts and crafts. They tend to enrich your life and are relatively non-consuming and non-polluting.

Minimize your energy losses – Repair leaky faucets and water pipes. Replace incandescent lights with fluorescent lights, insulate walls and roofs, and replace old appliances with energy-saving appliances.

Use ecologically friendly (green) materials – Consider using natural or recycled materials when buying new homes or remodeling your existing house. Buy only materials that meet environmental and safety

[27] The CIA Fact Book, (*http://www.cia.gov*).

standards.

Change your diet – Eat more vegetables and less meat. Consume organic instead of processed foods.

Reduce, reuse, and recycle – Reduce material consumption by extending products' useful lives. Don't buy new products unless you absolutely need them; share or repair whenever possible. When it is necessary, purchase only durable and energy-efficient devices. Replace disposable with reusable products. Manufacturers do not pay your energy bills. In fact, to reduce costs and expand their market base, they have every incentive to make their devices less efficient and less durable. Reuse items (for other applications) instead of buying new ones; upgrade to a newer model and give the old one to a needier owner. Recycle only if the options for repair, reuse, and upgrade are no longer available.

Educate yourself – Learn about people and their cultures, travel to foreign countries, and explore the beauty of their land, language, religion, and cultural heritage.

On the Community and National Levels

Promote sustainability – Restore the environment to the condition of equilibrium by planting new trees, cropping native species, protecting wildlife, and enhancing fisheries. Reward efficiency and not waste by cutting subsidies to polluting industries, and giving rebates to manufacturers, builders, and those who save material and energy.

Be proactive – Don't wait until the environment is so stressed that the damage done has become irreversible; look for signs of stress, and act to resolve the problem in a timely fashion.

Empower individuals – Encourage individuals to practice conservation. For example, city planners can build convenient bike routes and transit authorities can subsidize bikes costs. Companies can institute flex-time, allow employees to work from home, build shower facilities for bikers, and partially subsidize rent for employees who live nearby.

Deploy empowering technologies – These help increase productivity while preserving natural resources. Examples are those that improve irrigation, reduce pest damage, and encourage soil conservation and enrichment. They also accelerate research and development of renewable energy sources.

Revamp manufacturing practices – This can minimize energy and material use during the entire life of the product. Design for durability, reusability, recyclability, and for manufacturing practices that promote energy efficiency and allow for easy repair, dismantling, reuse of scrap materials, and disposal of wastes.

Eliminate subsidies – Subsidies to industries that promote inefficient and unproductive use of materials and energy and agricultural companies that produce products that degrade soil fertility, pollute the environment, or waste a large amount of water should be eliminated. Internalize the externalities so that the price of a product reflects the full cost including health costs and costs of cleaning the environment. The only exception is to subsidize those industries and products that are beneficial to society, but need help to overcome initial market barriers.

Reform tax laws – Reduce or eliminate income tax and substitute it with environmental and carbon auctions and carbon taxes. Carbon taxes are levied against individuals and corporations in proportion to the amount of petroleum they consume. In carbon auction, companies bid for the right to pollute. This makes sense because labor earns income from productive work, whereas polluting corporations and energy inefficient industries profit by being wasteful and non-sustainable. Industries should be taxed on the amount of raw material and energy they use and the toxic pollutants they release into the atmosphere in order to cover health costs, environmental cleanup, and restoration of the environment to pre-release conditions. Higher taxes will increase cost to polluting companies and result in a loss in sales. This forces companies to make their products safer, more efficient, and produce less waste.[28] Proceeds should be reinvested towards projects that promote sustainability – for example, use the carbon auction and carbon tax to build hydrogen infrastructure. The proceeds from pollution tax can be redirected as subsidies to clean and make industries efficient, or used to spur research and development in resource productivity. Income from parking rentals and toll roads can be used to pay for efficient public transportation, and savings from dematerialization can foster more efficient manufacturing. The interest on car loans can be increased while decreasing mortgage interests for community housing. Those with too many children can be penalized; the proceeds can help improve education, provide better health care, and to build playgrounds for children.

Participate in local and national politics – Support politicians who advocate and promote realistic sustainable environmental policies and short–term goals that are consistent with long-term objectives. Doing so also reduces political tension and instability in the world, which is often caused over control of foreign natural resources.

Educate – Teach the public environmental and energy literacy, the need for proactive practices and social activism. Universities should integrate energy and environmental courses in the liberal arts and

[28] This point has been refuted by some economists who argue that for a purely efficient market, a shift from income tax to pollution tax will have a net negative impact on the economy .Others argue that the conclusions are made based on false assumptions and that the net effect is indeed positive. [See for example, Economics and the Environment, 3rd Ed. by E. S. Goodstein, John Wiley & Sons, Inc., 2002, pp. 170-173.]

engineering disciplines. This book is aimed as a small step toward this direction.

On the Global and International Levels

Support democratic governments – When these governments work to reduce poverty, support human rights, provide education, protect the environment, and promote sustainable economic development, they should be given support.

Empower the grass root activists and microenterprises – Give loans and technical assistance to millions of tiny family-owned businesses or microenterprises with only a few employees. These small businesses are more flexible, lack the bureaucracy, have strong ties to their local community, can bring social benefits at only a fraction of the cost, and help more effectively protect their immediate environment and thus global ecology.

Oppose war and military intervention - Securing the flow of oil and other natural resources by force is neither the answer to our short-term energy needs, nor is it in our long-term interest. Military interventions will undoubtedly antagonize people, encourage extremism, foment terrorism, and cause political instability that eventually reverberates throughout the world and daunt us in our home.

Support the elimination of economic inequality – Forgive debts to poorer nations, transfer sustainable technologies to developing nations, and support globalization efforts, but only as much as it reduces the gap between industrial and poor nations, helps maintain peace, and encourages disarmament. Help to eliminate poverty by providing free education, jobs, and basic health care.

We're not passengers on Spaceship Earth: We're the crew. We're not resident; we're citizen. The difference, in both cases, is responsibility. ~ Rusty Schweikart, Apollo astronaut (1946-)

Concluding Remarks

In this text we covered different aspects of energy and problems associated with the uncontrolled consumption of raw materials and nonrenewable energy resources. Thus far, we have been abusing the environment and exploiting our natural resources in an unsustainable way. The choice to continue on this path or to take steps in modifying our living habits to follow a sustainable path is up to us. If we use history as evidence, unless we make a deliberate effort in reaching sustainability, we will have to fight many more wars over the control of natural resources -- land, water, mineral, and especially oil. What is clear is that the longer we wait to introduce meaningful changes in the way we are currently using our natural resources and treating our environment, the more difficult it will be to achieve a sustainable future. Technological advances can help sustainability for as long as the society as a whole makes a concerted effort to direct them toward this goal. Short of this commitment, technology will only serve to promote its implicit goal of unhindered growth at the expense of increasing stress on the environment and its eventual collapse. The major challenge of our times is how to improve the standard of living of the average person without putting undue pressure on the resources at our disposal.

The highly industrialized countries, the United States in particular, have the largest economies, the most access to technology, and vast intellectual and financial resources. They are also the biggest polluters and producers of wastes and have the moral responsibility to lead the world toward practices that enable our children to enjoy a sustainable future and a healthy environment for many generations to come.

Additional Information

Books

1. Hawkens, P., Lovins, A, and Lovins, L. H., Natural Capitalism: Creating the Next Industrial Revolution, Rocky Mountain Institute, 1999.
2. Meadows, D., Randers, J., and Meadows, D., Limit to Growth: 30-year Update, Chelsea Green Publishing, 2004.
3. Diamond, J., Collapse: How Societies Choose to Fail or Succeed, Penguin Group, USA, 2004.

Periodicals

1. *Journal of Political Ecology: Case Studies in History and Society,* JPE is produced at the Bureau of Applied Research in Anthropology, the University of Arizona Library, Tucson, Arizona. The journal covers research articles into the linkages between political economy and human environmental impact.
2. *World Watch Magazine* (*http://www.worldwatch.org*).

Non-Government Organizations and Websites

1. World Bank *(http://web.worldbank.org).*
2. United Nations Environment Program *(http://www.unep.org).*
3. Rocky Mountain Institute *(http://www.rmi.org).*
4. Greenpeace *(http://www.greenpeace.org).*
5. Green Seal *(http://www.greenseal.org).*
6. Nature Conservancy *(http://www.nature.org).*
7. The Sierra Club *(http://www.sierraclub.org).*
8. Friends of the Earth *(http://www.foe.co.uk).*
9. Women's Environment and Development Organization *(http://www.wedo.org).*
10. World Wide Fund *(http://www.panda.org).*
    Exercises

Exercises

I. Essay Questions:

1. How do neoclassical and ecological economists differ on what sustainability means? In what points do they agree?

2. What are various factors that limit world population?

3. Do the different views expressed by ecologists and economists represent their widely different value systems?

4. Why is GDP a bad measure of sustainability? List at least two problems. Does a better index of sustainability exist?

5. Compare the GDP and NNW data for your country and conclude whether consumption is on a sustainable path.

6. What does the word "development" mean to you? What do you mean by land development? Human development? Economic development? Are these concepts compatible with ecologically sustainable development? What is viable global development?

7. What is the significance of IPAT Equation? Define terms and give examples of how this equation is used.

8. Define "Sustainable Energy" and give some examples of sustainable and non-sustainable energy use.

9. What strategies governments and individuals could use to adapt to global climate change?

10. Visit the websites for the non-government organizations (NGO) listed above and make a summary of the mission and activities carried out by each organization:
    a. What are the mission, goals, and objectives of each organization?
    b. Where do their budgets come from?
    c. What have their major accomplishments been and how have they impacted world sustainability? Explain.
    d. Do any of these organizations appeal to you and your philosophy? Are you willing to perform some volunteer work or contribute financially?
    e. Do you know of any other international organization that has had significant success in helping stop environmental degradation?

II. Multiple Choice Questions:

1. What name is given to the global ecosystem?
    a. Atmosphere
    b. Biosphere
    c. Hydrosphere
    d. Lithosphere
    e. Mesosphere

2. Which of the following factors limits the ultimate population of the world?
    a. The size of the earth
    b. The area of the arable land
    c. The amount of available food
    d. Environmental factors such as global warming and pollution
    e. All of the above

3. Sustainability implies
    a. Using natural resources as slowly as possible
    b. Using only as much as is replaced by natural processes
    c. Not introducing new technology too quickly
    d. Discovering new resources to allow maximum economic growth
    e. All of the above

4. A sustainable society is not achievable unless
    a. There is public awareness of the fragility of the environment
    b. More citizens participate in political process.
    c. Individuals evaluate the impact of his/her actions on the world as a whole
    d. We consider the planet earth and not necessarily the country where we were born as our home
    e. All of the above

5. To make earth sustainable, we need to devise strategies that
   a. Keep human birth and death rates in balance
   b. Assure soil erosion does not exceed the natural rate of new soil formation
   c. Assure enough trees are planted to offset deforestation
   d. Use renewable energy at a rate that is slower than its sources are regenerated
   e. All of the above

6. Which type of light bulb has the highest efficiency?
   a. Halogen
   b. Incandescent
   c. Fluorescent
   d. All are equally efficient

7. The higher the efficiency of an electrical appliance,
   a. The faster it uses energy
   b. The more energy it uses
   c. The cheaper it is
   d. The shorter the payback period
   e. All of the above

8. According to the data published by the World Bank, what share of the world's income does the richest 20% of the population receive?
   a. Less than 10%
   b. 32%
   c. 55%
   d. 86%
   e. More than 95%

9. According to the data published by the World Bank, what share of the world's income does the poorest 20% of the population receive?
   a. About 1%
   b. 5%
   c. 10%
   d. 15%
   e. More than 15%

10. The best philosophy for dealing with the increasing volume of garbage is to
    a. Reduce material consumption as much as possible
    b. Recycle
    c. Reuse
    d. Increase the number of landfills
    e. Incinerate

11. Which of the following statements is true?
    a. Populations increase in the United States at a faster rate than in Europe.
    b. Population growth is higher in poorer countries.
    c. People with less education are likely to have more children.
    d. Africa has the largest percentage of population growth.
    e. All of the above.

12. The main reason why many US manufacturing firms relocate their plants to developing countries is because
    a. Immigration policies have practically shut off the flow of immigrant workers to the United States
    b. The environmental regulations are less stringent in developing countries
    c. Labor costs are lower in the developing countries
    d. a, b, and c
    e. b and c, but not a

II. True or False?

1. Roughly one quarter of the US GDP is wasted on nonproductive human activities.

2. Just-in-time manufacturing is the process by which a manufacturer can deliver their product without delay.

3. Design for manufacturing is the process by which material and energy consumed through the product life-cycle is minimized.

4. "Natural Capitalism" is the doctrine practiced in all capitalist countries.

5. Sustainable development requires new resources to be found as fast as they are consumed.

6. The neoclassical economists see population as an asset.

7. According to Lovins, human productivity, rather than resource productivity, will fuel the future economy.

8. The best approach for minimizing waste is recycling.

9. The IPAT equation states that the impact of technology on the environment will increase proportionally to the rate at which the technology grows.

10. The snob effect applies to those who live in Beverly Hills more than to those who live in the neighboring Hollywood.

III. Fill-in the Blanks

1. The largest number of humans that the earth can sustain indefinitely at the current rate of per capita consumption of natural resources is known as the earth's holding __________.

2. According to the World Resource Institute, by 2050 roughly one of every two persons will live in __________.

3. Being stylish by following latest fashion is one example of the ______________ effect.

4. The US produces roughly ____________ million tons of toxic wastes every year.

5. OEM is the acronym stands for the "Original ________________________."

6. Unlike __________ who consider population as the main cause of the environmental degradation, _______ economists see population as an asset and a necessary component for a sustainable economy.

7. The consumption of expensive material goods instead of the less expensive goods of similar utility is called _________________.

8. __________ can have either positive or negative impact on the environment.

9. Using CAD software to manufacture three-dimensional prototype of a device is called _________________.

10. ___________ lights are a lot more efficient than ___________ lights.

IV. Project I - Life Cycle Analysis: Electric versus Conventional Bikes

In this project you are asked to carry out a simple qualitative life-cycle (cradle-to-cradle) comparison of the energy expenditure between electric and conventional bicycles.

1. What assumptions are you making in carrying this comparison?
2. Which one consumes more energy per kilometer of travel? What form of energy is used by an electric bike? By an ordinary bike?
3. What are the differences if an individual pedals faster than an electric bike? If the rider travels twice as fast by an electric bike than a conventional bike?
4. Which one takes more energy to manufacture? Describe what each component does and what forms of energy it requires to operate.
5. What are the total energy costs to manufacture different batteries? Compare four types of batteries, Li-ion, NiCad, NiMH, and lead-acid. You can consult the paper by Rydh, C., "Life Cycle Inventory of Recycling Portable NiCad Batteries," *Resource Conservation, and Recycling*, March 2002, Vol 34, 289-309, to find energy costs to manufacture and ship various types of batteries. Which battery is the most energy-efficient?
6. What is the energy cost associated with the production of food necessary for delivery of muscle power to operate the conventional vehicle?
7. Does it make economical sense to buy an electric bicycle? In terms of energy efficiency? In terms of convenience, health benefits, etc.?

Project II - Biosphere II

In 1991, to simulate ecological processes on earth and asses the possibility of life within a closed system, eight scientists, four men, and four women, entered a tightly sealed glass and steel greenhouse structure named Biosphere II (Biosphere I refers to the earth itself). The structure, located near Oracle, Arizona, was 3.5 acres – the size of three football fields – and became home to the scientists for two years. The experiment was originally designed to last 100 years, with scientists rotating every few months or years. All air, water, and nutrients were recycled within the system, which was almost entirely isolated. Sunlight and electrical power were allowed into the dome, and every two weeks samples of soil, air, and other species were sent outside through an airlock for further tests and analysis.

The building housed a diversity of ecosystems or biomes including a desert, a tropical rainforest, a savanna, a marsh, a field for farming, and an ocean with coral reef. The scientists were accompanied by insects, pollinators, fish, reptiles, and mammals that were supposed to replicate the earth and its ecosystem.

In this project you are asked to research the results of these studies, the follow up experiments and the lessons learned by answering the following questions:

1. Why was the project called Biosphere II?
2. How much did the project cost?
3. What were the primary goals of the project? Did the project achieve its goals?
4. How did the scientists maintain their diets? Where did food and water come from?
5. What were the main complaints of the inhabitants during the two year incarceration?
6. What species survived the experiments and which perished? Why?
7. How different was the atmosphere of Biosphere II from Earth's atmosphere? Could the environment be maintained? How high did the level of carbon dioxide rise?
8. How did the quality of water, air, and soil change? Was the ecosystem capable of repairing and maintaining itself? How did plant and animal lives flourish under these conditions? Did any of its species become extinct?
9. Do you consider the experiment a success? Was it worth the amount of time and money spent?
10. What were the lessons learned?
11. What is the status of Biosphere II today?
12. Do you have suggestions for further experiments or a better utilization of the resources and facility?

Project III - Sustainable Village (Groups of 3-5 students)

A group of 10,000 families are planning to design a completely self-sufficient and sustainable community for themselves and their children. As a consultant, you are asked to develop a blueprint of how such a community would be built. You are going to advise the villagers on:

1. What source of energy they should use? Where should they find the resources, and what advantages and disadvantages will various energy options have?
2. What kind of food should be produced and how should the villagers produce it?
3. Assuming that the villagers have the option to devise their own tax laws and enact their own environmental regulations, what kind of tax structure do you recommend?
4. How should the community pay for public goods such as health care costs, cleanup cost, and salaries of police, teachers, etc.?
5. What would schools look like, and who would pay for the education?
6. What sport facilities and entertainment facilities do you recommend?
7. Are there going to be any restrictions on the type of activities they can take or the number of children they can have? What kind of incentive should they offer for family planning? For keeping the environment clean? How should they handle wastes?

8. Assuming the villagers have access to limited amount of oil and other fossil resources at reasonably cheap prices, how would you use these resources?
9. What would the villagers use for transportation between their homes and work? For taking long trips and for vacations?

Project IV - Sustainability and social responsibility

In this project you are asked to analyze the pattern of energy use in your household in order to investigate ways of reducing fossil fuel consumption and help the environment. First you are asked to estimate the annual energy consumption in your household by adding up the energy 1) spent to produce your food, 2) needed for lighting, 3) required to run your electrical and gas appliances, and 4) used to meet your transportation needs. Then you are asked to estimate the potential savings in fossil fuel consumption if you make slight modification in your lifestyle.

Please start by answering the following questions:

a. How many people live in your household? What are their ages and genders?
b. Is any member of your household a vegetarian?
c. List all electrical and gaseous appliances in your house. How many hours is each appliance used in a typical day?
d. How many lights are usually on? What kind are they?
e. How many cars do you have in your home? How many miles is each car driven? What are their fuel efficiencies?

Now calculate the following:

1. Total food calories consumed in your household each day.
2. Gallons of fossil fuel needed to grow the food you need. (It takes roughly 10 fossil fuel calories to produce each food calorie in average American diet. This number may vary widely in different countries.)
3. Energy used for lighting.
4. Total electrical energy used to operate the appliances and for lighting. Compare this to the actual energy used as indicated in your electricity bill.
5. Total thermal (natural gas) energy used in your household (cooking, heating, etc.). Compare this to actual energy used as indicated in your gas bill.
6. Total energy needed for transportation.
7. Add items 1-6 to find the total annual energy consumption in MJ. How many barrels of petroleum do you use to meet your annual energy needs? (1 bbl = 6.1×10^9 J)
8. Explain sensible approaches to cut your total energy use without significantly changing your lifestyle. You may consider some changes in diet, buying more energy efficient appliances, switching to alternative sources of energy, turning off the air conditioner, lights, and other appliances when not needed, carpooling, taking public transportation, replacing your cars with more fuel efficient vehicles, etc. Be careful to make adjustments that you can actually make with little effort, cause no discomfort, and requires no or only minor sacrifices in your standard of living.
9. How much fossil fuel can you save by making these changes? How much money can you save?
10. How much can the US save in fossil fuel imports if all Americans were to make similar adjustments? What would be the impact on the environment? On the US economy?

Project V - Sustainability and Religion: Was Jesus an Environmentalist?

In the 2008 US presidential election, there is a debate among the religious right groups on whether or not they should endorse the fight against the global warming and other environmental pollution.

One group argues that God created the Earth and expects that its inhabitants to protect the earth and its environment. If God saves humans, then he must also save the environment. It is thus a religious duty to do everything possible to sustainable the environment and all its species . Another group sees this only a tactical move by liberal groups to divert the debate away from issues such as abortion and gay marriage. After all the same science that promotes evolution over creationism, is also promotes environmental protection. Since science is wrong as how the world works, then it cannot be trusted in deciding what is good for the environment. What do you think?

Work Sheet for Project IV				
Food	Person 1	Person 2	Person 3	Total
Daily food calories Energy to grow food (kcal/d)				
Total annual food energy				MJ
Appliances				
TV Refrigerator Microwave oven Electric oven Gas range Others (list) ________				
Total annual energy by appliances				MJ
Lighting	Watts	Number	Hours on	Total
Incandescent Fluorescent Others				
Total annual energy for lighting				MJ
Transportation	Car 1	Car 2	Car 3	Total
Model Annual mileage Fuel efficiency (mpg or 100 km/lit) Gallons (liters) of gasoline				
Total annual transportation energy				MJ
Household total		**Barrels of oil**		**MJ**
US total		**Barrels of oil**		**MJ**
Savings by	Proposed action		MJ Saved	
a. Changing diet b. Energy efficient appliances c. Lighting d. Transportation e. Others (list) ________				
Total household saving		**Barrels of oil**		**MJ**
Total US savings		**Billion barrels of oil**		**Quad**
Impact on US economy	The US will save $ ________ billions from reduced dependence on the foreign oil.			
Environmental impact	Reduction in greenhouse gases: ________ million metric tons of CO_2			

A man travels the world in search of what he needs and returns home to find it. ~ George Augustus Moore (1852 - 1933)

index

D

E

S

T

Basic Information

Appendix A

Fundamental Constants

Avogadro's number = 6.023×10^{23}
Charge of the electron = 1.60×10^{-19} C
Density of air = 1.03 kg/m^3
Density of water = 1000 kg/m^3 = 1 g/cm^3 = 62.4 lb/ft^3
Heat capacity of water = 4.18 kJ/kg.K = 1 g-cal/g.°C =
1 Btu/lb.°F = 1.163 kWh/kg.K

Mass of an electron = 9.1×10^{-31} kg
Mass of a neutron = 1.675×10^{-27} kg
Mass of a proton = 1.672×10^{-27} kg

Mass of the moon = 7.4×10^{22} kg
Mass of the earth = 6×10^{24} kg
Mass of the sun = 2×10^{30} kg

Size of hydrogen atom (Bohr radius) = 5.29×10^{-11} m
Radius of the moon = 1,738 km
Radius of the earth (at equator) = 6,400 km
Radius of the sun = 700,000 km
Average distance between the earth and the sun = 1.495×10^{8} km
Average distance between the earth and the moon = 3.84×10^{5} km

Solar constant = 1,340 W/m^2
Speed of light (in vacuum) = 3.00×10^{8} m/s
Speed of sound in air (at STP) = 331 m/s
One light year = 9.46×10^{12} km
Stephan-Boltzmann's constant = 5.67×10^{-8} W/m^2.K^4
Acceleration of gravity on earth's Surface = g = 9.81 m/s^2
Gravitational constant = G = 6.67×10^{-11} m^3/kg-s^2 =
6.67×10^{-11} N-m^2/kg^2

Useful Arithmetic Skills

Powers of Ten

Addition and Subtraction:
Example: $2 \times 10^2 + 10^{-1} - 0.3 \times 10^1 = (2 \times 100) + (1 \times 0.1) - (0.3 \times 10) = 200 + .01 - 3 = 197.1$

Multiplication: $10^a . 10^b = 10^{a+b}$
Example: $10^2 . 10^3 = 10^5$; $(2 \times 10^{-3}).(4 \times 10^2) = 8 \times 10^{-3+2} = 8 \times 10^{-1} = 0.8$

Division: $10^a / 10^b = 10^{a-b}$
Example: $10^2/10^3 = 10^{2-3} = 10^{-1}$; $(2 \times 10^{-2})/(10^{-3}) = 2 \times 10^{-2+3} = 2 \times 10^1 = 20$

Power: $(10^a)^b = 10^{a.b}$
Example: $(10^2)^3 = 10^{2.3} = 10^6$; $(2 \times 10^2)^{-1.5} = 2 \times 10^{-2 \times 1.5} = 2 \times 10^{-3} = 0.002$

Power	Value	Name
10^{-12}	0.000000000001	Trillionth
10^{-9}	0.000000001	Billionth
10^{-6}	0.000001	Millionth
10^{-3}	0.001	Thousands
10^{0}	1	One
10^{3}	1,000	Thousand
10^{6}	1,000,000	Million
10^{9}	1,000,000,000	Billion
10^{12}	1,000,000,000,000	Trillion

Logarithms

Definition: $N = a^x$; $x = \log_a N$
for $a = e = 2.71828$; $N = e^x$; $x = \ln N$ (natural logarithm)
for $a = 10$; $N = 10^x$; $x = \log_{10} N = \log N$ (logarithm base 10)

Base other than 10: $\log_b a = \log a / \log b$
Example: $\log_3 5 + \log 5/\log 3 = 0.699 / 0.477 = 1.465$

Addition and Subtraction:
Example: $\log(3) + \log(2) = 0.477 + 0.301 = 0.778$

Multiplication: $\log(a.b) = \log a + \log b$
Example: $\log 6 = \log(3x2) = \log 3 + \log 2 = 0.477 + 0.301 = 0.778$

Division: $\log(a/b) = \log a - \log b$
Example: $\log(1.5) = \log(3/2) = \log 3 - \log 2 = 0.477 - .0301 = 0.176$

Power: $\log(a)^b = b. \log(a)$
Example: $\log(10^{0.3}) = 0.3 \log(10) = 0.3x1 = 0.3$; $\log(5^{-2}) = -2 \log(5) = -1.398$

N	$log_{10} N$	ln N	e^x	e^{-x}
0	--	--	1.000	1.000
1	0	0	2.718	0.368
2	0.301	0.693	7.389	0.135
3	0.477	1.099	20.086	0.050
4	0.602	1.386	54.598	0.018
5	0.699	1.609	148.41	$6.74x10^{-3}$
6	0.778	1.792	403.43	$2.48x10^{-3}$
7	0.845	1.946	1096.6	$9.12x10^{-4}$
8	0.903	2.079	2981.0	$3.35x10^{-4}$
9	0.954	2.197	8103.1	$1.23x10^{-4}$
10	1.000	2.303	22026	$4.54x10^{-5}$

Unit Conversion

Appendix B

Metric Units and Abbreviation

Quantity	Unit	Symbol
Time	second	s
Length	meter	m
Speed	meter/s	m/s
Volume	cubic meter	m^3
Mass	kilogram	kg
Weight (Force)	newton	$N = kg.m/s^2$
Torque	newton-meter	N.m = J
Energy	joule	J
Power	watt	W = J/s
Specific power	watt per kilogram	W/kg
Power density	watt per cu. meter	W/m^3

Conversion Tables

Time

1 year = 365.25 d = 8,766 h = 3.16×10^7 s

Length

1 micron =1 μm = 10^{-6} m
1 nanometer = 1 nm = 10^{-9} m
1 meter (m) = 1.094 yd = 3.281ft
1 kilometer (km) = 1,000 m = 0.62 mi = 3,281 ft
1 foot (ft) = 12 in = 0.305 m
1 mile (mi) = 1.609 km = 1,609 m = 1,760 yd = 5,280 ft
1 light year = 9.46×10^{15} m
1 astronomical unit (AU) (average distance between earth and sun) = 1.49×10^{11} m

Speed (Velocity)

1 m/s = 3.28 ft/s = 3.6 km/hr = 2.24 mph
1 km/h = 0.62 mph = 0.278 m/s
1 mph = 1.609 km/h = 1.467 ft/s

Area

1 square mile (mi^2) = 2.59 km^2 = 640 acre = 2.79×10^7 ft^2
1 square meter (m^2) = 10.76 ft^2
1 hectacre = 2.47 acre

Volume

1 gallon (U.S.) = 3.785 L
I gallon (Imperial) = 1.2 gallon (U.S.)
1 cubic meter (m^3) = 1,000 L = 264 U.S. gallons = 35.3 ft^3 = 6.29 bbl
1 barrel (bbl) = 42 US gallons = 0.159 m^3 = 159.1 L

Mass

1 kilogram (kg) = 1,000 g = 2.21 lb = 10^{-3} ton (metric)
1 pound (lb) = 0.453 kg = 4.536×10^{-4} ton (metric)
1 ton (metric) = 1,000 kg = 2,240 lbs
1 short ton = 2,000 lb = 907 kg

Density

1 g/cm^3 = 1 kg/L = 1,000 kg/m^3 = 62.4 lb/ft^3 = 8.35 lb/gallon
1 pound/gallon U.S.) = 0.12 kg/L

Pressure

1 pascal (Pa) = 1 N/m^2
1 atmosphere (atm) = 101 kPa = 760 mm Hg = 14.7 lb/in^2

Energy

1 quad = 10^{15} Btu = 1.055 EJ
1 Megajoule (MJ) = 10^6 J = 0.278 kWh = 948 Btu
1 Btu = 1,055 J = $2.93\text{x}10^{-4}$ kWh = 0.252 kcal
1 kWh = $3.6\text{x}10^6$ J = 3412 Btu
1 Calorie (food calorie) = 1 kcal = 4.184 kJ = 3.968 Btu
1 therm = 10^5 Btu = $1.055\text{x}10^5$ kJ
1 ton of TNT = $4.184\text{x}10^9$ J = 10^6 food Calorie
1 electron-volt (eV) = $1.602\text{x}10^{-19}$ J

Power

1 watt (W) = 1 J/s
1 kilowatt (kW) = 1.34 hp = 0.239 kcal/s = 0.948 Btu/s
1 TW = 30 quads/yr
1 horsepower (hp) = 0.746 kW = 0.178 kcal/s = 0.707 Btu/s
1 ton refrigeration = 12,000 Btu/hr = 3.517 kW

To Convert	Into	Multiply By
Btu	joule	1,055
Btu	kilowatt-hour	$2.93\text{x}10^{-4}$
Btu	calorie	242
Btu/hour	watt	0.293
Btu/hour	horsepower	$3.93\text{x}10^{-4}$
calorie	Btu	$3.968\text{x}10^{-3}$
cubic meter	cubic foot	35.31
cubic meter	gallon (U.S.)	264.2
cubic meter	liter	1,000
day	second	86,400
electron-volt	joule	$1.6\text{x}10^{-19}$
foot	inch	12
foot	meter	0.3048
foot/second	meter/second	0.3048
foot/second	mile/hour	0.68
gallon (U.S.)	cubic foot	0.1337
gallon (U.S.)	liter	3.785
horsepower	kilowatt	0.746
hour	second	3,600
joule	electron-volt	$6.24\text{x}10^{18}$
joule	Btu	$9.48\text{x}10^{-4}$
kilogram	pound	2.205
kilometer	meter	1,000
kilometer	mile	0.6214

To Convert	Into	Multiply By
kilometer/hour	meters/second	0.278
kilometer/hour	mile/hour	0.6214
kilometer/hour	knots	0.54
kilopascal	inch of water (4°)	4.015
kilowatt	horsepower	1.341
kilowatt-hour	Btu	3,413
kilowatt-hour	joule	3.6×10^6
knot	nautical mile/hr	1.000
knot	kilometers/hour	1.852
knots	miles/hour	1.151
liter	cubic centimeter	1,000
liter	cubic foot	0.0353
liter	cubic meter	0.001
liter	gallon (US)	0.264
meter	foot	3.28
meter	yard	1.094
meter/second	kilometer/hour	3.6
meter/second	mile/hour	2.237
metric ton	kilogram	1,000
mile	foot	5,280
mile (nautical)	mile	1.15
mile/hour	foot/second	1.467
mile/hour	kilometer/hour	1.609
minute	second	60
pound	kilogram	0.453
quad	Btu	10^{15}
square foot	square meter	0.093
square meter	square foot	10.76
square kilometer	acre	247
square kilometer	square mile	0.386
ton (metric)	kilogram	1,000
ton (short)	pound	2,000
watt	Btu/hour	3.413
watt	horsepower	1.341×10^{-3}
yard	meter	0.9144

Density of

Air @ STP	1.2 kg/m3 =
Water @ 4°C	1 kg/L = 1,000 kg/m^3 = 62.4 lb/ft^3
Natural gas	0.72 kg/m^3 = 44.9 lb/ft^3
Crude oil	0.86 kg/L = 136 kg/bbl = 7.16 lb/gallon

Fuel Equivalents

1 quad = 170 million barrels of crude oil = 0.273 cubic kilometer of oil = 40 million tons of coal = 28 billion m^3 natural gas

1 bbl of oil = 6.1 giga joule = 5.8 million Btu = 0.2 tons of coal = 164 cubic meters of natural gas

1 kg of U-235 fuel (3.5%) = 100 tons of coal = $2.6x10^{12}$ J

Fuel Efficiency Conversions			
MPG	**Miles/L**	**Km/L**	**L/100 km**
10	2.64	4.25	23.52
15	3.96	6.38	15.68
20	5.28	8.50	11.76
25	6.60	10.63	9.41
30	7.92	12.75	7.84
35	9.25	14.88	6.72
40	10.57	17.00	5.88
45	11.89	19.13	5.23
50	13.21	21.25	4.70
55	14.53	23.38	4.28
60	15.85	25.51	3.92

Energy Contents of Various Fuels

Fuel	Energy Density*			
Crude oil (kerosene)	38.5	MJ/L	6	MBtu/bbl
Gasoline	32.2	"	115	kBtu/gal
Diesel	35.6	"	128	"
Ethanol	21.2	"	76	"
Biodiesel	33.0	"	121	"
Hydrogen @ 35MPa	2.7	"	9.9	"
Standard coal	30.0	MJ/kg	12.9	kBtu/lb
Bituminous	20-30	"	8.6-12.9	"
Lignite	10-19	"	4.3-8.2	"
Natural gas @ STP	37.0	MJ/m^3	84	kBtu/gal
CNG @ 20 MPa	9.3	MJ/L	33	"
LPG @ 1.5 MPa	23.0	"	84	"
Methanol	16.7	"	53	"
Uranium (fissioned)	80	GJ/g	34	GBtu/lb
Uranium(0.5%)	400	GJ/kg	172	MBtu/lb

*Lower Heating Value (Net) except for hydrogen which is for HHV

Appendix C

Exponential Expiration Time

One of the pressing questions of our time is that for how long our nonrenewable resources last. When a quantity such as the rate of consumption of a resource is growing at a fixed percentage rate, r, after time t we have:

$$N = N_0 e^{rt} \qquad (C.1)$$

Where;
N_0 is the initial rate of consumption per year (t = 0),
N is the rate of consumption per year at a later time, and
r is the annual growth rate

The consumption rate will grow to twice the initial value $(N/N_0 = 2)$ after time

$$T_2 = \frac{\ln 2}{r} = \frac{0.693}{r} \cong \frac{70}{R} \qquad (C.2)$$

Where R is the percentage growth per year $(R = 100.r)$

The cumulative consumption of the resource between the present time $t = 0$ and time $t = T$ in the future is:

$$Q = \int_0^T N(t).dt = \frac{N_0}{r}(e^{rT} - 1) \qquad (C.3)$$

We can rearrange this equation to give the exponential expiration time, i.e. time it takes for the resource to be depleted.

$$EET = (1/r)\ln(rQ/N_0 + 1) \qquad (C.4)$$

In this equation, Q represents the ultimate size of the resource (given in barrels of oil, tons of coal, etc.), N_0 is the consumption rate (given in barrels of oil per year, tons of coal per day, etc.) at time t = 0.

Example C-1: The US energy consumption increased from 76.8 quads in 1984 to 100.4 quads in 2004 (See Table 1-3). Assuming that the consumption followed an exponential growth pattern, find

a. The annual growth rate,
b. The doubling-time,
c. Total US energy consumption during this period.

Solution: Substituting into Equations C-1 through C-3, we will have:

a. $100.4 = 76.8\, e^{20r}$; or
$r = 1/20 \ln (100.4/76.8) = 0.013$ (1.3% per year)

b. $T_2 = 70/1.3 = 52.2$ years

c. $Q = 76.8/0.013\, (e^{0.013x20} - 1) = 1{,}754$ quads

Example C-2: The total world petroleum endowment (that already produced and that will remain to be exploited) is estimated at 3,563 billion barrels (See Table 7-3). The same data indicates the cumulative production of 708 billion barrels and proven (remaining) reserves of 2,855 billion barrels. Assuming that the oil consumption was at 98.9 million barrels a day in 2000, calculate the remaining life expectancy (exponential expiration time) of the world's petroleum reserves.

Solution: Substituting for the total size of remaining reserves ($Q =$ 2,855 bbo), the annual consumption rate of $N_0 = 98.9x10^6x365 = 36.1x10^9$ barrels/year (36.1 bbo/year) at the turn of the century, the exponential expiration time is calculated for different rate of consumption growth as:

r (% per year)	EET (years)
1	58.3
2	47.4
3	40.5
4	35.7
5	32.0
6	29.1
7	26.8
8	24.9
9	23.3
10	21.9

Abbreviations and Acronyms

Appendix D

Acronyms

AQMD	Air Quality Management District
DNA	Defense Nuclear Agency
DOD	Department of Defense
DOE	Department of Energy
DOT	Department of Transportation
EPA	Environmental Protection Agency
FERC	Federal Energy Regulatory Commission
GAO	Government Accountability Office (Formerly General Accounting Office)
IAEA	International Atomic Energy Commission
IMF	International Monetary Fund
INPO	Institute of Nuclear Power Operation
NASA	National Aeronautics and Space Administration
NATO	North Atlantic Treaty Organization
NIST	National Institute of Standards and Technology
NRC	Nuclear Regulatory Commission
OPEC	Organization of Petroleum Exporting Countries
UN	United Nations
USSR	Union of Soviet Socialist Republics
WANO	World Association of Nuclear Operators
WHO	World Health Organization
WTO	World Trade Organization

Abbreviations

ABM	Anti Ballistic Missile
AQI	Air Quality Index
BACT	Best Available Control Technology
BEV	Battery-operated Electric Vehicle
BMI	Body Mass Index
BMR	Basal Metabolic Rate
Btu	British Thermal Unit
BWR	Boiling Water Reactor
CAA	Clean Air Act
CAD	Computer Aided Design
CAFÉ	Corporate Average Fuel Economy
CAFTA	Central American Free Trade Agreement
CAM	Computer Aided Manufacturing
CFC	Chlorofluorocarbon
CHP	Combined Heat and Power
CNG	Compressed Natural Gas
COLA	Cost of Living Adjustment
COP	Coefficient of Performance
CPI	Consumer Price Index
CWA	Clean Water Act

DNA	Deoxyribo Nucleic Acid
DU	Depleted Uranium
EER	Energy Efficiency Ratio
FFV	Flexible Fuel Vehicle
GDP	Gross Domestic Products
GNP	Gross National Products
HCFC	Hydrochlorofluorocarbon
HTGR	High Temperature Gas-Cooled Reactors
HWR	Heavy Water Reactor
ICE	Internal Combustion Engine
IOU	Investor Owned Utility
IPCC	Intergovernmental Panel on Climate Change
IPP	Independent Power Producers
ISO	Independent System Operators
ITER	International Thermonuclear Experimental Reactor
JET	Joint European Torus
LDC	Load Demand Curve
LEV	Low Emission Vehicle
LMFBR	Liquid Metal Fast Breeder Reactor
LNG	Liquefied Natural Gas
LWR	Light Water Reactor
MHD	Magneto Hydro Dynamics
MRI	Magnetic Resonance Imaging
MTBE	Methyl Tertiary Butyl Ether
NAAQS	National Ambient Air Quality Standards
NAFTA	North Atlantic Free Trade Agreement.
NMD	National Missile Defense System
NNW	Net National Welfare
NPV	Net Present Value
OEM	Original Equipment Manufacturer
OTEC	Ocean Thermal Energy Conversion
PNGV	Partnership for New Generation of Vehicles
PSC	Polar Stratospheric Cloud
PWR	Pressurized Water Reactor
PX	California Power Exchange Corporation
QF	Qualifying Facility
RAD	Radiation Adsorbed Dose
RNA	Ribonucleic Acid
SALT	Strategic Arms Limitation Treaty
SDI	Strategic Defense Initiative
SI	International System (of units)
SOC	State of Charge
START	STrategic Arms Reduction Treaty
TFTR	Tokomak Fusion Test Reactor
TNT	Trinitrotoluene
ULEV	Ultra Low Emission Vehicle
USCS	United States Customary System
VOC	Volatile Organic Compounds
ZEV	Zero Emission Vehicle

Glossary

A

absolute humidity: the mass of water vapor in a given volume of air (expressed in grams of water vapor per cubic meter) at atmospheric temperature and pressure.

absolute zero: the lowest possible temperature at which all atomic motion stops (-273.15° C).

acidic: a substance with pH lower than 7; the opposite of basic.

acid rain: rain with a pH less than 5.6.

adiabatic: the process in which no heat enters or leaves a system.

aerobic: occurring in presence of oxygen or air. Also see anaerobic.

albedo: the fraction of the sunlight that is reflected by earth, ice, and cloud back into space.

algae: plants which grow in moist or aquatic environments.

alkaline: basic, the opposite of acidic.

alpha particle: a particle consisting of two neutrons and two protons The nucleus of helium atom. Also see beta particle and gamma ray.

alternating current (AC): the flow of charge constantly changing direction. Almost all power produced by electric utilities in the United States is AC that shifts direction at a rate of 60 times per second. Also see direct current (DC).

alternative fuel vehicle (AFV): motor vehicles that run on fuels other than petroleum.

altitude: the distance above sea-level; the angle between the sun and the horizon. When the sun is on the horizon (sunrise and sunset), this angle is zero. Solar altitude is maximum at solar noon. The complement of solar altitude angle or the angle of the sun from a vertical line directly overhead is called zenith angle.

amino acid: the complex organic molecules that are the "building blocks" of proteins. Most organisms construct proteins from roughly 20 amino acids.

amperes (amp): the unit designating the amount of electricity flow through a conductor.

anaerobic: occurring in the absence of oxygen or air. Also see aerobic.

anemometer: a mechanical instrument used to measure wind speed and direction.

angle of incidence: the angle at which the sun's rays or insolation strike the earth's surface.

anion: an ion carrying a negative charge. Also see cation.

Antarctic Circle: latitude of 66.5° South. The northern limit of the area of the earth which experiences 24 hours of darkness or 24 hours of daylight at least once during the year. Also see Arctic Circle.

anthracite: hard coal found deep in the earth containing a high percentage of carbon and little volatile matter. It burns very hot, with little flame.

anthropocentric: human-centered.

anthropogenic: human made; usually used in the context of emissions that are produced as the result of human activities.

aquaculture: aquatic organism farming.

aquifer: a permeable rock layer that stores a significant amount of water or fluid.

ash: fine particles of solid or molten rock material ejected during a volcanic eruption or the non-organic, non-flammable substance left over after combustion.

associated gas: natural gas found in contact with oil in naturally-occurring underground formations. The gas may be present in a free state or dissolved in the petroleum. Also see non-associated gas and dissolved gas.

atmosphere: the layer of gases that surrounds the earth.

atom: the smallest unit of an element that still maintains its chemical characteristics.

atomic mass number: total number of an atom's protons and neutrons.

atomic number: the number of protons in the nucleus of an atom.

atomic mass: the combined weight of an atom's electrons, protons, and neutrons.

autotroph: an organism that produces its own food – it does not require outside sources of organic food energy for survival. Also see heterotroph, chemical autotrophs and photosynthetic autotrophs.

azimuth: the angle between a north-south line on the earth's surface and the horizontal projection of the sun's rays, measured from true south. By convention, solar azimuth is negative before noon and positive after noon.

B

back scattering: the fraction of solar radiation that is directed back into space as a result of particle scattering in the atmosphere.

bads: opposite of "goods." Bads are usually referred to anything we have to pay money to dispose of. Examples of bads are air pollutants, garbage, and toxic wastes.

baghouses: a series of fabric bags that act as filters to remove particulate matter from polluted air.

barometer: an instrument for measuring atmospheric pressure.

barrel: 42 U.S. gallons. One barrel of oil has an energy content of 6 million Btu.

base load: the minimum power production needed during a year.

basal metabolic rate (BMR): the number of calories a body burns at rest to maintain normal body functions. BMR changes with age, weight, height, gender, diet and exercise habits.

basic: a substance with a pH greater than 7. Also see acidic.

battery: a device that stores chemical energy and produces electric current. Batteries are typically built for specific purposes and they differ in construction accordingly. Broadly speaking, there are two applications that manufacturers build their batteries for: starting and deep cycle.

starter batteries have many thin lead plates which allow them to discharge a lot of energy very quickly for a short amount of time. However, they do not tolerate being discharged deeply, as the thin lead plates needed for starter currents degrade quickly under deep discharge and re-charging cycles. Most starter batteries will only tolerate being completely discharged a few times before being irreversibly damaged.

deep cycle batteries have thicker lead plates that make them tolerate deep discharges better. They cannot dispense charge as quickly as a starter battery. Some "marine" batteries are sold as dual-purpose batteries for starter and deep cycle applications. Such batteries should not be cycled deeply and should be avoided for deep-cycle applications unless space/weight constraints dictate otherwise.

bed: sedimentary structure that usually represents a layer of deposited sediment.

bedrock: a general term for any consolidated rock.

beta particle: an electron emitted from the nucleus of a radioactive isotope. Also see alpha particle and gamma ray.

Bhopal: the city in India where, in December 1984, a major leak at the Union Carbide pesticide facility killed thousands and injured tens of thousands more.

Big Bang: the theory of the formation of the universe. According to this theory about 15 billion years ago all of the matter and energy in the Universe was concentrated into an area smaller than the size of an atom. At one instance a huge explosion occurred that released an enormous amount of mass that began to expand at an incredible rate and matter, energy, space and time came into being. As the Universe expanded, matter began to coalesce into gas clouds, and then into stars and planets. Some scientists believe that this expansion is finite and will one day cease. After this point in time, the Universe will begin to collapse until a Big Crunch occurs.

biochemistry: the chemistry of life.

biodegradable: a matter that can be degraded, decomposed, or broken down by microorganisms into simple compounds such as water and carbon dioxide.

biodiesel: a fuel derived from vegetables and other organics matter having essentially identical composition to diesel fuel refined from petroleum.

biodiversity: refers to the diversity of different species (species diversity), genetic diversity among individuals within each species (genetic diversity), and the diversity of ecosystems (ecosystem diversity).

bioengineering: the genetic manipulation of species to produce new varieties and types of organisms.

biofuel: a fuel made from plants and other biomass.

biomass: living tissues accumulated over an area in a particular time interval.

biomass energy: energy resources derived from organic matter that include wood, agricultural waste, algae, sewage and other organic substances that can be burned to produce heat or transformed into other fuels through chemical processes.

biome: a large assemblage of animals and plants that includes many communities of a similar nature.

biosphere: the layer adjacent to the earth's surface where all life exists. The biosphere consists of all living things, plant and animal. Also called ecosphere.

biotic: referring to life, or influenced by living organisms.

bituminous coal: soft coal of medium quality containing relatively large amounts of carbonaceous matter. Bituminous coal has a heating value somewhat lower than anthracite, and is used primarily for the generation of electricity, production of coke, and space heating. Also see anthracite and lignite.

black body: a body that emits electromagnetic radiation, at any temperature, at the maximum possible rate. It also absorbs all electromagnetic radiation that is falling on it.

blackout: a power loss affecting many electricity consumers over a large geographical area for a significant period of time, perhaps due to a major breakdown.

body mass index (BMI): a widely used measurement for obesity calculated as the person's mass divided by the square of the height.

boiling water reactor (BWR): a nuclear power plant which is allowed to boil at the core, producing steam which is then used to drive electric turbines.

bottled gas: the liquefied petroleum gas stored in cylinders under pressure – mainly propane and butane.

bottoming cycle: a means of increasing the thermal efficiency of a steam electric generating cycle where some waste heat from the condenser is converted into electricity.

breeder reactor: a nuclear reactor that produces more fuel than it consumes. It is especially designed to actively convert non-fissionable isotopes of U-238 into fissionable isotopes of Pu-239 that can then be used as fuel.

brine: water which contains a large concentration of salt such as a salt lake or seawater.

British thermal unit (Btu): a unit for measurement of heat defined as the amount of energy required to raise the temperature of one pound of water one degree fahrenheit. One Btu is equal to 252 calories or 1055 joules.

brown coal: See lignite.

brownout: when the capacity of a power plant exceeds demand and the voltage to consumers drops by a few percent such that lights often dim.

buffer: a protective barrier. A substance capable of neutralizing acids and bases and thereby maintaining the original pH of the solution.

butane: a paraffin with a chemical composition of C_4H_{10}. It is used primarily for blending into high-octane gasoline, residential and commercial heating, and manufacturing of chemical and synthetic rubber.

C

California Energy Commission: a state agency responsible for energy policy in California.

California Public Utility Commission (CPUC): the administrative agency that regulates privately owned utilities, securing adequate service to the public at rates that are just and reasonable both to customers and shareholders of the utilities. The CPUC also provides electricity and natural gas forecasting, and analysis and planning of energy supply and resources.

calorie: a unit of heat energy. The amount of heat required to raise the temperature of 1 gram of water 1 degree celsius at standard atmospheric pressure. One heat calorie is equivalent to 4.2 joules. It is common to differentiate heat calories from food calories by using lowercase "c". Also see Calorie.

Calorie: or food calorie equivalent to 1,000 calories. It is common to differentiate food calories from heat calories by capitalizing the "C." Also see calorie.

capacity: the amount of electricity produced in an electricity generating station, or the total volume of natural gas that flows through a pipeline over a given amount of time. Capacity of a battery is the amount of usable energy it can store at a nominal voltage.

All other things equal, the greater the physical volume of a battery, the larger its total storage capacity. Capacity is expressed in amp-hour (Ah).

capacity factor: the fraction of a power plant's capacity that is utilized on average over time.

cap-and-trade system: in this system companies are allowed to continue polluting, but must buy the right to pollute more from companies that implement strategies that reduce their pollution.

carbohydrate: an organic compound composed of carbon, oxygen, and hydrogen atoms, such as sugars, starch, and cellulose.

carbon auction: companies are allowed to continue polluting, but are required to bid against other companies for a portion of the atmosphere they intend to use -- within guidelines that reduce overall pollution.

carbon capture and storage (CCS): The process of capturing and storing the CO_2 emission at the power plant instead of being released into the atmosphere.

carbon dating: the process of determining the age of certain archeological artifacts of a biological origin such as bone, cloth, wood and plant fibers.

carbon dioxide: molecules of a gas composed of one carbon and two oxygen atoms (chemical formula of CO_2). Carbon dioxide gas is transparent to solar light but absorbs radiation strongly in the long wave infrared band. This absorption causes the greenhouse effect. It is believed that the higher concentrations of carbon dioxide and other greenhouse gases resulting from human burning of fossil fuels are responsible for global warming.

carbon monoxide: a colorless, odorless, tasteless and highly poisonous gas made up of one carbon and one oxygen atom formed by the incomplete combustion of carbon or carbonaceous material, including gasoline. The chemical formula for carbon monoxide is CO.

carbon tax: tax levied on fossil fuels (or any fuels) in proportion to the amount of carbon emitted during combustion.

carcinogens: any chemical substance in the environment that can potentially cause cancer.

carnivore: an organism that consumes other living animals. Also see herbivore, omnivore, secondary consumers, and tertiary consumers.

Carnot efficiency: the highest theoretical efficiency a heat engine can have when operating between the two thermal energy reservoirs at temperatures T_L and T_H; h = $1 - T_L / T_H$.

carrying capacity: the maximum population of a species that a certain habitat can support.

cassandras: doom sayers, those who believe that the exponential growth in populations and consumption will ultimately totally degrade the environment. Cassandras are pessimists, opposite to cornucopians.

catalytic converter: a device that takes the pollutants emitted by automobile exhausts and smoke stacks (CO, NOx, and hydrocarbon) and converts them to less harmful substances (CO_2 and N_2).

catalytic cracking: the process by which the high-boiling fraction of crude petroleum is converted to lighter compounds, distillate, fuel oil and fuel gas using a catalyst and heat.

cation: an ion carrying a positive atomic charge. See also anion.

cell: the smallest self-functioning unit of living organisms. Some organisms such as bacteria consist of only one cell, but most of the organisms found on earth are made up of many cells.

cellulose: the fibrous woody material making up three-quarters of plant material.

Celsius scale: a temperature scale in which water boils at 100° and freezes at 0°.

Central American Free Trade Agreement (CAFTA): a trade agreement signed between the United States, Costa Rica, Honduras, Guatemala, and Nicaragua in 2005 under which these countries will remove all trade barriers, tariffs, and quotas between them. Also see North American Free Trade Agreement (NAFTA).

CFC: acronym for chlorofluorocarbon.

chain reaction: in nuclear reactions, when the fissioning of one atom releases neutrons that induce the fissioning of other atoms, and so forth. Similarly, in chemical reactions, the release of heat and other radicals results in a release of additional heat and radical species that makes the reaction self-sustaining.

chemical autotrophs: organisms that use the external energy found in chemical compounds to produce food molecules. The process used to produce food by these organisms is known as chemosynthesis. Also see photosynthetic autotrophs.

chemical energy: energy stored in molecular bonds. The energy generated when a compound undergoes chemical combustion, decomposes, or transforms to produce new compounds.

chemosynthesis: a process in which organisms extract inorganic compounds from their environment and convert them into organic nutrient compounds without the use of sunlight. Also see photosynthesis.

chlorofluorocarbons (CFC): artificially produced compounds composed primarily of carbon, fluorine, and chlorine that were introduced in the mid-1930s, used in refrigerants, solvents and insulating foam material. These chemicals were eventually implicated in the deterioration of the ozone layer and subsequently banned.

chlorophyll: the green pigment in plants and certain bacteria used to capture the energy in light through photosynthesis.

Clean Air Act: a federal statute enacted in 1963 that was the first of a series of acts and amendments that exerted increasing federal pressure on air polluters to clean up their emissions.

Clean Water Act: a federal statute enacted in 1972 that has been very successful in improving the water quality of lakes and rivers.

climate: the general pattern of weather conditions for a region over a long time period.

closed system: a system that transfers energy but not matter across its boundary to the surrounding environment. Our planet is often viewed as a closed system.

coal: the black or brown rock, formed under pressure from organic fossils in prehistoric times, that is mined and burned to produce heat energy. Depending on its carbon content coal is classified as anthracite, bituminous, or lignite.

coefficient of performance (COP): a non-dimensional measure of the ratio of the useful heating (or cooling) load delivered to the corresponding energy input.

cogeneration: a power plant which produces several types of energy simultaneously

– such as electricity and heat – that can be used locally, such as an electric generating station that uses waste heat from its gas turbines to produce steam for conventional steam turbines.

coke: a porous solid left over after the volatile constituents of coal are driven off by heat in the absence of or in a limited supply of air. It is used primarily in blast furnaces for smelting ores.

combined cycle plant: See cogeneration.

compressed natural gas (CNG): natural gas that has been compressed under high pressure, typically between 2,000-3,600 psi, and is held in a container.

condensation: the change in state of matter from vapor to liquid.

condensation nuclei: microscopic particles of dust, smoke or salt that act as nucleus for condensation of water vapor to water droplets in the atmosphere. Condensation normally occurs on these particles when relative humidity reaches 100 %.

condenser: a heat exchanger in which the refrigerant is heated to a high pressure and temperature and then condensed to liquid by rejecting heat.

conduction: the transfer of heat through a material by the motion of adjacent atoms and molecules without gross displacement of the particles. Also see convection and radiation.

conductivity: the ability of a material of a given thickness to transfer heat or electricity.

constant dollar: the dollar value in which the effect of changes in its purchasing power have been removed.

consumer: an organism who gets its organic nutrient from the consumption of tissues of producers and other consumers because it cannot synthesize the organic nutrients in requires. Different consumers are: carnivores, herbivores, and detritivores.

consumer price index (CPI): an index designed to measure the changes in the prices of goods and services as a ratio of the cost of a bundle of products and services at current prices to its cost at a base year.

control rod: a rod made of substances such as cadmium or boron with the ability to absorb neutrons used in a nuclear reactor to control or even halt the nuclear chain reaction.

convection: the transfer of energy by means of mass motions through a medium as a result of a temperature or density gradient. Also see conduction and radiation.

cooling load: the rate at which heat must be extracted from a space in order to maintain the desired temperature within the space.

cooling tower: a device for cooling water through direct contact with air.

core: a layer rich in iron and nickel found in the interior of the earth. It is composed of two sub-layers: the inner core and outer core. The core is about 7,000 kilometers in diameter.

coriolis force: an apparent force that, as a result of the earth's rotation, deflects objects to their right in the northern hemisphere and to the left in the southern hemisphere. Coriolis force is responsible for hurricanes, tornadoes, and twisters.

cornucopians: neoclassical economics who believe that advances in technology result in the growth in living standards which will offset loss of resources and degradation in the environment. Cornucopians are optimists, opposite to cassandras.

corporate average fuel economy (CAFE): average fuel consumption in miles per gallon, based on city and highway fuel economy measurements performed as part of the federal emissions test procedures.

criteria pollutants: pollutants that are mandated by Congress for regulation, including particulate matter, sulfur dioxide, nitrogen dioxide, ozone, lead, and carbon monoxide.

crude oil: simply called crude, is the petroleum found in the earth before it is refined into oil products.

cumulative production: the total amount of energy that has been produced so far.

curb weight: the weight of a vehicle including all standard equipment, spare tire and wheel, all fluids and lubricants to capacity, a full tank of fuel, and the weight of major optional accessories normally found on the vehicle.

current: a measure of how many electrons are flowing through a conductor. Current is usually measured in amperes (A). Current flow over time is defined as ampere-hours, a product of the average current and the amount of time it flowed.

D

deep mining: extraction of coal or minerals at great depths. Coal is usually deep-mined at no more than 1,500 feet.

degree day: a convenient way to measure the energy need for a given day and is calculated by taking the average of the day's high and low temperatures. The difference between this temperature and the design temperature (taken as 65°F) summed up over all days that require heating (colder than 65°F) is the number of heating degree-days. Similarly, the temperature difference summed up over all days that require cooling (warmer than 65°F) is the number of cooling degree-days.

demand: the rate at which energy is delivered.

demand-side management: in economics, the methods used to manage energy demand including energy efficiency, load management, fuel substitution and load leveling. See also supply-side management.

density: a measure of how tightly the atoms of a substance are packed; mass per unit volume. For gases, density is the number of atoms and molecules per unit volume. Also see specific gravity.

deoxyribonucleic acid (DNA): the basic hereditary molecule of life on Earth and material of which genes are composed. A form of nucleic acid that is organized into a double-helix molecule. DNA is used by most organisms to chemically code their genetics and to direct the development and functioning of cells.

depleted uranium (DU): mostly u-238, is a toxic and radioactive waste product of the enriching uranium in nuclear reactors. Because of its high density, DU has been used in manufacturing armored vehicles, missiles warheads, and various ammunitions.

depth of discharge (DOD): a measure of how deeply a battery is discharged. When a battery is 100% full, the DOD is 0%. Conversely, when a battery is 100% empty, the DOD is 100%. For example, starter batteries are not designed to be discharged deeply (no more than 20% DOD). Indeed, if used as designed, they hardly discharge at all; engine starters are very energy-intensive but their duration of use is very short.

deregulation: elimination of regulation from a previously regulated industry or sector of an industry.

detritivores: are organisms that recycle detritus (decomposing organic matter), returning it into the food chain.

deuterium: an isotope of hydrogen with a nucleus containing one proton and one neutron and an atomic mass number of 2.

diffused solar radiation: solar radiation received by the earth's atmosphere or surface that has been modified by atmospheric scattering.

diffusion: (1) mixing of one substance into another; (2) refraction of solar insolation in many directions.

diminishing returns, law of: as the supply of a resource declines, increasing efforts to extract the resource produce progressively smaller relative amounts of the resource.

dioxin: a group of extremely toxic compounds that are inadvertently synthesized in incinerators when trash and garbage are burned.

direct current: electricity flowing continuously in the same direction. See also alternate current.

direct energy conversion: direct production of electricity from an energy source without transferring the energy to a working fluid or steam. For example, photovoltaic direct solar electricity conversion.

dissociation: a chemical process where a compound or molecule breaks up into simpler constituents.

dissolved gas: natural gas dissolved in oil within a reservoir.

distillation: the desalination of salt water by evaporation so as to remove the dissolved salts.

diversity: the number of different species in a given region.

DNA: acronym for deoxyribonucleic acid.

dose: the amount of ionizing radiation absorbed per unit mass of irradiated material.

double glazing: windows made of two sheets of glass with an airspace between.

doubling time: the time it takes a population of a given size growing at a fixed rate, to double in size.

dry steam: the conventional type of geothermal energy used for electricity production. The principal dry steam resource area is the Geyser in Northern California which is one of only two known areas in the world for dry steam – the other being Larderello, Italy.

durable goods: material goods or products that are designed to last a relatively long period of time.

E

ecological footprint: total burden that various human activities puts on the earth.

ecology: the study of interrelationships of animals and plants to one another and to their environment.

economic efficiency: a term that refers to the optimal production and consumption of goods and services; occurs when prices of products and services reflect their marginal costs.

economics: the study of the production, distribution, and consumption of goods and services.

ecosystem: the interacting system of the biological community and its nonliving environment.

EER: acronym for energy efficiency ratio.

efficacy (lighting): the ratio of light from a lamp to the electrical power consumed, expressed as lumens per watt and including ballast losses.

efficiency: the ratio of what we get to what we put in. Applied to energy, it is the useful energy delivered by a device to the energy supplied to it over the same period or cycle of operation. The first law efficiency is the ratio of useful output to input. The second law efficiency is the ratio of actual output to output under ideal conditions for the same input.

electric vehicle: a vehicle powered by electricity, usually provided by batteries but may also be provided by photovoltaic (solar) cells or a fuel cell.

El Niño: the occasional development of warm ocean surface waters along the coast of Ecuador and Peru. El Niño normally occurs around Christmas and usually lasts for a few weeks to a few months. See also La Niña

electrical generator: a device that converts thermal, chemical or mechanical energy into electricity.

electric resistance heater: a device that converts electric energy to heat which can be transferred to a space by fans.

electrical energy: energy resulting from electric current created by a flow of charged particles (electrons).

electrolysis: the process of breaking a chemical compound down into its elements by passing a direct current through it. Electrolysis of water, for example, produces hydrogen and oxygen.

electromagnetic energy: energy stored in electromagnetic waves or radiation. Any object with a temperature above absolute zero (-273° C) emits this type of energy. The intensity of energy released is a function of the temperature of the radiating surface. The higher the temperature the greater the quantity of energy released. Electromagnetic radiation covers the entire range of wavelengths or frequencies extending from gamma rays to the longest radio waves and including visible light.

electron: an elementary particle consisting of a charge of negative electricity equal to about 1.602×10^{-19} coulombs and having a mass of about 9.109534×10^{-28} grams when at rest.

electrostatic precipitator: a device to remove charged particulates from smoke coming out of a smokestack.

emigration: the process of leaving one country to take up permanent or semi-permanent residence in another; the opposite process to immigration.

emission standard: the maximum amount of a pollutant legally permitted to be discharged from a single source.

emissivity: the ratio of total radiative flux from a body to that of a black body under the same environmental conditions.

emittance: the energy radiated by the surface of a body per second per unit area. Emittance values range from 0.05 for brightly polished metals to 0.96 for flat black paint.

energy: the capacity for doing work. Forms of energy include: thermal, mechanical, electrical and chemical.

energy capacity: the energy delivered in kilowatt hours by battery or an electrical generating station.

energy density: the energy contained in a given volume of a material (J/m^3)

energy efficiency: in transportation, the ratio of output to input energy expressed in miles traveled per gallon of fuel (mpg).

energy efficiency ratio (EER): the ratio of cooling capacity of an air conditioning unit in Btu per hour to the total electrical input in watts under specified test conditions.

energy intensity: in transportation, the ratio of energy input to a process to the useful output from that process; expressed in gallons of fuel per passenger-mile.

entropy: the measure of the disorder or randomness of energy and matter in a system.

Environmental Protection Agency (EPA): established in 1970, the U.S. EPA is charged with the task of repairing the damage done to the natural environment and with establishing new criteria for making a cleaner and healthier environment.

EPA: acronym for Environmental Protection Agency.

enzyme: proteins used to facilitate and regulate chemical reactions within living cells.

equator: the great circle of the celestial sphere whose plane is perpendicular to the axis of the earth. Location on the earth with a latitude of 0°.

equilibrium: a condition where one or more attributes of a system does not change with time.

equinox: either of the two times each year when the sun crosses the equator and day and night are of equal length everywhere. The autumnal equinox occurs on the first day of fall (around September 22). The vernal equinox occurs on the first day of spring (around March 21).

estimated ultimate recovery (EUR): the total amount of a resource that will have been produced from a reservoir when all production ceases. Also see cumulative production and proved reserves.

estuaries: transitional ecosystems between ocean and freshwater biomes; a partially enclosed coastal area at the mouth of a river where nutrient rich fresh water meets with salty ocean water.

ethanol: otherwise known as ethyl alcohol, alcohol, or grain-spirit is a colorless, volatile, flammable liquid C_2H_5OH that is the intoxicating agent in liquors and is also used as a solvent. In transportation, ethanol is used by itself as a vehicle fuel (E100 – 100% ethanol by volume), blended with gasoline (E85 – 85% ethanol by volume), or as a gasoline octane enhancer and oxygenate (10% by volume).

euphotic zone: The upper, illuminated zone of aquatic ecosystem (meaning well-lit in Greek).

EV: See electric vehicles.

evaporation: the change of state from the liquid to vapor phase.

evaporative cooling: cooling by exchange of latent heat from water sprays, jets of water, or wetted material.

evolution: a process in which the whole universe is a progression of interrelated phenomena . Biological evolution concerns with the process of change in living organisms during the history of life on earth.

externality: cost borne by people who are not party to the transaction that imposes upon them.

F

Fahrenheit scale: a scale for measuring temperature. In this scale, water boils at 212° and freezes at 32°.

fault: a crack or fracture in the earth's surface in which there has been movement of one or both sides relative to the other. Movement along the fault can cause earthquakes or, in the process of mountain-building, can release underlying magma and permit it to rise to the surface as a volcanic eruption.

Federal Energy Regulatory Commission (FERC): an independent regulatory commission within the U.S. Department of Energy that has jurisdiction over energy producers that sell or transport fuels for resale in interstate commerce; the authority to set oil and gas pipeline transportation rates and to set the value of oil and gas pipelines for rate making purposes; and regulates wholesale electric rates and hydroelectric plant licenses.

feedback loop: a process referring to a circle of causes and effects that connects an effect of a particular action to changes in its surroundings, which in turn influences further action. If the effect tends to reinforce that action, then feedback is positive. On the other hand if this cause and effect relationship tends to be self-limiting, then feedback is negative. An example of a positive feedback is the process of global warming. An example of a negative feedback is the operation of a thermostat.

ferrel cell: three-dimensional atmospheric circulation cell located at roughly 30 to 60° North and South of the equator.

fertile: in nuclear industry refers to nuclides that do not directly undergo induced fission but from which fissile material is generated. For example, uranium-238 is not fissionable itself, but can be fissionable by absorbing a fast neutron and convert to plutonium-239.

FFV: acronym for flexible fuel vehicle.

fissile: in nuclear industry refers to nuclides capable of sustained chain reaction and fission. The three most important fissile materials are uranium-233, uranium-235, and plutonium-239.

fission: the release of energy from the splitting of an atom's nucleus. This is the energy process used in conventional nuclear power plants to make the heat needed to run steam electric turbines. Also see fusion.

fissionable material: any material with atoms that can undergo nuclear fission. Uranium-238 is a fissionable material but not *fissile.*

flare gas: natural gas that is burned as it is released from an oil field. No energy is collected from this process.

flat plate collector: a device used to collect solar energy; made of a piece of metal painted black on the side facing the sun, to absorb the sun's heat. The heat can be transferred to a liquid and then used as desired (for instance, to heat a building).

flexible fuel vehicle (FFV): a vehicle that can operate on either alcohol fuels (methanol or ethanol) or regular unleaded gasoline or any combination of the two from the same tank.

flue gas: gas that is left over after fuel is burned and which is disposed of through a pipe or stack to the outer air.

fluidized bed combustion: a process where coal is pulverized (turned into powder) and then carried by a stream of air and burned. The process reduces sulfur dioxide emissions from coal combustion.

fog: droplets composed of water, ice crystals or smoke particles.

food chain: the movement of energy through the trophic levels of organisms. In most ecosystems, this process begins with photosynthetic autotrophs (plants) and ends with carnivores and herbivores.

food web: the complex patterns of energy flow in an ecosystem that describe the interrelationships by which organisms consume other organisms.

fossil: the remains of living organisms, such as hard and soft parts of plants and animals, tracks and burrows, whole organisms preserved intact in amber or tar, and fossilized dung.

fossil fuels: carbon based remains of organic matter that has been geologically transformed into coal, oil and natural gas.

freezing: the change in state of matter from liquid to solid, such as the formation of ice from liquid water.

Freon: See chlorofluorocarbons.

fuel: a substance that can be burned to produce heat.

fuel cell: a device that converts the chemical energy of a fuel directly into electricity. The principal components of a fuel cell are catalytically activated electrodes for the fuel (anode) and the oxidant (cathode) and an electrolyte to conduct ions between the two electrodes. Electrons are bypassed through an external wire which constitutes an electric current.

fuel reprocessing (nuclear): the means for obtaining usable, fissionable material from spent reactor fuel.

fuel rod (nuclear): a long slender tube that holds fissionable fuel. Fuel rods are assembled into bundles called fuel elements or assemblies, which are loaded individually into the reactor core.

fumarole: a vent or opening in a volcanic region from which hot gases and vapors escape.

fusion: the combining of isotopes of light elements to form a heavier element, the process of which releases heat and electromagnetic radiation energy. Also see fission.

fusion energy: a technology, presently under development, where atoms are combined under the most extreme heat and pressure. It is the energy process of the sun and the stars.

G

gamma radiation: ionizing radiation with a wavelength less than 0.03 nanometers that readily penetrates the body tissues of organisms.

gasohol: gasoline that contains 10 percent ethanol by volume.

GDP: See gross domestic product.

gene: organic material responsible for passing inheritance or traits. In most organisms these adaptations are coded through the organic molecule DNA.

genetic code: The sequence of rules by which DNA constructs the genes.

generator: See electric generator.

geostrophic wind: wind in the upper atmosphere that moves parallel to isobars.

geothermal energy: energy with its origin in the internal heat of the earth.

glacier: the accumulation of snow and ice developed over a long time period.

glazing: a covering of transparent or translucent material (typically glass or plastic) used for admitting light.

global warming: the warming of the earth's average global temperature because of an increase in the concentration of greenhouse gases.

glucose: the simple six-carbon sugar with chemical formula $C_6H_{12}O_6$.

GNP: See gross national product.

gradient: the steepness of a slope as measured in degrees, percentage, or the rise/run ratio.

gravity: the tendency of one mass to attract another mass. First proposed by Newton in 1686, the force of attraction is found to be proportional to the product of the objects' masses and inversely proportional to the distance that separates them.

greenhouse effect: the characteristic tendency of some transparent materials (such as glass) to transmit short wavelength solar radiation and block radiation of longer wavelengths (such as heat). This tendency leads to a heat build-up within the space enclosed by such a material. In the atmosphere, greenhouse effect refers to the warming of the lower atmosphere due to the accumulation of greenhouse gases that trap heat near the surface of the earth. See global warming.

greenhouse gases: gases responsible for the greenhouse effect. These gases include: carbon dioxide (CO_2); methane (CH_4); nitrous oxide (N_2O); chlorofluorocarbons (CFCs; and ozone (O_3). These gases allow visible and ultraviolet light (shortwave radiation) to pass through the atmosphere but block (long wave) infrared radiation from being re-emitted by the earth. In the last few centuries, the activities of humans have caused the concentration of the major greenhouse gases to increase, raising the earth's average temperature. As a result, temperature has already increased by 0.3 to 0.6° C, since the beginning of this century.

grid: a network of interconnected power lines and generators that is managed so that the generators are dispatched as needed to meet the requirements of the customers connected to the grid at various points.

The electric utility companies' transmission and distribution system that links power plants to customers through high power transmission line service (110-765 kilovolts [kV]); high voltage primary service for industrial applications and street rail and bus systems (23-138 kV); medium voltage primary service for commercial and industrial applications (4-35 kV); and secondary service for commercial and residential customers (120-480 V).

Grid can also refer to the layout of a gas distribution system of a city or town in which pipes are laid in both directions in the streets and connected at intersections.

gross domestic product (GDP): the total market value of all goods and services produced in a country in a given calendar year.

gross national product (GNP): the total market value of the goods and services produced by a nation before deduction or depreciation charges and other allowances for capital consumption which is widely used as a measure of economic activity. In the United States, GNP is calculated quarterly by the Department of Commerce.

gulf stream: the warm ocean current that originates in and around the Caribbean and flows across the North Atlantic to northwest Europe.

H

half-life: the time required for one half of the nuclei in a radioisotope to disintegrate. Half-lives for radioisotopes range from a few millionths of a second to several billion years.

halon: chemical compound containing fluorine, bromine, chlorine, or iodine used in fire fighting. Halons, like chlorofluorocarbons, are considered destructive of the earth's ozone shield.

hazardous waste: wastes that are particularly dangerous or destructive in one or more of the following ways: ignitable, corrosive, reactive, or toxic.

heat: a form of energy representing the aggregated internal energy of the motion of atoms and molecules in a body.

heat capacity: the amount of energy necessary to raise the temperature of a given mass by one degree. Heat capacity may be calculated by multiplying the mass by the specific heat.

heat engine: an engine that converts heat to mechanical energy. Examples are automobile engines, jet engines, and power plants.

heat gain: an increase in the amount of thermal energy in a space, resulting from direct solar radiation, heat flow through walls, windows, and other building surfaces, and the heat given off by people, lights, equipment, and other sources.

heat loss: a decrease in the amount of thermal energy in a space, resulting from heat flow through walls, windows, the roof and other building surfaces and from infiltration of cold outside air.

heat pump: a device for transferring heat from a low-temperature medium (outside air in the winter) to a high temperature one (air inside the house).

heat transfer: energy transferred from one object to another because of the temperature difference between them. Heat is commonly transferred by conduction, convection, and radiation.

heating load: the rate at which heat must be added to a space in order to maintain the desired temperature within the space.

heating value: the amount of heat given off by the complete combustion of a unit mass of fuel.

herbicide: a chemical substance used to kill plant weeds.

herbivore: animals that consume plants. Also known as a primary consumer. Also see omnivore and carnivore.

hertz: a unit of electromagnetic wave frequency equal to one cycle per second.

heterotroph: an organisms that cannot produce their own food. Also see autotrophs.

horsepower: a unit of power equal to 746 watts.

hot dry rock: an impermeable subsurface rock 5,000 meters or more below the earth's surface heated by geothermal energy.

hot spot: an area of exceptionally high species richness or concentration (e.g. near a power plant or smoke stack).

hybrid vehicle: a vehicle with two or more power sources, normally an internal combustion engine together with an electric propulsion system.

hydrocarbon: organic compound composed primarily of hydrogen and carbon atoms.

hydroelectric power: or hydropower is electricity produced by falling water that turns a turbine generator.

hydrosphere: oceans and other waters of the earth.

hydrothermal reservoir: an underground zone of porous rock containing hot water and dry steam.

I

IAEA: acronym for the International Atomic Energy Commission.

immigration: the process of entering one country from another to take up permanent residence. Opposite of emigration.

incineration: the burning of trash and garbage at high temperatures in a large furnace for its disposal and to generate electricity.

independent power producer (IPP): companies or individuals who construct electricity-generating plants and then sell the electricity to an electric utility at wholesale prices. The utility then resells this power to end-use customers. IPPs usually do not own transmission lines to transmit the power that they generate. Although IPPs generate power, they are not franchised utilities, government agencies or qualifying facilities.

independent system operator (ISO): a neutral operator responsible for maintaining instantaneous balance of the grid system. The ISO performs its function by controlling the dispatch of flexible plants to ensure that loads match resources available to the system.

indigenous energy resources: power and heat derived from sources native to a region. These include geothermal, hydro, biomass, solar and wind energy. The term usually is understood to include cogeneration facilities.

industrial smog: air pollution resulting from various industrial activities, consisting of a combination of sulfur dioxide, suspended droplets of sulfuric acid, and a variety of suspended solid particles. Also see photochemical smog.

infiltration: the uncontrolled leakage of air through cracks and gaps in the building around windows, doors and duct systems.

infrared radiation: low-energy, long-wavelength (between 0.7 and 100 micrometers) electromagnetic radiation perceived by humans as heat.

inner core: the inner region of the earth's core made of solid iron and nickel. The inner core has a diameter of about 1,200 kilometers.

insolation: the total amount of solar radiation (direct, diffuse, and reflected) that is received by the earth's surface.

instability: atmospheric condition where a parcel of air is warmer than that of the surrounding air, causing it to rise in the atmosphere.

installed capacity: also called the nameplate capacity is the total manufacturer-rated capacities of equipment such as turbines, generators, condensers, transformers, and other system components.

Intergovernmental Panel on Climate Change (IPCC): a large, international group of officials, scientists, and other researchers who, under the auspices of the United Nations, have been investigating the issue of global climate change, particularly potential future global warming.

internal combustion engine: an engine in which fuel is burned inside. Examples are cars' gasoline or rotary engines. Also see external combustion engine.

International Atomic Energy Commission (IAEA): an international agency under auspices of the United Nations, established in 1957. The agency is responsible for monitoring international nuclear activities and helping nations with safe, secure and peaceful nuclear technologies.

investor owned utility (IOU): a privately owned, for-profit company that provides a utility, such as water, natural gas or electricity, to a specific service area. A designation used to differentiate a utility owned and operated for the benefit of shareholders from municipally owned and operated utilities and rural electric cooperatives.

ion: an atom or group of atoms that carries either a positive (cation) or negative (anion) electrical charge.

ionizing radiation: the emission of alpha or beta particles or gamma rays from radioisotopes that can dislodge one or more electrons from atoms they strike. The free electrons can form charged ions in living tissue that can react with and damage cells or cause cancer.

ionosphere: a region in the atmosphere above 50 kilometers from the earth's surface where relatively large concentrations of ions and free electrons exist. The ionosphere is important in communications because it redirects AM radio transmissions, and thus extends the range that radio transmissions can travel.

IOU: acronym for investor owned utility.

IPP: acronym for independent power producer.

ISO: acronym for independent system operator.

isotope: a form of an element where the number of neutrons in its atomic nucleus is different than the number of protons.

isotopic dating: dating technique used to determine the age of a rock and mineral through the decay of radioactive elements. Also see carbon dating.

J

joule: a unit of work or energy equal to the amount of work done when the point of application of force of 1 newton is displaced 1 meter in the direction of the force. It takes 1,055 joules to equal a British thermal unit. It takes about 1 million joules to make a pot of coffee.

K

Kelvin scale: a scale for measuring temperature. In this scale absolute zero is 0 kelvins, water boils at 373.15 kelvins and freezes at 273.15 kelvins.

kerogen: the bituminous material in shale which yields oil when heated.

kerosene: a petroleum distillate suitable for use as an illuminant when burned in wick lamps.

kilocalorie (kcal): a unit of energy equal to 1,000 calories.

kilopascal (kPa): a unit of pressure equal to 1,000 pascals. 1 atm=100 kPa.

kilowatt (kW): a unit of electrical power equal to 1,000 watts. Typical homes, with central air conditioning and other equipment have a demand of 1-5 kW.

kilowatt-hour (kWh): the most commonly-used unit of electricity consumed over time, i.e., one kilowatt of electricity supplied for one hour.

kinetic energy: the energy of motion.

L

La Niña: very strong tropical Pacific trade winds in the central and eastern Pacific Ocean. Also see El Niño.

land breeze: surface winds blown from land to water during the night. Also see sea breeze.

landfill gas: gas generated by the natural decomposition of municipal solid waste by anaerobic microorganisms in sanitary landfills. The gases produced, carbon dioxide and methane, can be collected by a series of low-level pressure wells and can be processed into a medium Btu gas that can be burned to generate steam or electricity.

last mile delivery: the delivery (communication fiber, electrical wiring, or gas pipeline) from the public network to private houses and firms.

latent heat: the energy required or released when a substance changes phase without changing its temperature. Latent heat of fusion is the amount of energy absorbed during melting or released during freezing. Latent heat of vaporization is the amount of energy absorbed during vaporization or released during condensation.

latitude: a north-south measurement of position on the earth, measured in degrees, minutes, and seconds. A line connecting all places of the same latitude is termed a parallel. Latitude ranges from (0°) at the equator to 90° north and 90° south at the north and south poles.

lava: the molten magma released from a volcanic vent or fissure.

law of conservation of energy: this law (also called the first law of thermodynamics) states that energy can neither be created nor destroyed, but only transferred from one form to another. Thus, the total amount of energy available in the universe is constant.

leeward: the side facing away from the wind. Opposite of windward.

lever: a bar than can move freely around a fixed position. There are three kinds of levers. In the **first class lever,** the fulcrum is between the force and the weight (shovel, seesaw, and scissors). When the weight is between the fulcrum and the force, the lever is called a **second class lever** (tweezers, wheelbarrow). In **third class levers** the force is located between the fulcrum and the weight (baseball bat).

life cycle analysis: also called the cradle-to-grave analysis, analyzes the entire "life cycle" of a product, from procurement of the raw materials through its use, to its eventual disposal and the possible reuse or recycling of its components. Life cycle cost is the amount of money necessary to own, operate and maintain a building over its useful life.

life expectancy: the average number of years that a typical person at a certain age can expect to live.

light duty vehicles: combined automobiles and light trucks.

light water reactor (LWR): a nuclear power unit that uses ordinary water as the coolant and the moderator. A LWR may be a boiling water reactor or a pressurized water reactor.

light year: the distance that light travels in a vacuum in one year. It is approximately 9.7 trillion kilometers.

lignin: amorphous polymer making up about a quarter of plant material that provides rigidity and together with cellulose forms the woody cell walls of plants and the cementing material between them.

lignite: also called brown coal, is the coal with carbon content somewhere between that of bituminous coal and peat. The texture of the original wood is often visible in lignite.

limestone: the sedimentary rock composed of carbonate minerals.

liquefaction: the process of making synthetic liquid fuel from coal. The term is also used to refer to a method for making gasoline and heating oil from petroleum.

liquefied natural gas (LNG): a mixture of gaseous hydrocarbons, mainly propane and butane, that change into liquid form under moderate pressure.

liquefied petroleum gas (LPG): the same as propane. LPG is commonly used as a fuel in rural homes for space and water heating, as a fuel for barbecues and recreational vehicles, and as a transportation fuel.

lithosphere: the rigid crust of the earth.

LNG: See liquefied natural gas.

load: the amount of electric power supplied to meet one or more of end user's needs.

load factor: a percent telling the difference between the amount of electricity a consumer used during a given time span and the amount that would have been used if the usage had stayed at the consumer's highest demand level during the whole time. The term is also used to mean the percentage of capacity of an energy facility - such as a power plant or gas pipeline -- that is utilized in a given period of time. In transportation, it means the total passenger miles divided by total vehicle miles (or average number of passengers per vehicle).

load management: strategies taken to reduce power demand at peak load times or to shift some of the load to off-peak times. Load management may be pursued by persuading consumers to modify behavior or by using equipment that regulates electric consumption.

longitude: a west-east measurement of position on the earth measured in degrees, minutes, and seconds. A line connecting all places of the same longitude is termed a meridian. Measurements of longitude range from the prime meridian (0°) to 180° west and 180° east from this point.

low emission vehicles (LEV): vehicles certified by the California Air Resources Board, or another similar agency, to have emissions from zero to 50,000 miles no higher than 0.075 g/mi of non-methane organic gases, 3.4 g/mi of carbon monoxide, and 0.2 g/mi of nitrogen oxides. Emissions from 50,000 to 100,000 miles may be slightly higher.

LPG: See liquefied petroleum gas.

M

magma: the molten rock beneath the earth's crust. To extract energy from magma resources requires drilling near or directly into a magma chamber and circulating water down the well in a convection- type system.

magneto hydrodynamics (MHD): a means of producing electricity directly by moving liquids or gases through a magnetic field.

mantle: the zone of the earth below the crust and above the core.

marginal cost: in economics is the cost to the utility of providing the next increment of product or service.

market failure: when market prices do not reflect all the true costs of a product or service (externalities).

mass number: the total number of neutrons and protons in the nucleus of an atom. Approximate measure of the mass of an atom. Also see atomic number.

mean solar day: the time it takes to complete one earth rotation relative to the sun (for example, from midnight to midnight). This measurement takes 24 hours and is longer than a sidereal day because it includes the effect of the earth's revolution around the sun.

mechanical efficiency: the ratio of the mechanical energy output to the mechanical energy input.

mechanical work: the work associated a force acting in the direction of motion.

melting: the process of conversion of solids to liquids.

meridian: a circular arc passing through poles. All points on the same meridian have the same longitude.

metabolism: enzymatic reactions performed by the cells of a living organism.

methane: a light hydrocarbon and the main component of natural gas and marsh gas. It is the product of the anaerobic decomposition of organic matter, fermentation in animals and is one of the greenhouse gases. The primary sources for methane are from rice cultivation, domestic grazing animals, termites, landfills, coal mining, and oil and gas extraction. The chemical formula for methane is CH_4.

methanol: also known as methyl or wood alcohol is a colorless highly toxic liquid with essentially no odor and very little taste. Methanol is formed by steam reforming natural gas, by the distillation of wood, or the catalytic reaction of carbon monoxide (CO) with hydrogen (H_2), under high temperature and pressure. Methanol is the simplest alcohol and has the chemical formula of CH_3OH. In transportation, methanol is used as a vehicle fuel by itself (M100), or blended with gasoline (M85).

methyl alcohol: a common liquid hydrocarbon fuel that is also called methanol and is used in some gasoline blends. It has the chemical formula of CH_3OH.

methyl tertiary butyl ether (MTBE): a clean-burning oxygenate with high octane and low volatility added to unleaded gasoline to reduce carbon monoxide emissions. It is one of the primary ingredients in reformulated gasoline.

microwave: electromagnetic radiation with a wavelength between 0.1 to 100 centimeters. It falls between infrared and radio wavelengths on the electromagnetic spectrum.

midpoint: the point at which half of ultimate production has been consumed.

moderator: a substance like water, graphite, or beryllium used in a nuclear reactor to slow down fast neutrons.

molecule: a minute particle that consists of connected atoms of one or many elements.

monopoly: a situation where one firm controls market sales. A monopoly can produce a given level of output at a lower total cost than can any combination of multiple firms.

Montreal Protocol: an international agreement reached in 1987 at a meeting in Montreal, Canada, whereby a number of industrialized countries pledged to freeze CFC production at 1986 levels and then gradually decrease its production until it was eliminated 1999.

mountain breeze: the local thermal circulation pattern in mountainous areas, where surface winds blow from areas of higher elevation to valley bottoms during the night. The reverse happens during the day.

MTBE: acronym for methyl tertiary butyl ether.

municipal solid waste: locally collected garbage, which can be processed and burned to produce energy.

mutation: the change in the structure of a gene or chromosome.

N

NAFTA: See North American Free Trade Agreement.

natural gas: gas commonly found in the pores of sedimentary rocks of marine origin, composed mainly of methane, but also containing ethane, butane, propane and other gases.

neap tide: a tide that occurs every 14 to 15 days and coincides with the first and last quarter of the moon. This tide has a small tidal range because the gravitational forces of the moon and sun are perpendicular to each other. Contrasts with spring tide.

negative feedback: the change in the state of a system that counteracts the effect of the initial alteration.

net primary productivity: total amount of chemical energy fixed by the processes of photosynthesis minus the chemical energy lost through respiration.

neutron: an atomic sub-particle found in the nucleus of an atom. This particle is similar in mass to a proton but does not have a charge.

newton: the force required to accelerate a mass of 1 kg by 1 m/s^2.

nitric acid: acid formed as a result of the interaction of nitric oxides with water. It has the chemical formula HNO_3.

nitrogen fixation: the biological or chemical process where gaseous nitrogen is converted into solid forms of nitrogen. Biological fixation is done by specialized microorganisms like bacteria. Chemical fixation occurs at high temperatures. One natural process that can produce enough heat to fix atmospheric nitrogen is lightning.

nitrogen oxides (NOx): oxides of nitrogen produced by the burning of fossil fuels. Nitrogen oxides consist mainly of two gases: nitric oxide (NO) and nitrogen dioxide (NO_2). These gases are produced by bacterial action in the soil and by high temperature combustion. Both gases are chief components of air pollution and responsible for the production of photochemical smog.

node: a point in a network where electricity can be supplied to the network or withdrawn from it.

non-associated gas: natural gas that is not trapped with petroleum in the same reservoir. Also see associated gas.

non-ionizing radiation: a form of electromagnetic radiation that does not have enough energy to cause ionization of atoms in living tissue. Examples of this type of radiation include radio waves, microwaves, infrared light, and ordinary light. See also ionizing radiation.

non-renewable resource: a resource that is finite in quantity and is being used faster than its ability to regenerate itself. Fossil fuels are a prime example of nonrenewable energy sources.

North Atlantic Free Trade Agreement (NAFTA): trade agreement signed between the United States, Mexico, and Canada in 1992 in which the three countries agree to remove all trade barriers, tariffs, and quotas between them. Also See Central American Free Trade Agreement (CAFTA).

NOx: acronym for nitrogen oxides.

nuclear energy: energy stored in the nucleus of atoms. Power obtained by splitting heavy atoms (fission) or fusing light atoms (fusion). A nuclear energy plant uses a controlled atomic chain reaction to produce heat. The heat is used to produce steam which runs conventional turbine generators.

Nuclear Regulatory Commission (NRC): an independent federal agency that ensures strict standards of public health and safety, environmental quality and national security are adhered to by individuals and organizations possessing and using radioactive materials. The NRC is the agency that is mandated with licensing and regulating nuclear power plants in the United States. It was formally established in 1975 after its predecessor, the Atomic Energy Commission was abolished.

nucleic acid: an organic compound composed primarily of carbon, hydrogen, nitrogen, oxygen, and phosphorus. DNA or deoxyribonucleic acid, the genetic blueprint of life, is an example of a nucleic acid.

nucleus: the dense central portion of an atom that is composed of neutrons and protons.

O

octane: a rating scale used to grade gasoline according to its antiknock properties. A measure of gasoline's resistance to self-ignition too early in the engine cycle, which causes knocking. The higher the octane rating, the lower the chance of premature ignition.

OECD: acronym for the Organization for Economic Cooperation and Development. OECD is an international organization established in 1948 to stimulate growth among the member states in Europe, the United States, Canada, Mexico and Japan.

OEM: acronym for original equipment manufacturer.

ohm: a unit of measurement of electrical resistance. One volt can produce a current of one ampere through a resistance of one ohm.

oil: the hydrocarbon-based liquid commonly found in the pores of sedimentary rocks of marine origin.

oil shale: a type of rock containing organic matter kerogen that produces oil when heated to high temperatures.

OPEC: acronym for the Organization of Petroleum Exporting Countries.

open system: a system that allows transfer of both matter and energy across its boundary.

organic matter: matter that contains living organisms or non-living material derived from organisms.

organic molecule: Any molecule containing carbon and involved in forming life. Carbon dioxide and carbonate minerals are not considered to be organic.

organism: any form of life.

Organization of Petroleum Exporting Countries (OPEC): founded in 1960 to unify and coordinate petroleum polices of its members. OPEC headquarters is in Vienna, Austria. OPEC currently has 13 members which include Saudi Arabia, Iran, Venezuela, Libya, Indonesia, United Arab Emirates, Algeria, Nigeria, Ecuador, Gabon, Iraq, Kuwait, and Qatar.

original equipment manufacturer (OEM): one who manufactures complex systems from components usually bought from other manufacturers.

outage: See blackout.

outer core: the outer region of the earth's core which is 2,300 kilometers thick and composed of liquid molten metals, mostly iron but also some nickel. Temperature of the outer core is relatively uniform, and varies from 5,000°C to about 4,000°C at the edges of the outer core.

overhang: any horizontal projection that serves as a shading element for a window in summer.

oxidation: the process of combining with oxygen.

oxygenate: a method for blending oxygen-rich liquids into gasoline in order to convert carbon monoxide to carbon dioxide and to better oxidize volatile organic compounds. The increased oxygen content given by oxygenates promotes more complete combustion, thereby reducing tailpipe emissions. Also see reformulated gasoline.

ozone: a triatomic form of oxygen. It is formed naturally in the atmosphere by photochemical processes. Ozone is a poisonous gas, and in the lower atmosphere an air pollutant. In the upper atmosphere, ozone acts as a shield protecting life on earth from deadly ultraviolet radiation from space.

ozone hole: a sharp seasonal decrease in stratospheric ozone concentration that occurs over Antarctica and much of the world. First detected in the late 1970s, the ozone hole continues to appear as a result of a complex chemical reaction in the atmosphere that involves chlorine and bromine atoms.

ozone layer: a layer in the stratosphere containing ozone, most concentrated at an altitude between about 12 and 16 miles (20-25 km). This layer is important to life on the earth as it absorbs harmful ultraviolet radiation.

P

particulate matter (PM): small particles of either a solid or a liquid that form smoke or soot.

pascal: the unit of pressure defined as newtons per square meter (N/m^2). One atmosphere is approximately 100 kilopascals.

passive solar system: a solar heating or cooling system that uses no external mechanical power to move the collected solar heat. Examples of passive solar systems are common glass, solar roofs, and solar chimneys.

PCB: or polychlorinated biphenyls is a group of organic compounds used in the manufacture of plastics and formerly used as a coolant in electric transformers. In the environment, PCBs are highly toxic to aquatic life.

peak load: the amount of electricity needed at the time of highest demand. Daily electric peaks on weekdays occur in the late afternoon and early evening. Annual peaks occur on hot summer days.

peak load power plant: the power generating station used to produce the extra electricity during peak load times. Also called peaking unit.

peaking capacity: the capacity of generating equipment intended for operation during the hours of highest loads.

peat: partially decomposed remains of organic matter that have accumulated in a water-saturated environment over a very long period of time. Geologically, peat is considered a very young form of coal and has a heating value of 6,600 Btu/pound in situ.

periodic table: a table that arranges chemical elements to describe their chemical properties.

permafrost: the zone of permanently frozen water found in high latitude soils and sediments.

permeability: a measure of the ability of soil, sediments, and rock to transport water. Permeability is dependent on the porosity of the medium the water is flowing through. Some rocks like granite have very poor permeability. Sand is the most pervious, while clay has the lowest permeability; silt is somewhere in the middle.

Persian Gulf: the waterway bounded by the countries of Iran, Iraq, Bahrain, Kuwait, Qatar, Saudi Arabia, and United Arab Emirates.

petrochemicals: chemicals made from oil.

petroleum: a generic term applied to oil and oil products.

phases of matter: matter can exist in three different forms: gas, liquid, and solid.

pH scale: a scale that is used to measure acidity; a pH of 7.0 is neutral, lower numbers are acidic, higher numbers are basic (alkaline).

photocell: See photovoltaic.

photochemical pollutant: a pollutant produced when sunlight initiates chemical reactions among NOx, volatile organic compounds, and other substances found in the air. Also see secondary pollutant.

photochemical smog: a condition that develops when primary pollutants (oxides of nitrogen and volatile organic compounds created from fossil fuel combustion) interact under the influence of sunlight to produce a mixture of hundreds of different and hazardous chemicals known as secondary pollutants. Also see industrial smog.

photon: a discrete packet of radiant energy.

photosynthesis: the chemical process whereby green plants and some bacteria can capture the energy of the sun and synthesize organic compounds from water and carbon dioxide.

photovoltaic: a semiconductor device that converts light directly into electricity.

plankton: a collective term for a variety of marine and freshwater organisms that drift on or near the surface of the water.

plate tectonics: the theory that the earth's lithosphere is divided into numerous plates that are in motion relative to each other; the continents ride on the backs of the plates and thus move ("drift") over geologic time.

plutonium: a heavy element containing 94 protons. Fissionable isotopes of this element can be used as fuel in nuclear reactors and can also be manufactured into bombs.

pollutant standards index (PSI): an index defined as a measure of air quality. Index of 100 is assigned to moderate air quality. Higher numbers indicate poorer air quality.

population density: the number of individuals of a particular species per unit found in a specified area.

positive feedback: the process in which a system responds to change in a way that magnifies the initial change.

potential energy (gravitational): the stored energy of a body by virtue of its position.

power: the work performed in unit time, measured in watts or horsepower (h.p.).

power (electrical): the product of voltage and current, measured in watts. Power over time is usually defined in watt-hours (Wh), the product of the average number of watts and time. Your energy utility usually bills you per kilowatt-hour (kWh), one of which is 1,000 watt-hours.

power (mechanical): the product of force and velocity. Alternatively it can be calculated by dividing work by the time it takes to perform that work.

power (thermal): a measure of the amount of heat flow in unit time. It is often expressed in kJ/hour, kW, or Btu/hr.

power density: power per unit volume (W/m^3) or per unit area (W/m^2); in engines, it usually implies power per unit volume of engine or gas turbine. In wind power industry, it refers to power per unit rotor area. When we discuss biomass resource, we often refer to the power generated per unit area of the land mass.

power plant: a central station generating facility that converts some form of energy into electrical energy.

ppm (parts per million): a common unit of concentration of gases or vapor in air. For example, 1 ppm of a gas means that 1 unit of the gas is present for every 1 million units of air.

predator: a consumer organism who feeds on prey.

pressurized water reactor (PWR): a method of producing nuclear power, cooled by water at elevated pressure to keep it from boiling when it reaches high temperatures.

prevailing wind: the dominant direction that a wind blows from.

primary carnivore: See secondary consumer.

primary consumer: organisms that subsists on the producers (plants) for nourishment, usually herbivores.

primary pollutant: air pollutants that enter the atmosphere directly from the source such as tailpipes and smoke stacks. Also see secondary pollutant.

primary producer: organisms that occupy the first trophic level in the grazing food chain. These organisms are photosynthetic autotrophs.

producer: any of various organisms (like a green plant) which produce their own food by photosynthesis and may become food sources for other organisms.

propane: also known as LPG is a hydrocarbon gas molecule consisting of three atoms of carbon and eight atoms of hydrogen, and one of the constituents of natural gas.

protein: organic substances primarily composed of carbon, hydrogen, nitrogen, and some other minor elements which are arranged in about 20 different compounds known as amino acids. The various amino acids found in a protein are linked together by peptide bonds.

proton: a positively-charged elementary particle identical to the nucleolus of a hydrogen atom.

pumped hydroelectric storage: a system where excess power is used to pump water to a reservoir during off-peak periods. During peak periods, the reservoir is used to operate auxiliary hydroelectric generators.

PX: the California Power Exchange Corporation; a state chartered, non-profit corporation charged with scheduling and providing markets for energy and ancillary services. PX is independent of the ISO and all other market participants.

Q

quad: one quadrillion (10^{15} Btu), an amount of energy equal to 170 million barrels of oil.

qualifying facility (QF): non-utility power producers that generate electricity using renewable and alternative resources, such as hydro, wind, solar, geothermal or biomass (solid waste). Under federal law, QFs have the right to sell their excess power output to the public utilities.

R

rad (radiation adsorbed dose): a unit for measuring absorbed radiation. One rad equals 100 ergs of radiation energy per gram of absorbing material.

radiant energy: energy transferred by the exchange of electromagnetic waves from a hot or warm object to one that is cold or cooler. In some cases it refers to the radiation emitted from the sun.

radiation: the flow of energy via electromagnetic waves such as light.

radioactive decay: the natural decay of the nucleus of an atom.

radioactivity: a property of isotopes of atoms from a nucleus as it disintegrates. In the process some elementary particles (such as alpha and beta particles) and energy (such as gamma rays) are emitted.

radioisotope (radioactive isotope): an unstable isotope of an element.

radiometer: an instrument used to measure radiation over a range of wavelengths.

ramp: a slanted surface used to lift a weight.

Rankine cycle: the steam-cycle consists of a boiler, a steam turbine, a condenser, and a pump.

reformer: a device for on-board production of hydrogen.

reformulated gasoline: a cleaner-burning gasoline with the oxygenate additive MTBE.

refrigerant: a fluid such as R-134a or ammonia. Refrigerants are usually gas at room temperatures but turn into liquid at temperatures considerably colder than the ambient.

refuse: refers to both trash and garbage.

remediation: efforts to reduce effects of pollution after it has been released into the environment.

renewable energy sources: resources that will regenerate within a human time scale and are regarded as practically inexhaustible. These include solar, wind, hydro, ocean thermal, and wood. Geothermal is considered renewable by some and non renewable by others depending on the time the reservoir needs to recharge itself.

reserve capacity: the extra generating capacity of an electric utility to meet peak or abnormally high demands for power, or when other generating units are off-line for maintenance, repair or refueling.

reserves: a retainer or holder of a resource for use at a future time. Natural resources that have been discovered and can be exploited for profit. Proved reserves are reserves that are expected to be recovered from future explorations, and which can be produced economically through application of improved recovery techniques. Inferred reserves (or indicated reserves) are estimates of the amount of reserve growth.

residential sector: private households consuming energy primarily for space heating and cooling, water heating, refrigeration, lighting, cooking, etc.

resistance: the ability to resist the flow of heat or current. Resistance depends on the cross section of the conductor (the smaller the cross section, the greater the resistance) and its temperature (the hotter the cross section, the greater its resistance). Also see conductance.

resource: known or inferred available sources of energy supply that are not exploitable for profit. Conventional resources are discrete concentrations of resources that can be extracted using traditional methods. Examples of conventional resources are petroleum products such as gasoline and kerosene. Unconventional resources are resources whose technical development is in relatively early stages, and which require practices different from traditional production methods. Examples of unconventional resources are shale oils and tar-sands.

respiration: the process of the "burning" of food molecules in an organism to release the energy required for all metabolic processes. Also see photosynthesis.

restructuring: the separation of the various utility functions into several individually-operated and individually-owned entities (such as generation, distribution, and retail).

retail market: a market in which electricity and other energy services are sold directly to the end-use customer.

ribonucleic acid (RNA): a form of nucleic acid used by most organisms to read the genetic information found in DNA and to produce specific organic molecules used in the development and functioning of cells.

rocket: a device where a solid or liquid fuel and an oxidizer react in the combustion chamber producing high-pressure combustion gases that are then expanded in a nozzle, producing the thrust to propel the rocket.

rubbish: a general term that includes trash, garbage, and other items such as construction and demolition debris.

R-value: a unit of thermal resistance used for comparing insulating values of different materials. The higher the R-values, the greater the material's insulating properties and the slower heat flows through it.

S

salinity: concentration of dissolved salts in water.

scattering: the atmospheric process where small particles and gas molecules diffuse rays of the incoming solar radiation in random directions without changing their wavelengths.

screw: an inclined plane wrapped around a cylinder or a cone.

sea breeze: surface winds blown from water to land during the daytime. Also see land breeze.

second law of thermodynamics: See thermodynamics (laws of).

secondary consumer: animals that feed on the primary consumers.

secondary pollutants: pollutants not emitted directly from the source but formed after primary pollutants and other components of the air react, mainly with ozone. Also see primary pollutants.

secondary standards: an air-quality standard designed to protect property, vegetation, etc. as opposed to health.

sedimentary rock: a rock formed on or near the earth's surface by the settling or precipitation of materials.

shale oil: See oil shale.

shrub: a woody plant species that is smaller than a tree. Shrubs usually do not have a trunk.

sick building syndrome: syndrome resulting from indoor air pollutants in a building with symptoms such as headaches, nausea, and allergic reactions.

sidereal day: the time it takes the earth to make one complete rotation relative to the position of a fixed star. This measurement takes 23 hours and 56 minutes. Also see solar day.

silt: mineral particles with a diameter between 0.004 and 0.06 millimeters.

smog: a mixture of smoke and fog, mostly a brown haze of photochemical pollutants. Also see photochemical smog.

sin tax: euphemism for a tax levied on certain activities, which are subject of widespread disapproval - usually alcohol and tobacco.

SOC: See state of charge.

soil permeability: the rate at which a fluid moves vertically through the soil.

soil porosity: the ratio of the volume of voids to the total volume of the soil.

solar cell: a photovoltaic cell that can convert light directly into electricity. A typical solar cell uses semiconductors made from silicon.

solar collector: a device that absorbs solar energy to heat a fluid. The heated fluid can be used directly or used to heat water or space.

solar concentrator: the process of concentrating sunlight on a relatively small area.

solar day: the time it takes earth to complete one rotation relative to the sun. Also see sidereal day.

solar energy: thermal and light energy radiated from the sun.

solar irradiation: the amount of radiation, both direct and diffuse, that can be received at any given location.

solar noon: the time of the day when the sun is aligned with true north and true south. Half-way between sunrise and sunset.

solar power satellite: a proposed process of using satellites in geosynchronous orbit above the earth to capture solar energy with photovoltaic cells, convert it to microwave energy, beam the microwaves to earth where they would be received by large antennas and changed from microwaves into usable electricity.

solar radiation: electromagnetic radiation that originates from the sun. Also see insolation, direct solar radiation, and diffused solar radiation.

solid waste: garbage, trash, refuse, and rubbish that result from mining, agricultural, commercial, and industrial activities.

specific energy: energy per unit mass (J/kg). In batteries, it is defined as the rated energy capacity of a battery divided by the total battery system weight; measured in watt-hours per kilogram.

specific gravity: the ratio of the mass of a body to the mass of an identical volume of water at a specific temperature. Also see density.

specific heat: the amount of heat needed to raise the temperature of one gram of a substance one degree celsius. In US customary units, the quantity of heat, in Btu, needed to raise the temperature of one pound of material one degree fahrenheit.

spectrum: radiation emitted from a body at different wavelengths.

speed of light: the velocity of light in a vacuum (3×10^8 meters per second).

spring tide: a tide that occurs every 14 to 15 days and coincides with the new and full moons. During this time, tides have the largest range because the gravitational forces of the moon and the sun are complementary to each other. Also see neap tide.

state of charge: describes how full a battery is charged. The exact voltage to battery charge correlation is dependent on the temperature of the battery. Cold batteries will show a lower voltage when full than hot batteries.

Stirling engine: an external combustion engine which works by heating (expanding) and cooling (contracting) a captive gas such as helium or hydrogen.

strategic petroleum reserve: stockpiles of government-owned crude oil stored at various reserves that can be drawn on during emergencies and severe oil supply disruptions. A reserve of 750 million barrels of oil is available that can be released only by the order of the President of the United States.

stratosphere: the thermal layer of the atmosphere above the troposphere (11 to 50 kilometers above the earth's surface) in which temperature increases with altitude. The ozone layer exists within the stratosphere. Also see troposphere.

strip-mining: the surface mining of minerals and resources such as coal.

subatomic particles: extremely small particles that make up the internal structure of atoms.

sublimation: the change in phase of matter directly from solid to vapor without passing through the liquid state. Examples of substances undergoing sublimation at standard temperatures and pressures are dry ice and iodine.

subsidence: the sinking of the earth's surface.

substation: an intermediary facility that steps up or steps down the voltage in utility power lines. Voltage is stepped up where power is sent through long-distance transmission lines. It is stepped down where the power is to enter local distribution lines.

sulfur dioxide: a colorless gas consisting of two atoms of oxygen and one atom of sulfur (SO_2) found in volcanic eruptions, ocean spray, organic decomposition and the products of burning fossil fuels. Once in the air, in the presence of water vapor, it forms sulfuric acid mist that can be found in fogs or precipitated as rain or snow.

superconductor: a synthetic material that has very low or no electrical resistance. A major research effort is underway to find materials that behave as superconductors near room temperature. Once found, electrical transmission lines may be built that offer little or no resistance, thus conserving energy usually lost in transmission.

supply-side management: in economics, refers to all efforts required to find new resources and processes that improve supply of a natural resource. See also demand-side management.

surface waves: a special type of seismic wave that travels across the earth's surface, causing the earth's surface to roll or sway like waves on the ocean.

sustainability: means different thing to different people. Neoclassical economists equate it with economic progress, while ecologists define sustainability as meeting the needs of today's generation, without reducing the quality of life for future generations.

sustainable growth: development that focuses on making social, economic, and political progress to satisfy global human needs, desires, aspirations, and potential without damaging the environment; also known as sustainable development.

syncrude: synthetic crude oil made from coal or from oil shale. Also see synfuel.

synfuel: synthetic gas or synthetic oil. Fuel made by artificial means (in contrast to that which is found directly in nature). Also see syncrude.

T

temperature: a measure of the average speed of atoms and molecules.

therm: a unit of energy equal to 100,000 Btu, mostly used to express amounts of gas consumption.

thermal conductivity: See conductivity.

thermal efficiency: the ratio of work output to energy input.

thermal energy: See heat.

thermal power plant: any electrical generating facility using heat to boil water or another working fluid to turn it into vapor or steam used to run a generator. Examples of thermal power plants are oil, coal, gas, nuclear, and geothermal. Wind, hydroelectric, or solar photovoltaic electrical generating facilities are not considered thermal plants.

thermal storage: a material with high specific heat capable of storing energy. Typical thermal mass materials include concrete, rock, brick, and water.

thermocline: the transition layer of water between the mixed warm layer at the surface and the cold deep water.

thermodynamic equilibrium: a condition where all nonuniformities (temperature, pressure, density, electric potential, etc.) have smoothed out.

thermodynamics (laws of): the fundamental laws of nature dealing with quantitative and qualitative aspects of energy. The three laws of thermodynamics are: **the first law of thermodynamics**, which states that energy may be transformed from one form to another, but its total remains constant; the **second law of thermodynamics**, which states that heat cannot be converted completely to work; and the **third law of thermodynamics**, which states that heat capacity and entropy of every crystalline solid becomes zero at a temperature of absolute zero, which is impossible toreach.

thermodynamics: the study of the conversion of energy into other forms and of their practical applications.

tidal current: an ocean current created by the tidal rise and fall of the ocean surface.

tidal period: the time it takes for one tidal cycle (between two high tides).

tide: the cyclical rise and fall of the surface of the oceans caused by the gravitational attraction of the sun and moon on the earth.

ton of refrigeration: the cooling effect equal to 12,000 Btu/hour

transformer: an electronic device used to raise or lower one electric potential to another.

transportation sector: private and public vehicles including automobiles, buses, trucks, motorcycles, railroads, aircrafts, ships, barges, and natural gas pipelines.

trophic level: a group of organisms that occupy the same position in a food chain.

tropical rainforest: the plush forested biome near the equator.

tsunami: a large ocean wave formed as a result of underwater seismic activities or volcanic eruption. Its height is rather small, about 1 meter in open oceans, but reaches over 15 meters in height when it enters shallow coastal waters.

U

ultraviolet radiation: electromagnetic radiation with a wavelength between visible and x-ray (0.1-0.4 micrometers).

updraft: air current with upward velocity.

upper mantle: the layer of the earth's interior between lower mantle and crust.

uranium: a radioactive element, commonly used as fuel in fission reactors.

utility: an entity that owns or operates facilities for generation, transmission, distribution, or sale of electricity; a regulated, vertically-integrated electric company.

V

valley breeze: the wind pattern which develops as a result of non-uniform heating between the valley to areas of higher elevation.

vapor: the state of a compound at temperatures above its boiling point..

vapor pressure: the pressure exerted by vapor molecules at a given temperature.

vernal equinox: the time when the sun crosses the equator and day and night are of equal length everywhere, indicating the start of spring. It occurs around March 21 or 22.

viscosity: a measure of resistance to flow in a fluid due to intermolecular friction.

visibility: the distance that a large object can be seen with naked eyes.

visible radiation: the electromagnetic radiation visible to naked human eyes.

volatile organic compounds (VOCs): organic molecules mainly composed of carbon and hydrogen atoms.

volt: the unit of electromotive force. It is the amount of force required to drive a steady current of one ampere through a resistance of one ohm. The power delivered to homes and offices is 120 volts in the U.S. and 220 volts in Europe.

voltage: the potential to perform work through electricity.

W

watt: a unit of power in SI units, named after James Watt. 1W =1 J/s.

watt-hour: a metric unit of energy; defined as power expended over one hour.

wave: to swing back and forth or up and down. Waves are characterized by their amplitude (vertical distance between a wave's trough and crest), wavelength (distance between two successive wave crests or troughs), and period (the time it takes for a wave to travel the distance of one wavelength).

wedge: a moving inclined plane.

well: a shaft or hole sunk to obtain oil, brine, or gas. An exploratory well (or dry hole) is drilled to assess justification for completion of an oil or gas well. A development well is a well drilled within the proved area of an oil or gas reservoir. A service well is a well drilled for the purpose of supporting production in an existing field. Examples are gas, water, steam, or air injections, and salt-water disposal.

wind: motion of air relative to earth as a result of a pressure or temperature difference.

wind chill factor: the temperature that still air would have to be to result in the same sensation as a combination of wind and temperature on naked human flesh.

windward: the side facing the direction from which wind is blowing. See also leeward.

winter solstice: the shortest day of the year or first day of winter (around December 21 or 22).

work: the energy required to displace a force by a distance along the direction of force.

X

x-ray: electromagnetic radiation with a wavelength between gamma rays and ultraviolet radiation.

Y

Yucca Mountain: the proposed site in Nevada where a permanent repository for nuclear waste is being built. It is scheduled to open for operation in 2010.

Z

zenith: the highest point reached in the heavens by a celestial body. The complement of solar altitude angle or the angle of the sun from a vertical line directly overhead is called zenith angle. Also see altitude.

Zeroth Law of Thermodynamics: See Laws of Thermodynamics

Sources:

1. Encarta, Online Encyclopedia. *(http://encarta.msn.com).*
2. Wikipedia, The Free Encyclopedia *(http://wikipedia.org).*
3. Merriam Webster Encyclopedia *(http://www.m-w.com).*
4. British Britannica *(http://www.britannica.com).*
5. Department of Energy Websites *(http://www.doe.gov).*
6. Department of Transportation Website *(http://www.dot.gov).*
7. British Petroleum Website *(http://www.bp.com).*
8. Fundamentals of Physical Geography, Pidwirny, M., University of British Columbia Okanagan, Canada VIV 1V7. *(http://www.physgeog.net).*